Springer Tracts in Mechanical Engineering

T0171815

Springer Tracts in Mechanical Engineering (STME) publishes the latest developments in Mechanical Engineering - quickly, informally and with high quality. The intent is to cover all the main branches of mechanical engineering, both theoretical and applied, including:

- Engineering Design
- Machinery and Machine Elements
- Mechanical Structures and Stress Analysis
- Automotive Engineering
- Engine Technology
- Aerospace Technology and Astronautics
- Nanotechnology and Microengineering
- Control, Robotics, Mechatronics
- MEMS
- Theoretical and Applied Mechanics
- Dynamical Systems, Control
- Fluids Mechanics
- Engineering Thermodynamics, Heat and Mass Transfer
- Manufacturing
- Precision Engineering, Instrumentation, Measurement
- Materials Engineering
- Tribology and Surface Technology

Within the scope of the series are monographs, professional books or graduate textbooks, edited volumes as well as outstanding PhD theses and books purposely devoted to support education in mechanical engineering at graduate and post-graduate levels.

Indexed by SCOPUS, zbMATH, SCImago.

Please check our Lecture Notes in Mechanical Engineering at http://www.springer.com/series/11236 if you are interested in conference proceedings.

To submit a proposal or for further inquiries, please contact the Springer Editor **in your country**:

Dr. Mengchu Huang (China)
Email: mengchu.Huang@springer.com
Priya Vyas (India)
Email: priya.vyas@springer.com
Dr. Leontina Di Cecco (All other countries)
Email: leontina.dicecco@springer.com
All books published in the series are submitted for consideration in Web of Science.

More information about this series at http://www.springer.com/series/11693

Georg-Peter Ostermeyer · Valentin L. Popov ·
Evgeny V. Shilko · Olga S. Vasiljeva
Editors

Multiscale Biomechanics and Tribology of Inorganic and Organic Systems

In Memory of Professor Sergey Psakhie

 Springer

Editors
Georg-Peter Ostermeyer
Institute of Dynamics and Vibrations
Technische Universität Braunschweig
Braunschweig, Niedersachsen, Germany

Evgeny V. Shilko
Institute of Strength Physics and Materials
Science
Russian Academy of Sciences
Tomsk, Russia

Valentin L. Popov
Technische Universität Berlin
Berlin, Germany

Olga S. Vasiljeva
Department of Biochemistry and Molecular
and Structural Biology
Jozef Stefan Institute
Ljubljana, Slovenia

ISSN 2195-9862 ISSN 2195-9870 (electronic)
Springer Tracts in Mechanical Engineering
ISBN 978-3-030-60126-3 ISBN 978-3-030-60124-9 (eBook)
https://doi.org/10.1007/978-3-030-60124-9

This Springer imprint is published by the registered company Springer Nature Switzerland AG
The registered company address is: Gewerbestrasse 11, 6330 Cham, Switzerland

Preface

The monograph *Multiscale Biomechanics and Tribology of Inorganic and Organic Systems* is dedicated to the memory of Prof. Sergey Grigorievich Psakhie (02.03.1952–22.12.2018). The topic of the monograph reflects the broad range of scientific focus areas of Prof. Psakhie. In compiling this book, we attempted to represent the various aspects of his multifaceted research interests ranging from theoretical physics research and computer modeling for understanding materials at the atomic scale up to applied science for solving specific problems of the rocket and space industry, medicine or geotectonics. The authors of the monograph are the colleagues of S. G. Psakhie from the Institute, collaborators from his international network, his former students, participants of the Workshops he organized and his friends.

Braunschweig, Germany
Berlin, Germany
Tomsk, Russia
Ljubljana, Slovenia
June 2020

Georg-Peter Ostermeyer
Valentin L. Popov
Evgeny V. Shilko
Olga S. Vasiljeva

Acknowledgements

We thank the authors of individual chapters of the monograph for their contributions. We also would like to express a heartfelt thanks to Dr. Jasminka Starcevic for her immense support during the assembly and editing of this book. She is not among the authors of the book, but her decisive assistance in preparation of the monograph was her contribution to the memory of her colleague and friend, Prof. Psakhie.

We would like to acknowledge Technische Universität Braunschweig (Germany), Technische Universität Berlin (Germany), Institute of Strength Physics and Materials Science (Russia), and Jozef Stefan Institute (Slovenia) for financial support of the Open Access publication.

Braunschweig, Germany	Georg-Peter Ostermeyer
Berlin, Germany	Valentin L. Popov
Tomsk, Russia	Evgeny V. Shilko
Ljubljana, Slovenia	Olga S. Vasiljeva
June 2020	

Contents

In Memory of Sergey G. Psakhie

Evgeny V. Shilko, Valentin L. Popov, Olga S. Vasiljeva,
and Georg-Peter Ostermeyer

1 Scientific Biography of Professor Sergey Grigorievich Psakhie

E. V. Shilko
Institute of Strength Physics, Materials Science SB RAS, 634055 Tomsk, Russia

V. L. Popov (✉)
Technische Universität Berlin, 10623 Berlin, Germany
e-mail: v.popov@tu-berlin.de

O. S. Vasiljeva
Jozef Stefan Institute, Ljubljana 1000, Slovenia

G.-P. Ostermeyer
Technische Universität Braunschweig, 38106 Brunswick, Germany

© The Author(s) 2021
G.-P. Ostermeyer et al. (eds.), *Multiscale Biomechanics and Tribology
of Inorganic and Organic Systems*, Springer Tracts in Mechanical Engineering,
https://doi.org/10.1007/978-3-030-60124-9_1

Sergey G. Psakhie was born on March 2, 1952 in Tomsk (the main city of the Tomsk region in Siberia, USSR) in a family of school teachers Grigory A. Psakhie and Nadezhda A. Psakhie.

His father, **Grigory Abramovich Psakhie**, was one of the most famous and respected school teachers in the history of Tomsk Region. He has worked as a physics and astronomy teacher for more than half a century and was a school principal in Tomsk and Tomsk Region for about 40 years. Grigory Abramovich is rightfully called the Innovative Teacher. He was one of the first in the USSR to introduce the most advanced system of teaching physics and astronomy at that time, the Shatalov system, and has introduced a large number of other innovations in the school education system in the Tomsk Region. For many years of fruitful educational and enlightenment activities, Grigory Psakhie was awarded numerous orders, medals, and diplomas. The most valuable for him were the Order of the October Revolution and the Janusz Korczak medal "To a teacher who raised students" (the latter had a cult status among the teachers in the USSR). The Academic Lyceum in Tomsk is named after Grigory Psakhie.

Sergey's mother, **Nadezhda Alekseevna Psakhie**, devoted her whole life to preschool and primary school education for children. For more than half a century in the profession, Nadezhda Psakhie worked as a kindergarten teacher, head of a kindergarten, primary school teacher and educator. At all places of work, she was considered an exemplary teacher and educator. She was awarded numerous diplomas of the Departments of public education of Tomsk and the Tomsk Region. Grigory and Nadezhda Psakhie lived together for 65 years and were awarded a special certificate from the President of the Russian Federation Vladimir Putin as a family who lived a long, happy life together.

Grigory and Nadezhda Psakhie

Sergey Psakhie raised and taught two daughters (**Olga** and **Natalia**), and son **Ivan**. Olga (married name Vasiljeva), a Professor of Biochemistry at Jozef Stefan Institute (Slovenia) and an expert in protease biology and oncology, has been working with her father, Sergey Psakhie, on several interdisciplinary projects. Natalia Psakhie is a product manager working at top Silicon Valley (California) tech companies. Ivan Psakhie became a scientist in the molecular biology field.

Sergey Psakhie spent a significant period of his childhood in small villages New Vasyugan, Middle Vasyugan, Kargasok in the north of the Tomsk Region. In some of these villages there were no kindergartens, so Sergey became independent very early and learned to read as early as 3 years old. His parents devoted a lot of time to raising and educating their son, so when Sergey entered the school, the school administration offered to immediately transfer Sergey to the 2nd grade. Already at school, Sergey Psakhie manifested diverse interests, high intelligence, and leadership. He was seriously interested in physics and chess, and at the same time passionately engaged in sports and dances, participated in the performances of the Greek school theater, and wrote talented poems. At the same time, Sergey demonstrated excellent performance in almost all school subjects. In high school, Sergey dreamed of entering a flight school, but in the end, his father-instilled love of physics won.

Upon school graduation, Sergey Psakhie entered the **Physics Department of Tomsk State University**, from which he **graduated in 1976**.

In 1976–1979, Sergey Psakhie studied at the graduate school of Tomsk State University (TSU) and carried out scientific research under the guidance of Professor Viktor E. Panin. Viktor Panin became for Sergey not just a scientific supervisor, but a teacher, colleague, and like-minded person for many subsequent decades. In the graduate school of TSU, Sergey Psakhie chose as a scientific specialization a new, and only emerging at that time, direction: computer modeling of processes and phenomena in solids at the atomic level. This direction would later become one of the leading and most successful scientific areas of his future scientific school.

In 1981, Sergey Psakhie successfully defended his thesis for the degree of candidate of physical and mathematical sciences. The theme of his dissertation was "Investigation of the interaction between atoms of alloying elements and vacancies in diluted aluminum-based alloys".

From 1980 to 1984, Sergey Psakhie was a junior researcher at the Institute of Atmospheric Optics of the Siberian Branch of the Academy of Sciences of the USSR (now V. E. Zuev Institute of Atmospheric Optics of the Siberian Branch of the Russian Academy of Sciences) in the Department of solid-state physics and materials science. This department was headed by his scientific adviser Viktor Panin. The best qualities of Sergey Psakhie as a great scientist, such as scientific instinct, understanding of the physical nature of phenomena, the ability to formulate ideas and results, and extraordinary organizational talent, were clearly manifested already during this period.

In 1984, a new research institute was opened in Tomsk—the Institute of Strength Physics and Materials Science of the Siberian Branch of the Academy of Sciences of the USSR (after the collapse of the USSR in 1992 it was renamed to the Institute of Strength Physics and Materials Science of the Siberian Branch of the Russian

Academy of Sciences, ISPMS SB RAS). Professor Viktor E. Panin was the organizer and first director of the Institute. Together with him and a group of ambitious young scientists, Sergey Psakhie came to the new institute as a senior researcher. The Institute became for Sergey Grigorievich a scientific home for the rest of his life. ISPMS SB RAS is associated with all of his key scientific and career achievements.

In 1985, young ambitious scientist Sergey Psakhie founded a new Laboratory of automation (from 1998, the Laboratory of computer-aided design of materials) and became its head. The scientific formation of his best students have been working in this laboratory. Among them are Dr. Konstantin P. Zolnikov, Dr. Sergey Yu. Korostelev, Dr. Alexey Yu. Smolin, Dr. Andrei I. Dmitriev, Dr. Evgeny V. Shilko, Dr. Andrey V. Dimaki, and many others.

Sergey Psakhie (second from the left) with visitors in the Laboratory of computer-aided design of materials (ca. 1990)

In 1990, Sergey Psakhie successfully defended his dissertation for the degree of Doctor of Physical and Mathematical Sciences at the age of only 38 years and in **1991** received a Doctor of Science degree. The theme of his dissertation was "Interparticle interactions and nonlinear properties of metals under mechanical stress".

In 1991, Sergey Psakhie took the post of deputy director of the Institute for Research. In this position, he worked until 1993.

In 1994, Sergey Psakhie worked as a visiting professor at North Carolina State University (USA) by the invitation of one of the leading experts in high-rate processes and phenomena in solids, professor Yasuyuki (Yuki) Horie. Work at NCSU significantly changed the worldview of Sergey Psakhie. He fully realized that world-class science can only be successfully developed in close international cooperation. The foundations of such cooperation were laid in the American period of his career, and Sergey Psakhie was engaged in its development throughout his subsequent scientific career at ISPMS SB RAS. Close collaboration of Sergey Psakhie and his research team with leading American, European and Chinese scientists enabled the achievement of advanced fundamental and applied results in materials science, and led to international recognition in the form of highly cited papers in high ranked journals, international patents, and scientific contracts with leading European universities and international industrial corporations.

The period of work at the NCSU has become extremely fruitful scientifically. In collaboration with Prof. Yasuyuki Horie, Sergey Psakhie developed a new particle method, namely, movable cellular automaton (MCA) method. This method was originally developed to model mechanically activated chemical reactions in powder mixtures. In fact, it was a hybrid numerical technique, which combines the formalisms of discrete element and cellular automaton methods. Later, the formalism of this method was used to create an advanced implementation of the discrete element method, namely of the method of homogeneously deformable discrete elements.

In the same year 1994, Sergey Psakhie became a member of the New York Academy of Sciences and American Ceramic Society.

From 1995 to 2002, Sergey Psakhie continued his scientific work at ISPMS SB RAS as the head of the Laboratory of computer-aided design of materials in close international cooperation with Prof. Yuki Horie, Prof. Zongguang Wang (Institute of Metal Research CAS, China), Dr. Stanko Blatnik (Jožef Stefan Institute and INOVA d.o.o., Slovenia), Dr. Simon Zavsek (Velenje Coal Mine, Slovenia), Prof. Jože Pezdič (University of Ljubljana), Prof. Georg-Peter Ostermeyer (Technische Universität Berlin, Technische Universität Braunschweig after 2002) and Prof. Valentin L. Popov (Universität Paderborn, Technische Universität Berlin after 2002). Sergey Psakhie was closely related to these scientists not only by joint research, but also by many years of friendship.

Thanks to broad scientific collaboration, several new areas of the Sergey Psakhie scientific school were created during this period, and later became extremely successful:

- development of the formalism of hybrid cellular automaton method (the coupled discrete element–finite difference numerical technique) for the study of gas-saturated and fluid-saturated porous materials including coal, sandstone, bone tissues and so on;
- computational study of friction and wear in technical and natural tribounits at different scales using a particle-based approach;
- development of an approach to stress state prognosis and managing the displacement mode and seismic activity of tectonic faults.

The latter approach was co-developed in collaboration with prominent experts in geology and geophysics, Prof. Sergey V. Goldin and Dr. Valery V. Ruzhich. In addition to computational modeling, a key component of these studies was the long-term field research in the dynamics of the segments of tectonic faults and the Lake

Baikal ice cover, as a model block-structured medium. These studies included long-term monitoring of displacements and seismicity in combination with managing impacts (explosions, vibrations, fluid injection). An important result of these studies, which Sergey Psakhie was rightly proud of, is the Patent of Russian Federation (2006) "A method for controlling the displacement mode in fragments of seismically active tectonic faults".

Sergey Psakhie's colleague, Prof. Valentin L. Popov, has contributed to the development of the three aforementioned scientific areas. He initiated and participated in a large number of joint projects and is the co-author of dozens of joint papers. For the past two decades, Valentin Popov has been Sergey Psakhie's closest friend.

In 2002, Sergey Psakhie took a new decisive step. He became the director of his native Institute (ISPMS SB RAS), succeeding his teacher Academician Viktor E. Panin in this post. Sergey Psakhie made this decision after much deliberation, but subsequently he never regretted it. In his new post, Sergey Psakhie became one of the most famous and successful directors of research institutes in Russia and was able to execute the most ambitious projects in a wide spectrum of scientific fields from materials science and mechanical engineering to geotectonics and biomedicine. Under the leadership of Sergey Psakhie, the ISPMS SB RAS developed rapidly and in 2017 became one of the top research institutes in Russia. It is a first-rank institute and is included in the TOP-10 Russian scientific institutes based on the number of papers published in journals indexed in Web of Science.

Despite working at the research institute, Sergey Psakhie never lost touch with his Alma Mater. **Since 1992**, he worked as a professor at the Department of Strength and Design of Tomsk State University, and **in 2005** he took the post of head of the Department of High Technology Physics in Mechanical Engineering at Tomsk Polytechnic University. Working both at the research institute and at leading Tomsk universities, Sergey Psakhie made an invaluable contribution to the formation of Tomsk Consortium of Scientific and Educational Organizations enabling integration of Tomsk's educational and scientific systems.

Sergey Psakhie believed that the presence of scientific schools in Russia, including strong Tomsk scientific schools in the fields of theoretical physics, material science, and advanced medicine, is a key advantage of the education system in the Russian Federation. He always considered Tomsk as a place of attraction for the implementation of joint projects with world leading scientific organizations and did a lot to form such projects in the field of materials science. Examples are the project aimed to develop and create the production of advanced wound healing materials "VitaVallis", development of a new multi-beam electron beam technology for high-performance additive production of large-size metal products and structures for key industries in Russian Federation.

The high reputation of Sergey Psakhie as an outstanding organizer of science, determined his active and multifaceted work in the Russian Academy of Sciences.

From January **2004**, he combined presiding the ISPMS SB RAS with the post of the Deputy Chairman of the Presidium of the Tomsk Scientific Center of the SB RAS, and from **2006 to 2013** with the post of the Chairman of the Presidium of the Tomsk Scientific Center of the SB RAS.

Sergey Psakhie was one of the key leaders in the Siberian Branch of the Russian Academy of Sciences, a member of the Presidium of the SB RAS, and Deputy Chairman of the SB RAS contributing to innovative activities and the development of integrated scientific and educational systems in Siberia.

The merited recognition of Sergey Psakhie's high scientific and organizational achievements and scientific reputation led to his election as a Corresponding Member of the Russian Academy of Sciences **in 2011**.

Sergey Psakhie has contributed to scientific and organizational work in multiple governmental, municipal, and public organizations. He was a member of the Russian Foundation for Basic Research (RFBR) Council, the Russian National Committee on Theoretical and Applied Mechanics, the Russian National Committee on Tribology, the Interstate Coordination Council on physics of strength and plasticity of materials, the Innovation Council of the Siberian Branch of the Russian Academy of Sciences, multiple coordination councils (including international), the expert committee of the Russian Youth Prize in Nano-industry, the committee for the competition of technological projects, the Tomsk Region Administration, a number of councils for awarding prizes in the Tomsk Region, etc. Over the years, he was a member and co-chair of the organizing committees of many international conferences. He was a member of the editorial boards of three international journals.

For his tenure and achievements in science and the organization of science and education in Russia, Sergey Psakhie was awarded a large number of honors including the title of "Honorary Worker of Science and Technology of the Russian Federation", the medal of the Russian Cosmonautics Federation "For Merits", honorary badges of the SB RAS "Silver Sigma", the medal "Honored Veteran of the SB RAS", and the Order of Friendship, among others.

Despite the high workload of organizational activities, Sergey Psakhie always considered science as his main priority. His great fundamental knowledge, unusually wide erudition, sharp mind, and talent made it possible to achieve great results in various scientific fields. Sergey Psakhie has always actively supported interdisciplinary research conducted at the intersection of different sciences. He saw the prospect of such research in the fact that well-established methods and approaches from one field of science, after some modification, can be efficiently used to obtain breakthrough results in another scientific field. The range of scientific interests of Sergey Psakhie was unusually broad and covered the problems of theoretical research and computer modeling of complex nonlinear processes in technical and natural (biological and geological) materials.

In the last decade, Sergey Psakhie's main research interests have been related to the study of the interactions between hard matter and soft matter in multiphase contrast materials, the study of the features and anomalies of the behavior of solids under mechanical confinement (confined matter), the study of dusty plasma, etc. In particular, implementation of the method of particles (movable cellular automaton method) which he developed in cooperation with Prof. Y. Horie and Prof. V. Popov has been broadly used to solve fundamental and important practical problems in the field of mechanics of solids, fracture mechanics, mechanics and physics of friction and wear, geomechanics and tectonophysics, biomedicine.

In 2007, he initiated a new scientific direction connected with the theoretical study of plasma-dust crystals. The study of dusty plasma, which is often called the new state of matter, is one of the most promising scientific fields and is developed by collaborations of leading scientific groups around the world.

Since his student years and throughout his scientific career, Sergey Psakhie was actively involved in radiation materials science. His first publications in 1980s were dedicated to this topic. In the 2000s, Sergey Psakhie initiated scientific cooperation with A. A. Bochvar High-Technology Scientific Research Institute for Inorganic Materials (Moscow, Russia), the lead institution for the development of promising structural materials for nuclear power engineering. This extremely fruitful collaboration brought together several leading Russian scientific teams and continues to this day. The results of theoretical studies of Sergey Psakhie's scientific group are unique and important for the development of new materials for nuclear energy, as they shed light on the particular behavior of solids under conditions of high temperatures, pressures and radiation exposure that materials experience in the core of nuclear reactors.

A large place in the professional activity of Sergey Psakhie in recent years has been devoted to scientific and technical cooperation with S.P. Korolev Rocket and Space Corporation Energia (Korolev, Russia). Joint research was aimed at solving specific problems of the rocket and space industry. Among the largest projects implemented was the development of methods and equipment for non-destructive testing of welded joints, the efficient friction stir welding technology, 3D-printing technology of products from polymeric materials under zero-gravity conditions, a new method and equipment for repair and restoration of the surface of glass illuminators damaged as a result of the impact of micrometeorites, and others. A huge role in the implementation of these ambitious projects was played by Alexander G. Chernyavsky, Deputy Chief Designer of S. P. Korolev Rocket and Space Corporation Energia, who became not only a like-minded person, but also a friend of Sergey Psakhie.

Sergey Psakhie initiated and implemented innovative scientific direction related to the physics of nanoscale states of substances. Since the mid-2000s, the ISPMS SB RAS has become one of the leading nanotechnology innovation centers in Russia.

Sergey Psakhie made a great contribution to the study of the biological effect of low-dimensional metal oxide nanostructures and their use for biomedical applications, including the creation of new cancer treatment strategies. Latter studies have been performed in collaboration with his daughter, Prof. Olga Vasiljeva (Jozef Stefan Institute, Slovenia). As such, his publication with multidisciplinary international team "Ferri-liposomes as an MRI-visible drug-delivery system for targeting tumors and their microenvironment" published in Nature Nanotechnology journal, has more than 300 citations. In 2018, Sergey Psakhie and his team received a US patent for the use of such materials to suppress tumor growth (US Patent 10105318 "Low-dimensional structures of organic and/or inorganic substances and use thereof").

The staff of his laboratory and chair treated Sergey Psakhie not just as head and colleague, but as a Teacher and Scientist. Over the years, he mentored 5 Doctors of Sciences and about 20 Candidates of Science (Ph.D.), and published with co-authors more than 300 papers. Nowadays, students of his students are successfully defending scientific dissertations.

Sergey Psakhie suddenly passed away on **December 22, 2018**. On the morning of the day of his death, he met with colleagues and had a long list of planned events. The death of a talented Russian physicist, an outstanding organizer of science, a man who devoted himself to serving and promoting science, shocked the entire Tomsk scientific and educational community and the large number of his Russian and international colleagues and friends.

The loss of such a great scientist and person is shocking and irreparable and will remain so for many more years. Nonetheless, the principles of scientific activities created and applied by Sergey Psakhie, his scientific ideas and undertakings would allow the ISPMS SB RAS, Tomsk and Russian science to successfully develop further. Multiple colleagues, collaborators and mentees of Sergey Psakhie will continue to implement the initiatives of Sergey Psakhie and carry on his innovative work.

The name **Sergey G. Psakhie** is forever inscribed in gold letters in the history of Russian and international science.

Sergey Grigorievich Psakhie (02.03.1952–22.12.2018)

2   Georg-Peter Ostermeyer: Twenty Years of Friendship with Sergey Psakhie

G.-P. Ostermeyer is director of the Institute of Dynamics and Vibrations at the University of Braunschweig, Germany.

My wife Ulrike and I first met Professor Psakhie in 1999 in Berlin. There, at the TU Berlin, I established a Collaborative Research Center on friction and invited Prof. Psakhie several times in this context. Besides scientific discussions, we also explored Berlin together and went on extensive hikes in the nature around Berlin. We became friends very soon and he remained our very good friend until the end.

This was also because we were both interested in the philosophy of discrete methods in tribology. We both shared the opinion that mesoscopic simulation methods in particular are essential for tribology. We were in contact all the time and have developed together many ideas. One of these ideas was the simulation of the thermalization process on the mesoscale based on the Shannon theorem. The method of meso particles, which I developed in the late 1990s, and the method of movable cellular automata, developed by Sergei Psakhie, solved this problem in different ways, and the combination of both approaches was a very attractive idea.

It was always a pleasure to have discussions with Sergey Psakhie. With his very broad range of interests and his extensive experience in discrete methods, we would often return to the topic of numerical modeling of multiscale phenomena. Examples of such are material textures in friction processes, which react on slow time scales with surprising motion dynamics, to friction loads, or chemical reactions in the friction boundary layer, which are found on completely different time and size scales. We were fascinated by the idea of treating multiscale effects with different abstract, i.e. scale-independent, methods, whereby the interconnection of these methods alone represents the scale spread. My wife Ulrike, a mathematician, was always fascinated by the mathematical depth of his argumentation even on topics, which were quite far from his fields of research, and Sergey Psakhie enjoyed these inspiring discussions with her.

The last time we met personally was during the International Workshop "Advances in Tribology: Science, Technology and Education" in Karlovy Vary in 2015, organized by Sergey Psakhie together with Valentin Popov and myself. This meeting was like a throwback to the early days of our friendship. Once again, we were walking (and working) together in the nature. One time the three of us took a shortcut on the way back to the hotel - but the intensive discussions made sure that we didn't arrive any earlier. The plan to hold a joint conference in Tomsk was born and we made plans to try to organize this conference thematically and chronologically.

Unfortunately, these plans could not be realized because Sergei Psakhie left us suddenly and completely unexpectedly in December 2018. This was a great shock for all of us.

I am very glad that I can contribute to the memory of the great scientist and man Sergei Psakhie by co-editing this monograph and co-organizing the commemorative workshop on "Multiscale Biomechanics and Tribology of Inorganic and Organic

Systems" in Tomsk in October 2019. My wife and I experienced and got to know Tomsk, the home city of Sergey, up close for the first time. We visited his place of work, and also the nature around Tomsk, with the feeling that he was still very much with us.

At the German-Russian Workshop "Advances in Tribology: Science, Technology and Education", Karlovy Vary, Czech Republic, March 2, 2015. First row left: S. G Psakhie, rigth: G. P. Ostermeyer, behind them: Ulrike Ketterl-Ostermeyer. Other participants (from the left): E. Shilko, V. Aleshin, M. Popov, V. Popov, A. Smolin, A. Dmitriev, R. Pohrt, J. Starcevic, E. Kolubaev, A. Korsunsky, A. Dimaki

3   Valentin L. Popov: A Word of Sergey Psakhie

V. L. Popov is the head of Department of System Dynamics and Friction Physics at the Technische Universität Berlin.

My collaboration and close friendship with Sergey Grigorievich Psakhie started in 1997. Together with Prof. E. Santner, at that time the head of tribology department of the Federal Institute of Materials Research and Testing (BAM), we submitted a joint project devoted to simulation of tribological processes with the Method of Movable Cellular Automaton. This project became a great success and established for many years one of research directions at the BAM.

Sergey Psakhie was the most important partner of the Department of System Dynamics and Friction Physics of Technische Universität Berlin, which I headed since 2002. Our cooperation covered diverse areas to which tribology can be applied: molecular motors and earth tectonics, problems of material wear and damping of aerospace structures, active control of friction and, of course, numerical modeling methods in tribology. Most important for us was scientific collaboration in the field of application of mesoparticle approach, in particular Cellular Automata, to tribological problems. Among a large number of papers, I can highlight our very fundamental and highly cited programmatic joint paper, "Numerical Simulation Methods in Tribology", appeared in 2007, which contained the most important future developments in this field such as Method of Dimensionality Reduction, applications of MCA and stochastic differential equations.

The forms of cooperation were correspondingly manifold: annual German-Russian workshops (in fact international conferences), seismological field expeditions, multiple joint projects, invited professorships, international laboratories, visits of student groups, academic exchanges and finally the Double Degree Program with the Polytechnic University initiated by Prof. Psakhie in 2006. The program ran on the base of the Department of High Technology Physics in Mechanical Engineering headed by Sergey Psakhie. Since that time, about 100 German students studied in Tomsk thanks to these programs.

Sergey Psakhie was director of an academic institute. And he suited this position very well. He was a surprisingly harmonious and versatile person. Despite the fact that he devoted his life to materials, he was always attracted to biological topics. He believed that the twenty-first century is the century of biology, especially molecular biology. He was an expert in both fields. And not just an expert. He was a visionary, theorist and practitioner in one. Sergey Psakhie was a very erudite and well-read man. My wife Elena, a philologist with a double education (graduated from the Tomsk State University in Russian philology and from Paderborn University in German philology) had many times to urgently re-read works, heroes or conflicts of which Sergey mentioned.

In October of 2018, Sergey Psakhie and I, together with the rector of TSU, Eduard Vladimirovich Galazhinsky, discussed a new scientific direction - active biocontact mechanics, the purpose of which is the natural restoration of joints controlled by mechanical and medical means, instead of surgical replacement. Already in December 2018 the preparation of a joint multilateral project was in full swing. But when I received the sad news on Saturday morning, 22. December 2018, my thoughts were not about cooperation and not about laboratories. Because Sergey Psakhie was not only my colleague, but also a close friend of my whole family. The memory machine has started and cannot stop until now…

In 2005, we celebrated Christmas at the Baltic Sea, in Warnemünde. It is easy to imagine the weather in North Germany at the end of December. Nevertheless, Sergey jokingly suggested taking a swim next morning. Indeed, at 6.00 am in the morning, still in the dark, my sons, Nikita and Misha woke him up and pulled him to go swimming. There is a historical photograph capturing Sergey Psakhie with Nikita and Misha swimming in fur hats in the Baltic Sea on the 25th of December.

We often took vacations together and Sergey supplied us with ideas for movies to watch. One of these movies was "The Discreet Charm of the Bourgeoisie" by Luis Bunuel, which we watched many times.

Like many parents, I sometimes suffer from a critical attitude towards my children. Sergey constantly corrected me and told me how talented they are and what they fascinate him with.

Sergey introduced Russia to my German colleagues. I would like to mention just two names: Jasminka Starcevic, who was a devoted close friend of Sergey Psakhie; and, of course, Professor Ostermeyer with his wife Ulrike, with whom Sergey Psakhie had a family friendship.

Sergey Psakhie was a true friend who we regard as a member of our family. He entered our lives a long time ago and will remain in them forever.

Sergey Psakhie and Valentin Popov at the German-Russian Workshop on "Numerical simulation methods in tribology: possibilities and limitations" (Technische Universität Berlin, March 2005)

4   Lev B. Zuev: From My Memories of Sergey G. Psakhie

L. B. Zuev is the head of Laboratory of Strength Physics at the Institute of Strength Physics and Materials Science, SB RAS.

I met Sergey Grigorievich Psakhie in 1983. At that time, I headed the Department of Physics at the Siberian Metallurgical Institute in Novokuznetsk, but my moving to the Institute of Strength Physics and Materials Science in Tomsk (which had just formed) was under discussion. The future director of the institute, V. E. Panin, invited me to participate in a Session of the Scientific Council of the USSR Academy of Sciences on strength and ductility. The session was held in a pioneer camp in the village of Zavarzino, located close to the Tomsk Academic Township. Sergey Psakhie was responsible for organization of this Session. It was the fall of 1983 - four months before the official opening of the Institute.

I worked with Sergey Psakhie as his deputy for ten years, from 2002 to 2012. In these years, on his initiative, the research topics and the structure of the Institute have been substantially modernized. New laboratories and new research areas opened, often quite distant from the physics of strength, but ideally fitting into materials science in a broader sense. The technological base of the Institute has been significantly expanded and improved. This allowed rapid progress in our research at the highest international level.

I liked the style of the meetings Sergey Psakhie ran at the Institute. They were relatively brief. He usually listened to a number of points of view, which were presented by the participants, and then quickly made a decision. I want to note that the decision was made at the right time, not too early, and not too late. This contributed to the formation of an optimal point of view, but excluded a long debate that almost never leads to a useful result.

It seems important to me that Sergey Psakhie had a sense of what is now called "growth points", an ability to notice teams and people who have the potential for growth and development. He always considered it his duty to promote their development, which, in turn, contributed to the progress of the Institute as a whole.

I think that even organizational work gave Sergey Psakhie pleasure. He clearly sought to concentrate in his hands the work on opening new areas of research, networking, negotiations, making key decisions, choosing partners, etc.

I was twelve years older than Sergey Psakhie, and with such a difference in age we have not been friends. However, closer, what is called "human" contacts with him developed due to my health problems. His enormous decisive help in this respect did not belong to the scope of duties of director. When a medical treatment was necessary, he immediately picked up the phone, called medical doctors he knew, and even accompanied me to the hospital, which was not necessary at all. Sergey Psakhie considered such help a person's duty.

I am thankful to Sergey Psakhie for a lot, and especially for the editing of my book "Autowave plasticity. Localization and collective modes", which was published in 2018. He proposed the concise title of this monograph and wrote a short introduction to it. I still managed to present him a copy of the book…

I think that Sergey Psakhie was a man of integrity, and such people are always complex. Nevertheless, I am happy that I had the opportunity to work with him for a long time…

Meeting of the director board of ISPMS SB RAS (ca. 1985). Third from the left: S. G. Psakhie, second from the left L. B. Zuev. The meeting is chaired by the founder and first director of ISPMS SB RAS, V.E. Panin

5   Valery V. Ruzhich: On the "Earthquake Vaccine" Project of Sergey Psakhie

V. V. Ruzhich is the Principal Researcher of the laboratorium of Sesimology at the Institute of Earth's Crust, SB RAS, Irkutsk.

The eternal question of how to reduce the destructive consequences of strong earthquakes remains unresolved for many thousands of years. Humanity does not have means to affect the energetics of the deep tectonic processes leading to earthquakes. However, it is legitimate to pose the question: is it possible to find ways to reduce, disperse, or "streamline" the rampant play of the underground elements, manifested in the form of seismo-tectonic catastrophes? This issue was at the center of the project that emerged in the late 1990s and was realized under guidance of Sergey Psakhie. Journalists called the methods developed in the framework of this project "vaccination" from earthquakes.

Our first meeting with Sergey Psakhie took place in 1999 at the initiative of Academician Sergei Vasilievich Goldin at one of the interdisciplinary seminars. The ambitious idea of creating methods of controlled technogenic impacts on faults to mitigate and dissipate the destructive energy stored in the crust interested Sergey Psakhie, although it was very far from the subject of his main research. He belonged to a rare category of researchers with a huge arsenal of knowledge and own scientific developments, but at the same time very sensitive to innovation. He initiated a project with participation of specialists from 7 scientific organizations of the Russian Federation and Technische Universität Berlin. By 2003, at the Listvyanka landfill in the Angarsk seismically active fault zone, wells were drilled, strain and seismic regime monitoring was organized, and the seismic zone was modeled theoretically. In August 2004, original field experiments were conducted on a real seismic hazardous fault with vibrations, shocks and explosions in combination with injecting water solutions through wells into particular fault fragments. With this complex action, it was possible to induce an accelerated creep mode and to shift the fault banks on the length of about 100 m by 8–10 mm. In 2006, these unique results were recognized by the Siberian Branch of the Russian Academy of Sciences as a breakthrough in seismic research. In the same year, the authors obtained the patent of the Russian Federation "A method for controlling the displacement mode in fragments of seismically active tectonic faults".

Ice chronicle on Baikal. In 2005 Sergey Psakhie proposed a unique scientific project to study the conditions of deformation and dynamic fracture of the ice cover of Lake Baikal, related to the seismically active Baikal rift basin. Annually, many kilometers long main cracks occur in the ice cover, which are accompanied by dynamic phenomena called ice impacts - registered in the form of seismic tremors with an energy comparable to weak tectonic earthquakes. It was assumed that their study might facilitate understanding of similar processers in the lithosphere. An interdisciplinary team of researchers from several scientific institutions, under the guidance of Sergey Psakhie, collected information that contributed to improvement of methods for controlling the regimes of seismic emission generation.

Our last meeting with Sergey Psakhie took place in Tomsk in the fall of 2018, during the annual international conference. With his extensive scientific and organizational activities, Sergey subjected his health to extreme loads, reaching far beyond the functions of the director of a research institute. His regular trips to Moscow and Novosibirsk for solving numerous organizational and funding issues did not leave him any time for recreation. At our last lunch in the cafe of the Academic Township in Tomsk, we discussed the plan for a joint winter expedition in 2019 on the ice of Lake Baikal. When parting, I asked him about his health—he looked tired, but habitually lively. Sitting behind the steering wheel of his SUV, he cheerfully replied: "Fine, only the left lower leg aches a little…"

Two months later, on the day of my birth, friends from Tomsk sent a deafeningly deplorable message about the sudden tragic departure of Sergey Grigorievich Psakhie…

As a tribute of deep respect and in memory of Sergey Psakhie, it is necessary for all of us, his friends and followers to continue research on improving the "vaccine" that is in great demand by the international community against imminent seismic disasters.

On the ice of Lake Baikal. Prof. Ruzhich is the right in the front row. Prof. Psakhie is behind him

6 Most Important Publications of Prof. Sergey Grigorievich Psakhie

I. *Particle-based and continuum approaches to computational modeling of materials*

Development of particle-based and continuum approaches to computational modeling of the mechanical behavior of materials under complex loading conditions.

1. Grigoriev, A. S., Shilko, E. V., Skripnyak, V. A., **Psakhie, S. G.** (2019). Kinetic approach to the development of computational dynamic models for brittle solids, *International Journal of Impact Engineering*, 123, 4–25.
2. Grinyaev, Yu. V., Chertova, N. V., Shilko, E. V., **Psakhie, S. G.** (2018). The continuum approach to the description of semi-crystalline polymers deformation regimes: the role of dynamic and translational defects, *Polymers*, 10, 1155.
3. **Psakhie, S. G.**, Dimaki, A. V., Shilko, E. V., Astafurov, S. V. (2016). A coupled discrete element-finite difference approach for modeling mechanical response of fluid-saturated porous materials, *International Journal for Numerical Methods in Engineering*, 106, 623–643.
4. Shilko, E. V., **Psakhie, S. G.**, Schmauder, S., Popov, V. L., Astafurov, S. V., Smolin, A. Yu. (2015). Overcoming the limitations of distinct element method

for multiscale modeling of materials with multimodal internal structure, *Computational Materials Science*, 102, 267–285.

5. **Psakhie, S. G.**, Shilko, E. V., Grigoriev, A. S., Astafurov, S. V., Dimaki, A. V., Smolin, A. Yu. (2014). A mathematical model of particle–particle interaction for discrete element based modeling of deformation and fracture of heterogeneous elastic–plastic materials, *Engineering Fracture Mechanics*, 130, 96–115.

6. **Psakhie, S. G.**, Shilko, E. V., Smolin, A. Yu., Dimaki, A. V., Dmitriev, A. I., Konovalenko, Ig. S., Astafurov, S. V., Zavsek, S. (2011). Approach to simulation of deformation and fracture of hierarchically organized heterogeneous media, including contrast media, *Physical Mesomechanics,* 14, 224–248.

7. Grinyaev, Yu. V., **Psakhie, S. G.**, Chertova, N. V. (2008). Phase space of solids under deformation, *Physical Mesomechanics,* 11, 228–232.

8. **Psakhie, S. G.**, Smolin, A. Yu., Stefanov, Yu. P., Makarov, P. V., Chertov, M. A. (2004). Modeling the behavior of complex media by jointly using discrete and continuum approaches, *Technical Physics Letters*, 30, 712–714.

9. **Psakhie, S. G.**, Horie Y., Ostermeyer G.-P., Korostelev, S. Yu., Smolin, A. Yu., Shilko, E. V., Dmitriev, A. I., Blatnik, S., Spegel, M., Zavsek, S. (2001). Movable cellular automata method for simulating materials with mesostructured, *Theoretical and Applied Fracture Mechanics*, 37, 311–334.

10. Popov, V. L., **Psakhie, S. G.** (2001). Theoretical principles of modeling elasto-plastic media by movable cellular automata method. I. Homogeneous media, *Physical Mesomechanics*, 4, 15–25.

11. **Psakhie, S. G.**, Horie, Y., Korostelev, S. Yu., Smolin, A. Yu., Dmitriev, A. I., Shilko, E. V., Alekseev, S. V. (1995). Method of movable cellular automata as a tool for simulation within the framework of mesomechanics, *Russian Physics Journal,* 38, 1157–1168.

II. ***Contact interaction, friction and wear***

1. Dimaki, A. V., Dudkin, I. V., Shilko, E. V., **Psakhie, S. G.**, Popov, V. L. (2020). Role of Adhesion Stress in Controlling Transition between Plastic, Grinding and Breakaway Regimes of Adhesive Wear, *Scientific Reports*, 10, 1585.

2. Willert, E., Dmitriev, A. I., **Psakhie, S. G.**, Popov, V. L. (2019). Effect of elastic grading on fretting wear, *Scientific Reports*, 9, 7791.

3. Popov, V. L., Dimaki, A. V., **Psakhie, S. G.**, Popov, M. V. (2015). On the role of scales in contact mechanics and friction between elastomers and randomly rough self-affine surfaces, *Scientific Reports*, 5, 11139.

4. Li, Q., Dimaki, A., Popov, M., **Psakhie, S. G.**, Popov, V. L. (2014). Kinetics of the coefficient of friction of elastomers, *Scientific Reports*, 4, 5795.

5. **Psakhie, S. G.**, Popov, V. L., Shilko, E. V., Smolin, A. Yu., Dmitriev, A. I. (2009). Spectral analysis of the behavior and properties of solid surface layers. Nanotribospectroscopy, *Physical Mesomechanics*, 12, 221–234.

6. Popov, V. L., **Psakhie, S. G.** (2007). Numerical simulation methods in tribology, *Tribology International*, 40, 916–923.

7. Bucher, F., Dmitriev, A. I., Ertz, M., Knothe, K., Popov, V. L., **Psakhie, S. G.**, Shilko, E. V. (2006). Multiscale simulation of dry friction in wheel/rail contact, *Wear*, 261, 874–884.

8. Popov, V. L., **Psakhie, S. G.**, Dmitriev, A. I., Shilko, E. V. (2003). Quasi-fluid nano-layers at the interface between rubbing bodies: simulations by movable cellular automata, *Wear*, 254, 901–906.

9. Popov, V. L., **Psakhie, S. G.**, Shilko, E. V., Dmitriev, A. I., Knothe, K., Bucher, F., Ertz, M. (2002). Friction coefficient in "rail-wheel" contacts as a function of material and loading parameters, *Physical Mesomechanics*, 5(3-4), 17–24.

III. *Seismic activity of geological media and controlling displacement modes in fault zones*

1. Shilko, E. V., Dimaki, A. V., **Psakhie, S. G.** (2018). Strength of shear bands in fluid-saturated rocks: a nonlinear effect of competition between dilation and fluid flow, *Scientific Reports*, 8, 1428.

2. Ruzhich, V. V., **Psakhie, S. G.**, Chernykh, E. N., Shilko, E. V., Levina, E. A., Dimaki, A. V. (2018). Baikal ice cover as a representative block medium for research in lithospheric geodynamics, *Physical Mesomechanics*, 21, 223–233.

3. Grigoriev, A. S., Shilko, E. V., Astafurov, S. V., Dimaki, A. V., Vysotsky, E. M., **Psakhie, S. G.** (2016). Effect of dynamic stress state perturbation on irreversible strain accumulation at interfaces in block-structured media, *Physical Mesomechanics*, 19, 136–148.

4. **Psakhie, S. G.**, Dobretsov, N. L., Shilko, E. V., Astafurov, S. V., Dimaki, A. V., Ruzhich, V. V. (2009). Model study of the formation of deformation-induced structures of subduction type in block-structured media. Ice cover of Lake Baikal as a model medium, *Tectonophysics*, 465, 204–211.

5. **Psakhie, S. G.**, Shilko, E. V., Astafurov, S. V., Dimaki, A. V., Ruzhich, V. V., Panchenko, A. Yu. (2008). Model study of the formation and evolution of deformation induced structures of the subduction type in the ice cover of Lake Baikal, *Physical Mesomechanics*, 11, 55–65.

6. **Psakhie, S. G.**, Ruzhich, V. V., Shilko, E. V., Popov, V. L., Astafurov, S. V. (2007). A new way to manage displacements in zones of active faults, *Tribology International*, 40, 995–1003.

7. Dobretsov, N. L., **Psakhie, S. G.**, Ruzhich, V. V., Popov, V. L., Shilko, E. V., Granin, N. G., Timofeev, V. Yu., Astafurov, S. V., Dimaki, A. V., Starcevic, Ya. (2007). Ice cover of Lake Baikal as a model for studying tectonic processes in the Earth's crust, *Doklady Earth Sciences*, 413, 155–159.

8. **Psakhie, S. G.**, Ruzhich, V. V., Shilko, E. V., Popov, V. L., Dimaki, A. V., Astafurov, S. V., Lopatin, V. V. (2005). Influence of the state of interfaces on the character of local displacements in fault-block and interfacial media, *Technical Physics Letters*, 31, 712–715.

9. **Psakhie, S. G.**, Ruzhich, V. V., Smekalin, O. P., Shilko, E. V. (2001). Response of the geological media to dynamic loading, *Physical Mesomechanics*, 4(1), 63–66.

IV. *Vortices as a fundamental mechanism of material response to loading*

1. Shilko, E. V., Astafurov, S. V., Grigoriev, A. S., Smolin, A. Yu., **Psakhie, S. G.** (2018). The fundamental regularities of the evolution of elastic vortices generated in the surface layers of solids under tangential contact loading, *Lubricants*, 6, 51.

2. Shilko, E. V., Grinyaev, Yu. V., Popov, M. V., Popov, V. L., **Psakhie, S. G.** (2016). Nonlinear effect of elastic vortexlike motion on the dynamic stress state of solids, *Physical Review. E,* 93, 053005.

3. **Psakhie, S. G.,** Shilko, E. V., Popov, M. V., Popov, V. L. (2015). The key role of elastic vortices in the initiation of intersonic shear cracks, *Physical Review. E,* 91, 063302.

4. **Psakhie, S. G.,** Zolnikov, K. P., Dmitriev, A. I., Smolin, A. Yu., Shilko, E. V. (2014). Dynamic vortex defects in deformed material, *Physical Mesomechanics,* 17, 15–22.

5. **Psakhie, S. G.,** Smolin, A. Yu., Shilko, E. V., Korostelev, S. Yu., Dmitriev, A. I., Alekseev, S. V. (1997). About the features of transient to steady state deformation of solids, *Journal of Materials Science and Technology,* 13, 69–72.

6. **Psakhie, S. G.,** Smolin, A. Yu., Korostelev, S. Yu., Dmitriev, A. I., Shilko, E. V., Alekseev, S. V. (1995). The study of establishing the steady mode of deformation of solids by the method of movable cellular automata, *Pis'ma Zh. Tech. Phys,* 21(20), 72.

V. *Atomic mechanisms of inelastic deformation of crystalline solids*

1. **Psakhie, S. G.,** Zolnikov, K. P., Kryzhevich, D. S., Korchuganov, A. V. (2019). Key role of excess atomic volume in structural rearrangements at the front of moving partial dislocations in copper nanocrystals, *Scientific Reports,* 9, 3867.

2. Korchuganov, A. V., Tyumentsev, A. N., Zolnikov, K. P., Litovchenko, I. Yu., Kryzhevich, D. S., Gutmanas, E., Li, S. X., Wang, Z. G., **Psakhie, S. G.** (2019). Nucleation of dislocations and twins in fcc nanocrystals: Dynamics of structural transformations, *Journal of Materials Science and Technology,* 35, 201–206.

3. Zolnikov, K. P., Korchuganov, A. V., Kryzhevich, D. S., Chernov, V. M., **Psakhie, S. G.** (2019). Formation of Point Defect Clusters in Metals with Grain Boundaries under Irradiation, *Physical Mesomechanics,* 22(5), 355–364.

4. Korchuganov, A. V., Zolnikov, K. P., Kryzhevich, D. S., **Psakhie, S. G.** (2017). Primary Ion-Irradiation Damage of BCC-Iron Surfaces, *Russian Physics Journal,* 60, 170–174.

5. Korchuganov, A. V., Zolnikov, K. P., Kryzhevich, D. S., Chernov, V. M., **Psakhie, S. G.** (2016). MD simulation of plastic deformation nucleation in stressed crystallites under irradiation, *Physics of Atomic Nuclei,* 79(7), 1193–1198.

6. Zolnikov, K. P., Korchuganov, A. V., Kryzhevich, D. S., Chernov, V. M., **Psakhie, S. G.** (2015). Structural changes in elastically stressed crystallites under irradiation, *Nuclear Instruments & Methods in Physics Research Section B-Beam Interactions with Materials and Atoms,* 352, 43–46.

7. Dmitriev, A. I., Nikonov, A. Yu., **Psakhie, S. G.** (2011). Atomistic mechanism of grain boundary sliding with the example of a large-angle boundary $2 = 5$. Molecular dynamics calculation, *Physical Mesomechanics*, 14, 24–31.

8. **Psakhie, S. G.,** Zolnikov, K. P., Dmitriev, A. I., Konovalenko, Iv. S. (2009). Kinematic properties of nanostructures based on bilayer nanocrystalline films, *Physical Mesomechanics*, 12, 112–116.

9. **Psakhie, S. G.,** Zolnikov, K. P., Kryzhevich, D. S. (2007). Elementary atomistic mechanism of crystal plasticity, *Physics Letters A*, 367, 250–253.

10. **Psakhie, S. G.,** Zolnikov, K. P., Kryzhevich, D. S., Lipnitskii, A. G. (2006). On structural defect generation induced by thermal fluctuations in materials with a perfect lattice under dynamic loading, *Physics Letters A*, 349, 509–512.

11. **Psakhie, S. G.,** Zolnikov, K. P. (1998). Possibility of a vortex mechanism of displacement of the grain boundaries under high-rate shear loading, *Combustion Explosion and Shock Waves*, 34(3), 366–368.

12. **Psakhie, S. G.,** Zolnikov, K. P. (1997). Anomalously high rate of grain boundary displacement under fast shear loading, *Technical Physics Letters*, 23, 555–556.

13. **Psakhie, S. G.,** Korostelev, S. Y., Negreskul, S. I., Zolnikov, K. P., Wang, Z. G., Li, S. X. (1993). Vortex mechanism of plastic-deformation of grain-boundaries—computer-simulation, *Physica Status Solidi B-Basic research*, 176(2), K41–K44

14. Zolnikov, K. P., **Psakhie, S. G.,** Panin, V. E. (1986). Alloy phase-diagrams using temperature, concentration and density as variables, *Journal of Physics F—Metal Physics*, 16, 1145–1152.

15. **Psakhie, S. G.,** Panin, V. E., Chulkov, E. V., Zhorovkov, M. F. (1980). Pseudopotential theory calculation of bounding energy of zinc atom with vacancy in aluminum, *Fizika Metallov i Metallovedenie*, 50, 620–622.

16. **Psakhie, S. G.,** Panin, V. E., Chulkov, E. V., Zhorovkov, M. F. (1980). Calculation of the bond-energy of Mg and Zn impurities with vacancies in Al-alloys, *Izvestiya Vysshikh Uchebnykh Zavedenii Fizika*, 8, 99–104.

VI. ***Biological and medical applications***

1. Sharipova, A., Gotman, I., **Psakhie, S. G.,** Gutmanas, E. Y. (2019). Biodegradable nanocomposite Fe-Ag load-bearing scaffolds for bone healing, *Journal of the Mechanical Behavior of Biomedical Materials*, 98, 246–254.

2. Lerner, M. I., Mikhaylov, G., Tsukanov, A. A., Lozhkomoev, A. S., Gutmanas, E., Gotman, I., Bratovs, A., Turk, V., Turk, B., **Psakhie, S. G.,** Vasiljeva, O. (2018). Crumpled aluminum hydroxide nanostructures as a microenvironment dysregulation agent for cancer treatment, *Nano Letters*, 18, 5401–5410.

3. Sharipova, A., Swain, S. K., Gotman, I., Starosvetsky, D., **Psakhie, S. G.,** Unger, R., Gutmanas, E. Y. (2018). Mechanical, degradation and drug-release behavior of nano-grained Fe-Ag composites for biomedical applications, *Journal of the Mechanical Behavior of Biomedical Materials*, 86, 240–249.

4. Tsukanov, A. A., **Psakhie, S. G.** (2017). From the soft matter-hard matter interface to bio-self-organization and hybrid systems, *Physical Mesomechanics*, 20, 43–54.

5. Tsukanov, A. A., **Psakhie, S. G.** (2016). Energy and structure of bonds in the interaction of organic anions with layered double hydroxide nanosheets: A molecular dynamics study, *Scientific Reports*, 6, 19986.

6. Lozhkomoev, A. S., Glazkova, E. A., Bakina, O. V., Lerner, M. I., Gotman, I., Gutmanas, E. Y., Kazantsev, S. O., **Psakhie, S. G.** (2016). Synthesis of core-shell AlOOH hollow nanospheres by reacting Al nanoparticles with water, *Nanotechnology*, 27, 205603.

7. Bakina, O. V., Glazkova, E. A., Svarovskaya, N. V., Lozhkomoev, A. S., Lerner, M. I., **Psakhie, S. G.** (2015). The influence of precursor disaggregation during synthesis of low-dimensional AlOOH structures on their morphology, *Russian Physics Journal*, 57, 1669–1675.

8. Mishnaevsky, L. Jr., Levashov, E., Valiev, R. Z., Segurado, J., Sabirov, I., Enikeev, N., Prokoshkin, S., Solov'yov, A. V., Korotitskiy, A., Gutmanas, E., Gotman, I., Rabkin, E., **Psakhie, S. G.**, Dluhoš, L., Seefeldt, M., Smolin, A. (2014). Nanostructured titanium-based materials for medical implants: Modeling and development, *Materials Science and Engineering: R: Reports*, 81, 1–19.

9. Mikhaylov, G., Mikac, U., Magaeva, A. A., Itin, V. I., Naiden, E. P., Psakhye, I., Babes, L., Reinheckel, T., Peters, C., Zeiser, R., Bogyo, M., Turk, V., **Psakhie, S. G.**, Turk, B., Vasiljeva, O. (2011). Ferri-liposomes as an MRI-visible drug-delivery system for targeting tumours and their microenvironment, *Nature Nanotechnology*, 6, 594–602.

VII. *Dusty plasma as a special state of particulate matter*

1. **Psakhie, S. G.**, Zolnikov, K. P., Abdrashitov, A. V. (2010). On the formation of structural states in dusty plasmas & *Physical Mesomechanics*, 13(5–6), 275–282.

2. Abdrashitov, A. V., Zolnikov, K. P., **Psakhie, S. G.** (2010). Effect of the Anisotropy of Confining Field on the Structure of Dusty Plasma Clusters, *Technical Physics Letters*, 36(10), 910–913.

3. **Psakhie, S. G.**, Zolnikov, K. P., Abdrashitov, A. V. (2009). Studying the response of a Coulomb ball of charged dust particles to external pulsed loads &*Technical Physics Letters*, 35(2), 120–122.

4. **Psakhie, S. G.**, Zolnikov, K. P., Skorentsev, L. F., Kryzhevich, D. S., Abdrashitov, A. V. (2008). Structural features of bicomponent dust Coulomb balls formed by the superposition of fields of different origin in plasma, *Physics of Plasmas,* 15, 053701.

5. **Psakhie, S. G.**, Zolnikov, K. P. (2008). Structure of binary dust Coulomb balls in confining fields of different origin, *Physical Mesomechanics*, 11(3-4), 144–148.

6. **Psakhie, S. G.**, Zolnikov, K. P., Skorentsev, L. F., Kryzhevich, D. S., Abdrashitov, A. V. (2008). Structural features of two-component dusty plasma Coulomb balls, *Technical Physics Letters*, 34(4), 319–322.

Biomechanical and Tribological Aspects of Orthopaedic Implants

Irena Gotman

Abstract Orthopaedic and dental implant treatments have allowed to enhance the quality of life of millions of patients. Total hip/knee arthroplasty is a surgical replacement of the hip/knee joint with an artificial prosthesis. The aim of joint replacement surgery is to relieve pain improve function, often for sufferers of osteoarthritis, which affects around a third of people aged over fifty. Nowadays, total hip and knee replacement (THR) surgeries are considered routine procedures with generally excellent outcomes. Given the increasing life expectancy of the world population, however, many patients will require revision or removal of the artificial joint during their lifetime. The most common cause of failure of hip and knee replacements is mechanical instability secondary to wear of the articulating components. Thus, tribological and biomechanical aspects of joint arthroplasty are of specific interest in addressing the needs of younger, more active patients. The most significant improvements in the longevity of artificial joints have been achieved through the introduction of more wear resistant bearing surfaces. These innovations, however, brought about new tribocorrosion phenomena, such as fretting corrosion at the modular junctions of hip implants. Stiffness mismatch between the prosthesis components, non-physiological stress transfer and uneven implant-bone stress distribution are all involved in premature failure of hip arthroplasty. The development of more durable hip and knee prostheses requires a comprehensive understanding of biomechanics and tribocorrosion of implant materials. Some of these insights can also be applied to the design and development of dental implants.

Keywords Total joint replacements · Low friction arthroplasty · Stress shielding · Wear of polyethylene · Hard-on-hard articulations · Fretting wear · Tribocorrosion

I. Gotman (✉)
Department of Mechanical Engineering, ORT Braude College, Karmiel, Israel
e-mail: irenag@braude.ac.il

G.-P. Ostermeyer et al. (eds.), *Multiscale Biomechanics and Tribology of Inorganic and Organic Systems*, Springer Tracts in Mechanical Engineering, https://doi.org/10.1007/978-3-030-60124-9_2

1 Introduction to Orthopaedic and Dental Implantable Devices

Orthopaedic and dental implants are surgical components that replace or interface with the bone. The most important implantable orthopaedic devices are total joint replacements, primarily total hip (THR) and total knee replacements (TKR), Fig. 1(left).

The aim of joint replacement surgical procedure (arthroplasty) is to relieve pain, improve function, and enhance quality of life, often for sufferers of osteoarthritis, which affects around a third of people aged over 50. Worldwide, more than one million THR surgeries and about two million TKR surgeries are performed every year, approximately 50% of which are done in the US. The rapidly aging population and a high prevalence of degenerative bone conditions in the elderly drive the demand for joint replacements even higher. According to American Academy of Orthopaedic Surgeons, the number of primary THRs in the US is projected to reach 635,000 in the year 2030 (171% increase vs. year 2014) and 1.23 million in the year 2060 (330% increase). Similarly, the projections for primary TKR are 1.28 million in the year 2030 (189% increase vs. year 2014) and 60 million in the year 2060 (382% increase).

Total joint arthroplasty is considered one of the most successful surgical interventions performed today. The reported survivorship after 15 years is above 90% for total hip replacements [1] and ranges between 82 and 98% for total knee replacements [2]. Despite this success, failures of joint replacements do occur, in which case the patient is required to undergo a revision surgery to replace the failed implant. In fact, the number of revision procedures is increasing faster than the number of primary arthroplasties. The projections of American Academy of Orthopaedic Surgeons for the US are 72,000 revision THRs (142% increase vs. year 2014) and 120,000 revision TKRs (190% increase) in the year 2030 and 110,000 revision THRs (219% increase vs. year 2014) and 253,000 revision TKRs (400% rise) in the year 2060. The rising rates

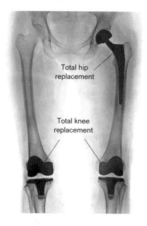
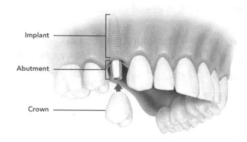

Fig. 1 Total joint replacements (left) and implant-supported dental prosthesis (right)

of failure are not surprising given the fact that patients currently receiving THR/TKR are ~20% heavier, more physically active, and live more than 25% longer compared with several decades ago. According to the latest study, approximately 42% of THRs will not last more than 25 years [3, 4]. This means that some artificial joints will need to be replaced at least once during a patient's lifetime. For younger patients (under 60 years of age), the lifetime risk of revision is around 30%, with the highest revision rates reported for males between the ages of 50–55 [5]. These clinical data indicate that the ultimate goal of joint arthroplasty—long-term pain-free function for the rest of the patient's life—has not yet been achieved.

Another important class of bone-interfacing prostheses are dental implants, Fig. 1(right). A dental (or endosseous) implant is a surgical replacement of the tooth root that interfaces with the with the bone of the jaw or skull to support a dental prosthetic (e.g. crown or bridge). The main objective of implant supported dental restoration is the permanent replacement of missing teeth. Every year, more than 800,000 dental implants are placed in the United States and more than 1.8 million in the European Union. As population is aging, the demand for dental implants will rise significantly in the next decade. Projection models suggest that among the US adults missing teeth, the prevalence of dental implant use could reach as high as 23% by the year 2026 [6]. Dental implants are designed to last a lifetime and have, according to different sources, a success rate of over 90% at 10–15 years follow-up [7]. However, failures do happen, especially in medically compromised patient population. It has been suggested that failure patterns and mechanisms behind bone loss around dental implants have much in common with joint replacements, and that biomechanical under- or overloading and synergy between friction, wear and corrosion are responsible for the majority of dental and orthopaedic implant system failures [8–10]. Therefore, a total hip replacement will be used throughout this paper as it is a representative example of an orthopaedic implant and because the knowledge from the discipline of orthopedics can be applied to oral implants.

2 Tribology of Total Hip Replacement

2.1 Charnley Low Friction Arthroplasty

The hip is one of the body's largest weight-bearing joints. This geometrically simple "ball-in-socket" joint consists of the head (top of the thigh bone, or femur) articulating inside the acetabular socket of the pelvis. The layer of articular cartilage covering the bone surfaces lubricated by the synovial fluid provides the joint with exceptional tribological properties. Total hip arthroplasty consists of replacing both the acetabulum and the femoral head with artificial components. The first total hip replacements were performed by Wiles (1938) and McKee (1951) [11]. In those early designs, both bearing surfaces (acetabular cup and femoral head) were made of stainless steel and were fixed to the bone with screws and bolts. These historical implants experienced high incidence of early failure associated with the component loosening, typically of the acetabular cup. The unsatisfactory clinical performance was primarily due to elevated friction, jamming and wear within the bearings. It turned out that the main limitation of the early THR designs was that they mimicked normal hip joint anatomy. Large femoral heads coupled with inconsistent manufacturing tolerances generated high frictional torque (turning force) on the articulating surfaces leading to high shear stresses and loosening at the acetabular cup-bone interface.

Realization of the tribological nature of failure of early hip replacement designs brought Sir John Charnley to introduce his revolutionary concept of "low-friction arthroplasty" (LFA) [12]. LFA follows the principle of low-frictional torque based on the largest possible difference between the radius of the femoral head and that of the outer aspect of the acetabular component [13]. Charnley and his colleagues concluded that in order to minimize frictional torque and protect the cup-bone interface, the head diameter should be not greater than half of the external diameter of the cup. Consequently, the head diameter of THR was reduced from the earlier used 41.5–22.2 mm, less than half anatomical femoral head diameter (48–55 mm on the average). Charnley also recognized that in addition to the low frictional torque design, it is important that the acetabular cup material had a low friction coefficient against the material of the femoral head (stainless steel in Charnley prosthesis). The first material used was a self-lubricating polymer polytetrafluoroethylene (PTFE). PTFE sockets, however, wore out disappointingly fast causing "intense foreign body reaction" to wear debris and gross destruction of bone. The next polymeric material used—ultra-high molecular weight polyethylene (UHMWPE) [14] proved much more successful. UHMWPE had excellent wear resistance, low friction and high impact strength, and no problems were observed with metal-on-UHMWPE (M-PE) bearings in the early years post-implantation. Thus, the biomechanical concept of low friction arthroplasty combined with the use of a low-wear acetabular cup material (UHMW polyethylene) started a new era in joint replacement surgery. Very soon, a more biocompatible and corrosion resistant cobalt-chromium alloy (CoCr) came to replace stainless steel in the femoral component of Charnley prostheses. From then on, arthroplasty has known considerable evolution, but metal-on-UHMWPE (CoCr-PE) articulation remains the gold standard for artificial hips and other artificial joints, including the knee and shoulder. It is definitely Sir John Charnley to be credited with advancing our understanding of tribological effects as they apply in orthopaedics, and the significance of friction, wear and lubrication of implant materials for their longevity and function, and particularly the body's reaction to the particulate debris produced as a result of implant wear [15].

Another breakthrough made by John Charnley was the introduction of a self-curing acrylic resin (bone cement) as a grouting agent to secure the implant components to bone. The cement is injected as a dough-like mass and hardens around the implant to ensure its anchorage in the bone [16]. The use of bone cement allowed for the firm fixation of hip replacements unachievable with the previously used screws and bolts. Charnley cemented THA rapidly gained widespread popularity and became one of the most successful orthopaedic procedures with reported survivorship rates greater than 90% at 15–20 years. In young, physically active patients, however, failure rates were significantly higher. The main mode of failure was aseptic loosening (loss of fixation) of one or both implant components, secondary to periprosthetic osteolysis—resorption of bone surrounding the implant [17, 18]. At the time, osteolysis was thought to be caused by biological reaction to bone cement described as "cement disease". This led to the erroneous conclusion that the problem of aseptic loosening can be solved if the use of bone cement is avoided. Consequently, the innovative concept of cementless fixation was developed. In the cementless approach, implant components are stabilized within the bone by bone ingrowth into the porous surface layer or by bone ongrowth onto the textured surface. The material of choice for cementless prostheses are titanium alloys (mostly Ti6Al4V) due to their superior biomechanical compatibility: Ti6Al4V is capable of osseointegration (establishing direct contact with bone) and has a low modulus of elasticity, half that of CoCr alloy (110 vs. ~230 GPa) [19]. The latter is important for minimizing periprosthetic bone resorption due to stress shielding caused by stiffness mismatch between the implant and the bone (elastic modulus of ~20 GPa) [20]. Against early expectations, cementless fixation did not eliminate the problem of aseptic loosening of hip replacements and the outcomes were no better than with cemented THA.

2.2   Wear of Polyethylene—The Main Culprit of Aseptic Loosening

Once it became clear that not "cement disease" was the problem, the proposition that aseptic loosening is related to particulate wear debris from the UHMWPE acetabular component was put forward. The hard metal femoral head can produce wear of the polyethylene surface during articulation through both abrasive and adhesive mechanisms. Positive asperities on the hard counterface can abrade the polyethylene surface, which is relatively softer, producing abrasive wear debris. Friction between the articulating surfaces shears off particles, producing adhesive wear debris. Hard particles present in the joint space (cement, metal, bone) can enter between the articulating surfaces, embed in the polyethylene and abrade the metallic counterface—"third body" wear mechanism. In total knee replacements, the dominant form of wear is delamination of polyethylene which occurs as a result of cyclic compressive-tensile loading that leads to subsurface cracking. Wear particles migrate into tissues and are phagocytosed by macrophages which become activated and release pro-inflammatory cytokines that stimulate bone resorption (osteolysis) around the implant leading to prosthesis loosening [21–23]. Typically, billions of submicron UHMWPE wear particles (average diameter of 0.3–0.5 μm) per year are released into periprosthetic fluids. Two main factors affecting the volumetric wear of UHMW polyethylene are diameter and material of the femoral head that articulates against the polyethylene. CoCr alloy is a metal traditionally used for the femoral head. Meanwhile, titanium alloys exhibit poor tribological behavior under abrasive and adhesive wear and should not be used for manufacturing femoral heads. Titanium is much softer than CoCr and is easily scratched by hard "third body" particles that intrude between the articulating surfaces which results in increased friction and abrasive wear of the polyethylene. For CoCr heads articulating against UHMWPE, larger heads are associated with greater volumetric wear of polyethylene and high revision rates. The popular 28 mm diameter is a compromise between wear performance and risk for dislocation of the implant.

3 Alternative Bearing Surfaces

Since the inflammatory response to wear debris was established as the main cause of aseptic loosening, efforts at extending joint replacement longevity have focused primarily on development of more wear resistant bearing surfaces. Two major directions included (i) improving the quality of UHMWPE and (ii) avoiding the use of polyethylene bearing altogether. Alternatively, attempts were made to improve the wear resistance of metallic components by providing them with a hard, wear resistant surface.

Prior to being introduced into clinical practice, new artificial joint materials must be submitted to realistic preclinical tests. The tribological performance of novel THA bearing couples is tested in hip joint simulators designed to mimic the biomechanics of hip joint in a simulated physiological environment. Despite the reported discrepancies between in vitro simulation results and wear data from explanted devices, joint simulators are instrumental in predicting clinical wear performance of new bearing surfaces and identifying the risk of clinically relevant wear [24].

3.1 Cross-Linked Polyethylene

Radiation cross-linking significantly decreased the wear rate of UHMWPE against CoCr in simulation studies, Fig. 2.

In total hip arthroplasty, this has translated into better long-term outcomes and a significant reduction in the rate of revision for younger patients [25]. Furthermore, highly cross-linked polyethylene may allow use of large-diameter femoral heads without concern about increased polyethylene wear [26]. The biomechanical rationale for using large-diameter femoral heads is that they allow for a greater range of motion and limit the risk of dislocation by increasing jump distance—distance the head has to "jump" before leaving the acetabular cup [27]. For total knee arthroplasty, however, clinical evidence has been inconclusive, and no distinctive improvement was observed when highly cross-linked polyethylene (HXLPE) was used compared to conventional UHMWPE [28–30]. This could be attributed to distinct biomechanical environments and different relative contributions of polyethylene wear mechanisms in the two types of joints: adhesive and abrasive wear in the hip versus fatigue wear and delamination in the knee. In addition, the reduced fatigue strength and toughness

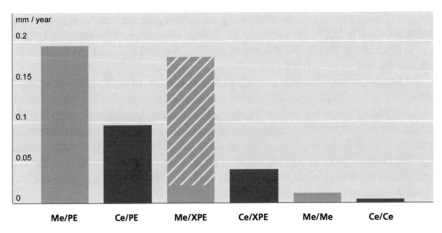

Fig. 2 In vivo linear wear rates of different bearing couples.: Me—metal, PU—UHMW polyethylene, Ce—ceramic, XPE—cross-linked polyethylene. https://www.ceramtec.com/biolox/clinical-experience/wear-osteolysis/

of cross-linked polyethylene may cause the cracking of HXLPE knee replacement components.

3.2 Ceramic-on-Polyethylene Articulation

The wear of polyethylene acetabular components has been further reduced by replacing the femoral head material from the CoCr metal to a ceramic. The first ceramic used was aluminum oxide (alumina). Alumina is oxidation resistant, biocompatible, very hard and scratch resistant. Furthermore, the fine grain structure allows alumina to be polished to a very low surface roughness (Ra < 0.05 μm) resulting in reduced wear of the polyethylene countersurface. In addition, the high wettability of alumina positively affects the lubricating film thus decreasing the coefficient of friction [31]. Due to the brittle nature of alumina, however, a few (but potentially devastating) fracture failures of the early ceramic heads (Biolox® forte, CeramTec, GmbH) were reported. To reduce the risk of brittle fracture of femoral heads, an alternative ceramic material—yttria stabilized zirconia (Y-TZP) was introduced. Y-TZP is a metastable material that exhibits an extremely high (for a ceramic) fracture toughness due to a unique transformation toughening mechanism. High fracture toughness combined with excellent tribological behavior against polyethylene made zirconia femoral heads very popular in the last decade of the past century. It turned out, however, that when exposed to body fluids, the metastable tetragonal phase may transform to the stable monoclinic structure [32, 33]. This aging process occurs in vivo on the surface of zirconia heads, leading to their roughening and microcracking. The problem became apparent in the year 2000 when an unusually large amount of failures was reported following a change in the manufacturing process of zirconia heads. In 2001, the company St. Gobain Desmarquest (Vincennes, France) issued a voluntary recall, and the use of zirconia femoral heads in hip arthroplasty came to an end.

An important step towards enhancing the resistance of ceramic femoral heads to brittle fracture was the introduction of zirconia-toughened alumina—a composite ceramic (Biolox delta) whose fracture toughness is twice that of Biolox forte [34]. The volumetric wear of polyethylene acetabular cups articulating against ceramic femoral heads is several times lower than that of their metal-on-polyethylene counterparts, Fig. 2. The absence of allergic reaction to alumina makes ceramic-on-polyethylene articulation especially suitable for patients suffering from immune hypersensitivity to metals such as nickel, chromium and cobalt.

3.3 Hard-on-Hard Articulations

Notwithstanding the improved wear resistance of cross-linked polyethylene, the most spectacular reductions in volumetric wear of articulating joints (by one-two orders of magnitude) are achieved with hard-on-hard (metal-on-metal and ceramic-on-ceramic) bearings, eliminating altogether the soft polyethylene component, Fig. 2.

3.3.1 Ceramic-on-Ceramic Bearings

Ceramic-on-ceramic (CoC) articulations exhibit by far the lowest wear rates and are considered a viable option for young, active patients. Excellent mid-term clinical outcomes are reported for modern CoC hip replacements using zirconia-toughened alumina (Biolox delta), and component fractures are extremely rare [35, 36]. One well-recognized and annoying complication of ceramic-on-ceramic THA is squeaking—a high pitched, audible sound that occurs during movement, often related to a specific activity [37, 38]. The reported incidence of squeaking in CoC THA lies between 0.5% and >20%. A likely cause of squeaking is adverse tribological conditions caused by the loss of fluid film lubrication and high friction between the ceramic components. Friction generates forced vibrations that cause the metallic parts to resonate and convert vibrational energy into an audible noise. The incidence of squeaking is strongly affected by implant- and patient-specific factors.

3.3.2 Metal-on-Metal Bearings

Metal-on-metal (MoM) total hip arthroplasty almost totally abandoned in the mid-1970's in favor of Charnley's metal-on-polyethylene THA, made its comeback in the very beginning of the twenty-first century [39–41]. By that time, it became clear that the failures of fist generation metal-on-metal implants were not due to the bearing surface material but were mainly caused by design errors and inadequate manufacturing. Survivorship analysis of hip replacements implanted between 1965 and 1973 revealed a surprisingly great longevity among some of the original all-metal designs. At long-term follow-up, the wear of the long-lived metal-on-metal McKee–Farrar prostheses was by at least one order of magnitude smaller compared to the metal-on-polyethylene Charnley prostheses. Novel CoCr-on-CoCr devices designed with the standard small femoral head diameter (22 and 28 mm) exhibited very low wear rates in hip simulator tests, only slightly higher than those of ceramic-on-ceramic couples. At the same time, they possessed an obvious advantage of not being brittle. An important observation from the simulation tests was that more effective fluid film lubrication and correspondingly low wear rates were achieved with larger diameter femoral heads. Based on these results, large diameter ($\geq$36 mm) MoM articulations were developed and quickly gained a big share of the market, both in the USA (approx. 30% in 2006–2007) and worldwide. It took only a few years to realize that MoM THA was associated with higher revision rates and lower patient satisfaction. In addition to the well-known phenomenon of aseptic loosening), a new mode of failure was observed—adverse local tissue reaction (ALTR) [42, 43]. ALTR included periprosthetic soft tissue inflammation, soft tissue necrosis, and pseudotumor formation. Some patients were asymptomatic but those presenting with pain and elevated metal blood levels had to be revised.

The final blow for metal-on-metal designs came in 2010 when DePuy, J&J's orthopaedic branch voluntarily recalled its ASR MoM hip system due to an unacceptably high failure rate (~13% after 5 years) [44]. The use of metal-on-metal devices declined rapidly to less than 1% of all the THR systems being implanted today. The analysis of failed implants revealed that wear particles from MoM articulations are approximately 50 nm in size, much smaller than the 0.3–0.5 μm UHMWPE debris particles. Thus, despite the low volumetric wear of MoM bearings, the actual number of released particles is considerably higher than for conventional metal-on-PE bearings. Moreover, these metallic nanoparticles are more biologically active and corrode rapidly in the body fluids releasing large amounts of potentially toxic cobalt and chromium ions. The unfavorable outcomes of large diameter metal-on-metal hip replacements are the result of their complex and not well-understood tribology. From the biomechanical and tribological point of view, MoM articulations were found to be extremely unforgiving to positioning and manufacturing mistakes: slight deviations from the optimal alignment, sphericity and radial clearance could lead to adverse lubrication conditions and excessive wear. To date, there are no FDA-approved metal-on-metal total hip replacement devices marketed for use in the US, the only available options being ceramic-on-ceramic, ceramic-on-polyethylene and metal-on-polyethylene bearings.

4   Bearing Materials in Total Knee Replacement

In total knee arthroplasty (TKR), bearing surface options have been much more limited. Practically all state-of-the-art knee replacements use a CoCr alloy femoral component articulating on polyethylene. The knee joint has a complex nonconformal geometry and is subjected to high contact stresses. Polyethylene is sufficiently compliant to accommodate stress concentration caused by misalignment or surface-to-surface contact of asperities. The rigid nature of ceramics makes ceramic-on-ceramic articulation much less forgiving of surface irregularities and slight malposition thus leading to increased risk of brittle fracture. Therefore, transferring the benefits of excellent tribological properties of ceramics to the complex geometry of knee prostheses remains challenging. At the present time, all-ceramic knee endo-prosthesis is not a feasible option. Several TKA designs having a ceramic femoral component articulating on polyethylene are available for clinical use or are under clinical trial [45–47]. These devices represent a promising alternative for patients with a known hypersensitivity to metals, but it is still early to draw conclusions regarding their long-term outcomes in terms of longevity, wear damage and incidence of brittle fracture.

5   Surface-Modified Bearing Materials

The desired alternative to existing articulating materials for joint implants would combine the fracture toughness of metals with the wear performance of ceramics [48]. One approach for achieving this is to deposit or overlay a ceramic coating onto a metallic substrate. The bond created between the deposited coating and the substrate is only physical (rather than chemical) resulting in relatively weak coating adhesion. Given the significantly different mechanical properties of the ceramic film and the underlying metal, adhesive failure between the two materials occur under load or during articulation. Diamond like carbon (DLC) and titanium nitride (TiN) are the most extensively studied wear resistant coatings for artificial joints. Despite their high hardness and excellent biocompatibility, delamination and spalling of such coatings has been observed in clinical trials and some wear simulation tests. Insufficient adhesion and inadequate load bearing capacity of the underlying softer metallic substrate are believed to be the major obstacles on the way to successful implementation of hard coatings into clinical practice [48]. It is believed that these shortcomings of externally applied ceramic layers can be alleviated by diffusional surface hardening—reactively diffusing a non-metallic element into the substrate at elevated temperatures thus transforming the surface from metal to ceramic.

Oxidized zirconium (OXINIUM, Smith & Nephew Orthopaedics) was developed for orthopedic applications to provide improvements over CoCr alloy for resistance to roughening, frictional behavior, and biocompatibility without the mechanical limitations of brittle monolithic ceramics [49]. The ceramic surface is formed by heating a zirconium alloy in air to allow oxygen to diffuse into the substrate and to transform the metal surface to zirconium oxide (zirconia) ceramic. Despite the consistently lower wear of polyethylene components articulating against the ceramic surfaced OXINIUM in knee and hip joint simulator tests, clinical studies have shown no statistically significant differences in mid-term implant survivorship between OXINIUM and CoCr components [50–52]. Additional research is needed for the clinical performance of OXINIUM to better understand long-term outcomes. Meanwhile, OXINIUM contains no detectable nickel or chromium, which makes such implants a safer choice for patients with metal allergies.

6   Fretting Wear Damage of Total Joint Replacements

6.1   Modular Connections of Hip Prostheses

In early hip replacement devices, the femoral stem and head were produced as a single-piece, monolithic component—a so-called monobloc design, Fig. 3a. Nowadays, almost all hip joints are modular and consist of a separate femoral head that fits on the stem, Fig. 3b, c. The reliable joining of modular components of total joint replacements is based on the concept of a Morse taper, i.e. that of the cone in the cone [53, 54]. The two components of the Morse taper form a firm fit that relies on friction and mechanical interlocking. Modularity provides many advantages, such as greater intraoperative flexibility allowing the surgeon to restore the patient's anatomy and

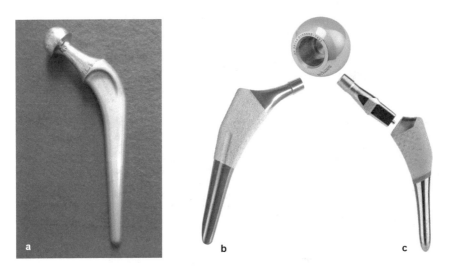

Fig. 3 Different designs of femoral components of total hip replacement: monoblock (Charnley) (**a**) and modular (**b, c**): b—monolithic stem, c—dual-taper stem

to adjust leg length, decreased implant inventory and potential ease of revision by exchanging only the failed component. Furthermore, modularity allows the combination of head and stem made of different materials with specific properties thus optimizing the clinical performance of the whole assembly. For example, a stem from titanium alloy that is most suitable for cementless fixation but has an inadequate wear resistance can be combined with a hard, wear resistant Co alloy or ceramic head.

Despite its benefits, the modular design has been associated with higher revision rates due to adverse tissue reaction, neck fracture and femoral head disassociation. Modularity creates additional mechanical junctions (neck-head and neck stem interfaces having a crevice-like geometry) that become weak points where micromotion and wear can occur [55]. The hip joint is subject to cyclic stresses from gait loading amounting to more than one million cycles a year. As a result of cyclic loading, a low amplitude oscillating relative motion occurs at the taper junction of femoral components made of dissimilar materials and having different rigidity. This leads to the tribological process of fretting causing surface damage of the fitting contact surfaces. The process is often referred to as "mechanically assisted crevice corrosion" (MACC) and can be briefly described as follows [56–59], Fig. 4a–d. Rubbing between the taper surfaces under stress leads to mechanical disruption of the protective oxide film and corrosion followed by rapid regeneration of the oxide layer (repassivation). This is accompanied by oxygen consumption, metal ion release and hydrolysis, and voltage drop. As the mechanical damage to the oxide is continuously repeated, oxygen in the crevice is depleted while the liberated hydrogen ions acidify the fluid to the point where repassivation becomes impossible. Given that the corrosion resistance

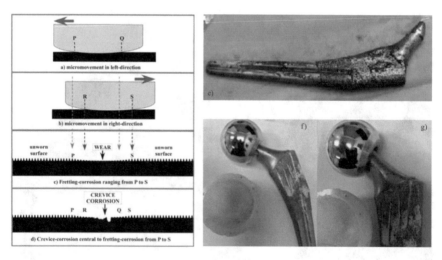

Fig. 4 a–d Schematic of mechanically assisted crevice corrosion (MACC); **e** retrieved Accolade stem with marked fretting wear of the neck taper; **f** retrieved Accolade stem with instability and dissociation of the head-neck junction. Reproduced from P. Walker et al., Reconstructive Review 2016, September; 6(3):13–18

of both cobalt and titanium alloys relies on passivity, the inability to rebuild the passive oxide film results in active corrosion. An additional feature of MACC is hydrogen gas accumulation below the modular neck and hydrogen embrittlement on the surface of titanium components [60, 61]. MACC at the femoral head-neck interface of modular hip replacements is commonly referred to as trunnionosis.

Wear debris and metal ions produced at taper junctions can lead to elevated blood metal ion concentrations and adverse local tissue reactions. The histological appearance of periprosthetic tissues surrounding corroded trunnions is similar to tissues surrounding failed hip replacements with metal-on-metal bearings. Some patients will remain asymptomatic; others will develop adverse clinical symptoms that require revision: necrosis, pseudotumors, pain, etc. In heavy patients, cases of catastrophic fatigue fracture of titanium male stem tapers initiated at notch-like irregularities of the fretted surface were reported [62, 63].

Several modular hip systems are available on the market, differing in design and femoral stem/neck/head material. As discussed above, the head is typically made of the wear-resistant cobalt-chromium alloy or ceramic whereas the stem can be either cobalt-chromium or titanium alloy. The stem can be monolithic (one-piece design), Fig. 3b, or can have an exchangeable neck (dual-taper stem), Fig. 3c. In the latter case, an additional stem/neck interface is introduced which only increases the concern regarding the occurrence of mechanically assisted crevice corrosion. Ti alloys are the common material choice for cementless stems due to their superior osseoconductivity and low stiffness preventing bone resorption secondary to stress-shielding. In this respect, beta-titanium alloys whose elastic modulus is significantly lower than that of the standard Ti-6Al-4V alloy are most favorable candidate materials [64]. However, combining beta-titanium with the high-modulus Co-Cr alloy components has proven disastrous and led to several major Hip Replacement recalls. The beta-titanium involved is a proprietary Ti-12Mo-6Zr-2Fe (TMZF®) alloy having the elastic modulus of around 75 GPa that was developed by Stryker Orthopaedics.

When a monolithic TMZF stem (Accolade I stem design) was used in combination with a Co-Cr alloy head (LFIT V40) supposed to minimize the risk of dislocation, high incidence of failures due to taper wear and adverse local tissue reaction was reported [65–68]. The patients experienced pain and restricted motion requiring a revision surgery. In some cases, tapers were damaged to such a significant level that head dissociation from the stem occurred. Examples of such gross stem taper failure (GTF) and head disassociation are shown in Fig. 4e–g. It is hypothesized that micromotion and fretting corrosion at the taper junction results in the widening of the gap between the head and the neck which allows the head to turn and move on the stem. The harder Co-Cr head abrades the softer titanium alloy neck to such an extent that the head can easily disassociate from the stem. Importantly, GTF has occurred recurrently with stems made of the low-modulus TMZF alloy, very rarely with stems from the standard Ti-6Al-4V alloy and never with the high modulus Co-Cr alloy stems suggesting the influence of the material stiffness. Indeed, numerical modelling has shown that the deformation and micromotion at the Co-Cr head-stem taper interface was significantly larger for the TMZF-alloy stem compared to the CoCr and even to Ti-6Al-4V alloy stems. The phenomenon of head disassociation was most often observed with large diameter Co-Cr heads (36 mm and larger) and these were voluntarily recalled by the manufacturer in 2016. Other femoral head sizes, as well as ceramic heads remain on the market however problems with the device continue to be of concern as failures in implant sizes outside of the recall are being reported. The Accolade I stem made of the low-modulus TMZF beta titanium alloy was never recalled but its use declined rapidly. In 2012 Stryker replaced Accolade I stem with the standard Ti-6Al-4V alloy stem (Accolade II).

The most well-known case of tribocorrosion of modular hip replacements is the failure of dual-taper Rejuvenate and ABG II stems launched by Stryker Orthopaedics in 2009 [69–72]. Both designs combined a low-modulus TMZF femoral stem and an exchangeable cobalt-chromium alloy neck. Since Co-Cr is harder and stiffer than titanium, it was suggested that this would allow safer and long-term use of the modular neck. The femoral heads were either Co-Cr or ceramic articulating on a UHMW polyethylene acetabular cup. Extremely high revisions rates secondary to tribocorrosion at the taper connection were reported for both designs, reaching, for Rejuvenate stem, 65% three years post-implantation. Due to these unacceptable failure rates and the ensuing FDA investigation, Stryker was forced to issue a voluntary recall of both products in 2012. Similarly to the case of Accolade II, the culprit in the failure of the dual-taper mixed-metal stem was the low elastic modulus of the TMZF alloy. Following the recall of the Rejuvenate and ABG II, Stryker discontinued the use of the low-stiffness beta-titanium and replaced it with the standard Ti-6Al-4V alloy. Current recommendations regarding modular hip replacements include avoiding femoral stems with low flexural rigidity and reducing the number of modular junctions, e.g. by using fixed neck stems. Also, substituting Co-Cr alloy heads with the chemically inert ceramic heads is expected to reduce MACC since corrosion processes will only occur on the metallic stem taper and will not be accelerated by galvanic coupling with a dissimilar metal [73, 74]. From the biomechanical point of view, smaller diameter heads are typically reported to produce less fretting damage since they generate smaller head-neck moment arm and correspondingly smaller torsional forces at the head-neck taper junction [75, 76]. The use of small diameter heads (36 mm and less), however, is associated with an increased risk of dislocation of hip replacement and is not willingly accepted by orthopaedic surgeons.

It follows from the above discussion that even if all the guidelines are followed, tribocorrosion at modular junctions of orthopaedic and dental implants exposed to cyclic loading cannot be fully eliminated. To overcome the problem of head-neck taper degradation, different surface engineering approaches are being investigated. One possibility is to coat a titanium alloy stem taper with a hard, wear-resistant film. The results of in vitro evaluation of TiN and TiN/AlN coatings suggest that these coatings provide superior fretting and fretting corrosion resistance to the tapered interfaces of the Co-Cr-Mo and Ti-6Al-4V alloy components [77–79]. Increasing the interfacial bond strength between the coating and the substrate could improve the fretting and corrosion resistance even more.

6.2 Stem-Cement Interface

Another joint replacement zone prone to tribocorrosion is the stem-cement inter-face of cemented prostheses [80–83]. Under physiological loading, this interface experiences a low amplitude oscillatory micromotion. During relative sliding, hard radiopacifier particles (e.g., ZrO_2) within the cement abrade the polished surface of the femoral stem and induce a tribocorrosive interaction. The effect is most pronounced for titanium alloy stems that experience larger flexural deflections and are more easily abraded than the stiffer and harder CoCr stems [84]. Fretting wear damage results in the formation of gaps/crevices between the cement mantle and the titanium stem leading to crevice corrosion of the metallic surface. Both the surface damage and the immunological reaction to released particles and ions compromise the stem stability and may lead to premature failure of the cemented joint prosthesis. Similarly to the taper junctions of modular implants, the low elastic modulus of tita-nium here is a drawback rather than an advantage. Flexural deflections of femoral stems lead to the cracking of the cement mantle and debonding at cement-stem interface, and both phenomena are much more pronounced for the low-stiffness tita-nium stems. These biomechanical and tribocorrosion problems make Ti alloy stems a much less popular option for cemented hip replacements. Although cemented Ti alloy stems are still available on the market, it is believed by many that the use of titanium stems in cemented THA should be abandoned [85].

7 Tribocorrosion in Dental Implants

Despite high success rates, 5–11% of implant-supported dental restorations fail within 10–15 years and must be removed [86]. The dominant failure mode of dental implants is peri-implantitis (inflammation of tissues surrounding the implant) and the associated loss of supporting bone. Tribocorrosion at internal connections between the prosthesis parts (implant, abutment and crown) and at implant-bone interface is among the major contributors to peri-implantitis [87–90]. The implant and the abutment are made of titanium or titanium alloy, whereas the crown is usually made of porcelain or zirconia ceramic. Despite very precise machining, there always is a microgap between the implant and the prosthetic connector. During mastication, micromotion occurs at the implant-abutment and abutment-crown interface leading to fretting and crevice corrosion. The situation is obviously similar to mechanically assisted crevice corrosion at taper connections of total hip replacements. The released degradation products (metal ions and metal oxide particles) initiate inflammatory tissue response that can eventually result in peri-implant bone loss. Furthermore, material loss enlarges the microgap between the abutment and the implant allowing for rapid gap colonization by oral microorganisms and subsequent bacterial infec-tion. The combined action of microbiological, mechanical and tribocorrosion factors promotes peri-implantitis and the associated bone loss eventually leading to implant failure.

8 Summary

With the growing demand for orthopaedic and dental implants and expectations of longer device lifetimes for the younger patients, wear and corrosion of articulating surfaces and modular junctions of these implantable devices are a prime concern. Aside from the patient's activity and physiological state, the biotribological performance of a prosthesis depends on the mechanical design and the materials used (metals, ceramics, polymers). Despite the substantial improvements achieved in both directions, the ultimate goal of total joint replacements—long-term pain-free function for the rest of the patient's life—has not yet been achieved. Currently there are gaps in our understanding of biomechanics and wear behavior of medical implants. Closing these gaps will help guide future research in this field and improve the longevity of orthopaedic and dental implantable devices.

References

1. Ferguson RJ, Palmer AJR, Taylor A, Porter ML, Malchau H, Glyn-Jones S (2018) Hip replacement. Lancet 392:1662–1671
2. Price AJ, Alvand A, Troelsen A, Katz JN, Hooper G, Gray A, Carr A, Beard D (2018) Knee replacement. Lancet 392:1672–1682
3. Cook R, Davidson P, Martin R (on behalf of NIHR Dissemination Centre) (2019) More than 50% of hip replacements appear to last 25 years. Journal BMJ 367:l5681
4. Evans JT, Evans JP, Walker RW, Blom AW, Whitehouse MR, Sayers A (2019) How long does a hip replacement last? A systematic review and meta-analysis of case series and national registry reports with more than 15 years of follow-up. Lancet 393:647–654
5. Bayliss LE, Culliford D, Monk AP, Glyn-Jones S, Prieto-Alhambra D, Judge A, Cooper C, Carr AJ, Arden NK, Beard DJ, Price AJ (2017) The effect of patient age at intervention on risk of implant revision after total replacement of the hip or knee: a population-based cohort study. Lancet 389:1424–1430
6. Elani HW, Starr JR, Da Silva JD, Gallucci GO (2018) Trends in dental implant use in the U.S., 1999–2016, and projections to 2026. J Dental Res 97(13):1424–1430
7. Moraschini V, da C. Poubel LA, Ferreira VF, dos SP Barboza E (2015) Evaluation of survival and success rates of dental implants reported in longitudinal studies with a follow-up period of at least 10 years: a systematic review. Int J Oral Maxillofac Surg 44:377–388
8. Albrektsson T, Becker W, Coli P, Jemt T, Mölne J, Sennerby L (2019) Bone loss around oral and orthopedic implants: an immunologically based condition. Clin Implant Dentistry Related Res 21(4):786–795
9. Coli P, Sennerby L (2019) Is peri-implant probing causing over-diagnosis and over-treatment of dental implants? J Clin Med 8(8):1123. https://doi.org/10.3390/jcm8081123
10. Souza JCM, Henriques M, Teughels W, Ponthiaux P, Celis JP, Rocha LA (2015) Wear and corrosion interactions on titanium in oral environment: literature review. J Bio- Tribo-Corrosion 1(13). https://doi.org/10.1007/s40735-015-0013-0

11. Gomez PF, Morcuende JA (2005) Early attempts at hip arthroplasty–1700s to 1950s. Iowa Orthop J 25:25–29
12. Wroblewski BM, Siney PD, Fleming PA (2006) The Charnley Hip replacement—43 years of clinical success. Acta Chirurgiae Orthopaedicae et Traumatologiae Cechoslovaca 73(1):6–9
13. Wroblewski BM, Siney PD, Fleming PA (2009) The principle of low frictional torque in the Charnley total hip replacement. J Bone Joint Sur Br Vol 91-B:855–858
14. Sobieraj MC, Rimnac CM (2009) Ultra high molecular weight polyethylene: mechanics, morphology, and clinical behavior. J Mech Behav Biomed Mater 2(5):433–443
15. Sonntag R, Beckmann NA, Reinders J, Kretzer JP (2015) Materials for total joint arthroplasty—biotribology of potential bearings. Imperial College Press, London, pp 15–40
16. Vaishya R, Chauhan M, Vaish A (2013) Bone Cement. J Clin Orthop Trauma 4(4):157–163
17. Iannotti JP, Balderston RA, Booth RE, Rothmann RH, Cohn JC, Pickens G (1986) Aseptic loosening after total hip arthroplasty. J Arthroplasty 1:99–103
18. Sundfeldt M, Carlsson LV, Johansson CB, Thomsen P, Gretzer C (2006) Aseptic loosening, not only a question of wear: a review of different theories. J Acta Orthop 77(2):177–197
19. Geetha M, Singh AK, Asokamani R, Gogia AK (2009) Ti based biomaterials, the ultimate choice for orthopaedic implants—a review. Prog Mater Sci 54:397–425
20. Sumner DR (2015) Long-term implant fixation and stress-shielding in total hip replacement. J Biomech 48(5):797–800
21. Abu-Amer Y, Darwech I, Clohisy JC (2007) Aseptic loosening of total joint replacements: mechanisms underlying osteolysis and potential therapies. Arthritis Res Therapy 9(1):6
22. Harris WH (1994) Osteolysis and particle disease in hip replacement. A review. Acta Orthop Scand 65:113–123
23. Wooley PH, Schwarz EM (2004) Aseptic loosening. Gene Ther 11:402–407
24. Medley JB (2016) Can physical joint simulators be used to anticipate clinical wear problems of new joint replacement implants prior to market release? Proc Inst Mech Eng Part H J Eng Med 230(5):347–358
25. Dumbleton JH, D'Antonio JA, Manley MT, Capello WN, Wang A (2006) The basis for a second-generation highly cross-linked UHMWPE. Clin Orthop Relat Res 453:265–271
26. Allepuz A, Havelin L, Barber T, Sedrakyan A, Graves S, Bordini B, Hoeffel D, Cafri G, Paxton E (2014) Effect of femoral head size on metal-on-HXLPE hip arthroplasty outcome in a combined analysis of six national and regional registries. J Bone Joint Surg 96(1):12–18
27. Banerjee S, Pivec R, Issa K, Kapadia BH, Khanuja HS, Mont MA (2014) Large-diameter femoral heads in total hip arthroplasty: an evidence-based review. Am J Orthop (Belle Mead, N.J.) 43(11):506–512
28. Brown TA, van Citters DW, Berry DJ, Abdel M (2017) The use of highly crosslinked polyethylene in total knee arthroplasty. Bone Joint J 99B(8):996–1002
29. Wilhelm SK, Henrichsen JL, Siljander M, Moore D, Karadsheh M (2018) Polyethylene in total knee arthroplasty: where are we now? J Orthop Surg 23(6):1–7
30. de Steiger RN, Muratoglu O, Lorimer M, Cuthbert AR, Graves SE (2015) Lower prosthesis-specific 10-year revision rate with crosslinked than with non-crosslinked polyethylene in primary total knee arthroplasty. Acta Orthop 86(6):721–727
31. Cash D, Khanduja V (2014) The case for ceramic-on-polyethylene as the preferred bearing for a young adult hip replacement. HIP Int 24(5):421–427
32. Habermann B, Ewald W, Rauschmann M, Zichner L, Kurth AA (2006) Fracture of ceramic heads in total hip replacement. Arch Orthop Trauma Surg 126:464–470
33. Chevalier J (2006) What future for zirconia as a biomaterial? Biomaterials 27(4):535–543
34. Kurtz SM, Kocagöz S, Arnholt C, Huet R, Ueno M, Walter WL (2014) Advances in zirconia toughened alumina biomaterials for total joint replacement. J Mech Behav Biomed Mater 31:107–116
35. Reuven A, Manoudis G, Aoude A, Huk OL, Zukor D, Antoniou J (2014) Clinical and radiological outcome of the newest generation of ceramic-on-ceramic hip arthroplasty in young patients. Adv Orthop Surg. https://doi.org/10.1155/2014/863748

36. Kuntz M, Usbeck S, Pandorf T, Heros R (2011) Tribology in total hip arthroplasty (Knahr K, ed). Springer, Berlin, pp 25–40

37. Brockett CL, Williams S, Zhongmin J, Isaac GH, Fisher J (2013) Squeaking hip arthroplasties. J Arthrop 28(1):90–97

38. Mai K, Verioti C, Ezzet KA, Copp SN, Walker RH, Colwell Jr CW (2010) Incidence of 'squeaking' after ceramic-on-ceramic total hip arthroplasty. Clin Orthop Relat Res 468(2):413–417

39. Triclot P (2011) Metal-on-metal: history, state of the art (2010). Int Orthop 35:201–206

40. Drummond J, Tran P, Fary C (2015) Metal-on-metal hip arthroplasty: a review of adverse reactions and patient management. J Funct Biomater 6(3):486–499

41. Delaunay C, Petit I, Learmonth ID, Oger P, Vendittoli PA (2010) Metal-on-metal bearings total hip arthroplasty: the cobalt and chromium ions release concern. Orthoaed Traumatol Surg Res OTSR 96(8):894–904

42. Matharu GS, Eskelinen A, Judge A, Pandit HG, Murray DW (2018) Acta Orthop 89(3):278–288

43. Engh CA, MacDonald SJ, Sritulanondh S, Korczak A, Naudie D, Engh C (2014) Metal ion levels after metal-on-metal total hip arthroplasty. J Bone Joint Surg 96(6):448–455

44. Fernández-Valencia J, Gallart X, Bori G, Ramiro SG, Combalía A, Riba J (2014) Assessment of patients with a DePuy ASR metal-on-metal hip replacement: results of applying the guidelines of the Spanish Society of Hip surgery in a tertiary referral hospital. Adv Orthop. https://doi.org/10.1155/2014/982523

45. Bergschmidt P, Bader R, Kluess D, Zietz C, Mittelmeier W (2012) The all-ceramic Knee endoprosthesis—the gap between expectation and experience with ceramic implants. Seminars Arthrop 23:262–267

46. Meier E, Gelse K, Trieb K, Pachowsky M, Henning FF, Mauerer A (2016) First clinical study of a novel complete metal-free ceramic total knee replacement system. J Orthop Surg Res 11(21). https://doi.org/10.1186/s13018-016-0352-7

47. Solarino G, Piconi C, De Santis V, Piazzolla A, Moretti B (2017) Ceramic total knee arthroplasty: ready to go? Joints 5(4):224–228

48. Gotman I, Gutmanas EY, Hunter G (2017) Comprehensive biomaterials II (Ducheyne P, ed). Elsevier, Amsterdam, pp 165–203

49. Hunter G, Jones WM, Spector M (2005) Total knee arthroplasty (Bellemans J, Ries MD, Victor JMK, ed). Springer-Verlag, Heidelberg, pp 371–377

50. Vertullo CJ, Lewis PL, Graves S, Kelly L, Lorimer M, Myers P (2017) Twelve-year outcomes of an oxinium total knee replacement compared with the same cobalt-chromium design: an analysis of 17,577 prostheses from the Australian Orthopaedic Association National Joint Replacement Registry. J Bone Joint Surg Am 99(4):275–283

51. Heyse T, Haas SB, Efe T (2012) The use of oxidized zirconium alloy in knee arthroplasty. J Expert Rev Med Dev 9(4):409–421

52. Piconi C, De Santis V, Maccauro G (2017) Clinical outcomes of ceramicized ball heads in total hip replacement bearings: a literature review. J Appl Biomater Funct Mater 15(1):e1–e9

53. Vierra BM, Blumenthal SR, Amanatullah DF (2017) Modularity in total hip arthroplasty: benefits, risks, mechanisms, diagnosis, and management. Orthopedics 40(6):355–366

54. Hernigou P, Queinnec S, Henri C, Lachaniette F (2013) One hundred and fifty years of history of the Morse taper: from Stephen A. Morse in 1864 to complications related to modularity in hip arthroplasty. Int Orthop 37(10):2081–2088

55. Falkenberg A, Drummen P, Morlock MM, Huber G (2019) Determination of local micromotion at the stem-neck taper junction of a bi-modular total hip prosthesis design. Med Eng Phys 65:31–38

56. Urish KL, Giori NJ, Lemons JE, Mihalko WM, Hallab N (2019) Trunnion corrosion in total hip arthroplasty-basic concepts. Orthop Clin North Am 50:281–288

57. Weiser MC, Lavernia CJ (2017) Trunnionosis in total hip arthroplasty. J Bone Joint Surg Am 99:1489–1501

58. Pourzal R, Lundberg HJ, Hall DJ, Jacobs JJ (2018) What factors drive taper corrosion? J Arthrop 33(9):2707–2711

59. Mistry JB, Chughtai M, Elmallah RK, Diedrich A, Le S, Thomas M, Mont MA (2016) Trunnionosis in total hip arthroplasty: a review. J Orthop Traumatol 17(1):1–6
60. Rodrigues DC, Urban RM, Jacobs JJ, Gilbert JL (2009) In vivo severe corrosion and hydrogen embrittlement of retrieved modular body titanium alloy hip-implants. J Biomed Mater Res B Appl Biomater 88(1):206–219
61. Weber AE, Skendzel JG, Waxman DL, Blaha JD (2013) Symptomatic aseptic hydrogen pneumarthrosis as a sign of crevice corrosion following total hip arthroplasty with a modular neck: a case report. JBJS Case Connector 3(3):e76
62. Morlock MM, Bünte D, Ettema H, Verheyen CC, Hamberg A, Gilbert J (2016) Primary hip replacement stem taper fracture due to corrosion in 3 patients. J Acta Orthop 87(2):189–192
63. Fokter SK, Rudolf R, Molicnik A (2014) Titanium alloy femoral neck fracture—clinical and metallurgical analysis in 6 cases. Acta Orthop 87(2):197–202
64. Niinomi M, Nakai M (2011) Titanium-based biomaterials for preventing stress shielding between implant devices and bone. Int J Biomater, Article ID 836587. https://doi.org/10.1155/2011/836587
65. Swann RP, Webb JE, Cass JR, Van Citters DW, Lewallen DG (2015) Catastrophic head-neck dissociation of a modular cementless femoral component. JBJS Case Connector 5(3):e71. https://doi.org/10.2106/JBJS.CC.N.00179
66. Urish KL, Hamlin BR, Plakseychuk AY, Levison TJ, Kurtz S, DiGioia AM (2017) Letter to the Editor on "Trunnion failure of the recalled low friction ion treatment cobalt chromium alloy femoral head." J Arthrop 32(9):2857–2863
67. Martin AJ, Jenkins DR, Van Citters DW (2018) Role of corrosion in taper failure and head disassociation in total hip arthroplasty of a single design. J Orthop Res 36(11):2996–3003
68. Matsen Ko L, Chen AF, Deirmengian GK, Hozack WJ, Sharkey PF (2016) The Journal of bone and joint surgery. American 98(16):1400–1404
69. Molloy DO, Munir S, Jack CM, Cross MB, Walter WL, Walter WK (2014) Fretting and corrosion in modular-neck total hip arthroplasty femoral stems. J Bone Joint Surg 96(6):488–493
70. Walsh CP, Hubbard JC, Nessler JP, Markel DC (2015) Revision of recalled modular neck rejuvenate and ABG femoral implants. J Arthrop 30(5):822–826
71. Nawabi DH, Do HT, Ruel A, Lurie B, Elpers ME, Wright T, Potter HG, Westrich GH (2016) Comprehensive analysis of a recalled modular total hip system and recommendations for management. J Bone Joint Surg 98:40–47
72. Di Laura A, Hothi HS, Henckel J, Kwon YM, Skinner JA, Hart AJ (2018) Retrieval findings of recalled dual-taper hips. J Bone Joint Surg Am 100(19):1661–1672
73. Kurtz SM, Kocagöz SB, Hanzlik JA, Underwood RJ, Gilbert JL, MacDonald DW, Lee GW, Mont MA, Kraay MJ, Klein GR, Parvizi J (2013) Do ceramic femoral heads reduce taper fretting corrosion in hip arthroplasty? A retrieval study. Clin Orthop Relat Res 471(10):3270–3282
74. Bull SJ, Moharrami N, Langton D (2017) Mechanistic study of the wear of ceramic heads by metallic stems in modular implants. J Bio Tribo-Corrosion 3:7
75. Triantafyllopoulos GK, Elpers ME, Burket JC, Esposito CI, Padgett DE, Wright TM (2016) Otto Aufranc award: large heads do not increase damage at the head-neck taper of metal-on-polyethylene total hip arthroplasties. Clin Orthop Relat Res 474(2):330–338
76. Morlock MM (2015) The taper disaster—how could it happen? HIP Int 25(4):339–346
77. Goldberg JR, Gilbert JL (2003) In vitro corrosion testing of modular hip tapers. J Biomed Mater Res Part B Appl Biomater 64(2):78–93
78. Zachary JC, Coury JG, Cohen J (2017) Taper technology in total hip arthroplasty. JBJS Rev 5(6):e2
79. Shenhar A, Gotman I, Radin S, Ducheyne P, Gutmanas EY (2000) Titanium nitride coatings on surgical titanium alloys produced by a powder immersion reaction assisted coating method: residual stresses and fretting behavior. Surf Coat Technol 126(2–3):210–218
80. Shearwood-Porter N, Browne M, Milton JA, Cooper MJ, Palmer MR, Latham JM, Wood RJK, Cook RB (2017) Damage mechanisms at the cement–implant interface of polished cemented femoral stems. J Biomed Mater Res Part B Appl Biomater 105(7):2027–2033

81. Choi D, Park Y, Yoon Y-S, Masri BA (2010) In vitro measurement of interface micromotion and crack in cemented total hip arthroplasty systems with different surface roughness. Clin Biomech 25:50–55

82. Zhang H, Brown LT, Blunt LA, Jiang X, Barrans SM (2009) Understanding initiation and propagation of fretting wear on the femoral stem in total hip replacement. Wear 266(5–6):566–569

83. Bader R, Steinhauser E, Holzwarth U, Schmitt M, Mittelmeier W (2004) A novel method for evaluation of the abrasive wear behaviour of total hip stems at the interface between implant surface and bone cement. Proc Inst Mech Eng, Part H 218(4):223–230

84. Thomas SR, Shukla D, Latham PD (2004) Corrosion of cemented titanium femoral stems. J Bone Joint Surg (BR) 86B:974–978

85. Hallan G, Espehaug B, Furnes O, Wangen H, Hol PJ, Ellison P, Havelin LI (2012) Is there still a place for the cemented titanium femoral stem? Acta Orthop 83(1):1–6

86. Noronha Oliveira M, Schunemann WVH, Mathew MT, Henriques B, Magini RS, Teughels W, Souza JCM (2018) Can degradation products released from dental implants affect peri-implant tissues? J Periodontal Res 53:1–11

87. Cruz HV, Souza JCM, Henriques M, Rocha LA (2011) Biomedical tribology (Paulo Davim J, ed). Nova Science Publishers, Hauppauge (New York), pp 2–33

88. Corne P, De March P, Cleymand F, Geringer J (2019) Fretting-corrosion behavior on dental implant connection in human saliva. J Mech Behav Biomed Mater 94:86–92

89. Alrabeah GO, Knowles JC, Petridis H (2018) Reduction of tribocorrosion products when using the platform-switching concept. J Dent Res 97:995–1002

90. Fransson C, Wennström J, Tomasi C, Berglundh T (2009) Extent of peri-implantitis-associated bone loss. J Clin Periodontol 36:357–363

A New Method for Seismically Safe Managing of Seismotectonic Deformations in Fault Zones

Valery V. Ruzhich and Evgeny V. Shilko

Abstract The authors outline the results of long-term interdisciplinary research aimed at identifying the possibility and the methods of controlling tangential displacements in seismically dangerous faults to reduce the seismic risk of potential earthquakes. The studies include full-scale physical and numerical modeling of P-T conditions in the earth's crust contributing to the initiation of displacement in the stick-slip regime and associated seismic radiation. A cooperation of specialists in physical mesomechanics, seismogeology, geomechanics, and tribology made it possible to combine and generalize data on the mechanisms for the formation of the sources of dangerous earthquakes in the highly stressed segments of faults. We consider the prospect of man-caused actions on the deep horizons of fault zones using powerful shocks or vibrations in combination with injecting aqueous solutions through deep wells to manage the slip mode. We show that such actions contribute to a decrease in the coseismic slip velocity in the fault zone, and, therefore, cause a decrease in the amplitude and energy of seismic vibrations. In conclusion, we substantiate the efficiency of the use of combined impacts on potentially seismically hazardous segments of fault zones identified in the medium-term seismic prognosis. Finally, we discuss the importance of the full-scale validation of the proposed approach to managing the displacement regime in highly-stressed segments of fault zones. Validation should be based on large-scale tests involving advanced technologies for drilling deep multidirectional wells, injection of complex fluids, and localized vibrational or pulse impacts on deep horizons.

Keywords Seismically active fault · Earthquake · Friction · Healing · Field experiment · Vibro-pulse impact · Deep drilling · Fluid injection · Shear stress relaxation · Slip control

V. V. Ruzhich (✉)
Institute of the Earth's Crust, SB RAS, 664033 Irkutsk, Russia
e-mail: ruzhich@crust.irk.ru

E. V. Shilko
Institute of Strength Physics and Materials Science, SB RAS, 634055 Tomsk, Russia
e-mail: shilko@ispms.tsc.ru

© The Author(s) 2021
G.-P. Ostermeyer et al. (eds.), *Multiscale Biomechanics and Tribology of Inorganic and Organic Systems*, Springer Tracts in Mechanical Engineering, https://doi.org/10.1007/978-3-030-60124-9_3

1 Introduction

The methods developed by the world community for countering natural seismic disasters are insufficient since they do not allow an efficient reduction of the almost annual losses caused by strong earthquakes [1–3]. To date, one has come to understand that difficulties in finding solutions to this problem are concerned with the lack of reliable information on the deep tribophysical and geochemical characteristics of earthquake preparation processes in fault zones. These geological processes are hidden at inaccessible seismic focal depths of the lithosphere (7–30 km) and occur at elevated temperatures and pressures of hundreds of megapascals. An important role in the preparation of earthquakes belongs to the migration of fluids of various compositions [4, 5]. The results of a long- and medium-term prognosis make it possible to identify foci with indications of the final stage of the preparation of strong earthquakes with a probability of the order of $P = 0.6$–0.7. In most cases, these are the same segments of the zones of interplate and intracontinental seismic faults, where "hot spots" of dangerous or catastrophic events have already arisen for many tens or hundreds of thousands of years. The modern possibilities of making a short-term forecast of strong and catastrophic earthquakes are still limited. Moreover, such a forecast cannot efficiently ensure seismic safety, since it does not prevent the large-scale negative consequences of destruction [6]. The world scientific community recognizes the need to search for new ways to more reliably ensure seismic safety.

The chapter is devoted to a review of the results of a joint 20-year interdisciplinary research led by Professor Sergey Grigorievich Psakhie, the corresponding member of the Russian Academy of Sciences. These studies were aimed at exploring the safe methods of managing deformations (displacements) in the seismically dangerous segments of active faults [7, 8]. Specialists from various scientific institutions were involved, including the Institute of the Earth's Crust SB RAS (IEC SB RAS), the Institute of Strength Physics and Materials Science SB RAS (ISPMS SB RAS), the Institute of Geosphere Dynamics RAS (IDG RAS), Technische Universität Berlin (TU Berlin), as well as other scientific institutions from Ukraine, Mongolia, and China. The success achieved was largely related to the use of diverse scientific approaches and methods including computer simulation, laboratory physical modeling, and field experiments in rock massifs and regions of the block-structured ice cover of Lake Baikal. Such a variety of techniques and objects of study made it possible to take a fresh look at the nature and mechanisms of destruction processes in the earth's crust and reveal a set of criteria governing the occurrence of seismically dangerous dynamic phenomena in zones of deep faults.

In the chapter, we outline key aspects of this multidisciplinary research.

In particular, we describe the results of a detailed geological, geophysical, and petrological study of the structure and physicochemical transformations of rocks in earthquake foci in the deep horizons of faults. Such information was obtained through examining the deeply denudated blocks of rocks in the fault zones, which were exhumed to the earth's surface from depths of 10–20 km. Using new methods of analysis, a number of representative deep segments of fault zones (deep paleoseismic dislocations) were studied in detail. The main attention was paid to segments, which were the centers of earthquakes in past epochs of geological evolution.

Significant efforts of the collaboration participants were directed at conducting field experiments to study the complex mechanisms of the formation of seismic radiation sources. The advantages of such experiments over laboratory tests on small samples of artificial and natural materials are the large size of the studied objects (segments of fault zones tens or hundreds of meters long), the interrelation between the studied objects and the surrounding rocks, the possibility of using broadband seismic stations and strain gages, as well as the possibility of modeling various impacts using high-power jacks, pile drivers and weak explosions. Emphasis was placed on revealing the effect of the impacts on the features of changing the displacement mode in the studied fault segments from slow creep to accelerated creep and coseismic slip. These features largely determine the generation of seismic oscillations of a wide amplitude-frequency range.

Based on a comprehensive analysis of the results of geological and geophysical studies of deep paleoseismic dislocations, field experiments, and numerical modeling, we formulated the concept of the response of faults to external actions and the approach to controlling and managing deformations in fault segments using man-made vibro-pulse and hydraulic wave impacts. Due to the limited effectiveness of numerical modeling and low-scale field experiments in the near-surface part of the earth's crust, the authors cooperated with deep drilling specialists, which have high competence in the exploration and exploitation of hydrocarbon deposits in the Late Proterozoic crystalline formations of the Siberian Craton [9, 10]. The combination of dynamic mechanical impacts with the injection of solutions into fault segments through deep multidirectional wells significantly expands the possibilities of the developed approach to managing displacements in seismically dangerous fault segments. Here we also discuss the prospects of hydraulic wave impacts on deep seismically active segments by means of drilling horizontal, inclined, and vertical wells up to 5–7 km deep, hydraulic fracturing of rock massifs, and injection of solutions of various compositions.

Finally, we consider the difficulties of implementing a complex impact on the deep fault segments and provide a brief overview of current and planned studies by different scientific groups.

2 Methodological Basis

Of the many earthquake source focal models known in the world, the "stick–slip" model is considered to be the optimal and most general model [11, 12]. It is based on the fundamental tribological laws of sliding friction in seismic-generating fault zones, which have a textured roughness, specific mineralization, and fluid saturation. This model uses the main achievements of tribology and, in particular, takes into account the influence of such frictional parameters as the ratio of shear and normal resistance forces (characterized by stiffness ratio), roughness, the variation of sliding velocity, the presence of lubricant, wear of the sliding surfaces, vibrations, etc. [13]. Detailed geological and experimental study of fracture processes in rock massifs has shown that the fundamental tribological patterns of creep or stick-slip sliding can be successfully used to determine the causes and mechanisms of strong earthquakes in fault segments [14, 15].

To clarify the conditions of earthquake preparation, we studied the zones of seismic dislocations that formed not only with modern earthquakes but also with paleo-earthquakes. Traces of such ancient earthquakes are captured in the seismic focal horizons of the denudated crust. During the long geodynamic evolution of the lithospheric shell of the Earth, these horizons were on the Earth's surface in the orogeny regions after kilometers-long denudation. They are currently available for visual geological exploration. To obtain a more in-depth understanding of the real structure of the deep segments of the sources of strong earthquakes, these studies were conducted in the Baikal rift zone and the collision zones in Mongolia. The results of these studies are discussed below.

From the viewpoint of practical geodynamic applications of the methods of physical mesomechanics, it is advisable to consider the lithospheric shell of our planet as the upper hierarchically organized geosphere in which the seismotectonic destruction process occurs at various scales according to well-known geomechanical laws. In particular, various tribochemical and geomechanical processes manifest themselves at various hierarchical scales of the earth's crust. According to the well-known kinetic concept of the strength of solids by S. N. Zhurkov and his followers [16, 17], multiscale fracture-related processes manifest themselves in stage-by-stage accumulation of small discontinuities and their coalescence into longer ones due to the fracture of the separating "jumpers" (structural barriers). This can be observed in geological destructive processes. Hierarchical transitions within the self-organized process of destruction of the earth's crust reach the highest level in the form of the formation of intraplate and interplate seismic faults with a length of hundreds to thousands of kilometers.

The basic geomechanical model of a seismically active fault is based on the idea that a focus of a very strong earthquake with $M = 7.0$–9.0 is formed as a result of successive long and multi-stage coalescence of faults ranging from the first tens of kilometers to longer ones (70–120 km). This is clearly illustrated by many recent strong earthquakes, such as the Gobi-Altai earthquake of 1957 ($M = 7.9$). To find out the conditions for the formation of different-scale foci of earthquakes (including foci of super-earthquakes) and mechanisms of fracture of these interface zones, the authors use a multiscale approach. It is based on the data of many years of field experiments, geological and structural studies, and numerical modeling using the method of movable cellular automata. On each scale, specific tribochemical and geomechanical factors that determine the features of contact interactions, including the frictional regimes, can be distinguished. The friction features, in turn, affect the amplitude-frequency and energy parameters of the generation of seismic vibrations

[18]. Further development of this concept is associated with an analysis of the geomechanical consequences of fracture of structural barriers and contact interactions of
different scale asperities on the glide planes of the fault limbs.

Pore fluids play an important and, in some cases, determining role in petrochemical
processes in zones of highly stressed and seismically dangerous faults. Interstitial
fluid largely determines the slip modes, and therefore, the modes and features of
the generation of seismic vibrations [19, 20]. Traditionally, the term "fluids" include
liquids and gases that fill void spaces arising from deformation in rock masses.
Fluids are also magmatic melts, which themselves are sources of hydrothermal fluids
ascending to the earth's surface through conducting faults [21]. The migration of
fluid solutions affects the shear resistance of the faults through the lubrication and
hydraulic effects on the walls of the fractured rock. Therefore, the pore fluid should
be taken into account among key factors that control the nucleation and dynamics of
coseismic displacements in discontinuities of different hierarchical levels.

3   Geological Study of Exhumed Seismic Dislocations of Paleo-Earthquakes in the Southeastern Boundary of the Siberian Craton

To determine the deep-seated conditions for the modern earthquakes, a collaboration
of scientists from IEC SB RAS and IDG RAS carried out a geological and geophysical
study of the exhumed deep segments of the Primorsky section of the marginal zone
on the southeastern marginal seam of the ancient Siberian craton. Due to natural
exhumation from the middle depths of the earth's crust, deep fault segments are now
available for study. Here, the traces of different age foci of paleo-earthquakes (ancient
coseismic slips) formed at elevated temperatures and pressures are preserved. We note
that similar P-T conditions currently exist at depths of 10–25 km, where the seismic
focal horizon is located and most of the modern earthquake foci of various energy
scales occur (including the strongest events with $M = 7.0$–7.9). Many of them are
confined to the zone of the Primorsky riftogenic fault with a length of about 200 km.
This fault is located on the northwestern coast of the Baikal rift depression (Fig. 1) and
was formed for 65 Ma. It is important to note that the deeply denudated (exhumed)
Pribaikalsky segment of the fault allows the visual study of the structure and P-T
conditions for the formation of ancient earthquakes of different ages (including those
with an age of the order of 700–450 Ma) at activation depths of 12–18 km [22].

Ancient sliding mirrors, formed during the dynamic interaction of irregularities,
were sampled in this fault segment (Fig. 2). The samples were subjected to petrological studies using the polished sections. Tectonites (rocks transformed by the friction of rocks) taken in the zone of later earthquakes have indications of gneissing,
cataclase, and milonitization in the Early Paleozoic period. To elucidate the natural
conditions of tribochemical processes in the deep foci of paleo-earthquakes and

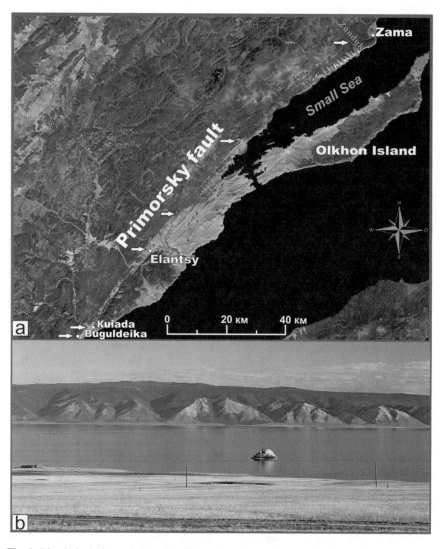

Fig. 1 Morphological expression of the Primorsky riftogenic zone: view from space (**a**) and one of the sections on the shore of the Small Sea (**b**). The arrows in (a) indicate the test sites where a comprehensive study of ancient coseismic displacements was carried out. These displacements occurred hundreds of millions of years ago in the centers of paleo-earthquakes at depths of about 15–20 km, at high temperatures and pressures

the structure of contact spots with traces of coseismic slip, we carried out a petrological study of changes in the material composition of the mineral coating on slip planes. Coseismic slip and the structure of the surface layers were analyzed by sliding mirrors containing indications of high-rate displacements, including pseudotachylyte (products of frictional heating in faults under stress metamorphism at elevated

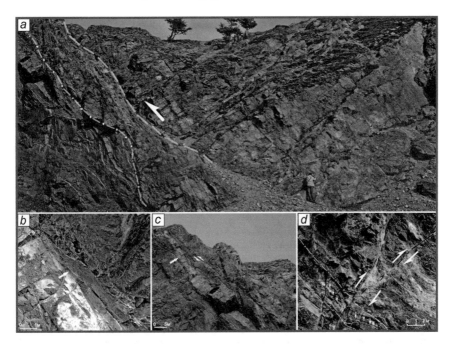

Fig. 2 Coseismic fractures of different ages containing multiscale asperities in the centers of paleo-earthquakes at seismic focal depths from 5 km to 12–18 km: **a** the oldest coseismic slip zone at a depth of about 18 km; **b** low-scale roughness in a small coseismic rupture; **c** longer coseismic rupture and "macroasperity" shown by the arrow; **d** intersecting coseismic ruptures that occurred at different times and different depths and PT conditions, as shown by minerals deposited from fluids. The dash-dot line in (a) shows the oldest coseismic slip zone at a depth of about 18 km. The dashed line and the arrow to the right (slip direction) denote the younger zone of coseismic slip. Slip mirrors and newly formed mineralization with traces of thermal heating during friction were taken from this zone

temperatures, pressures, and the participation of fluids). Recently, G.A. Sobolev and his colleagues demonstrated the possibility of efficient application of modern physical methods for determining strains and stresses in fine ground nanocrystals of various minerals, including quartz, albite, chlorite, epidote, and other varieties found on sliding mirrors from the exhumed segment of the Vilyui deep fault [23, 24]. Consequently, tectonic glide mirrors identified at fault exhumation sites, as well as pseudotachylyte, can be used to recognize and analyze coseismic planes. This makes it possible to obtain important information about the kinematics and tribophysical processes in the deep fault segments at the moments of the generation of seismic events by the examples of past tectonic eras.

The analysis of the slip mirrors showed the finest brittle attrition of minerals. This is typical for high-velocity sliding. The important role of fluids in eliminating (healing) the consequences of fracture is illustrated through the presence of fine-grained aggregates of quartz grains and feldspars with a subordinate amount of sericite, chlorite, and iron ore minerals. It is important to note that such mineral

Fig. 3 The cross-section of a sample of an ancient sliding mirror taken from the zone of the exhumed segment of the Dolinozersky Fault (still highly seismic hazardous) in southern Mongolia. It captures the consequences of two acts of coseismic slip (the arrow at the upper edge) with the "healing" by the epidote and subsequent quartz mineralization

neoformations have a reduced shear resistance and can play the role of a mineral lubricant that facilitates subsequent sliding acts and prevents reverse closure of crack walls. Sliding of rough walls of the fault is accompanied by crack opening (dilatancy), and mineralized fluids are sucked into the vacuum gap. This ensures filling the cavities with new minerals, including ore. This is the hydrothermal mechanism of mineral "healing" of fault segments in the upper part of the earth's crust at relatively low pressures and temperatures [25].

The collected information on the dating of identified coseismic planes and the P-T-f conditions for the formation of slip mirrors and pseudotachylyte was used to make a qualitative and quantitative estimation of the exhumation of the Primorsky section of the margin, as well as of the structure of the seismic focal horizon. In particular, the oldest absolute dating established by the $^{40}Ar/^{39}Ar$ method is 673 ± 4.8 Ma. This corresponds to one of the stages of the Neoproterozoic era, i.e., the time of break-up of the Rodinia supercontinent and the separation of the Siberian craton along the margin. 415.4 ± 4.1 Ma refers to a later stage in the evolution of the margin during the collision of the Olkhon terrane with the Siberian craton [22]. The obtained dating made it possible to roughly estimate the occurrence depths of the identified strong paleo-earthquakes. They amounted to about 18 km in the earlier (Neoproterozoic) period, and about 12 km in the later (Middle Paleozoic) stage [22]. Similar results were obtained in the neighboring section of the Primorsky fault (5 km to the northeast). In the coseismic slip plane, the temperature of frictional heating during the occurrence of pseudotachylyte reached approximately 900 °C, and the pressure reached 700–800 MPa. After dynamic slip, the pressure rapidly decreased to a "natural" (lithostatic) value, which was estimated as ≤150–200 MPa in the considered case. This corresponds to depths of the order of 5–7 km [26].

Figure 3 shows a photo of a very ancient sliding mirror taken from the exhumed zone of the Dolinozersky Fault in southern Mongolia. A devastating Gobi-Altai earthquake ($M = 7.9$) occurred here in December 1957.

The polished section shows the internal structure of the coseismic rupture zone with the shearing of asperities on the slip plane and the transfer of fragments of basaltic anorthosite from an ancient volcano. Large and small fragments of the parent rock move within the mass of newly formed chlorite and calcite minerals deposited from the flowing fluid. These minerals healed the fracture cavity. Mineral "healing" products also fill wing cracks that formed under the condition of large friction in combination with the fracture of asperities. Wear debris carries over when sliding in the direction of the white arrows. This is the widespread mechanism of the formation of sources of strong earthquakes accompanied by the occurrence of wing cracks.

Therefore, the geological study of deep segments of fault zones opens up a unique opportunity to reconstruct the conditions for the nucleation of coseismic slips in the areas of the most severe contact interaction of different scale asperities in fault zones with the participation of fluids. The mineralization of the fluid determines the mineral composition of crack fillers. Mineral composition in many respects affects the frictional sliding of the fault walls during subsequent activations. Although the processes of stress-metamorphic transformation of rocks in the considered fault zones at seismic focal depths occurred in past eras of seismotectonic activation, it is clear that the same mechanisms act in modern earthquake foci. Therefore, the database of ancient earthquakes is fundamentally important for creating adequate models of modern earthquake sources [14].

4   Features of the Response of Tectonic Fault Segments to Man-Caused Impacts

Below we describe the most important results of field experiments carried out in different years on the segments of the Angarsky and Primorsky faults in the Baikal rift zone.

The first one was carried out in 2004–2006 on the segment of the Angarsky fault zone near the Listvyanka village (southwestern coast of Lake Baikal), Irkutsk Region. The Angarsky fault has a northwestern strike and is characterized by faults of the reverse-strike-slip type. The studies were conducted with the participation of scientific groups of several institutes of the Russian Academy of Sciences, as well as the Technische Universität Berlin. At this test site, the effect of high strain sensitivity of the principal slip zone to natural and man-made impacts was revealed [27]. The most important result of the research is related to the demonstration of the ability to manage the shear displacement mode in the fault zone. In order to carry out man-made impacts on a selected fault segment 100 m long, several wells were drilled to a depth of 30 m for 14 days, and water solutions were injected into the fault zone (Fig. 4a). Then local dynamic impacts with weak explosions at a depth of about 15 m

Fig. 4 General view of the test site in the Angarsky fault zone section (**a**), a graph of shear displacements in the fault zone at different stages of the experiment (**b**) and results of numerical modeling [7] showing the dependence of the shear resistance on the relative tangential displacement at a constant displacement velocity and various additional actions. Figure (a) shows a part of the test site with three wells for water injection and explosive actions, a platform for a falling pile and locations for strain gages and seismic sensors. The numbers in Figure (b) are as follows: stress tests with a falling pile before the experiment (1); well drilling, water injection, and explosions in the well (2); displacement "jumps" at the stage of return deformation (3). The numbers in Figure (c) are: "dry" fault zone, no additional impacts (1); "dry" fault zone, additional vibration (2); watered fault zone, no additional impacts (3); watered fault zone, additional vibration (4)

in watered wells were carried out at the next stage. To study the effect of the impacts on the stress state of the fault segment, we performed mechanical stress tests on the surface with a falling pile weighing 100 kg. Such tests were performed before and during the whole course of the experiment.

The fault responded in the form of initiated tangential displacements in the principal slip zone. This was recorded by strain and seismic sensors. The vibrations during drilling and subsequent watering of the fault zone, accompanied by weak cumulative explosions in watered wells, have led to a qualitative change in the displacement mode. In particular, the creep velocity increased by several orders of magnitude: from "natural" values of 0.65–3.3 μm/day (0.24 mm/year ÷ 1.2 mm/year) to 0.5 mm/day.

Although the creep acceleration was of a short-term nature (less than a month), the total amplitude of the initiated tangential displacements in the principal slip zone reached 8–10 mm (Fig. 4b). This corresponds to 10 years of accumulation of displacements in the natural mode. Note that in the months after the experiment, a partial recovery (reverse displacement) occurred in the slip plane, and the residual displacement amplitude decreased to 5–6 mm. This is the result of the elastic "return" of the surrounding rock mass. It is noteworthy that the return motion was accompanied by seismic pulses with amplitudes comparable with seismic amplitudes during initiated (accelerated) creep. A similar elastic-plastic partial return with significant "residual" displacements is also observed in seismic dislocations after strong earthquakes. This indicates the generality of the described effect for the zones of tectonic faults.

Field experiments and numerical simulations using the method of movable cellular automata [7] showed that the injection of aqueous solutions into the fault zone in combination with vibroimpulse impact trigger local relaxation of "excess" shear stresses in the fault segment (Fig. 4c). Stress relaxation is manifested particularly in the transition from natural to accelerated creep, followed by its gradual deceleration to natural values determined by regional tectonic motion. Displacement monitoring on the studied segment of the Angarsky fault over 14 years shows slow creep and extremely weak microseismicity. Moreover, the magnitude of deformation and seismic response of the fault segment to the standard dynamic test impacts of the falling pile is one order of magnitude lower than before the experiment. This confirms a considerable decrease in the level of shear stresses. Thus, the field experiment confirmed the statement about the possibility of a controlled change in the slip mode in the segments of fault zones to safely reduce the level of shear stresses and seismic activity [7, 28]. The developed method of shear stress relaxation in fault segments is protected by a patent of the Russian Federation [29].

From 2015 to 2018 joint research was conducted by the IEC RAS and IDG RAS research teams in the coastal segment of the Primorsky Fault zone on the Baikal coast (near Olkhon Island). The studies included geophysical observation and field experiments. Figure 5 shows a general view of the testing ground. The purpose of the experiments was to study the influence of quasistatic (carried out by jacks) and dynamic (explosive) effects on the displacement mode during contact interaction with a single artificial asperity. A hydraulic flat jack was installed in a large crack in the central part of the fault zone. The crack walls were pulled apart with a force of 43 tons. A 5 cm thick granite gneiss ore was inserted into the cavity to study the fracture of the rock under conditions of high local compressive stress and additional (man-made) impacts. After the jack was removed, the ore was compressed by the crack walls with the force of the crack extension (~ 43 tons).

The processes in the studied segment of the fault zone after the insertion of a single macroasperity were recorded by strain gages and seismometers. After several days of monitoring the natural deformations and the seismic background, a series of weak explosions were performed (the mass of one powder charge was 0.1 kg). The explosion chamber was located in the large plate within the fault segment at a distance of 3 m from the seismic sensors. The recorded amplitude of seismic

Fig. 5 Studied segment of the Primorsky fault zone with measuring equipment (**a**) and artificial asperity (granite gneiss ore) with cracks resulting from experiments (**b**). The numbers in Figure (a) are as follows: jacks (1); pump (1a); displacement sensors (2); seismic sensors (3). The inclined arrow in (b) shows the direction of the natural normal faulting displacement

acceleration during the explosion was 219 cm/s^2, and the frequency was about 70–75 Hz. Figure 6 shows a seismic recording for 330 min after the explosion. It reflects the result of contact interaction during the explosion, in particular, the initiated slip

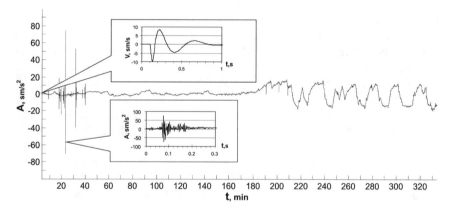

Fig. 6 Response to explosive action in one of the cracks in the Primorsky fault zone. The main figure shows the general view of the seismic record within 330 min after the explosion. The upper inset shows the oscillation velocity of the rock mass for 1 s and its upper value of 0.7 mm/s. The bottom inset shows the record of the first induced seismic response (the largest seismic pulse) that occurred 22 min after the explosion (amplitude 70 cm/s^2, frequency 160–175 Hz)

in the contact pair "crack wall—ore". Within 40 min after the explosion (with some delay) a series of strong seismic pulses (response to the explosion) were formed. Pulse amplitudes reached several tens of percent of the acceleration amplitudes during the explosion.

The measurements and visual observation showed that stress relaxation and redistribution in the surrounding massif after the explosion promote reaching the critical stresses at the contacts of the artificial asperity with the crack walls and cracking of the ore after some time (tens of minutes) after the explosion. Cracks propagated in the ore in the dynamic regime and were accompanied by the generation of a series of large seismic pulses. In Fig. 6, these pulses are shown on the left side (time interval of 15–40 min after the moment of explosion).

An analysis of the record of displacements in the main crack showed moderate compressive deformations of the crack with characteristic amplitudes of 25–35 μm in combination with reverse faulting during the period of the subsequent "calm". The largest seismic impulse was associated with the fracture of the artificial asperity and characterized by the following parameters: amplitude 70 cm/s^2, frequency 160–175 Hz. The next phase of the response began 190 min after the explosion and lasted about 140 min. It is characterized by low-frequency "pulses" with an average period of about 18 min and acceleration amplitudes of 15–20 cm/s^2. The generation of such extremely slow seismic pulses is a deformation response of the rock massif to a dynamic disturbance of a local stress state. The response manifests itself in the form of a return slip and can be described by viscoelastic models.

This example demonstrates the generation of high-velocity elastic and subsequent slowed-down oscillations of the rock massif caused by the intermittent slip mode at the contact with the large-scale asperity. It is noteworthy that the unloading of the surrounding rock mass manifested itself in the form of intermittent slip described by the stick-slip model, with the generation of slow deformation waves. In subsequent similar tests, the contact area of the ore and the crack walls was wetted. This led to the initiation of small cracks in the artificial asperity at the contact point accompanied by numerous packets of audible seismic-acoustic pulses. Over the subsequent months of application of natural static pressure by the rock mass, the ore gradually split into several large fragments.

The described experiment showed that an explosive (high-velocity) impact causes a strong viscoelastic-plastic return accompanied by high-amplitude seismic pulses and subsequent slow motions. These slow oscillations can presumably be regarded as being analog to slow waves associated with intermittent sliding of contacting surfaces in interplate faults [30, 31]. In this case, the observed propagation phenomena of slow waves are apparently associated with the phenomena of self-organization of ensembles of cracks in the rock mass (cracks are highly sensitive to the rate and amplitudes of external dynamic impacts). Note that slow motions were clearly recorded in other (larger-scale) experiments including technological explosions with multi-ton explosive charges in the quarry of the Udachnaya diamond pipe.

Based on the results of the above field experiments, as well as many other tests, we can draw an important conclusion. Safe stress relaxation in highly stressed (seismically hazardous) segments of tectonic fault zones can be achieved using local impacts with low impact velocities and energies [32]. An additional and extremely effective relaxation stimulation is the saturation of contact slip surfaces with aqueous solutions, especially when the physicochemical properties of the liquid determine the important role of the Rehbinder effect [33, 34].

5 Prospects for the Implementation of Controlled Impacts on Fault Segments Through Deep Wells

The examples discussed above consider methods of managing impacts on small near-surface areas of seismically active faults. The question remains whether these experiments can be extended to the scales of the long fault segments and deep regions characterized by high pressures and temperatures, and complex fluid circulation. To answer this question, we consider the possibility of implementing control actions on segments of deep sections of fault zones through wells up to 3–5 km deep. Here we can take the experience of drilling such wells within the ancient Siberian platform for injecting solutions and hydraulic fracturing. Such work is carried out, particularly, at hydrocarbon deposits in the Republic of Yakutia (Russia), where hydrocarbon reservoirs are at relatively shallow depths of about 2 km.

Deep drilling and hydraulic fracturing allow to open the natural fluid fracture-pore reservoirs containing oil, gas, and saltwater. It should be noted that the magnitude of the pressure in such reservoirs is often close to rock pressure. The application of this technology allows the use of fluid injection at a given pressure through a network of inclined wells as a hydraulic action on the walls of the principal slip plane, ensuring their opening or closing [35]. The hydraulic effect of the injected fluid can be compared to a first approximation with the action of a powerful hydraulic jack (the role of the jack is played by the fluid pumped under pressure). The effect of fluid saturation in combination with hydraulic fracturing reduces friction and changes the slip mode of the fault limbs at the injection sites.

Man-made pressure effects transmitted through the reservoir fluid system allow the fractured reservoir (within the hydraulic influence region) to be transferred to another stress state. These effects include the influence on the value of shear resistance in the segments of the drilled sections of the fault zones. When the seam pressure is reduced to a certain level, the permeability of the rock decreases, cracks close, and, therefore, friction and shear resistance increase. To re-open the cracks, it is necessary to re-increase the pressure of the fluid system to an appropriate level. This makes it possible to control the stress-strain state of fault segments and the mode of their seismic activity [36]. It is also important to take into account the natural ratio of the angles of inclination of fault zones with the vertical direction of gravity.

It is known that vertical cracks in the strike-slip fault zones have greater openness than inclined ones since the rock pressure on the walls of vertical cracks is much less than on the walls of inclined (and especially mildly sloping) cracks. Accordingly, the shear resistance of vertical cracks is also significantly less. In particular, during the drilling of reservoirs within the zones of subvertical faults, seam pressure can sharply drop to the value of pore pressure in the surrounding rock mass [37]. Another situation occurs with the subhorizontal inclination of the principal slip zone of a thrust fault. Here, the natural fluid system can be at extremely high pressures due to push-up conditions. In such cases, the use of standard drilling technologies can lead to strong dynamic responses of fault zones, especially in the absence of pore fluids. Injection of drilling fluid into the zones of subhorizontal fractures can lead to a sharp decrease in shear resistance and activation of coseismic displacements.

Injection of solutions into the fault at a controlled rate allows managing pressure drop and the frictional resistance of the fault zone, as was shown by experimental studies. Such technological impacts on highly stressed fault segments can be supplemented by the use of proppants to change the slip mode. Natural proppant-fixed cracks are colmataged, which prevents them from closing. Another technique is the use of additives in drilling fluids, for example, granules of copolymers (alpine drill beads). Such additives contribute to a decrease in frictional resistance in the contact spots of asperities in the fault zone and to a decrease in the critical density of accumulated elastic strain energy at which seismic vibrations are generated. The latter leads to a decrease in the amplitude of seismic oscillations.

Local hydrodynamic and tectonic conditions must be taken into account during the drilling process. For example, natural filtering fracture simplifies fluid injection and pellet delivery into the fault through drilled fan lateral branches in horizontal fishbones. At a subhorizontal slope of the principal slip zone of the thrust fault, the natural fluid system can be at extremely high pressures due to push-up conditions. This contributes to the creation of frictional instabilities in the fault zone. Upon identifying such features by geological and geophysical data, special measures to reduce seismic risk have to be developed.

Deep drilling of injection wells to depths of 5–7 km in the fault zones shows that the injection pressure and fluid saturation are able to propagate to much deeper horizons due to hydraulic connection with existing natural fluid systems. This indicates a real possibility of man-made impacts on seismically hazardous deep fault segments through deep wells since such impacts can reach the level of the seismic focal layer in the continental crust.

The example shown below is one of the options for drilling a well-studied segment of the Tunka fault in the Baikal rift zone near the Arshan settlement (Irkutsk Region, Russia). Some volcanic apparatuses and coseismic fractures of 4 destructive earthquakes ($M_w \sim 7.5$, the average recurrence period is 3.9 ± 0.6 thousand years) were found in the vicinity of this segment. The latest paleoseismic dislocation in this area is about 1 thousand years old. We also note the recent strong earthquake of August 22, 1814, in the neighboring segment of the Tunka fault zone. The energy of the tremors was about 9 balls. Currently, there is a seismic gap in the Arshan segment of the fault, which is a warning sign of the preparation of the next stage of activation in the coming years or the first decades. Based on these assumptions, the seismic situation in the Arshan segment of the Tunka fault can be assessed as potentially dangerous.

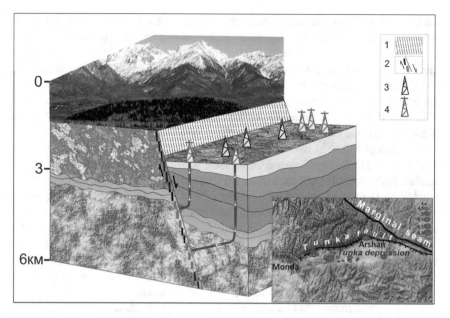

Fig. 7 The scheme of the proposed placement of two groups of inclined injection and monitoring wells for sequential hydrodynamic vibro-pulse impacts on the Arshan segment of the Tunka fault. The inset shows the central part of the Tunka rift basin, the Arshan segment of the Tunka fault zone, and the epicenters of earthquakes $M_w = 4.0$–6.9 over the past 70 years

Therefore, this segment is an appropriate place to validate the method of preventive relaxation of shear stresses using a hydraulic injection of fluids with controlled pressure into the fracture zone through a network of deep inclined or horizontal multilateral wells [38, 39]. Below is a simplified block diagram of the planned test site in the Arshan segment of the Tunka rift faulting zone (Fig. 7).

The advantages of using modern technology for driving deep wells into fault zones are determined by the possibility to control the volume and modes of injection of different solutions. Nevertheless, it is necessary to take into account the complex additional factors (geophysical, seismotectonic and geostructural), including the spatial orientation of principal slip zones with respect to the main axes of tectonic stresses, the length and thickness of fault zones, the values of dip angles, triple junctions, etc. It is also important to conduct preliminary test actions to assess the features of the "macroscopic" response of the fault zone.

6 Discussion

When discussing the results of the described interdisciplinary studies aimed at developing an approach to the man-made management of displacement modes in fault zones, it is important to consider the similar experience of foreign scientific groups. In particular, it is necessary to take into account the known results of studying the consequences of injecting solutions through wells 3–5 km deep. The watering of fault zones sometimes led to the initiation of many weak earthquakes. Hypocenters of earthquakes extended in depth to distances commensurate with the length of the wells themselves. This indicates fluid saturation of rock masses in much deeper horizons. The authors of the monograph [40] studied indirect seismological evidence of high and rapidly changing permeability of the middle and lower crust from data on the systematic emergence of hypocenters of aftershock sequences of strong earthquakes. Such a notion can be considered as an indication of the presence of a hydraulic connection between fluids of the lower and middle crust with hydrological systems in the surface layers of the earth's crust. Thus, fluid injected through insufficiently deep wells (3–5 km) can hydraulically act on deeper fluid-saturated layers of the earth's crust down to the depths of preparation of the sources of strong earthquakes. This way is assumed in the framework of the proposed approach.

Laboratory tribological studies indicate that many classical laws of friction are valid for zones of active tectonic faults and determine the regimes of seismic activity in deep fault horizons [13, 41, 42]. This means that the regimes of radiation of seismic vibrations during tectonic creep or coseismic slip should be considered as the results of visco-elastic–plastic contact interactions between randomly rough sliding surfaces. Classical tribological laws also hold upon transitions to higher hierarchical levels of seismotectonic destruction in the lithosphere. This refers, for example, to transient frictional slip regimes from aseismic to accelerated ones, or dynamic failure in the final stage of earthquake preparation. These regimes, in turn, affect the amplitude-frequency and energy parameters of the radiated seismic vibrations. The injection of aqueous solutions into fault zones not only reduces the strength characteristics of rocks and shear resistance in slip planes but also promotes a hydraulic (wedging apart) effect. The latter causes hydraulic fracturing or slow opening of natural cracks and hence reduces the contact area and the friction coefficient. The experience gained during the operation of underground hydrocarbon storage, the extraction of geothermal resources, and the development of gas and oil fields in seismically active regions show that aqueous solutions injected through wells under high pressure can reactivate existing deep fault segments and provoke dangerous earthquakes with $M_w \geq 4.0$–5.2 [43–46]. Many authors are of the opinion that the unintentional activation of faults caused by fluid injection should be used instead to manage the relaxation of excess stresses in fault zones. To implement this in practice, it is necessary to identify and use tribological mechanisms that control slip modes and generation of seismic vibrations in fault zones [47]. It was shown above that such knowledge helps to exclude the acts of strong seismic activation of fault segments at significant distances from injection wells.

Many representatives of the scientific community believe that even energetically moderate technogenic impacts including hydraulic fracturing by injecting drilling fluids can induce strong earthquakes even in aseismic areas. Examples are geological exploration and exploitation of underground geothermal water resources in Texas

(USA) and Switzerland [48]. However, experience of deep drilling in the exploration and development of hydrocarbon deposits in aseismic regions (including the Siberian platform), indicates the absence of such dangerous phenomena. Note that even underground nuclear explosions with megaton charges were not able to trigger multiscale transitions to merge small faults into a long seismically active fault and initiate destructive earthquakes with $M_w \geq 7$ [49]. Nevertheless, we cannot exclude induced low-energy seismicity in the form of a series of moderate earthquakes. The most striking example is the recent Bachat man-provoked seismic event in Kuzbass (Russia) on 18.06.2013 ($M_w = 6.1$) with a hypocenter depth of 4 km [50]. It was initiated after long-term mining operations in coal mining at depths of up to 2 km. This extraordinary earthquake is the result of regular powerful explosions, as well as the excavation of large volumes of rock mass over an area of 100 km^2.

The magnitude of the most well-known (and rather rare) technologically induced moderately hazardous earthquakes induced by mining operations ranges within 2.8–6.1. There are some examples of the initiation of strong earthquakes with $M_w = 7$ as a result of the uncontrolled long-term withdrawal of huge volumes of oil and gas from deep-seated deposits. This kind of phenomena took place near the city Gazli or the city Neftegorsk. A huge number of such examples show that the increasing power of anthropogenic impact on the upper layers of the earth's crust leads to the unintentional provocation of dangerous seismodynamic phenomena. This makes it necessary to mobilize the efforts of the scientific community to prevent such kind of phenomena. The proposed approach can be efficiently used to solve this problem.

7   Conclusion

Uncontrolled long-term high-energy man-made impacts on highly stressed areas of the earth's crust are increasingly leading to unintended seismic disasters. We showed the possibility to prevent or neutralize such disasters by implementing proactive measures including controlled vibration-pulse actions in combination with controlled injections of solutions through deep wells. Such kind of technology is based on the approach, which integrates advanced geological and geophysical methods in combination with advanced deep drilling technologies and numerical and physical modeling. This will make it possible, in the near future, to provide a more reliable solution to the global problem of ensuring seismic safety. The results considered above provide the basis for conclusions about the feasibility of projects aimed at effectively reducing seismic risk [51].

Professor S. G. Psakhie was among the drivers of the development of the approach to managing the displacement mode and seismic activity of highly stressed fault zones. At various seminars, we discussed the prospects of the implementation of large-scale field tests using vibro-pulse and hydraulic methods of influencing segments of seismically dangerous fault zones through drilled deep wells. The most challenging problems are appropriate funding, government approvals, and the preparation of test sites in areas, which are at a safe distance from settlements. Such test sites are available, for example, in the poorly populated territory of the Mongolian People's Republic, where seismic and geological studies have revealed a system of deep seismic hazardous faults. To create a technology aimed at solving the described multidisciplinary problem, combined efforts of the world community and the support of governments are required. The next immediate task for the implementation of such an important project is to organize a series of large-scale field tests in the segments of seismically dangerous faults using advanced deep drilling technologies.

Acknowledgements The authors are grateful to the great number of participants in the many years of research. The authors express especially grateful memory and deep gratitude to Professor S. G. Psakhie as the organizer, teacher, author, and mastermind of the studies aimed at solving the global problem of ensuring the seismic safety of the human community. The work was performed according to the Government research assignments for IEC SB RAS (V.V.R.), and for ISPMS SB RAS (E.V.S.).

References

1. Ogata Y (2017) Statistics of earthquake activity: models and methods for earthquake predictability studies. Annu Rev Earth Planet Sci 45:497–527. https://doi.org/10.1146/annurev-earth-063016-015918
2. Booth E (2018) Dealing with earthquakes: the practice of seismic engineering 'as if people mattered.' Bull Earthq Eng 6:1661–1724. https://doi.org/10.1007/s10518-017-0302-8
3. Panza FG, Kossobokov VG, Peresan A, Nekrasova K (2014) Earthquake hazard, risk and disasters. Academic Press, Cambridge, UK, pp 309–357 (Why are the Standard Probabilistic Methods of Estimating Seismic Hazard and Risks Too Often Wrong). https://doi.org/10.1016/B978-0-12-394848-9.00012-2
4. Chen X, Nakata N, Pennington C, Haffener J, Chang JC, He X, Zhan Z, Ni S, Walter JI (2017) The Pawnee earthquake as a result of the interplay among injection, faults and foreshocks. Sci Rep 7:4945. https://doi.org/10.1038/s41598-017-04992-z
5. Miller S (2013) The role of fluids in tectonic and earthquake processes. Adv Geophys 54:1–46. https://doi.org/10.1016/B978-0-12-380940-7.00001-9
6. Ruzhich VV, Psakhie SG, Shilko EV, Vakhromeev AG, Levina EA (2019) On the possibility of development of the technology for managing seismotectonic displacements in fault zones. AIP Conf Proc 2051:020261. https://doi.org/10.1063/1.5083504
7. Psakhie SG, Ruzhich VV, Shilko EV, Popov VL, Astafurov SV (2007) A new way to manage displacements in zones of active faults. Tribol Int 40:995–1003. https://doi.org/10.1016/j.triboint.2006.02.021
8. Ruzhich VV, Psakhie SG, Chernykh EN, Shilko EV, Levina EA, Dimaki AV (2018) Baikal ice cover as a representative block medium for research in lithospheric geodynamics. Phys Mesomech 21:223–233. https://doi.org/10.1134/S1029959918030062
9. Vakhromeev AG, Ivanishin VM, Sverkunov SA, Polyakov VN, Razyapov RK (2019) Deep well as a facility for on-line hydraulic studies of the stress state of the rock mass in fluid-saturated fractured reservoirs. Geodyn Tectonophys 10(3):761–778. https://doi.org/10.5800/GT-2019-10-3-0440
10. Pang X-Q, Jia C-Z, Wanf W-Y (2015) Petroleum geology features and research developments of hydrocarbon accumulation in deep petroliferous basins. Petrol Sci 12:1–53. https://doi.org/10.1007/s12182-015-0014-0
11. McGarr A (2012) Relating stick-slip friction experiments to earthquake source parameters. Geophys Res Lett 39:L05303. https://doi.org/10.1029/2011GL050327

12. Mclaskey GC, Yamashita F (2017) Slow and fast ruptures on a laboratory fault controlled by loading characteristics. J Geophys Res Solid Earth 122:3719–3738. https://doi.org/10.1002/2016JB013681

13. Popov VL (2010) Contact mechanics and friction. Physical principles and applications. Springer-Verlag Berlin Heidelberg, Berlin

14. Ruzhich VV, Kocharyan GG (2017) On the structure and formation of earthquake sources in the faults located in the subsurface and deep levels of the crust. Part I. Subsurface level. Geodyn Tectonophys 8(4):1021–1034. https://doi.org/10.5800/GT-2017-8-4-0330

15. Nielsen S (2017) From slow to fast faulting: recent challenges in earthquake fault mechanics. Philos Trans R Soc A Math Phys Eng Sci 375(2103):20160016. https://doi.org/10.1098/rsta.2016.0016

16. Zhurkov SN (1984) Kinetic concept of the strength of solids. Int J Fract 26:295–307. https://doi.org/10.1007/BF0096296.1

17. Petrov YuV, Karihaloo BL, Bratov VV, Bragov AM (2012) Multi-scale dynamic fracture model for quasi-brittle materials. Int J Eng Sci 61:3–9. https://doi.org/10.1016/j.ijengsci.2012.06.004

18. Filippov AE, Popov VL, Psakhie SG, Shilko EV (2006) Converting displacement dynamics into creep in block media. Tech Phys Lett 32:545–549. https://doi.org/10.1134/S1063785006060 60290

19. Scuderi MM, Collettini C (2016) The role of fluid pressure in induced vs. triggered seismicity: insights from rock deformation experiments on carbonates. Sci Rep 6:24852. https://doi.org/10.1038/srep24852

20. Cornelio C, Spagnuolo E, Di Toro G, Nielsen S, Violay M (2019) Mechanical behaviour of fluid-lubricated faults. Nat Commun 10:1274. https://doi.org/10.1038/s41467-019-09293-9

21. Loreto MF, Düşünür-Doğan D, Üner S, İşcan-Alp Y, Ocakoğlu N, Cocchi L, Muccini F, Giordano P, Ligi M (2019) Fault-controlled deep hydrothermal flow in a back-arc tectonic setting, SE Tyrrhenian Sea. Sci Rep 9:17724. https://doi.org/10.1038/s41598-019-53696-z

22. Ruzhich VV, Kocharyan GG, Travin AV, Saveleva VB, Ostapchuk AA, Rasskazov SV, Yasnygina TA, Yudin DS (2018) Determination of the PT conditions that accompanied a seismogenic slip along a deep segment of the marginal suture of the Siberian Craton. Dokl Earth Sci 481:1017–1020. https://doi.org/10.1134/S1028334X18080081

23. Sobolev GA, Vettegren' VI, Ruzhich VV, Ivanova LA, Mamalimov RI, Shcherbakov IP (2015) A study of nanocrystals and the glide-plane mechanism. J Volcanol Seismol 9:151–161. https://doi.org/10.1134/S0742046315030057

24. Vettegren VI, Ponomarev AV, Sobolev GA, Shcherbakov IP, Mamalimov RI, Kulik VB, Patonin AV (2017) Structural changes in the surface of a heterogeneous nanocrystalline body (sandstone) under the friction. Phys Solid State 59:588–593. https://doi.org/10.1134/S1063783417030313

25. Medvedev VY, Ivanova LA, Lysov BA, Ruzhich VV, Marchuk MV (2014) Experimental study of decompression, permeability and healing of silicate rocks in fault zones. Geodyn Tectonophys 5(4):905–917. https://doi.org/10.5800/GT-2014-5-4-0162

26. Ruzhich VV, Kocharyan GG, Saveleva VB, Travin A (2018) On the structure and formation of earthquake sources in the faults located in the subsurface and deep levels of the crust. Part II. Deep level. Geodyn Tectonophys 9(3):1039–1061. https://doi.org/10.5800/GT-2018-9-3-0383

27. Ruzhich VV, Truskov VA, Chernykh EN, Smekalin OP (1999) Neotectonic movements in fault zones of the Baikal region and their origin mecha-nisms. Russ Geol Geophys 40(3):356–368

28. Astafurov SV, Shilko EV, Psakhie SG, Rhuzich VV (2008) Effect of local stress on the interface response to dynamic loading in faulted crust. Russ Geol Geophys 49(1):52–58. https://doi.org/10.1016/j.rgg.2007.12.007

29. Psakhie SG, Popov VL, Shilko EV et al (2006) A method for controlling the displacement mode in fragments of seismically active tectonic faults. RF Patent 2273035, 27 Mar 2006

30. Bürgmann R (2018) The geophysics, geology and mechanics of slow fault slip. Earth Planet Sci Lett 495:112–134. https://doi.org/10.1016/j.epsl.2018.04.062

31. Michel S, Gualandi A, Avouac J-P (2019) Similar scaling laws for earthquakes and Cascadia slow-slip events. Nature 574:522–526. https://doi.org/10.1038/s41586-019-1673-6

32. Ostapchuk AA, Pavlov DV, Ruzhich VV, Gubanova AE (2019) Seismic-acoustics of a block sliding along a fault. Pure Appl Geophys. https://doi.org/10.1007/s00024-019-02375-1

33. Rebinder PA, Shchukin ED (1973) The surface phenomena in solids during the course of their deformation and failure. Soviet Physics Uspeki 15(5):533–554

34. Traskin VYu (2009) Rehbinder effect in tectonophysics. Izv Phys Solid Earth 45:952. https://doi.org/10.1134/S1069351309110032

35. Vakhromeev AG, Sizykh VI (2006) The role of nappe tectonics in the development of abnormally high formation pressure and economic metalliferous brines: a case study of the southern Siberian craton. Dokl Earth Sci 407:209–212. https://doi.org/10.1134/S1028334X06020115

36. Vakhromeev AG, Sverkunov SA, Ivanishin VM, Razyapov RK, Danilova EM (2017) Geodynamic aspects in the study of complex mining and geological conditions for drilling into oil-and-gas reservoirs in the riphean carbonate rocks: an overview of the problem as exemplified by the deposits in the Baikit petroliferous district. Geodyn Tectonophys 8(4):903–921. https://doi.org/10.5800/GT-2017-8-4-0323

37. Vakhromeev AG, Sverkunov SA, Siraev RU, Razyapov RK, Sotnikov AK, Chernokalov KA (2016) The primary method of drilling a horizontal hole in a fracture type of oil and gas saturated carbonate reservoir under conditions of abnormally low reservoir pressure. RF Patent 2602437, 20 Nov 2016, Bull. No. 32 (in Russian)

38. Ruzhich VV, Psakhie SG, Shilko EV, Vakhromeev AG, Levina EA (2018) On the possibility of development of the technology for managing seismotectonic displacements in fault zones. AIP Conf Proc 2051:020261. https://doi.org/10.1063/1.5083504

39. Ma T, Chen P, Zhao J (2016) Overview on vertical and directional drilling technologies for the exploration and exploitation of deep petroleum resources. Geomech Geophys Geo-Energy and Geo-Resour 2:365–395. https://doi.org/10.1007/s40948-016-0038-y

40. Rodkin MV, Rundquist DV (2017) Geofluidogeodynamics. Application to seismology, tectonics, and processes of ore and oil genesis. Publishing House "Intellect", Dolgoprudny (RU)

41. Popov VL, Grzemba B, Starcevic J, Popov M (2012) Rate and state dependent friction laws and the prediction of earthquakes: what can we learn from laboratory models? Tectonophysics 532–535:291–300. https://doi.org/10.1016/j.tecto.2012.02.020

42. Kocharyan GG, Novikov VA (2016) Experimental study of different modes of block sliding along interface. Part 1. Laboratory experiments. Phys Mesomech 19:189–199. https://doi.org/10.1134/S1029959916020120

43. Sibson RH (1973) Interactions between temperature and pore fluid pressure during an earthquake faulting and a mechanism for partial or total stress relief. Nat Phys Sci 243:66–68. https://doi.org/10.1038/physci243066a0

44. Bachmann CE, Wiemer S, Woessner J, Hainzl S (2011) Statistical analysis of the induced Basel 2006 earthquake sequence: introducing a probability-based monitoring approach for Enhanced Geothermal Systems. Geophys J Int 186(2):793–807. https://doi.org/10.1111/j.1365-246x.2011.05068.x

45. Guglielmi Y, Cappa F, Avouac J-P, Henry P, Elsworth D (2015) Seismicity triggered by fluid injection-induced aseismic slip. Science 348(6240):1224–1226. https://doi.org/10.1126/science.aab0476

46. Weingarten M, Ge S, Godt JW, Bekins BA, Rubinstein JL (2015) High-rate injection is associated with increase in U.S. mid-continent seismicity. Science 348(6241):1336–1340. https://doi.org/10.1126/science.aab1345

47. Rutqvist J, Rinaldi AP, Cappa F, Jeanne P, Mazzoldi A, Urpi L, Guglielmi Y, Vilarrasa V (2016) Fault activation and induced seismicity in geologic carbon storage—lessons learned from recent modeling studies. J Rock Mech Geotech Eng 8(6):789–804. https://doi.org/10.1016/j.jrmge.2016.09.001

48. Rinaldi AP, Rutqvist J, Cappa F (2014) Geomechanical effects on CO_2 leakage through fault zones during large-scale underground injection. Int J Greenhouse Gas Control 20:117–131. https://doi.org/10.1016/j.ijggc.2013.11.001

49. Tarasov NT, Tarasova NV (1995) Earthquakes induced by underground nuclear explosions. Environmental and ecological problems. NATO ASI Series (2. Environment), vol 4. Springer-Verlag Berlin, Heidelberg, Berlin, pp 215–223 (Response of seismoactive medium to nuclear explosions)
50. Emanov AF, Emanov AA, Fateev AV, Leskova EV, Shevkunova EV, Podkorytova VG (2014) Mining-induced seismicity at open pit mines in Kuzbass (Bachatsky earthquake on June 18, 2013). J Min Sci 50:224–228. https://doi.org/10.1134/S1062739114020033
51. Mirzoev K, Nikolaev AV, Lukk AA, Yunga SL (2009) Induced seismicity and the possibilities of controlled relaxation of tectonic stresses in the earth's crust. Izv Phys Solid Earth 45:885–904. https://doi.org/10.1134/S1069351309100061

Particle-Based Approach for Simulation of Nonlinear Material Behavior in Contact Zones

Evgeny V. Shilko, Alexey Yu. Smolin, Andrey V. Dimaki, and Galina M. Eremina

Abstract Methods of particles are now recognized as an effective tool for numerical modeling of dynamic mechanical and coupled processes in solids and liquids. This chapter is devoted to a brief review of recent advances in the development of the popular particle-based discrete element method (DEM). DEM is conventionally considered as a highly specialized technique for modeling the flow of granular media and the fracture of brittle materials at micro- and mesoscopic scales. However, in the last decade, great progress has been made in the development of the formalism of this method. It is largely associated with the works of the scientific group of Professor S. G. Psakhie. The most important achievement of this group is a generalized formulation of the method of homogeneously deformable discrete elements. In the chapter, we describe keystones of this implementation of DEM and a universal approach that allows one to apply various rheological models of materials (including coupled models of porous fluid-saturated solids) to a discrete element. The new formalism makes possible qualitative expansion of the scope of application of the particle-based discrete element technique to materials with various rheological properties and to the range of considered scales form microscopic to macroscopic. The capabilities of this method are especially in demand in the study of the features of contact interaction of materials. To demonstrate these capabilities, we briefly review two recent applications concerning (a) the effect of adhesive interaction on the regime of wear of surface asperities under tangential contact of bodies and (b) the nonmonotonic dependence of the stress concentration in the neck of the human femur on the dynamics of hip joint contact loading.

Keywords Discrete element method · Deformable element · Movable cellular automata · Many-body interaction · Plasticity · Poroelasticity · Surface adhesion · Adhesive wear · Bone tissue · Pore fluid

E. V. Shilko (✉) · A. Yu. Smolin · A. V. Dimaki · G. M. Eremina
Institute of Strength Physics and Materials Science SB RAS, 634055 Tomsk, Russia
e-mail: shilko@ispms.tsc.ru

© The Author(s) 2021
G.-P. Ostermeyer et al. (eds.), *Multiscale Biomechanics and Tribology of Inorganic and Organic Systems*, Springer Tracts in Mechanical Engineering,
https://doi.org/10.1007/978-3-030-60124-9_4

1   Introduction

Starting with the classic works of Cauchy and Navier [1, 2], the development of the formalism of the discrete representation of the medium at a continuum ("super-atomic") scale is considered among the fundamental problems for the mechanics of solids. In the framework of this representation, a matter is described by an ensemble of interacting particles. Each particle models a sufficient number of atoms or molecules to describe the state and response of the particle in terms of thermodynamic parameters and classical mechanical models (Fig. 1). A discrete description of solids and liquids was initially considered as a way to fill the gap between molecular mechanics and continuum mechanics [3]. However, the rapid development of the formalism of particle methods in the last two decades has made it possible to apply them to study the mechanical behavior of diverse solids as well as various mechanically assisted or activated processes in the entire spectrum of spatial scales from atomic to macroscopic.

The traditional approach to the numerical study of the behavior of materials on the "above-atomic" spatial scales is based on the methods of continuum mechanics such as finite element, finite difference and boundary element methods (FEM, FDM and BEM) [4–9]. The formalism of these methods allows easy implementation of various linear and nonlinear (including coupled thermomechanical and poroelastic) rheological models. Moreover, advanced implementations of FEM, FDM and BEM include the ability to directly model fracture [10, 11]. Despite the well-known advantages of continuum numerical methods, their fundamental limitation is difficulty in modeling of complex fracture-related problems including development of multiple fractures, contact interaction of the initial and newly formed surfaces, wear of surface layers and a change in surface roughness, flow of granular media, etc.

Mentioned limitation is not inherent in particle-based methods [12–19]. The most relevant and efficient particle-based method for numerical modeling of the above-said complex mechanical processes in solids is the discrete element method (DEM) [18, 19]. The constantly growing interest in this numerical technique is determined by the ability to solve a variety of complex and non-linear contact problems, where the processes of fragmentation and mass transfer of fragments play a key role. This

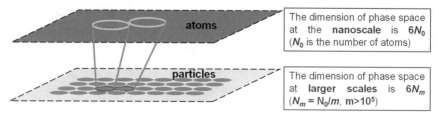

Fig. 1 Representation of the material by ensemble of particles as an extension of atomistic representation to higher spatial (and temporal) scales

chapter is devoted to the analysis of achievements in the development of DEM for computer simulation of the mechanical behavior of consolidated solids.

Creation of this method is attributed to Cundall [20, 21]. In the framework of original implementation of DEM, a discrete element is treated as a finite part of a solid body (or a particle in particulate/granular material) bounded by clearly defined (exact) surface. The latter qualitatively distinguishes this method from the methods of quasiparticles with "fuzzy" surface. Elements can be either chemically bonded if they model a consolidated material, or contact if the contact interaction of the fragments is modeled. Changing the type of bond of elements (chemically bonded ↔ contact ↔ noninteracting) is governed by the applied criteria of fracture, bond formation and contact loss [22, 23]. The stress state of a discrete element is determined by the mechanical load of the surrounding elements on the surface of the element. The absence of a constraint in the form of the continuity equation makes the DEM extremely attractive for numerical studying complex processes in contact zones.

The term "discrete element method" is now used as a generic name for a large group of numerical techniques based on these general principles for representing the medium [18, 21, 24, 25]. Various representatives of this group differ in several key features: the principle of local or global force equilibrium (explicit or implicit formulation); approximation of the shape of the volume modeled by a discrete element; approximation to the description of the deformability of a discrete element.

Explicit DEM is the most popular and is widely used to solve fracture- and contact interaction related dynamic problems. It implies the formulation of equations of motion for each discrete element and the parallel solution of the system of these equations with an explicit time marching scheme (Euler, Verlet or other integration algorithm). Since a discrete element simulates a finite volume of material, the mechanical interaction of such finite volumes should lead not only to their translational motion, but also to rotation determined by the moments of interaction forces. The form of the dynamics equations for the rotational degrees of freedom of an element (Euler's equations) is determined by the shape of an element. In the general case, Euler's equations are written in integral form using the inertia tensor (tensor of inertia, in turn, is formulated as an integral) [18]. Integral Euler equation is cumbersome and computationally costly, that is why elements with complex nonequiaxial geometry (polygonal [26], superquadric [27] elements) are used only for modeling granular or fragmented (block-structured) materials. At the same time, a consolidated material can be much more efficiently modelled by an ensemble of bonded equiaxial elements, the shape of which is approximated by equivalent sphere (3D problem) or a disk of a given height (2D problem) [18, 22, 28]. The efficiency of an approximation of an equivalent disk/sphere is determined by the simplicity of the Newton–Euler motion equations for a discrete element:

$$
\begin{cases}
m_i \frac{d^2 \vec{r}_i}{dt^2} = m_i \frac{d\vec{v}_i}{dt} = \vec{F}_i = \sum_{k=1}^{N_i} \vec{F}_{ik} = \sum_{k=1}^{N_i} \left(\vec{F}_{ik}^c + \vec{F}_{ik}^t \right) \\
J_i \frac{d\vec{\omega}_i}{dt} = \vec{M}_i = \sum_{k=1}^{N_i} \vec{M}_{ik}
\end{cases}
, \qquad (1)
$$

where i is the number of discrete element, $\vec{r}_i$, $\vec{v}_i$ and $\vec{\omega}_i$ are the radius-vector, velocity vector and angular velocity pseudovector respectively, m_i is the mass of the element i, J_i is moment of inertia of an equivalent disk or sphere, $\vec{F}_i$ and $\vec{M}_i$ are the total force and torque acting on the element i by the neighbors, $\vec{F}_{ik}$ is the force of interaction of the considered element i with the neighbor k, $\vec{F}_{ik}^c$ and $\vec{F}_{ik}^t$ are the central (along the line connecting mass centers of the elements i and k) and tangential (in transverse plane) components of the force $\vec{F}_{ik}$, $\vec{M}_{ik}$ is the moment of interaction forces (includes tangential moment of $\vec{F}_{ik}^t$ and twisting moment [22]), N_i is the number of neighbors of the element i.

One can see that the approximation of the equivalent disk/sphere allows the use of Euler's equations in the most trivial and computationally efficient form. Another important consequence of this approximation is the formal independence of the central and tangential interactions: the force $\vec{F}_{ik}^c$ does not cause acceleration in the plane of the tangential interaction, and the force $\vec{F}_{ik}^t$ does not cause acceleration along the line connecting the centers of mass of the elements.

The forces $\vec{F}_{ik}^c$ and $\vec{F}_{ik}^t$ of interaction of discrete elements are traditionally represented as the sum of the potential ($\vec{F}_{ik}^{cp}$ and $\vec{F}_{ik}^{tp}$) and viscous ($\vec{F}_{ik}^{cv}$ and $\vec{F}_{ik}^{tv}$) constituents [20, 22]. From the point of view of the rheological description, viscous forces have a meaning similar to the physical meaning of the damper in the Kelvin-Voight viscoelastic model. A key component of constructing a discrete-element model of a material is the determination of the structural type and coefficients of the potential interaction forces.

In the framework of the traditional DEM implementation, the central and tangential potential forces of interaction of equivalent balls/disks (F_{ik}^{cp} and F_{ik}^{tp}) are calculated in the pair-wise approximation. From the physical point of view, pair-wise potential corresponds to the approximation of a non-deformable (rigid) element (a system of springs or rods). Here, the term "rigid" is used in the sense that the interaction of the element i with the neighbor k does not change its volume and shape and, therefore, does not cause a change in the forces of interaction of the element i with other neighbors (these neighbors "feel" the result of the interaction in the pair $i - k$ only indirectly, through the motion of the element i). This approximation is widely used for micro- and mesoscopic description of the processes of damage accumulation and fracture of brittle materials, contact interaction of elastic bodies, including the dynamics of block-structured media. In this case, element-element interaction is traditionally modelled using harmonic interaction potentials (such an interaction is schematically represented by connecting the centers of the elements with two springs oriented in the central and tangential directions) [22, 28, 29]. Some models also use nonlinear (elastic–plastic or viscoelastic Maxwell type) formulations of pair-wise interaction forces F_{ik}^{cp} and F_{ik}^{tp} [30, 31] for simulation of granular media and porous structures with non-linear/ductile rheological properties of the material of the skeleton walls or granules. However, such potentials make possible adequate description of the mechanical behavior of porous systems only at a "low" scale (the scale of discontinuities or granules).

The key problems that strongly limit the range of application of the traditional implementation of DEM with pair-wise interaction forces are well known. They are (i) dependence of the macroscopic properties of an ensemble of discrete elements on the type of packaging and size distribution of elements, (ii) incorrect description of the plastic strain of an ensemble of elements (for example, plastic deformation of a sample may be accompanied by an uncontrolled change in its volume), and other related problems.

Various approaches to solving these fundamental difficulties within the framework of the concept of non-deformable elements have been proposed in last decades. In particular, stochastic dense packing of non-uniform-sized circular (2D) or spherical (3D) elements [22, 28, 29] is used to solve the problems of packing-induced anisotropy of the elastic response and packing-dependent ratio of elastic modules of an ensemble of elements. An alternative approach is to use the formalism of spring network model (lattice model [32, 33]) to build relationships for the forces of interaction of regularly packed uniform-sized elements [34–36]. The lattice model is based on the postulation of the form of interaction potential (harmonic potential is usually used for both central and angular interactions) and equalization of elastic strain energy stored in a unit cell of volume to the associated elastic strain energy of the modelled continuum. The material parameters derived from this equality are included in the relationships for the forces of element-element interaction. The above approaches made it possible to adequately describe the mechanical (and thermomechanical [36]) behavior of brittle materials under complex loading conditions. At the same time, they do not allow solving the key problem of incorrect modeling of nonlinear (and/or inelastic) mechanical behavior of materials with complex rheological properties (including rubber-like viscoelastic materials as well as metallic and polymer materials, whose macroscopic plasticity is not related to discontinuities).

The problem of correct modeling of nonlinear mechanical behavior of consolidated materials by the method of discrete elements can be generally solved only by using the approximation of the deformable element. In turn, the deformability of an element can be realized only within the framework of a many-body interaction of elements. This means that the potential interaction force must depend not only on the relative motion of the elements in the pair, but also on the interaction of each of them with other neighbors. The formulation of the general structural form of the potential interaction force and its specific realizations for materials with various rheological properties has been among the critical challenges for the DEM until recently.

2   Distinct Element Method with Deformable Elements

A meaningful contribution to the development of the formalism of DEM was done by Professor Sergey G. Psakhie and his team. Professor Psakhie was a founder of the new particle-based method, namely, the method of movable cellular automata (MCA) [37, 38].The basic principles of this method were developed in collaboration with Professor Yuki Horie (North Carolina State University, Los Alamos National

Lab). Originally, the MCA method was designed as a hybrid technique to model mechanically activated chemical reactions in powder mixtures [39, 40]. This original implementation combined the formalisms of discrete elements and cellular automata, in which the mechanical response of the particle was described using the DEM formalism, while the non-mechanical thermodynamic aspects of particle–particle interaction (including melting and mechanically activated chemical reaction) were modelled on the basis of the concept of cellular automata.

The most important achievement of S. G. Psakhie in the development of numerical particle-based modelling techniques is the proposed general formalism of the method of homogeneously (simply) deformable discrete elements.

The keystones of this formalism were laid in the framework of collaboration with Professor Valentin L. Popov (Technische Universität Berlin) and his scientific group. In a joint work of Professors Psakhie and Popov [41], the basic principles for describing the mechanical behavior of a discrete element (movable cellular automaton) as a deformable area of the medium were formulated. For a special case of an ensemble of close packed elements of the same equivalent radius, which models an isotropic two-dimensional continuum, a relation was proposed for the potential force of the central interaction of elements in the many-body approximation:

$$F_{ik}^{cp} = E^* \delta L_{ik} = E^* \left(\delta r_{ik} + D \sum_{j=1}^{N} \delta r_{ij} + D \sum_{m=1}^{N} \delta r_{km} \right). \tag{2}$$

Here, the symbol δ denotes the difference between the current and initial values of the corresponding parameter, $r_{ik} = |\vec{r}_i - \vec{r}_k|$ is the distance between the centers of mass of the elements i and k, L_{ik} is effective distance between the elements, N is the number of neighbors in the first coordination sphere. The coefficients E^* and D are expressed in terms of the elastic constants of the material and the element packing parameters. Derivation of these coefficients is based on the condition for ensuring the required values of Young's modulus and Poisson's ratio of the material [41]. This model is based on the same principles as the classic spring network models, but it has a fundamental difference. The force of the central interaction of the two elements is represented in the form of a superposition of the pair-wise component $E^* \delta r_{ik}$ and "hydrostatic" components. The latter are proportional to the change in the volumes of the interacting elements (here, we use the particular form of expression for the element's volume change in regular packing). The reasonableness of this formulation is confirmed by the linear relationship of the diagonal components of the stress tensor and the volume strain in the vast majority of macroscopic rheological models of materials (linear and non-linear elasticity, viscoelasticity, plasticity).

The proposed formalism actually uses the approximation of homogeneously deformable elements. It was further developed to describe plastic flow of the elements based on constitutive equations of the macroscopic continuum theory of defects [41]. Despite the clear advantages of the proposed formalism, it has the same key limitations as traditional lattice-based models. Among them, are the absence of the tangential interaction of elements (shear resistance force), packing-dependent artificial anisotropy of the integral response of the ensemble of elements at a significant distortion of the initial symmetry of the lattice, and the lack of a general and simple algorithm of implementation of complex rheological material models.

A generalized formulation of the method of homogeneously deformable discrete elements was proposed later in the works of Professor S. G. Psakhie with co-authors. It applies the concept of many-body interaction for the ensemble of arbitrarily packed different-size elements and is based on the following principles:

1. Approximation of equivalent disks/spheres (Fig. 2). Within the framework of this approximation, the dynamics of elements is described by Eq. (1), and the forces of central and tangential interaction are assumed to be formally unrelated to each other. Elements interact with each other through flat contact areas. The geometry and squares of these areas are determined by the local packing and sizes of the elements [22, 42, 43]. The potential interaction of the two elements is conveniently described in terms of specific forces of interaction (normal σ_{ik} and tangential τ_{ik} contact stresses):

$$\begin{cases} F_{ik}^{cp} = \sigma_{ik} S_{ik} \\ F_{ik}^{tp} = \tau_{ik} S_{ik} \end{cases}. \tag{3}$$

2. A discrete element is assumed to be homogeneously deformable, i.e., its stress–strain state is characterized by tensors of stress and strain (hereinafter called

Regular packing   **Stochastic packing**

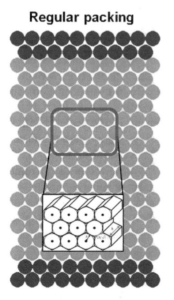

Fig. 2 Typical examples of 2D samples modelled by the ensembles of regularly and stochastically (dense) packed discrete elements. Figures show equivalent disks

average stresses $\overline{\sigma}_{\alpha\beta}$ and strains $\overline{\varepsilon}_{\alpha\beta}$, where α, $\beta = x$, y, z). To determine the components of average stress tensor, we use well-known classical relations for average stresses and "local" values of traction vectors at the contact areas of the element with neighbors (stress homogenization) [22, 42, 44]:

$$\sigma^i_{\alpha\beta} = \frac{R_i}{\Omega^0_i} \sum_{k=1}^{N_i} S^0_{ik} (\vec{n}_{ik})_\alpha \left(\vec{\Sigma}_{ik} \right)_\beta = \frac{R_i}{\Omega^0_i} \sum_{k=1}^{N_i} S^0_{ik} (\vec{n}_{ik})_\alpha \left(\sigma_{ik} (\vec{n}_{ik})_\beta + \tau_{ik} (\vec{t}_{ik})_\beta \right),$$

(4)

where R_i is the radius of equivalent sphere approximating the element i, Ω^0_i is the volume of unstrained element, S^0_{ik} is the contact square in unstrained pair $i - k$, $\vec{\Sigma}_{ik}$ is the traction vector at the area of contact of elements i and k (normal and tangential contact stresses are the components of this vector), $\vec{n}_{ik}$ is the unit normal vector directed along the line connecting the mass centers of the elements, $\vec{t}_{ik}$ is the unit tangent vector directed in the tangential plane, $(\vec{W})_\beta$ is the projection of some vector $\vec{W}$ onto the β-axis. We emphasize the generality of definition (4), which is applicable for arbitrary local packing of elements of various sizes (packing and sizes determine the values of the direction cosines and contact areas).

3. A consequence of the deformability of an element is the need to divide the spatial parameters of its interaction with a neighbor (pair overlap and relative tangential displacement) into two components, namely, the contributions of both elements:

$$\begin{cases} \Delta r_{ik} = \Delta q_{ik} + \Delta q_{ki} = R_i \Delta \varepsilon_{ik} + R_k \Delta \varepsilon_{ki} \\ \Delta l^{sh}_{ik} = R_i \Delta \gamma_{ik} + R_k \Delta \gamma_{ki} \end{cases},$$

(5)

where q_{ik} and q_{ki} are the distances from the mass centers of the interacting elements i and k to the central point of the contact area (they are equal to equivalent radii R_i and R_k respectively for the case of unstrained elements), ε_{ik} and ε_{ki} are central pair strains of discrete elements i and k, l^{shear}_{ik} is the value of relative tangential displacement of the elements (it is calculated with taking into account the element rotations [22, 39, 42]), γ_{ik} and γ_{ki} are the shear angles of discrete elements i and k (contributions to the total shear angle). In the general case, $\varepsilon_{ik} \neq \varepsilon_{ki}$ and $\gamma_{ik} \neq \gamma_{ki}$. Relations (5) are given in the incremental form [hereinafter the symbol Δ denotes an increment of some parameter over the time step of numerical integration of the motion equations (1)] because this form is convenient for the numerical implementation of complex rheological models. Note that the strains ε_{ik} and γ_{ik} are the components of the "local" strain vector, which is used in the definition of $\overline{\varepsilon}^i_{\alpha\beta}$ by the analogy to (4).

4. In the framework of approximation of deformable element, the specific normal and tangential forces (contact stresses) σ_{ik} and τ_{ik} are interpreted as the components of the specific force of mechanical response of the element i to the mechanical loading by the neighboring element k. These stresses are the functions of the

i-th element strains ε_{ik} and γ_{ik} in the pair $i-k$. We proposed the general form of these functions, which assumes homogeneous deformability of the element and linear relation between volume strain and mean stress (or pressure) [42–44]:

$$\begin{cases} \sigma_{ik} = \sigma_{ik}^{pair}(\varepsilon_{ik}, \dot{\varepsilon}_{ik}) + B_i\overline{\sigma}_{mean}^i = \sigma_{ik}^{pair}(\varepsilon_{ik}, \dot{\varepsilon}_{ik}) - B_i\overline{P}_i \\ \tau_{ik} = \tau_{ik}^{pair}(\gamma_{ik}, \dot{\gamma}_{ik}) \end{cases}. \tag{6}$$

Here, the upper index "$pair$" denotes pair-wise function, $\overline{\sigma}_{mean}^i = -\overline{P}_i = (\overline{\sigma}_{xx}^i + \overline{\sigma}_{yy}^i + \overline{\sigma}_{zz}^i)/3$, B_i is the material parameter. The first relation in (6) suggests that the normal (compressive/tensile) resistance of an element is determined by both the strain of this element ε_{ik} along the loading axis and the hydrostatic (liquid-like) component. The second relation is written in the pair-wise approximation, which ideologically corresponds to the relations connecting the off-diagonal components of the stress and strain (or strain rate) tensors in most rheological models of solids. The specific form of the pair-wise components σ_{ik}^{pair} and τ_{ik}^{pair} as well as the values of the material coefficients are determined by the applied rheological model of the material modeled by a discrete element. The necessity to satisfy Newton's third law ($\sigma_{ik} = \sigma_{ki}$ and $\tau_{ik} = \tau_{ki}$) leads to the following systems of equations, which are used to calculate the current value of interaction forces $\vec{F}_{ik}^{cp}$ and $\vec{F}_{ik}^{tp}$ for the motion equation (1):

$$\begin{cases} \sigma_{ik}^{pair}(\varepsilon_{ik}, \dot{\varepsilon}_{ik}) + B_i\overline{\sigma}_{mean}^i = \sigma_{ki}^{pair}(\varepsilon_{ki}, \dot{\varepsilon}_{ki}) + B_i\overline{\sigma}_{mean}^i \\ \Delta r_{ik} = \Delta q_{ik} + \Delta q_{ki} = R_i\Delta\varepsilon_{ik} + R_k\Delta\varepsilon_{ki} \end{cases}, \tag{7}$$

$$\begin{cases} \tau_{ik}^{pair}(\gamma_{ik}, \dot{\gamma}_{ik}) = \tau_{ki}^{pair}(\gamma_{ki}, \dot{\gamma}_{ki}) \\ \Delta l_{ik}^{sh} = R_i\Delta\gamma_{ik} + R_k\Delta\gamma_{ki} \end{cases}. \tag{8}$$

The solutions to each pair of equations are the values of pair strains. These strains are then used to calculate the current values of the forces of interaction of elements according to (6).

5. A pair of elements modeling a part of a consolidated material is assumed to be chemically bonded (linked). The central interaction of linked elements includes resistance to compression and tension, and the tangential interaction typically takes into account shear and bending resistance [42]. In the framework of the discrete element method, the elementary act of fracture at the considered spatial scale is the breaking of the chemical bond between the two elements. The condition of linked-to-unlinked transition is governed by a specified fracture criterion for a pair. This criterion is determined as a fracture condition at the contact area. Most fracture criteria in the mechanics of a deformable solid are formulated in force-like form in terms of the invariants of the stress tensor (Mises, Mohr–Coulomb, Drucker-Prager and other failure criteria). We proposed an approach to implementation of such kind of criteria within the framework of the formalism of deformable elements [42, 44]. It is based on determining the local stress tensor on the contact area of the linked pair of elements and

calculating its invariants. The local stress tensor $\sigma_{\alpha'\beta'}^{ik}$ is determined in the local coordinate system of the pair $i - k$. The specific forces σ_{ik} and τ_{ik} are used as the diagonal and off-diagonal components of this tensor. The missing 4 components are determined on the contact surface by linear interpolation of the corresponding components of average stress tensors in the interacting elements: $\sigma_{\alpha'\beta'}^{ik} = \left(\overline{\sigma}_{\alpha'\beta'}^{i} q_{ki} + \overline{\sigma}_{\alpha'\beta'}^{k} q_{ik}\right)/r_{ik}$. Here the accent means that average stresses are considered in the local coordinate system.

A bond break leads to a change in the interaction in a pair of discrete elements: the central interaction includes only compression resistance, and the dry [22] or viscous friction force is usually used as the tangential force. A pair becomes non-interacting if the value of the central force becomes equal to zero. We also note that the deformability of elements leads to a generalized formulation of contact detection condition, which takes into account a change in the linear dimensions of elements [42, 44].

6. Contact interaction of unlinked discrete elements is traditionally treated as non-adhesive. However, in many real systems, the adhesion of surfaces is an important factor determining the laws of friction and wear [45]. To adequately model the adhesive contact of surfaces, we assume attractive normal force acting between elements even after they are debonded [46, 47]. This force varies with separation of the surfaces of interacting elements according to prescribed model of adhesion (Dugdale's, Van-der-Vaals or other interaction potential). The value of separation is determined with taking into account deformation of elements along the normal $\vec{n}_{ik}$. The tangential force of interaction of unlinked and noncontact elements is assumed to be zero.

At large values of surface energy, the mechanical contact of chemically clean and smooth surfaces can be accompanied by the formation of a chemical bond (this effect is often called cold welding). This effect is taken into account in the method of deformable discrete elements based on the use of the criterion of unlinked-to-linked transition for contacting pairs of elements. The pair of elements becomes linked if the criterion is satisfied. We proposed some formulations of such kind of criteria including critical values of the contact normal stress and plastic work of deformation [42].

3 Principles of Implementation of Rheological Models

The most important advantage of the proposed general formulation of element-element interaction (6)–(8) is the possibility of simple implementation of various rheological models of solids. One can see from relation (4) that the components of average stress tensor are linearly related to the forces of interaction of the element with its neighbors. In turn, the interaction forces are linearly related to the components of average stress tensor by relation (6). A fully similar interconnection takes place between average strains and pair strains of the element. We first showed that the similarity of the relationship between average and local (contact) stress and strain parameters of the element inevitably leads to the fact that the specific formulation of relations (6) should be similar to the formulation of the constitutive relations for the material of a discrete element. In other words, the relation for the specific central force σ_{ik} of the ith element response to the mechanical loading by the neighbor k has to be formulated by means of direct rewriting of the constitutive equation for the diagonal components of the stress tensor. The specific force of tangential response τ_{ik} is formulated by direct rewriting the corresponding constitutive relation for off-diagonal stress components. Using these principles, we implemented macroscopic mechanical models of elasticity, viscoelasticity, and plasticity and applied them to study the behavior of various materials, including metals, ceramics, composite materials, rocks, rubbers, and even bone tissues [38, 42–44].

Below there is an example of the relations (6) for the quite general case of locally isotropic viscoelastic (described by the Prony series) material of the element. These relations are written in an incremental form:

$$
\begin{cases}
\Delta\sigma_{ik} = 2G_{i,\Sigma}\Delta\varepsilon_{ik} + \left(1 - \dfrac{2G_i}{3K_i}\right)\Delta\bar{\sigma}^i_{mean} \\[2mm]
\quad - 2\sum_p \dfrac{G_{i,Mp}}{\eta_{i,Mp}}\Delta\sigma_{ik,Mp} + 2\dfrac{\Delta\bar{\sigma}^i_{mean}}{3K_{i,\Sigma}}\sum_p \dfrac{G_{i,Mp}K_{i,Mp}}{\eta_{i,Mp}}. \\[3mm]
\Delta\tau_{ik} = 2G_{i,\Sigma}\Delta\gamma_{ik} - 2\sum_p \dfrac{G_{i,Mp}}{\eta_{i,Mp}}\Delta\tau_{ik,Mp}
\end{cases}
\tag{9}
$$

Here $G_{i,\Sigma} = G_{i,K} + \sum_p G_{i,M}$ is the instant shear modulus of viscoelastic material ($G_{i,K}$ is the shear modulus of the Kelvin element, $G_{i,Mp}$ is the shear modulus of the pth Maxwell element in a series), $K_{i,\Sigma}$ is the total bulk modulus determined in a similar way, $\eta_{i,Mp}$ is the dynamic viscosity of the p-th Maxwell element, $\sigma_{ik,Mp}$ and $\tau_{ik,Mp}$ are the contributions of the pth Maxwell element to the total specific force. It is easy to show that substitution of the relations (9) to the definition of average stress tensor (4) leads to rigorous fulfilment of the constitutive equation for the viscoelastic material in terms of average stresses and strains. The particular case of (9) is the linear-elastic model (the only Kelvin element), which is typically used to numerically study the elastic behavior of brittle and ductile materials.

It is well known that although the mechanisms of plasticity can qualitatively differ for the materials of various natures (defects of the crystal lattice in metals and alloys and discontinuities in microscopically brittle materials), their macroscopic inelastic behavior is adequately described on the basis of similar models based on the principles of the classical theory of plastic flow. In the framework of this theory, plasticity is described as instantaneous relaxation of "excess" stresses when the plasticity criterion (which is traditionally expressed in terms of the stress tensor invariants) exceeds a specified threshold value (yield shear stress).

Macroscopic inelastic (ductile) behavior of materials is conventionally modelled using associated (for metals and polymers) and non-associated (for ceramic materials, rocks and bone tissues under mechanical confinement) plastic flow models.

The most popular way to implement these models within explicit numerical methods of continuum mechanics is the use of radial return algorithm of Wilkins [48]. We were the first to show that this algorithm can be easily adapted to the formalism of deformable discrete elements. By the analogy with elasticity, the prescribed law of mapping the average stresses $\overline{\sigma}^i_{\alpha\beta}$ is rigorously satisfied when applying the stress correcting expressions to the specific response forces σ_{ik} and τ_{ik}. Using this way, we numerically implemented the widely used macroscopic model of plasticity of metals with von Mises yield criterion [42, 43] and non-associated plastic flow model of Nikolaevsky [44] (the macroscopic rock plasticity model with von Mises-Schleicher yield criterion [49]). Other models of elasticity and plasticity can be implemented within the formalism of deformable discrete elements using this direct way. In particular, recently we developed a numerical model of inelastic deformation and fracture of brittle materials, which takes into account the finite incubation time of structural defects and applies principles of the structural-kinetic theory of strength [50].

The generality of the developed formalism allows one to implement not only mechanical but also coupled (thermomechanical, poromechanical and so on) material models. One of the most significant recent achievements of Professor S. G. Psakhie and his scientific team is the development of a hybrid (coupled) DEM-based technique to model the mechanical behavior of "contrast" materials, namely the porous materials with solid skeleton and interstitial liquid [38, 51]. Well-known examples of materials with locally contrasting mechanical properties are watered porous rocks and rubbers, bone and soft tissues. The importance of adequate consideration of the liquid phase in the contrast materials is determined by the fact that such kind of materials possesses strongly nonlinear behavior and non-stationary mechanical characteristics (even in the case of elastic-brittle skeleton) due to redistribution of mobile interstitial fluid in the pore space. Note that the formalism of DEM-based hybrid method was developed in collaboration with Dr. S. Zavsek (Velenje Coal Mine) and Professor J. Pezdic (University of Ljubljana).

Within the framework of this hybrid technique, the discrete element is considered as porous and permeable solid. The mechanical behavior of both solid skeleton and interstitial fluid and their mutual influence are taken into account. In particular, using Biot's linear model of poroelasticity [52] and Terzaghi's concept of effective stresses, we developed the coupled macroscopic model of permeable brittle materials. The main constitutive equations of this model are:

1. Dependence of the pore volume of an element on the hydrostatic component of the external load and the pore pressure

$$\frac{\Omega_p - \Omega_{p0}}{\Omega} = \phi - \phi_0 = \frac{a}{K}\sigma_{mean} + \left(\frac{1}{K} - \frac{1+\phi}{K_s}\right)P_{pore}, \qquad (10)$$

where Ω_p and Ω_{p0} are the current and initial (in an undeformed element) values of the pore volume, φ and φ_0 are the corresponding values of porosity, Ω is the volume of the element, P_{pore} is the pore pressure in the volume of the element.

2. Constitutive equation of linear compressible liquid in the pore space

$$P_{pore} = P_{pore}^0 + K_{fl}\left(\frac{\rho_{fl}}{\rho_{fl}^o} - 1\right) = P_{pore}^0 + K_{fl}\left(\frac{m_{fl}}{\rho_{fl}^o V_p} - 1\right), \quad (11)$$

where ρ_{fl}^0 and P_{pore}^0 are the equilibrium values of the density and pressure of the fluid under atmospheric conditions (in the absence of a mechanical confinement), ρ_{fl} is the current value of the density of the liquid in the pore space of the element, K_{fl} is the bulk modulus of the liquid.

3. Hooke's law for poroelastic material of the element

$$\Delta\sigma_{\alpha\beta} = 2G\left(\Delta\varepsilon_{\alpha\beta} - \delta_{\alpha\beta}\frac{a\Delta P_{pore}}{K}\right) + \delta_{\alpha\beta}\left(1 - \frac{2G}{K}\right)\Delta\sigma_{mean}, \quad (12)$$

where $a = 1 - K/K_s$ is a coefficient of poroelasticity.

4. Modified formulations of yield and fracture criteria in terms of Terzaghi's effective stresses $\overline{\sigma}_{\alpha\beta}^{eff} = \overline{\sigma}_{mean} + P_{pore}$ [51].

5. The equations of motion of discrete elements (1) are supplemented by the classical equation of transport of interstitial fluid in the pore space of the material [53]. The transport equation is solved on an ensemble of discrete elements by the finite volume method [51].

Shown below examples demonstrate the capability of the developed formalism to implement various complex material models. This allows qualitative expansion of the range of simulated materials and the spatial scales under consideration.

4 Recent Applications of the Formalism of Deformable Elements

DEM is particularly efficient technique to study various aspects of the contact problems (including mechanisms of wear) in technical and natural friction pairs. One of the main principles of Professor Psakhie was a diversification of research activity and the use of the developed mathematical tools in a variety of scientific fields. The section is devoted to a brief outline of some recent results of computer study of contact problems in technical and biological systems. These studies were initiated and supervised by S. G. Psakhie in close collaboration with Professor V. L. Popov.

4.1 Surface Adhesion as a Factor Controlling Regimes of Adhesive Wear

Over the past two decades, the authors have carried out numerical studies of the laws of friction of rough surfaces of ductile and brittle materials. The key results of the studies are the features of the formation of the third body (quasi-liquid nano-layer) [54], the obtained generalized functional dependence of the friction coefficient on dimensionless combinations of material parameters and loading parameters [54, 55], the formulated principles of nanotribospectroscopy as a promising nondestructive-testing technique for assessment of nanostructured coating and surface layer damage [56].

Recent joint research in this field has focused on studying the "elementary" wear mechanisms (that is, the modes of fracture of individual asperities) and analyzing the effect of attractive (adhesive) force between spatially separated surfaces on the involved mechanism of asperity wear under the condition of low-angle collision. Note that adhesive wear of the surface layers of contacting bodies is a widely studied but still poorly predicted phenomenon [45]. A key factor determining the regime and the rate of wear is the adhesive interaction of surfaces in the contact spots of asperities. Here, the term "adhesive interaction" includes (a) attractive interaction between detached surfaces and (b) effect of "cold welding" (chemical bonding of contacting surfaces) for chemically clear surfaces with high surface energy. The first well-known attempt to generalize the patterns of adhesive wear is the analytical model of Rabinowicz [57, 58]. He examined two qualitatively different mechanisms of asperity wear (plastic smoothing and breakaway) and showed that the involving of a specific mechanism is determined by the size parameter, which is a combination of specific material parameters, including shear strength, elastic constants, and specific surface energy. In recent years, a number of scientific groups carried out extensive numerical studies of the laws of interaction of single asperities [59, 60]. These studies have shown that the dependence of the power of asperity wear on material parameters generally has a significantly more complex nonlinear form, and the spectrum of realized mechanisms is not limited to those considered by Rabinowicz. However, the vast majority of theoretical results were obtained for nanoscale asperities using atomistic simulation.

We were the first to make a systematic numerical analysis of typical modes of asperity wear on higher (micro- and mesoscopic) scales. Due to the capabilities of the developed formalism of deformable discrete elements, such a study was carried out for ductile and brittle materials.

The main result of the study is that we revealed and substantiated two dimension-less material parameters that control the regime of asperity wear [46].

The first one is the ratio of attractive stress σ_0 between the detached surfaces (the adhesion stress) to the shear strength of the material σ_j. We showed that the range of values of this dimensionless parameter could be divided into two intervals with its own wear regime in each of them. The border value of the ratio $\left(\sigma_0^*/\sigma_j\right)$ and the specific involved mechanisms (specific modes) of asperity wear are determined by the value of the second dimensionless parameter.

The second one is the dimensionless parameter a, which characterizes shear strength sensitivity to the applied normal stress (or mean stress). It can be expressed in terms of the ratio of material strength values under different loading conditions. Note that shear strength sensitivity parameter effectively characterizes fracture toughness and is closely related to material brittleness: $a \approx 1$ for highly ductile materials,

$1 < a < 1.5$ for moderately ductile materials, a reaches 5–10 for elastic-brittle solids. We showed that increase in the ratio of adhesion stress to shear strength is accompanied by the transitions:

1. from slipping (wear at the atomic scale) to grinding-based wear regime for the case of highly and moderately ductile materials;
2. from breakaway (separation of the asperity from the foundation) to grinding-based wear regime for the case of materials with a limited ductility or brittle.

We have built the qualitative map of asperity wear regimes in terms of dimensionless material parameters σ_0/σ_j and a [46]. Figure 3 shows a rough schematic representation of this classification of wear regimes, which agrees with the results of atomistic studies by other researchers, supplements and generalizes them to higher scales of surface roughness.

An important result of the DEM-based study is the determined dependence of the position of the boundary between the wear regimes (the border value σ_0^*/σ_j) separating "low adhesion" and "high adhesion" wear modes) on asperity size L. We showed that the value is determined not by the absolute value of L, but its relation to the length d of action of attractive potential between spatially separated surfaces [47]. This dependence has a nonlinear increasing profile with reaching a saturation level at the scale $\sim 10^4$ of the ratio L/d. For larger asperities $(L/d > 10^4)$, the results of the analysis of the wear regimes are scale invariant under the condition of scale invariance of the mechanical characteristics of the material.

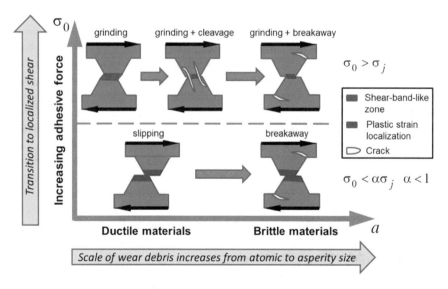

Fig. 3 Schematic classification of wear regimes of asperities for the case of tangential dry contact. The coefficient α is close to 1 for highly ductile materials and about 0.5–0.6 for brittle materials $\left(\alpha = \sigma_0^*/\sigma j\right)$ [46].

So, the surface adhesion stress is a criterion that determines the wear regime of asperities under the condition of tangential contact. The obtained map of asperity wear (Fig. 3) together with the revealed asperity size (scale) effect have both fundamental and practical significance as they allow forecasting the dominating mode of asperity wear for quite different materials from brittle to highly ductile.

4.2   Influence of Interstitial Fluid on the Sensitivity of the Femur to the Rate of Contact Loading

Contact loading is capable of determining not only surface wear and structural modification of surface layers, but also structural changes in the volume of contacting bodies. This is particularly relevant for biological (bone and cartilage) tissues. Functioning of these tissues is largely determined by the redistribution of interstitial fluid. Pore fluid has a complex nonlinear effect on the state and behavior of these biological materials. There are two key aspects of this influence. The first one is fluid flow in the pore space. Fluid flow provides the transfer of nutrients and oxygen and serves as a prerequisite for cell proliferation and tissue regeneration [61, 62]. The second aspect is the mechanical effect of pore pressure. It causes local tensile stresses in the skeleton and contributes to local fracture and gradual degradation of bone tissue. At extreme values of pore pressure it makes a significant contribution to the formation of cracks. The aforesaid argues the existence of optimal distributions of pore fluid (that is, optimal maximum local values and their gradient), which on the one hand provide sufficient fluid flows to ensure normal (regenerative) tissue activity, and on the other hand do not cause local fracture. Such distribution is formed under certain ("optimal") condition of the mechanical (contact) loading of the analyzed organ.

This problem is especially important in application to elements of the human musculoskeletal system (joints) because functioning of bone tissues in these regions strongly depends on the normal contact load and mode of tangential contact interaction. One of the topical problems in this field is the analysis of stress evolution and redistribution of interstitial fluid in the femur near the hip joint under dynamic loading. The developed formalism of permeable fluid-saturated discrete elements is an efficient tool for such a dynamic analysis up to the stage of macroscopic crack formation. Note that despite the topicality of the problem, there are practically no works devoted to the numerical study of the dynamic mechanical response of femur within the consideration of bone tissue as a multiphase fluid-saturated solid. The preliminary results of the numerical DEM-based study presented below are, therefore, pioneering in some way.

The aim of the study is to determine the loading rate sensitivity of healthy bone tissue of the proximal femur and bone tissue at different stages of osteoporosis. The main attention was paid to the regions of the femur in which the volume stresses are positive. This is due to the fact that bone is a brittle material and is characterized by

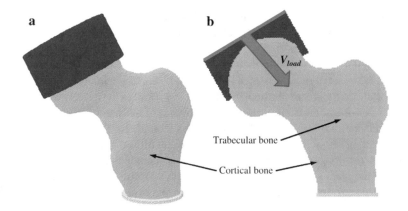

Fig. 4 General view of the model of the proximal femur (**a**) and its cross section (**b**) with differently colored cortical and spongy (trabecular) bone tissues

significantly lower tensile strength than compressive one. First of all, we analyzed the femoral neck (the typical region of the origin of the cracks).

The femur was modeled as a heterogeneous (shell) structure containing two fluid-saturated porous layers (Fig. 4): an inner spongy bone with low stiffness and an outer layer of cortical bone tissue. The latter plays the role of a hard shell and determines constrained conditions for the deformation of the soft spongy bone. The standard CAD model of the femur was used as the geometric basis of the DEM model [63]. We modelled compression of the proximal part of the femur along its main physiological direction (at an angle of about 45° relative to the orientation of the stem, Fig. 4b). The load was applied by setting the constant velocity V_{load} to the outermost elements of a special "cap" on the bone head. The base of the model was fixed. The model was loaded until the resistance force $F_{max} = 10$ kN was reached (10 kN, is about 70% of the critical load [64]). A velocity range from 1 to 8 m/s was considered. This range covers various modes of motor activity up to the "extreme" one.

Various authors previously showed that the strain rate dependence of the mechanical characteristics of heterogeneous (block-structured) brittle materials under confined loading conditions has not just a non-linear, but non-monotonic profile with a local minimum [65]. The reason for the non-monotonic nature of the dependence is the competition of two factors: pore fluid flow, which affects the redistribution of local stresses in the skeleton, and change in the equilibrium linear dimensions of the material due to change in pore pressure. The non-monotonic nature of the influence of interstitial fluid is especially pronounced under conditions of constrained deformation, when a change in linear dimensions causes a corresponding change in the degree of constraint. The biomechanical system under consideration (femur) also has a heterogeneous structure and is constrained by a hard cortical bone "shell". The shell causes confined deformation of a much softer inner region. This gives reason to suggest that under dynamic loading of such a system, the dependence of local stress values (including stresses in the femoral neck) on the strain rate can also be nonmonotonic with a local minimum. The magnitude of corresponding "optimal" loading velocity should be related to the characteristics of the porosity and permeability of the bone.

The simulation results confirmed these assumptions and, moreover, allowed us to estimate the characteristic values of the "optimal" loading velocity for healthy bone and bone at various stages of osteoporosis. The key result of the numerical study is the revealed non-linear and nonmonotonic dependence of the stress concentration in the femoral neck on the loading rate. Figure 5 shows examples of the distribution of mean and equivalent stresses in a healthy femur sample at different loading velocities. The simulation results show that an increase in V_{load} is accompanied by an increase in the values of the parameters characterizing the stress concentration in the upper part of the femoral neck. In particular, the volume of the region of maximum stresses and the magnitude of the maximum stress in the neck increase. However, when approaching the "optimal" value of the loading velocity ($\sim$3 m/s for the healthy bone), stress concentration decreases, then reaches minimum value at the "optimal" velocity, and then (with further increase in V_{load}) increases monotonously again. This effect takes place both for mean and equivalent stresses. The relative magnitude of the reduction in peak stresses in the neck reaches 10–20%, and the maximum decrease in the volume of the region of stress concentration amounts to 10%. Note that the loading velocities 3–4 m/s correspond to the regime of training motion of a human.

The described effect of reducing the heterogeneity of the stress state of bone tissue (particularly in the femoral neck) in the vicinity of the "optimal" loading velocity is directly related to the influence of pore fluid pressure and redistribution. Special simulations for the "dry" femur showed a monotonic dependence of the stress

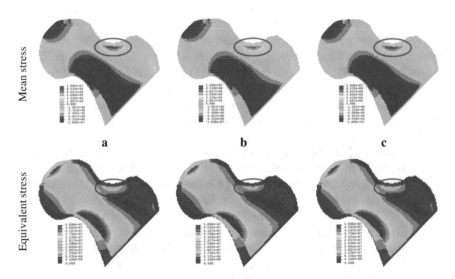

Fig. 5 Distributions of mean stress (upper row) and equivalent stress (lower row) in the proximal femur at different values of loading velocity: **a** 1 m/s; **b** 3 m/s; **c** 5 m/s. All pictures correspond to the same value of applied load $F_{max} = 10$ kN

concentration in the sample (including the femoral neck) on the loading rate in the entire considered velocity range.

Preliminary numerical studies have also shown that a change in bone tissue parameters corresponding to the successive stages of osteoporosis leads to a systematic increase in the values of "optimal" loading velocity. Bone tissue suffering from osteoporosis is characterized by the reduced strength and higher brittleness. Therefore, our results indicate that the selection of the optimal regime of motor activity for such a femur is questionable.

Finally, we note the importance of the revealed effect of the "optimal" loading rate. Despite the relatively small decrease in the maximum tensile stresses in the critical region of the bone (upper part of the neck), this effect can have a significant positive value in the long term (with multiple repetitions of the load). In addition to reducing the risk of microdamage nucleation/accumulation in the upper part of the neck, a decrease in the magnitude of positive volume stresses contributes to a more intensive and uniform circulation of interstitial fluid in this region of the bone. The consequence is a more complete and rational flow of substances necessary for the normal functioning of bone tissue at optimal loading intensity. So, the results obtained are critically important for determining the optimal modes of motor activity of people, as well as for developing strategies for treating osteoarthritis and preventing the negative consequences of osteoporosis.

5 Conclusion

A formalism of homogeneously deformable elements, which was developed by Professor Sergey G. Psakhie and his colleagues, made possible qualitative enhancement of the capabilities of the particle-based discrete element method. The main advantage of DEM is the ability to correctly describe the mechanical behavior of various materials with taking into account the accompanying thermal, hydromechanical and other effects. Deformability of elements is of particular importance when studying various aspects of dynamic contact interaction, for example, stick-to-slip transition in technical and geological contact zones, mechanisms of friction and wear, redistribution of pore fluid in the surface layers of geological and biological joints, etc.

Two key factors that identified the advantages of the formalism of deformable elements should be especially noted. The first one is the postulation of the many-particle form of relations for the element-element interaction forces. In this regard, DEM has much in common with molecular dynamics method. Pair-wise interatomic potentials are able to catch basic properties of crystal lattice, but do not describe many fundamentally important effects in the bulk and on the surface of solids, including those determining plasticity and phase transformations. Note that the approximation of a homogeneously deformable element is an efficient alternative to more computationally expensive combined DEM-FEM technique. The second factor is the proposed universal method for determining the specific form (and the values of constants) of the interaction forces. This method is based on the replication of the corresponding rheological relationships of the applied mechanical model. It "circumvents" the fundamental problem of traditional DEM, namely, the need to find a vector analogue of constitutive equations written in tensor form. The particular cases of rheological models mentioned in this chapter are related to locally isotropic materials, however, models of elasticity, ductility and local fracture of anisotropic materials can be implemented in a similar way.

At present time, the formalism of deformable DEM is actively developed and applied in various fields and in the wide range of spatial scales. Moreover, international scientific teams leaded by well-recognized scholars adopt these ideas and develop them in their own way. DEM with deformable elements becomes not just an addition to the classical continuum numerical methods, but to some extent their competitive even in the areas of mechanics, where they traditionally dominate.

Acknowledgements The work was performed according to the Government research assignment for ISPMS SB RAS.

References

1. Cauchy AL (1823) Recherches sur l'equilibre et le mouvement interieur des corps solides ou fluides, elastiques ou non lastiques. Bulletin des sciences par la Societe Philomatique de Paris 9–13
2. Navier CL (1823) Sur les lois de l'equilibre et du mouvement des corps solides elastiques. Bulletin des sciences par la Societe Philomatique de Paris 177–181
3. Kocsis A, Challamel N (2018) Generalized models and non-classic approaches in complex materials 1. Springer, Berlin, pp 451–486 (On the foundation of a generalized nonlocal extensible shear beam model from discrete interactions)
4. Zienkiewicz OC, Taylor RL, Fox DD (2014) The finite element method for solid and structural mechanics, 7th edn. Butterworth-Heinemann, Oxford
5. Balokhonov RR, Romanova VA, Kulkov AS (2020) Microstructure-based analysis of deformation and fracture in metal-matrix composite materials. Eng Fail Anal 110:104412. https://doi.org/10.1016/j.engfailanal.2020.104412
6. Moczo P, Kristek J, Galis M (2014) The finite-difference modelling of earthquake motions: waves and ruptures. Cambridge University Press, Cambridge (UK)
7. Garavand A, Stefanov YP, Rebetsky YL, Bakeev RA, Myasnikov AV (2020) Numerical modeling of plastic deformation and failure around a wellbore in compaction and dilation modes. Int J Numer Anal Meth Geomech 44(6):1–28. https://doi.org/10.1002/nag.3041
8. Cheng AHD, Syngellakis S (eds) (2019) Boundary elements and other mesh reduction methods XLI. WIT Press, Southampton (UK)
9. Popov VL, Pohrt R, Li Q (2017) Strength of adhesive contacts: influence of contact geometry and material gradients. Friction 5(3):308–325. https://doi.org/10.1007/s40544-017-0177-3
10. Yazid A, Abdelkader N, Abdelmajid H (2009) A state-of-the-art review of the X-FEM for computationalfracture mechanics. Appl Math Model 33(12):4269–4282. https://doi.org/10.1016/j.apm.2009.02.010
11. Pohrt R, Popov VL (2015) Adhesive contact simulation of elastic solids using local mesh-dependent detachment criterion in boundary elements method. Facta Univ Mech Eng 13(1):3–10

12. Liu MB, Liu GR (2010) Smoothed particle hydrodynamics (SPH): an overview and recent developments. Arch Comput Methods Eng 17:25–76. https://doi.org/10.1007/s11831-010-9040-7
13. Li X, Zhao J (2019) An overview of particle-based numerical manifold method and its application to dynamic rock fracturing. J Rock Mech Geotech Eng 11(3):684–700. https://doi.org/10.1016/j.jrmge.2019.02.003
14. Nabian MA, Farhadi L (2016) Multiphase mesh-free particle method for simulating granular flows and sediment transport. J Hydraul Eng 143(4):04016102. https://doi.org/10.1061/(ASCE)HY.1943-7900.0001275
15. Munjiza A, Smoljanović H, Živaljić N, Mihanovic A, Divić V, Uzelac I, Nikolić Ž, Balić I, Trogrlić B (2019) Structural applications of the combined finite–discrete element method. Comput Part Mech. https://doi.org/10.1007/s40571-019-00286-5
16. Rodriguez JM, Carbonell JM, Cante JC, Oliver J (2016) The particle finite element method (PFEM) in thermo-mechanical problems. Int J Numer Meth Eng 107(9):733–785. https://doi.org/10.1002/nme.5186
17. Cerquaglia ML, Deliege G, Boman R, Papeleux L, Ponthot JP (2017) The particle finite element method for the numerical simulation of bird strike. Int J Impact Eng 109:1–13. https://doi.org/10.1016/j.ijimpeng.2017.05.014
18. Jing L, Stephansson O (2007) Fundamentals of discrete element method for rock engineering: theory and applications. Elsevier, Amsterdam (NL)
19. Bicanic N (2017) Encyclopaedia of computational mechanics, 2nd edn. Wiley, Hoboken, pp 1–38 (Discrete element methods)
20. Cundall PA, Strack ODL (1979) A discrete numerical model for granular assemblies. Geotechnique 29(1):47–65
21. Cundall PA, Hart RD (1992) Numerical modelling of discontinua. Eng Comput 9(2):101–113. https://doi.org/10.1108/eb023851
22. Potyondy DO, Cundall PA (2004) A bonded-particle model for rock. Int J Rock Mech Min Sci 41:1329–1364. https://doi.org/10.1016/j.ijrmms.2004.09.011
23. Ivars DM, Pierce ME, Darcel C, Reyes-Montes J, Potyondy DO, Young RP, Cundall PA (2011) The synthetic rock mass approach for jointed rock mass modelling. Int J Rock Mech Min Sci 48(2):219–244. https://doi.org/10.1016/j.ijrmms.2010.11.014
24. Mustoe GGW (1992) A generalized formulation of the discrete element method. Eng Comput 9(2):181–190. https://doi.org/10.1108/eb023857
25. Hatzor YH, Ma G, Shi G-H (2017) Discontinuous deformation analysis in rock mechanical practice. CRC Press, Boca Raton (Florida, US)
26. Schneider B, Ramm E (2019) Conceptual experiments and discrete element simulations with polygonal particles. Granular Matter 21(91). https://doi.org/10.1007/s10035-019-0930-6
27. Podlozhnyuk A, Pirker S, Kloss C (2017) Efficient implementation of superquadric particles in discrete element method within an open-source framework. Comput Part Mech 4:101–118. https://doi.org/10.1007/s40571-016-0131-6
28. Sinaie S (2017) Application of the discrete element method for the simulation of size effects in concrete samples. Int J Solids Struct 108:244–253. https://doi.org/10.1016/j.ijsolstr.2016.12.022
29. Potyondy DO (2015) The bonded-particle model as a tool for rock mechanics research and application: current trends and future directions. Geosyst Eng 18(1):1–28. https://doi.org/10.1080/12269328.2014.998346
30. Nosewicz S, Rojek J, Pietrzak K, Chmielewski M (2013) Viscoelastic discrete element model of powder sintering. Powder Technol 246:157–168. https://doi.org/10.1016/j.powtec.2013.05.020
31. Rojek J, Lumelskyj D, Nosewicz S, Romelczyk-Baishya B (2019) Numerical and experimental investigation of an elastoplastic contact model for spherical discrete elements. Comput Part Mech 6:383–392. https://doi.org/10.1007/s40571-018-00219-8
32. Wang G, Al-Ostaz A, Cheng AH-D, Mantena PR (2009) Hybrid lattice particle modeling: theoretical considerations for a 2D elastic spring network for dynamic fracture simulations. Comput Mater Sci 44(4):1126–1134. https://doi.org/10.1016/j.commatsci.2008.07.032

33. Puglia VB, Kosteski LE, Riera JD, Iturrioz I (2019) Random field generation of the material properties in the lattice discrete element method. J Strain Anal Eng Des 54(4):236–246. https://doi.org/10.1177/0309324719858849
34. Pieczywek PM, Zdunek A (2017) Compression simulations of plant tissue in 3D using a mass-spring system approach and discrete element method. Soft Matter 13:7318–7331. https://doi.org/10.1039/C7SM01137G
35. Zabulionis D, Rimsa V (2018) A lattice model for elastic particulate composites. Materials 11(9):1584. https://doi.org/10.3390/ma11091584
36. Rizvi ZH (2019) Lattice element method and its application to multiphysics. Dissertation in fulfilment of the requirements for the degree "Dr.-Ing." of the Faculty of Mathematics and Natural Sciences at Kiel University, Christian-Albrechts-Universität, Kiel
37. Psakhie SG, Horie Y, Ostermeyer GP, Korostelev SYu, Smolin AYu, Shilko EV, Dmitriev AI, Blatnik S, Špegel M, Zavšek S (2001) Movable cellular automata method for simulating materials with mesostructured. Theoret Appl Fract Mech 37(1–3):311–334
38. Psakhie SG, Shilko EV, Smolin AY, Dimaki AV, Dmitriev AI, Konovalenko IS, Astafurov SV, Zavshek S (2011) Approach to simulation of deformation and fracture hierarchically organized heterogeneous media, including contrast media. Phys Mesomech 14(1–5):224–248. https://doi.org/10.1016/j.physme.2011.12.003
39. Psakhie SG, Horie Y, Korostelev SYu, Smolin AYu, Dmitriev AI, Shilko EV, Alekseev SV (1995) Method of movable cellular automata as a tool for simulation within the framework of mesomechanics. Russ Phys J 38(11):1157–1168. https://doi.org/10.1007/BF00559396
40. Psakhie SG, Shilko EV, Smolin AY, Dmitriev AI, Korostelev SY (1996) Computer aided study of reaction-assisted powder mixture shock compaction at meso-scale. New computational technique. In: Proceedings of US-Russian workshop "Shock induced chemical processing", Saint-Petersburg, 23–24 June, 1996
41. Popov VL, Psakhie SG (2001) Theoretical principles of modeling elastoplastic media by movable cellular automata method. I. Homogeneous media. Phys Mesomech 4(1):15–25
42. Psakhie S, Shilko E, Smolin A, Astafurov S, Ovcharenko V (2013) Development of a formalism of movable cellular automaton method for numerical modeling of fracture of heterogeneous elastic-plastic materials. Frattura Ed Integrità Strutturale 24(7):26–59. https://doi.org/10.3221/IGF-ESIS.24.04
43. Shilko EV, Psakhie SG, Schmauder S, Popov VL, Astafurov SV, Smolin AYu (2015) Overcoming the limitations of distinct element method for multiscale modeling of materials with multimodal internal structure. Comput Mater Sci 102:267–285. https://doi.org/10.1016/j.commatsci.2015.02.026
44. Psakhie SG, Shilko EV, Grigoriev AS, Astafurov SV, Dimaki AV, Smolin AYu (2014) A mathematical model of particle–particle interaction for discrete element based modeling of deformation and fracture of heterogeneous elastic–plastic materials. Eng Fract Mech 130:96–115. https://doi.org/10.1016/j.engfracmech.2014.04.034
45. Wen S, Huang P (2017) Principles of tribology, 2nd edn. Wiley, London (UK)
46. Dimaki AV, Shilko EV, Dudkin IV, Psakhie SG, Popov VL (2020) Role of adhesion stress in controlling transition between plastic, grinding and breakaway regimes of adhesive wear. Sci Rep 10:1585. https://doi.org/10.1038/s41598-020-57429-5
47. Dimaki AV, Dudkin IV, Popov VL, Shilko EV (2019) Influence of adhesion force and strain hardening coefficient of the material on the rate of adhesive wear in a dry tangential frictional contact. Russ Phys J 62(8):1398–1408. https://doi.org/10.1007/s11182-019-01857-y
48. Wilkins ML (1999) Computer simulation of dynamic phenomena. Springer, Berlin
49. Nikolaevsky VN (1996) Geomechanics and fluidodynamics with application to reservoir engineering. Kluwer Academic Publishers (Springer), Berlin
50. Grigoriev AS, Shilko EV, Skripnyak VA, Psakhie SG (2019) Kinetic approach to the development of computational dynamic models for brittle solids. Int J Impact Eng 123:14–25. https://doi.org/10.1016/j.ijimpeng.2018.09.018
51. Psakhie SG, Dimaki AV, Shilko EV, Astafurov SV (2016) A coupled discrete element-finite difference approach for modeling mechanical response of fluid-saturated porous materials. Int J Numer Meth Eng 106(8):623–643. https://doi.org/10.1002/nme.5134

52. Detournay E, Cheng AH-D (1993) Comprehensive rock engineering: principles, practice and projects. Pergamon Press, Oxford (UK), pp 113–171 (Fundamentals of poroelasticity)
53. Basniev KS, Dmitriev NM, Dmitriev NM, Chilingar GV (2012) Mechanics of fluid flow. Wiley, London (UK)
54. Popov VL, Psakhie SG, Dmitriev AI, Shilko E (2003) Quasi-fluid nano-layers at the interface between rubbing bodies: simulations by movable cellular automata. Wear 254(9):901–906. https://doi.org/10.1016/S0043-1648(03)00244-8
55. Bucher F, Dmitriev AI, Ertz M, Knothe K, Popov VL, Psakhie SG, Shilko EV (2006) Multiscale simulation of dry friction in wheel/rail contact. Wear 261(7–8):874–884. https://doi.org/10.1016/j.wear.2006.01.046
56. Psakhie SG, Shilko EV, Popov VL, Starcevic J, Thaten J, Astafurov SV, Dimaki AV (2009) Assessment of nanostructured ceramic coating damage Nanotribospectroscopy. Russ Phys J 52(4):380–385. https://doi.org/10.1007/s11182-009-9242-3
57. Rabinowicz E (1958) The effect of size on the looseness of wear fragments. Wear 2(1):4–8
58. Popova E, Popov VL, Kim DE (2018) 60 years of Rabinowicz' criterion for adhesive wear. Friction 6(3):341–348. https://doi.org/10.1007/s40544-018-0240-8
59. von Lautz J, Pastewka L, Gumbsch P, Moseler M (2016) Molecular dynamic simulation of collision-induced third-body formation in hydrogen-free diamond-like carbon asperities. Tribol Lett 63:26. https://doi.org/10.1007/s11249-016-0712-9
60. Molinari J-F, Aghababaei R, Brink T, Frérot L, Milanese E (2018) Adhesive wear mechanisms uncovered by atomistic simulations. Friction 6:245–259. https://doi.org/10.1007/s40544-018-0234-6
61. Christen P, Ito K, Ellouz R, Boutroy S, Sornay-Rendu E, Chapurlat RD, van Rietbergen B (2014) Bone remodelling in humans is load-driven but not lazy. Nat Commun 5:4855. https://doi.org/10.1038/ncomms5855
62. Wittkowske C, Reilly GC, Lacroix D, Perrault CM (2016) In vitro bone cell models: impact of fluid shear stress on bone formation. Front Bioeng Biotechnol 4:87. https://doi.org/10.3389/fbioe.2016.00087
63. Cheung G, Zalzal P, Bhandari M, Spelt JK, Papini M (2004) Finite element analysis of a femoral retrograde intramedullary nail subject to gait loading. Med Eng Phys 26(2):93–108. https://doi.org/10.1016/j.medengphy.2003.10.006
64. Todo M (2018) Biomechanical analysis of hip joint arthroplasties using CT-image based finite element method. J Surg Res 1:34–41
65. Shilko EV, Dimaki AV, Psakhie SG (2018) Strength of shear bands in fluid-saturated rocks: a nonlinear effect of competition between dilation and fluid flow. Sci Rep 8:1428. https://doi.org/10.1038/s41598-018-19843-8

A Tool for Studying the Mechanical Behavior of the Bone–Endoprosthesis System Based on Multi-scale Simulation

Alexey Yu. Smolin, Galina M. Eremina, and Evgeny V. Shilko

Abstract The chapter presents recent advances in developing numerical models for multiscale simulation of the femur–endoprosthesis system for the case of hip resurfacing arthroplasty. The models are based on the movable cellular automaton method, which is a representative of the discrete element approach in solid mechanics and allows correctly simulating mechanical behavior of a variety of elastoplastic materials including fracture and mass mixing. At the lowest scale, the model describes sliding friction between two rough surfaces of TiN coatings, which correspond to different parts of the friction pair of hip resurfacing endoprosthesis. At this scale, such parameters of the contacting surfaces as the thickness, roughness, and mechanical properties are considered explicitly. The next scale of the model corresponds to a resurfacing cap for the femur head rotating in the artificial acetabulum insert. Here, sliding friction is explicitly computed based on the effective coefficient of friction obtained at the previous scale. At the macroscale, the proximal part of the femur with a resurfacing cap is simulated at different loads. The bone is considered as a composite consisting of outer cortical and inner cancellous tissues, which are simulated within two approaches: the first implies their linear elastic behavior, the second considers these tissues as Boit's poroelastic bodies. The later allows revealing the role of the interstitial biological fluid in the mechanical behavior of the bone. Based on the analysis of the obtained results, the plan for future works is proposed.

Keywords Bone · Endoprosthesis · Mechanical behavior · Friction · Computer simulations · Particle-based method

A. Yu. Smolin (✉) · G. M. Eremina · E. V. Shilko
Institute of Strength Physics and Materials Science of Siberian Branch of Russian Academy of Sciences, Tomsk, Russian Federation
e-mail: asmolin@ispms.ru

Tomsk State University, Tomsk, Russian Federation

G. M. Eremina
e-mail: anikeeva@ispms.ru

E. V. Shilko
e-mail: shilko@ispms.tsc.ru

© The Author(s) 2021
G.-P. Ostermeyer et al. (eds.), *Multiscale Biomechanics and Tribology of Inorganic and Organic Systems*, Springer Tracts in Mechanical Engineering,
https://doi.org/10.1007/978-3-030-60124-9_5

1   Introduction

Endoprosthetics is an effective way to treat such common degenerative diseases of large human joints as osteoporosis and arthritis. Wear in a pair of friction of the constituent elements of the joint endoprosthesis design has a significant impact on its operational resource. Primarily, this refers to the prostheses of the hip and knee joints. The key role in the wear process belongs to the structure of the surface layers of the contacting elements. To improve the tribological characteristics of metal endoprostheses, which are currently the most widely used, reinforcing coatings are used. In practice, titanium is usually used as a metal and titanium nitride (TiN) as a coating.

Since endoprostheses are intended for a long stay in the human body, they should not lead to any discomfort. Therefore, an important stage in the development of endoprostheses is their testing. Endoprosthesis tests are divided into several stages; these are preclinical and clinical trials. Clinical trials are conducted through the installation of an endoprosthesis in a living human body. Using this test, the final result is determined. However, when conducting clinical trials, there is a danger that a poorly or improperly selected endoprosthesis may adversely affect the patient's health. Therefore, much more attention is paid to preclinical trials when developing endoprostheses. Talking about the mechanical behavior of the endoprosthesis, its preclinical studies can be divided into experimental and theoretical. Experimental studies are tests using special equipment that simulates real dynamic loading experienced by an endoprosthesis in the body. In particular, standard methods, such as instrumented indentation, scratching, and three-point bending, are used to determine the tribological properties of the surface layers of materials. It is worth noting that conventional experimental studies are carried out under standard atmospheric conditions and their results can differ greatly from the behavior of materials in real conditions inside a living body. Therefore, part of the experimental studies is carried out in a medium simulating a living body (in vitro), or on test samples placed into the animal (in vivo). However, for obvious reasons, the number and possibilities of such studies are very limited.

Theoretical studies of the mechanical behavior of the endoprosthesis are most often carried out using computer simulation. It should be noted that modern numerical methods of solid mechanics allow a detailed study of the mechanical behavior of endoprostheses, which takes into account the influence of a variety of factors. However, to fully utilize all the features of this approach, it is necessary to use special software.

In the last decade, the development of computer-aided tools for studying the mechanical behavior of endoprostheses has been the subject of extensive work by leading world scientific groups in such well-known centers as McGill University (Canada), Liverpool John Moores University (Great Britain), Iowa State University (USA) and many others. The high relevance of this problem is also confirmed by numerous publications on this topic in leading scientific journals.

However, tools taking into account the influence of structural features of surface layers on the wear of contacting surfaces, as well as the effect of wear on the mechanical behavior of the endoprosthesis as a whole, have not yet been developed. The second important factor that has to be considered in modeling the mechanical behavior of living bone is the presence of interstitial fluid in it. The availability of such tools would be very relevant in the study of the durability of a new class of endoprostheses made from new composite materials, including titanium alloys with a nanostructured TiN coating.

2 State-of-the-Art

There are two methods of hip arthroplasty: total hip replacement and hip resurfacing. Total hip replacement suggests cutting the femur head and inserting a prosthetic implant into the bone. In the case of hip resurfacing only a very thin layer of the femur head is replaced by a metal cap, which is hollow and shaped like a mushroom. Despite the more complicated procedure of the operation, the latter method is preferable for young people who lead an active life. Previously, only metals were used for the friction pair of the resurfacing endoprosthesis, in particular, the CoCrN alloy. This led to the fact that during life after the surgical operation, severe wear was observed in the friction pair with a large release of metal particles, which led to necrosis of cells and small phagocytes, and severe allergic vasculitis was also observed [1]. Therefore, in recent years there has been active scientific research on the use of titanium alloys with superthin ceramic coatings in resurfacing endoprosthetics. Moreover, titanium is usually used as a metal and titanium nitride (TiN) as a coating [2, 3]. The structure of the surface layer of the coating plays a key role in the wear process and is determined by the methods and regimes of its application. There are several ways to apply this coating: vacuum spraying (PVD), chemical vapor deposition (CVD), powder nitriding (PIRAC-powder immersion reaction assisted coating nitriding). The coatings obtained by PVD and CVD have insufficient adhesion to the substrate and are prone to delaminating under dynamic loads, while the adhesion of the coating obtained by the PIRAC method is significantly higher. Therefore, the use of metal implants with a PIRAC coating is very promising. Compounds of carbides, borides, nitrides are used as coating materials applied by the PIRAC method [4, 5].

Computer simulation allows one to study the mechanical behavior of endoprostheses, taking into account the influence of a variety of factors on it. Therefore, in preclinical studies, computer simulation is widely used to predict the mechanical behavior of prostheses. For example, modeling based on numerical methods of continuum mechanics allows a detailed study of the behavior of the coating and surface layer under contact loading. Thus, the influence of coating thickness on the load–displacement curve of instrumented indentation of ceramic coatings on a metal substrate was studied in [6–8]. The role of the surface roughness of the ceramic coating in the features of the mechanical behavior of the coating-substrate system during instrumented indentation was numerically studied in [9–12]. The authors of

[13–17] simulated the mechanical behavior of the coating–substrate system in instrumented indentation and scratching and analyzed stress fields in contact areas, as well as features of the load–displacement curves.

There is practically no theoretical work on the study of wear in a friction pair of a resurfacing endoprosthesis of a hip joint. Most of the work have been devoted to wearing in a friction pair of endoprostheses for total hip replacement. In theoretical studies of the wear process in a friction pair of an endoprosthesis, two approaches are used: with explicit consideration of wear debris and its implicit consideration. In the numerical study of wear in a friction pair by the finite element method, an implicit consideration of wear particles is used. First of all, this circumstance is associated with large computational costs and the complexity of explicit modeling of material failure. Most of the research [18, 19] on the theoretical study of wear in the friction pair of the endoprosthesis are based on the technique proposed in [20], where the ratio of Archard/Lancaster is used to describe the wear via calculating the wear coefficient for adhesive or abrasive wearing. This ratio defines linear wear as a function of contact pressure, loading rate, and wear coefficient, which in turn depends on the tribological characteristics of two contacting surfaces, such as roughness and various physical and mechanical characteristics of the materials of the contacting pair. The surface roughness is taken into account implicitly [21–23].

Most of the works on numerical modeling of the mechanical behavior of the bone–endoprosthesis system is performed using the methods of continuum mechanics. Thus, research of Kuhl and Balle [24], the influence of the type of prosthetics on the stress state in the system was investigated; establishing that the stress field of the system during resurfacing endoprosthetics is close to the stresses in a healthy joint. The results of a numerical study of the mechanical behavior of the bone–endoprosthesis system depending on the geometric characteristics of the implant and its design features are presented in [25], and on the implant material in [26].

Work on studying the mechanical behavior of bone tissue with biological fluid started long ago [27]. At the same time, the Biot model of poroelasticity was most widely used to describe the mechanical behavior of bone tissues. In papers [28–30] the Biot model of the isotropic poroelastic medium was used to describe bone tissues in the framework of continuum mechanics methods.

Taken together, this analysis of the literature shows that for the correct prediction of the mechanical behavior of the bone–endoprosthesis system, the development of numerical models that can describe the bone within the framework of a poroelastic body taking into account possible fracture of bone tissues, as well as wear in the friction pair of the endoprosthesis, is in demand.

3 The Problem Statement

The aim of this chapter is to present recent advances in development of numerical models for multi-scale simulation of the femur–endoprosthesis system for the hip resurfacing arthroplasty. The models are based on the movable cellular automaton method, which is a representative of the discrete element approach in solid mechanics. According to the chosen multi-scale approach, the model of the lowest scale (mesoscale) describes sliding friction between two rough surfaces of TiN coatings, which correspond to different parts of the friction pair of hip resurfacing endoprosthesis. At this scale, we consider explicitly such parameters of the endoprosthesis friction pair as the thickness, roughness, and mechanical properties of the corresponding TiN coatings. The next scale of the model is a resurfacing cap rotating in the artificial acetabulum insert with an explicit account of sliding friction based on the effective coefficient of friction obtained at the previous scale. At the macroscale, we consider compression of the proximal part of the femur with a resurfacing cap. This macro-model considers the bone as a composite consisting of outer cortical and inner cancellous tissues. Here we use two approaches: the first implies simple linear elastic behavior of both tissues, the second considers these tissues as Boit's poroelastic bodies with accounting for the role of the interstitial biological fluid in the mechanical behavior of the bone. For comparison, we also consider the macroscopic model for healthy bone in compression.

4 Description of the Modeling Method

For simulating the mechanical behavior of the materials for the bone and prosthesis, we use the particle-based method of movable cellular automata (MCA), which was proposed by professors Sergey Psakhie and Yasuyuki Horie in 1994 and firstly published in 1995 [31]. Since that time, the method has been being actively developed in Prof. Psakhie's lab and its latest description can be found, for example, elsewhere in papers [32, 33] as well as in book chapters [34, 35].

MCA is a representative of so-called discrete element methods (DEM). The main principles of the method are as follows. A simulated body is represented by an ensemble of bonded equiaxial discrete elements of the same size (called movable cellular automata), which spatial position and orientation, as well as state, can change due to local interaction with nearest neighbors. Automata interact with each other through their contacts. The initial value of the contact area, as well as the automaton volume, is determined by the size of automata and their packing. The main advantage of the MCA-method in comparison with DEM is the generalized many-body formulas for central interaction forces acting between the pair of elements, similar to the embedded atom force filed used in molecular dynamics. It is based on computing components of the average stress and strain tensors in the bulk of automaton according to the homogenization procedure described in [32]. Use of many-body interaction forces allows correct simulation within discrete element approach of such important features of the mechanical behavior of solids like Poisson effect and plastic flow.

When describing the kinematics and dynamics of an automaton motion, its shape is approximated by an equivalent sphere. This approximation is the most widely used in the discrete element method and allows one to consider the forces of central and tangential interaction of automata as formally independent. This makes also possible to use the simplified Newton–Euler equations of motion to govern translational motion and rotation of the movable automata.

$$
\begin{cases}
m_i \dfrac{d^2 \mathbf{R}_i}{dt^2} = \displaystyle\sum_{j=1}^{N_i} \mathbf{F}_{ij}^{\text{pair}} + \mathbf{F}_i^{\Omega}, \\[2ex]
\hat{J}_i \dfrac{d\boldsymbol{\omega}_i}{dt^2} = \displaystyle\sum_{j=1}^{N_i} \mathbf{M}_{ij}
\end{cases}
\tag{1}
$$

where $\mathbf{R_i}$, $\boldsymbol{\omega}_i$, m_i and $\widehat{J}_i$ are the location vector, rotation velocity vector, mass and moment of inertia of ith automaton respectively, $\mathbf{F}_{ij}^{\text{pair}}$ is the interaction force of the pair of ith and jth automata, $\mathbf{F}_i^{\Omega}$ is the volume-dependent force acting on ith automaton and depending on the interaction of its neighbors with the remaining automata. In the latter equation, $\mathbf{M}_{ij} = q_{ij}(\mathbf{n}_{ij} \times \mathbf{F}_{ij}^{\text{pair}}) + \mathbf{K}_{ij}$, here q_{ij} is the distance from the center of ith automaton to the point of its interaction (contact) with jth automaton, $\mathbf{n}_{ij} = (\mathbf{R}_j - \mathbf{R}_i)/r_{ij}$ is the unit vector directed from the center of ith automaton to the jth one and r_{ij} is the distance between automata centers, $\mathbf{K}_{ij}$ is the torque caused by relative rotation of automata in the pair.

Movable automata are treated as deformable. Strains and stresses are assumed to be uniformly distributed in the volume of each automaton. Within the framework of this approximation, the values of averaged stresses in the automaton volume may be calculated as the superposition of forces applied to different parts of the automaton surface. In other words, averaged stress tensor components are expressed in terms of the interaction forces with neighbors [32]:

$$
\overline{\sigma}_{\alpha\beta}^i = \frac{R_i S_{ij}^0}{\Omega_i^0} \sum_{j=1}^{N_i} \left[f_{ij}\left(\vec{n}_{ij}\right)_\alpha + \tau_{ij}\left(\vec{t}_{ij}\right)_\beta \right]
\tag{2}
$$

where i is the automaton number, $\overline{\sigma}_{\alpha\beta}^i$ is the component $\alpha\beta$ of the averaged stress tensor, α, $\beta = x, y$, z (XYZ is the global coordinate system), Ω_i^0 is the initial volume of the automaton i, S_{ij}^0 is the initial value of the contact area between the automata i and j, R_i is the radius of the equivalent sphere (semi-size of automaton i), f_{ij} and τ_{ij} are specific values of central and tangential forces of interaction between the automata i and j, $\left(\vec{n}_{ij}\right)_\alpha$ and $\left(\vec{t}_{ij}\right)_\alpha$ are the projections of the unit normal and unit tangent vectors onto the α-axis , N_i is the number of interacting neighbors of automaton i.

Invariants of the averaged stress tensor $\overline{\sigma}_{\alpha\beta}^i$ are used to calculate the central interaction forces $\left(f_{ij},\ \tau_{ij}\right)$ and the criterion of an inter-element bond breaking (local fracture). The components of the averaged strain tensor $\overline{\varepsilon}_{\alpha\beta}^i$ are calculated in increments using the specified constitutive equation of the simulated material and the calculated increments of mean stress.

In [32] it is shown that the relation for the force of central interaction of automata can be formulated based on the constitutive equation of the material for the diagonal components of the stress tensor, while the force of tangential interaction can be formulated on the basis of similar equations for non-diagonal stress components. When implementing the linear elastic model, the expressions for specific values of the central and tangential forces of the mechanical response of the automaton i to mechanical action from the neighboring automaton j are written as follows:

$$\begin{cases} \Delta f_{ij} = 2G_i \Delta \varepsilon_{ij} + D_i \Delta \sigma_i^{\text{mean}} \\ \Delta \tau_{ij} = 2G_i \Delta \gamma_{ij} \end{cases}, \tag{3}$$

where the symbol Δ means increment of the corresponding variable during time step Δt of the numerical scheme of integration of the motion equations, $\Delta \varepsilon_{ij}$ and $\Delta \gamma_{ij}$ are the increments of normal and shear strains of the automaton i in pair $i - j$, G_i is the shear modulus of the material of the automaton i, K_i is the bulk modulus, $D_i = 1 - 2G_i/K_i$.

Due to the necessity of the third Newton's law $\left(\sigma_{ij} = \sigma_{ji} \text{ and } \tau_{ij} = \tau_{ji} \right)$, the increments of the reaction forces of the automata i and j are calculated based on the solution of the following system of equations.

$$\begin{cases} \Delta f_{ij} = \Delta f_{ji} \\ R_i \Delta \varepsilon_{ij} + R_j \Delta \varepsilon_{ji} = \Delta r_{ij} \\ \Delta \tau_{ij} = \Delta \tau_{ji} \\ R_i \Delta \gamma_{ij} + R_j \Delta \gamma_{ji} = \Delta l_{ij}^{\text{sh}} \end{cases}, \tag{4}$$

where Δr_{ij} is the change in the distance between the centers of the automata for a time step Δt, $\Delta l_{ij}^{\text{sh}}$ is the value of the relative shear displacement of the interacting automata i and j. The system of equations (4) is solved for finding the increments of strains. This allows calculation of the increments of the specific interaction forces. When solving the system (4), the increments of mean stress and the values of specific forces in the right-hand sides of relations (3) are taken from the previous time step or are evaluated and further refined within the predictor–corrector scheme.

A pair of automata can be in one of two states: bound and unbound. Thus, in MCA fracture and coupling of fragments (crack healing, microwelding etc.) is simulated by the corresponding switching of the pair state. Switching criteria depend on physical mechanisms of material behavior [32, 33]. Note, that knowing stress and strain tensor in the bulk of an automaton, makes possible direct application of conventional fracture criteria written in the tensor form, herein we used von Mises criterion based on a threshold value of the equivalent stress.

5   Results and Discussion

5.1   Modeling Friction Pair of the Hip Resurfacing

We start with the simulation of the friction pair of the hip resurfacing endoprosthesis, which consists of two contacting bodies made of titanium alloy with TiN coating. First, we choose the materials parameters and validate the materials models to compare the simulation results for instrumented indentation with available experimental data. After that, we can simulate friction of the contacting bodies at the mesoscale to get an estimation of the coefficients of friction, which was used in the macroscopic model for rotation a resurfacing cap in the artificial acetabulum insert.

5.1.1   Materials Characterization

The values for the main physico-mechanical properties of the titanium alloy Ti6Al4V that we chose from the literature [36] are shown in Table 1 (corresponding Young's modulus $E = 110$ GPa). Geometric features of the TiN coating and its physicomechanical properties are determined by the deposition modes at PIRAC forming [37]. So, at a deposition temperature of 700 °C and a treatment time of 48 h, the coating on the titanium substrate has a thickness of 1.3 μm with an average roughness height of 0.15 μm (mode 1) and elastic modulus, depending on the deposition regime, was equal to $E_1 = 258$ GPa; at a deposition temperature of 800 °C and a treatment time of 4 h the coating on the titanium substrate has a thickness of 1.4 μm with an average roughness height of 0.132 μm (mode 2) and an elastic modulus of $E_2 = 258$ GPa; at a deposition temperature of 900 °C and a treatment time of 2 h the coating on the titanium substrate has a thickness of 1.5 μm with an average roughness height of 0.265 μm (mode 3) and an elastic modulus of $E_3 = 321$ GPa. According to this information, we chose the values for the material properties of the TiN coating that are presented in Table 1. Data for yield stress σ_y and ultimate strength σ_b and strain ε_b were obtained using reverse analysis of the load–displacement curve for instrumented indentation [38].

Table 1 Properties of the model materials for the friction pair

Material	Bulk modulus, K, GPa	Shear modulus, G, GPa	Density, ρ, kg/m^3	Yield stress, σ_y, GPa	Ultimate strength, σ_b, GPa	Ultimate strain, ε_b
Ti6Al4V	92	41	4420	0.99	1.07	0.100
TiN mode 1	173	104	5220	4.50	5.50	0.075
TiN mode 2	173	104	5220	4.50	5.50	0.075
TiN mode 3	205	129	5220	4.50	5.50	0.075

5.1.2   Validation of the Models for Materials

The first numerical test, which we performed for validation of the models used for further simulations, was instrumented indentation [39]. General view of the model geometry for this test is shown in Fig. 1a. The model specimen was a parallelepiped consisting of titanium substrate, interface and TiN coating. The loading was simulated by moving all the automata of the Berkovich indenter with constant velocity $V_z = -1$ m/s until the required penetration depth is reached, then we apply them the velocity $V_z = -1$ m/s for unloading (Fig. 1b).

Using the procedure proposed by Oliver and Pharr [40] applied to processing the simulation results we plot the dependence of the material hardness on the penetration depth (Fig. 2) for three kinds of coatings obtained in different regimes. It can be seen from Fig. 2 that the hardest is the coating obtained by mode 3. All the curves correspond to experimental data from [37]. This allows us to conclude that the models of the materials behavior are validated for the normal contact loading conditions and can be used in further steps of our work.

The model specimen for scratch testing was also a parallelepiped consisting of titanium substrate, interface and TiN coating but elongated along the axis Y (Fig. 3). To simulate the force acting on the indenter along axis Z in the experiment, in our calculations we set the velocity of indenter automata to $V_Z = -0.5$ m/s (Fig. 3b) until the indenter was immersed into a predetermined penetration depth and after that the vertical velocity was set to zero. Here we considered penetration of the indenter only up to the interface layer; the possibility of the coating to detach from the substrate was not allowed. To move the indenter along the sample surface, a constant velocity of the indenter automata along the axis Y was set to $V_Y = 1$ m/s.

Based on the results of our simulations, images of the deformed sample were created, and the values of the critical force characteristics at certain stages of fracture were obtained. It was found that the value of the critical force for the onset of fracture

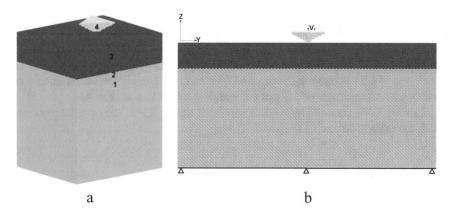

a   b

Fig. 1 The model specimen for indentation (**a**) and its cross-section with loading parameters (**b**) represented by automata packing (the numbers indicate the model materials: 1—titanium, 2—interface, 3—coating, 4—diamond)

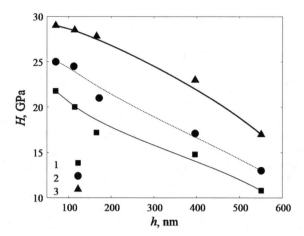

Fig. 2 Dependence of the hardness on the depth of penetration (the numbers indicate the modes of coating deposition)

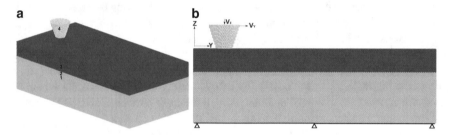

Fig. 3 The model specimen for scratch testing (**a**) and its cross-section with the loading parameters (**b**) represented by automata packing (the numbers indicate the model materials: 1—titanium, 2—interface, 3—coating, 4—diamond)

and cracking is largest for the coating thickness of 1.4 μm and roughness of 0.132 μm, and smallest for the thickness of 1.5 μm and roughness of 0.265 μm. At delamination of the coating, the maximum of critical force is typical for the specimen with a coating thickness of 1.5 μm and roughness of 0.265 μm, and the minimum for the specimen with a coating thickness of 1.3 μm and roughness of 0.15 μm (Fig. 4).

Critical loads obtained by modeling scratch testing are in very good agreement with the experimental data [41].

5.1.3 Modeling of Sliding Friction at the Meso-scale

At the next step, we developed a meso-model of sliding friction of the contacting surfaces of the endoprosthesis that explicitly accounts for the surface roughness, mean height of which is determined by the regime of the coating deposition [42].

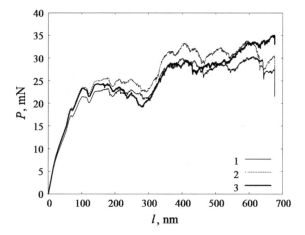

Fig. 4 The force acting on the indenter from the coating P versus the scratching path l of the indenter (the numbers indicate the modes of coating deposition)

Because at the macroscale we need just effective value of the friction coefficient, whereas the substrate and interface were not considered in this model.

Geometrically, the model for studying friction consists of two bodies in the form of parallelepipeds (Fig. 5). Each parallelepiped is usually divided into two parts: one is the contact surface, and the other is its base. The different roughness relief of the two contacting surfaces was set in the preprocessor of the MCA_3D software package. To simulate the environment of the developed meso-scale simulation box during sliding friction, periodicity conditions along the X and Y axes were set on its side faces.

In the initial state, the interacted bodies are separated in spaced so that the asperities of the rough surfaces do not touch each other. For their approaching and the beginning of contact interaction, the velocity V_Z along the Z-axis was applied to the automata of the top and bottom layers of the sample. After touching the upper and lower parts of the sample, traction forces (pressure $P = 0.75\sigma_y$) directed along the Z-axis, as well as horizontal velocity $V_y = 1$ m/s along the Y-axis, were applied to the automata of the top and bottom layers of the sample (Fig. 5b). The specified loading conditions at the top and bottom of the model had symmetrical character and the same magnitude, but the directions of the action were opposite. To avoid too hard loading in the friction zone, a gradual increase in the applied loads was implemented. To imitate the length of the real bodies along the vertical direction, it is necessary to damp the elastic waves arising in the friction zone and propagating along this direction. To this end, a special viscous damping force along the Z-axis was introduced for the automata of both loading regions.

According to the simulation results obtained, the dependence of the friction coefficient on the calculation time was plotted for different combinations of the roughnesses and reliefs (Fig. 6). Figure 6a shows that at the moment of contact of two

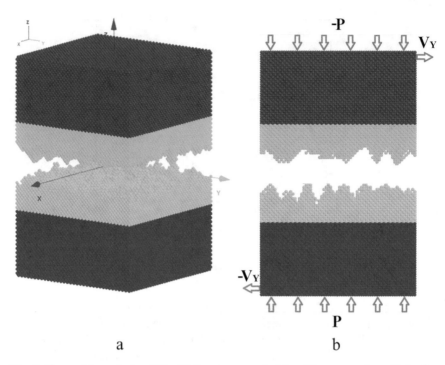

Fig. 5 The model sample for sliding friction at mesoscale (**a**) and its cross-section with loading parameters (**b**), the colors indicate different geometry blocks of the model

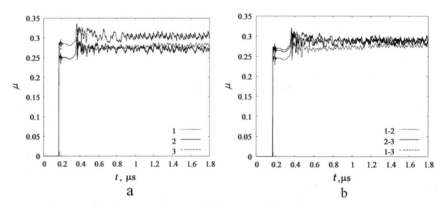

Fig. 6 Plots of the friction coefficient in the model at meso-scale versus the calculation time with the same (**a**) and different (**b**) modes of coating deposition (the numbers indicate these modes)

rough surfaces of different relief, but of the same roughness, the peak of the friction coefficient occurs, then the process goes to the stationary mode and the friction coefficient becomes almost constant. The plot shows that the greatest coefficient of friction $\mu_3 = 0.3$ is observed on the surfaces of the TiN coating in the first and third deposition modes, which corresponds to the experimental data [43], and the minimal $\mu_2 = 0.25$ in the second deposition mode. The thickness of the quasi-liquid mixing layer for the first and second regime is 0.08 μm, for the third regime—0.12 m.

It is known that an increase in the hardness of materials in contact with friction leads to a decrease in the friction coefficient, and an increase in the surface roughness leads to an increase in the friction coefficient [44, 45]. However, the materials used for the manufacture of endoprostheses should have increased hardness. Therefore, we have aimed to identify the most effective materials used in a friction pair. For this purpose, the calculations were carried out for the samples with different roughness and mechanical characteristics. According to the results of these numerical calculations, it was found that in the friction pair of the second and third modes $\mu_{2-3} = 0.27$; for the first and third modes $\mu_{1-3} = 0.28$ (Fig. 6b). The results indicate the promising use of materials obtained under different application conditions for different parts of the friction pair. At the modes 1–3 and 2–3, the thickness of the mixing layer is 0.1 μm, and for the pair of the modes 1–2—0.8 μm.

5.1.4 Modeling Friction in the Rotating Friction Pair

The next step in the numerical study of the friction in the hip joint with resurfacing endoprosthesis was the development of the macro-model of friction between the endoprosthesis casing cap rotating in the artificial insert of acetabular cup [46].

In the case of a one-component material, the geometric model consisted of a hemisphere simulating a casing cap for the femur head with an external diameter D_ext_cap = 36 mm and an interior diameter D_int_cap = 33 mm; a hemisphere simulating an acetabular cup of the hip prosthesis with an outer diameter of D_ext_insert = 41 mm and an interior diameter of D_int_insert = 38 mm; and also a conical "shell" for the cup imitating the surrounding bone tissue (Fig. 7b). In the case of a two-component (coated) material, the hollow hemispheres were additionally specified for a casing cap with an outer diameter of 35.9 mm and an inner diameter of 33.1 mm (Fig. 7a).

The load was applied by specifying the translational and rotational velocities for the automata of the resurfacing cap. These velocities corresponded to rotation of the cap as perfectly rigid body around the axis of symmetry of the corresponding sphere, which in our case was parallel to the axis OX. The value of the corresponding rotational velocity gradually increases from 0 to 10 s^{-1}. The bottom layer of the automata of the conical shell of the bone tissue was rigidly fixed (Fig. 7b).

When simulating a single rotational cycle, the maximum reaction force was not greater than 3 kN, which corresponds to the load of a walking man, and the angle of rotation of the resurfacing cap was 120°, which is typical for standard daily physical activities for a healthy person.

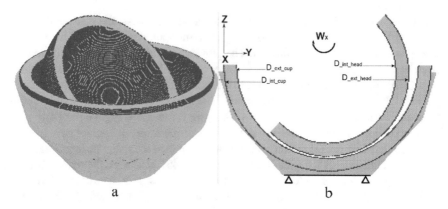

Fig. 7 General view of the model for friction pair of hip resurfacing endoprosthesis (**a**) and its cross section with loading parameters (**b**), represented by automata packing

Table 2 Properties of the model cortical bone

Material	Bulk modulus, K, GPa	Shear modulus, G, GPa	Density, ρ, kg/m^3	Ultimate strength, σ_b, GPa
Cortical bone	14	3.3	1850	0.12

Friction of Bone-Bone Pair

For comparison, based on the developed model we also studied friction between two cortical bone hemispheres (assuming them as a healthy joint) [47]. Cortical bone was considered as a linear elastic brittle material with the properties taken from [48] and shown in Table 2.

The simulation results for healthy friction pair of bone tissues showed that in the contact interaction zone of the acetabulum and femur head and behind this zone at extreme positions (the edge of the acetabulum), large compressive stresses aroused in the head and in the acetabulum with a maximum value not reaching 10 MPa. Such a load does not exceed the strength and, therefore, reduces the likelihood of premature wear of the femoral joint (Fig. 8).

Friction of Ti–Ti Pair

Then we considered stress fields in the simulated specimens for the titanium endoprosthesis joints without any coating, which are depicted in Figs. 9 and 10. The simulation results showed that in the case of a friction pair of a homogeneous metal material a large tensile stress with a maximum value not reaching 990 MPa appeared in the cap. Namely, it was observed at extreme positions in the zone of contact interaction of the acetabulum insert and resurfacing cap and behind it (edge of the

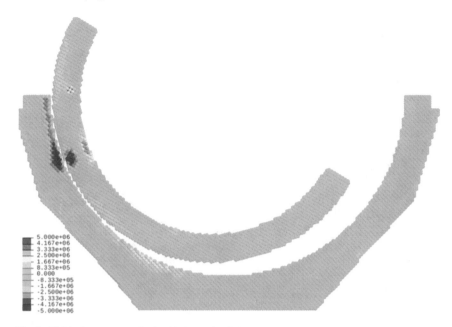

Fig. 8 Field of mean stress in the friction pair of the healthy hip joint (bone)

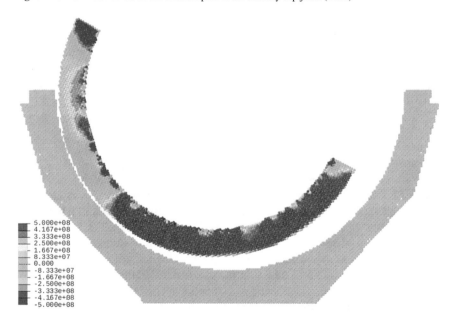

Fig. 9 Field of mean stress in the friction pair of the hip resurfacing endoprosthesis made of titanium alloy

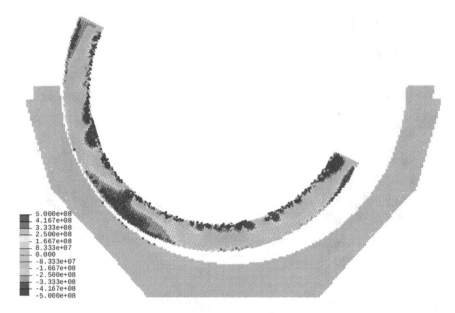

Fig. 10 Field of mean stress in the friction pair of the hip resurfacing endoprosthesis made of titanium alloy with TiN coating

acetabulum). Such a load exceeds the yield strength and, therefore, can lead to rapid wear of the surface of the resurfacing cap of the femur head of the joint (Fig. 9). These results are consistent with the data on the stress distribution in the metal head obtained in [49]. At the same time, the stress in the acetabulum insert did not exceed 100 MPa.

Friction of Ti–NiTi Pair

In the case of a coated endoprosthesis, a zone of tensile stresses with a maximum value of 1.1 GPa was observed in the contact zone, but this area was significantly smaller and concentrated mainly in the coating (Fig. 10). In addition, when using two-component materials in the friction pair, there was no noticeable increase in stress values in the cap when it was in the extreme positions. Consequently, the use of titanium alloys with a ceramic coating allows avoiding premature wear at the extreme positions of the femur head in the acetabulum. In the insert consisting of titanium alloy and coating, the value of compressive stresses reached 300 MPa. It should be noted that the magnitude of such stresses is not critical for the coating.

Thus, the presented results of the numerical simulations and their analysis suggest that the use of titanium alloy coated by TiN in the friction pair of hip resurfacing endoprosthesis can help avoiding premature wear of the endoprosthesis.

5.2 Modeling Bone–Endoprosthesis System

Finally, we developed a macromodel of hip resurfacing endoprosthesis with proximal part of the femur bone shown in Fig. 11. The geometry of the model was based on the so-called 3rd generation composite femur [50], which provides geometries of the cortical and cancellous bones as different solid bodies. A numerical model of the bone-endoprosthesis system was constructed for the resurfacing endoprosthesis with real geometric parameters. A CAD model from [50] was taken as a tubular femur, according to the parameters of which a personalized solid model for the endoprosthesis was created using FreeCAD software. For our purpose, we cut the top part of the bone geometry, added the resurfacing endoprosthesis (colored in cyan, and its coating colored in blue) and special loading part (colored in red) (Fig. 11b). Based on these solid models of the femur and endoprosthesis, mesh models were constructed in STL format, which then were imported into the MCA preprocessor as it shown [39].

Based on the developed model we simulated compression of the proximal part of the femur with resurfacing endoprosthesis, which can have a hardening coating or have not. The loading was applied by setting constant velocity in both horizontal and vertical directions as shown in Fig. 11b up to reaching the critical values of the resisting force of 3 and 10 kN [51]. The simulation results are presented in Fig. 12 as fields of mean stress.

It is reported in [52] that strength limit for the cancellous bone of healthy people reaches about 10 MPa in compression and 5 MPa in tension; and may be lower than 5 MPa and even 3 MPa for some diseases, respectively. Analysis of the stress field in the model system showed that maximum tensile stress was observed near the endoprosthesis, and did not reach critical values. Therefore, herein we tried to analyze the compression stress in the scale up to 5 MPa. From Fig. 12 one can see

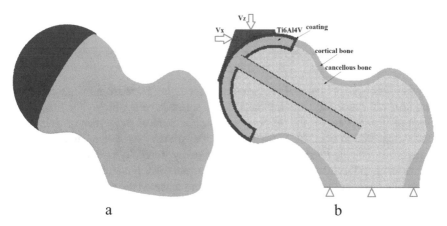

Fig. 11 The model of hip resurfacing endoprosthesis (**a**) and its cross-section with the scheme of loading (**b**)

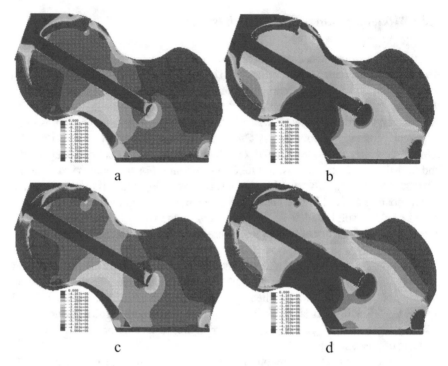

Fig. 12 Fields of mean stress in the cross-section of the system "bone-endoprosthesis" loaded by the force of 3 kN (**a, c**) and 10 kN (**b, c**) for titanium endoprosthesis (**a, b**) and titanium with TiN coating (**c, d**)

that maximum compression stress, which that can lead to fracture, is observed in the femoral neck.

5.3 Modeling of Biomaterials Based on Poroelastic Approach

It is well known, that the main difference of the living bone from one stored for a long time after removal from the body is the presence of the biological fluid. This is one of the reasons for the difference in the mechanical properties of living and "dead" bones. Moreover, this biological fluid may transfer across the whole bone and dramatically change its mechanical behavior under dynamic loading. That is why the next step in the further development of the numerical model described above is taking into account the biological fluid and its influence on the mechanical response of the femur.

5.3.1 Modification of the MCA Method to Enable Simulating Fluid-Saturated Materials

Automata that model fluid-saturated material are considered as porous and permeable. Pore space of such an automaton can be saturated with liquid. The characteristics of the pore space are taken into account implicitly through the specified integral parameters, namely, porosity ϕ, permeability k, and the ratio $a = 1 - K/K_S$ of the macroscopic value of bulk modulus K to the bulk modulus of the solid skeleton K_S. The mechanical influence of the pore fluid on the stresses and strains in the solid skeleton of an automaton is taken into account on the basis of the linear Biot's model of poroelasticity [53, 54]. Within this model, the mechanical response of a "dry" automaton is assumed linearly elastic and is described based on the above-shown relations. The mechanical effect of the pore fluid on the automaton behavior is described in terms of the local pore pressure P^{pore} (fluid pore pressure in the volume of the automaton). In the Biot model, the pore pressure affects only the diagonal components of the stress tensor. Therefore, it is necessary to modify only the relations for the central interaction forces in Eq. (3):

$$\Delta f_{ij} = 2G_i \left(\Delta \varepsilon_{ij} - \frac{a_i \Delta P_i^{\text{pore}}}{K_i} \right) + D_i \Delta \sigma_i^{\text{mean}} \tag{5}$$

Interstitial fluid is assumed to be linearly compressible. The value of fluid pore pressure in the volume of an automaton is calculated based on the relationships of Biot's poroelasticity model with the use of the current value of the pore volume. The pore space of the automata is assumed to be interconnected and provides the possibility of redistribution (filtration) of the interstitial fluid between the interacting elements. A pore pressure gradient is considered as the "driving force" of filtration. The fluid redistribution between automata is carried out by numerical solution of the classical equation of the fluid density transfer [55]. This equation is numerically solved using the finite volume method adopted for the ensemble of automata.

5.3.2 Choosing Poroelastic Parameters for Bone Tissues

The aim of this section is to choose the correct values of the model for both cortical and cancellous bone tissues. The first who considered bones like poroelastic bodies was Cowin [27]. In paper [56] he with co-authors provided the values of the main parameters of poroelastic body for cortical and cancellous tissues of the human bone. These parameters are as follows: Young's modulus, Poisson's ratio, the permeability, Biot's coefficient (or bulk modulus of the solid phase), the porosity, densities of the solid grain and the fluid. Later, several authors have made experimental and theoretical studies aimed to get the values of poroelastic parameters for some specific bones of humans and animals [48, 57–60]. Based on data published in the literature, one may conclude that there is a large scatter of the main poroelastic properties of the bone tissues. For example, geometric permeability estimates span across several

Table 3 Poroelastic properties of the model bone tissues

Bone tissue	Bulk modulus of the solid phase, K_S, GPa	Bulk modulus of the matrix, K, GPa	Shear modulus of the matrix, G, GPa	Density of the matrix, ρ, kg/m^3	Porosity ϕ	Geometric permeability, k, m^2
Cortical	17.0	14.0	5.55	1850	0.04	1.0×10^{-16}
Cancellous	17.0	3.3	1.32	600	0.80	3.5×10^{-13}

orders of magnitude $\left(10^{-25} - 10^{-10} \text{ m}^2\right)$, and values for Young's modulus vary from 1 up to 25 GPa [48]. That is why it is of special interest to study the peculiarities of the mechanical behavior of small model specimens in compression depending on the variation of these properties.

Basic values of the poroelastic properties chosen for our numerical models for both cortical and cancellous bone tissues are shown in Table 3. The fluid in both bone tissues is assumed to be the same and equivalent to salt water, with a bulk modulus $K_f = 2.4$ GPa, and density $\rho_f = 1000$ kg/m^3.

The developed model was applied to study the dynamic mechanical behavior of the fluid-saturated porous materials under uniaxial compression at a constant speed. We studied and analysed the dependences of the effective Young's modulus of fluid-saturated materials on the strain rate and the characteristic time of fluid redistribution in the pore space. In our calculations, the material parameters and the strain rate varied within wide limits: the permeability of the material varied within 4 orders of magnitude, the viscosity of the fluid varied within 2 orders of magnitude, the sample size changed within the order of magnitude, and the strain rate varied within 3 orders of magnitude [61].

We simulated uniaxial compression of 3D cubic specimens along the vertical axis (Z). The size of the automata for all models in this study was equal to 2 mm. The base size of the cubic specimens was chosen to be 5 cm. The initial pore pressure of interstitial fluid was assumed to be equal to atmospheric. Fluid could freely flow out from the compressed specimen through the side surface.

Analysis of the simulation results showed that under compression, the values of the mechanical characteristics of the fluid-saturated material are determined by the balance of two competing processes [62, 63]:

- deformation of the solid skeleton, providing compression of the pore space and a corresponding increase in the pore pressure of the interstitial fluid;
- outflow of the interstitial fluid through the side surface, which leads to the inverse effect of lowering pore pressure.

We revealed a key control parameter that determines the specific dynamic value of the mechanical characteristics of fluid-saturated materials, namely the dimensionless Darcy number:

$$D_a = \frac{T_{\text{Darcy}}}{T_{\text{load}}} \sim \frac{\eta_{\text{fl}} L^2}{k \Delta P} \dot{\varepsilon}_{\text{def}}, \tag{6}$$

where T_{Darcy} is the characteristic time of fluid filtration (Darcy time scale), $T_{\text{load}} \sim 1/\dot{\varepsilon}_{\text{def}}$ is the time scale of the specimen deformation, $\dot{\varepsilon}_{\text{def}}$ is the specimen strain rate, η_{fl} is the dynamic viscosity of the pore fluid, L is the characteristic length of the pore pressure difference ΔP (a half of length of the cubic side in the considered case). The parameter D_a characterizes the ratio of the timescales of deformation of the fluid-saturated porous specimen and filtration of the pore fluid.

The simulation results showed that the effective Young's modulus of fluid-saturated bone tissues non-linearly depends on the strain rate $\dot{\varepsilon}_{\text{def}}$, the specimen size L, the dynamic viscosity of pore fluid η_{fl}, and the permeability of solid skeleton k. In particular, Young's modulus of a fluid-saturated sample is minimal (equal to E_{min}) at infinitely small strain rates and tends to the maximum value (Young's modulus of undrained sample E_{max} [54]) at large ones. The key result is the established ability to build single "gauge" dependence applicable to specimens of porous materials of various sizes, characterized by different permeability of solid skeleton, different fluid viscosities, and deformed at different strain rates. An argument of such a "master curve" is the dimensionless Darcy number D_a (Fig. 13):

$$E = E_{\text{max}} + \frac{E_{\text{min}} - E_{\text{max}}}{1 + (D_a/D_{a_0})^p}, \tag{7}$$

where E_{min} corresponds to $D_a \to 0$ ("dry" specimen), E_{max} corresponds to $D_a \to \infty$ (undrained specimen), D_{a_0} and p are the fitting constants. This master curve has a logistic (sigmoidal) form but the fitting values of D_{a_0} and p (as well as E_{min} and E_{max}) are different for cortical and cancellous bones.

Based on the presented results we chose the values of the poroelastic parameters for both bone tissues applied to the model proximal femur corresponding to Darcy

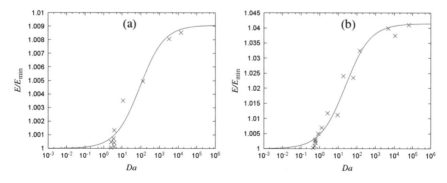

Fig. 13 Dependences of the normalized Young's moduli of the modeled specimens of cortical (**a**) and cancellous (**b**) bone tissues with different poroelastic properties on the Darcy number. Crosses show numerically obtained values, lines are approximating curves

numbers about 50, i.e. the middle of the plots range shown in Fig. 13. This means that at the loading rate of the model femur the effect of interstitial fluid flow is expected to be well pronounced.

5.3.3 Validation of the Materials Model

During life, the properties of the human bones change considerably; usually they become fragile. In [64], the concept of bone fragility was formulated as follows: bone tissue can adapt its shape and size in response to mechanical stress through a remodeling mechanism, during which the bones are formed or rearranged under the independent action of osteoclasts and osteoblasts. Remodeling is a process that supports the mechanical resistance of the skeleton, allowing you to selectively restore and replace damaged bone tissue. During the period of growth, these processes form a structure capable of adapting to the loads and maintaining strength. With age, the natural remodeling processes slow down under the influence of such factors as decreasing muscle mass and physical activity, malnutrition. Consequently, bone embrittlement occurs. Osteoporosis is a systemic skeletal disease characterized by low bone mass and impaired microarchitecture of bone tissue, which leads to increased bone fragility (reduced bone density) and a tendency to fractures. Researchers and clinicians continue to persist in finding ways to prevent the 8 million osteoporotic fractures occurring annually around the world [65]. An early and accurate assessment of the risk of fracture, with timely initiated treatment—this approach seems to be the most appropriate for reducing this number of fractures and associated personal and social losses.

There are many methods for the direct assessment of the structural strength and the material composition of the bone, such as testing of whole bone, bone mass, strength assessment using microindentation [66]. However, all of these methods are used for in vitro or in vivo research and are therefore not suitable for clinical practice. Recently, however, indentation has become positioned as a procedure for clinical use with minimal invasiveness [67]. In the indentation method, the bones are penetrated using an indenter tip with a depth sensor. Among the advantages of this method, there is an ability to measure material properties and microstructural features and identify local changes in bone material caused, for example, by disease or medication [68]. However, the method remains invasive and extremely local, therefore, at present, the development of methods for non-invasive diagnostics of the strength properties of local bone tissue is actively underway.

One of the new directions in this area is the combination of methods for visualizing the structure of bone tissue (CT and MRI) with computer modeling [69]. Therefore, the development of numerical models of the mechanical behavior of bone tissue during micro and macroindentation is relevant.

Here we describe a numerical model of the mechanical behavior of fluid-saturated cancellous tissue during indentation used for validation of the developed model of fluid-saturated bone tissues [70].

Geometrically, the bone tissue indentation model was a parallelepiped with a Berkovich pyramid (Fig. 1). During indentation, the counter-body was set as the rigid non-deformable indenter, which motion was set through the velocity in the vertical direction. This velocity was set to $V_z = -1$ m/s until the indenter was penetrated at a given depth, after which the velocity $V_z = 1$ m/s was set to simulate unloading. The simulation results were processed using the Oliver-Farr method [40].

The mechanical properties for the cancellous bone were chosen from Table 3, and the fluid in bone tissue is assumed to be equivalent to salt water as before.

It is known that bone tissue, when indented, behaves like a soft material, and gives an error in determining the elastic modulus using a standard experimental procedure; thus, indentation with a time delay of indentation depth (holding) looks to be better for this material [71]. Therefore, one of the first experiments to test the developed model was aimed to establish the fact of the influence of holding time on the mechanical properties of bone tissue. Three variants of loading were considered: (1) without holding $t_{hold} = 0$; (2) holding time corresponded to loading time $t_{hold} = t_{loading}$; (3) holding time was two times longer than loading time $t_{hold} = 2 \cdot t_{loading}$ (Fig. 14a).

When analyzing the load–displacement curves (Fig. 14b) using the Oliver-Farr method, the following values of recoverable characteristics were obtained: the hardness of 47 MPa for all loading regimes, the elastic modulus 3.7 GPa for no holding, $E = 3.5$ GPa for $t_{hold} = t_{loading}$, $E = 3.33$ GPa for $t_{hold} = 2 \cdot t_{loading}$. The obtained results showed that the elastic modulus is determined correctly at $t_{hold} = t_{loading}$ and corresponds to the specified input value. In addition, these calculations suggest that at constant deformation, stress relaxation occurs, which indicates that the model specimen of cancellous bone tissue possesses viscoelastic properties.

At the next stage, the effect of interstitial fluid flow in the porous skeleton of the material was investigated. According to the simulation results (Fig. 15a), it was established that the hardness of the undrained bone tissue specimen was 55 MPa, while the hardness of the fluid-saturated bone was 47 MPa, and the elastic moduli

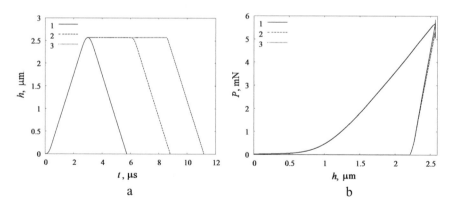

a

b

Fig. 14 **a** The indenter penetration depth versus time; **b** load–displacement curve for different holding times; numbers indicate loading regimes: (1) $t_{hold} = 0$, (2) $t_{hold} = t_{loading}$, (3)$t_{hold} = 2 \cdot t_{loading}$

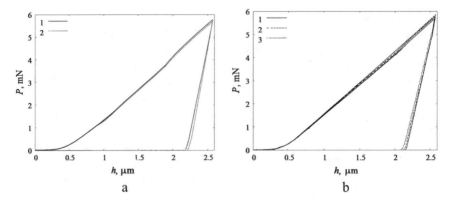

Fig. 15 Load–displacement curves for undrained (**a**), numbers indicate loading rates: (1) undrained, (2) fluid-saturated and fluid-saturated model specimens (**b**), numbers indicate loading rates: (1) $v = 1$ m/s, (2) $v = 5$ m/s, (3) $v = 10$ m/s

were 3.9 and 3.6 GPa, respectively. The results obtained correspond qualitatively to the data on the influence of fluid on the mechanical response of bone tissue by other authors [72–74].

Further calculations were carried out to study the effect of porosity value of the bone tissue and its permeability on the mechanical response of the bone in indentation. However, the results obtained indicate that with the chosen values of porosity (80%), the above factors do not have a significant effect on the mechanical response of the fluid-saturated bone tissue.

The viscoelastic behavior of the bone tissue is also characterized by the dependence of the mechanical characteristics on the loading rate; therefore, at the next stage of testing the numerical model development, experiments were performed on the indentation of a model specimen of cancellous bone tissue with different loading rates. The study of the effect of loading rate on the mechanical response of fluid-saturated bone tissue showed that an increase in loading rate leads to an increase in the values of recoverable characteristics (Fig. 15b), thus at the speed of $V = 1$ m/s we got $H = 47$ MPa, $E = 3.5$ GPa, at the speed of $v = 5$ m/s we got $H = 53$ MPa, $E = 3.7$ GPa, and at the speed of $v = 10$ m/s we got $H = 56$ MPa, $E = 3.95$ GPa. The obtained values, again, corresponds qualitatively to the results of other authors [75–77].

5.3.4 Modeling the Bone Compression

Similar as in the previous section, the geometry of the bone is based on the 3rd generation composite femur [50], which consists of the cortical and cancellous parts as different solid bodies. General view of the model represented as fcc packing of automata and its cross-section are shown in Fig. 16.

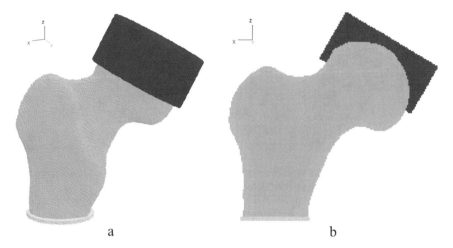

a   b

Fig. 16 General view of the model for proximal femur (**a**) and its cross-section (**b**)

At the bottom of the model, we place a disk with properties of the cortical bone; the automata of this disk are fixed. For compressing the femur, we place a special cylindrical "cap" on the femur head. This "cap" is shown in blue color in Fig. 16. Automata of the "cap" have the properties corresponding to cartilage. Loading is applied by setting the constant velocity $V = 1$ m/s to the automata of the upper face of the "cap". The velocity vector is directed along the face normal. Note, that this loading results in both compression and small bending of the bone.

To reveal the role of the interstitial fluid flow in the model femur under compression, we studied three different cases. In the first case, all bone tissues contained no fluid, i.e. were "dry" (so-called drained test). In the second case, we used the chosen poroelastic parameters from Table 3. In the third case, all pores containing fluid were assumed to be closed, which means no permeability of the materials (undrained test). Then we analyzed the distributions of mean stress, equivalent stress and pore fluid pressure at the final point at the total strain about 2%.

Fields of mean stress in the cross section of the model femur for all three cases are shown in Fig. 17. One can see that fluid filtration cause the increase in mean stress in the cortical part of the femur head, especially in the area of its contact with the loading "cap". At the same time, the drained and undrained tests do not differ considerably from each other. However, fields of equivalent stress in the cross section of the model for these cases shown in Fig. 18, clearly demonstrate that shear stress in the same area is much smaller for the undrained test, while for the two other cases are practically the same.

It can be clearly seen from Figs. 16 and 18 that the maximum stresses occur in the cortical part of the femur. Figure 19 shows the 3D view of both equivalent and mean stresses distribution in the model (particularly, the outer part of the cortical bone). It is obvious that the main stresses are localized in the femoral neck. The shear stresses propagate along the loading direction into the main part of the bone up to the

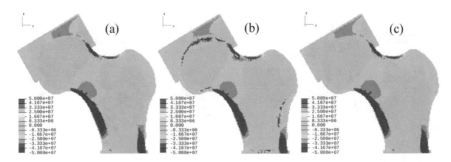

Fig. 17 Distribution of mean stress in cross-section of the model femur at drain test (**a**), test with chosen poroelastic parameters (**b**), and undrained test

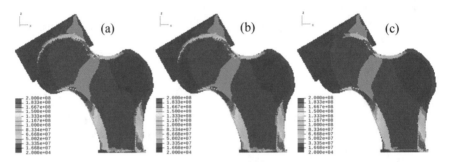

Fig. 18 Distribution of equivalent stress in cross-section of the model femur at drained test (**a**), test with chosen poroelastic parameters (**b**), and undrained test

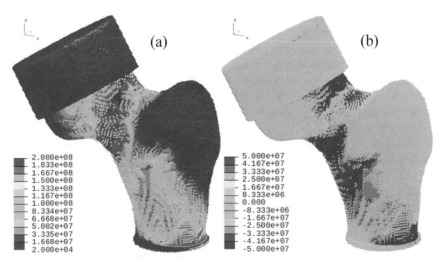

Fig. 19 Distribution of equivalent stress (**a**) and mean stress (**b**) in the model femur with chosen poroelastic parameters

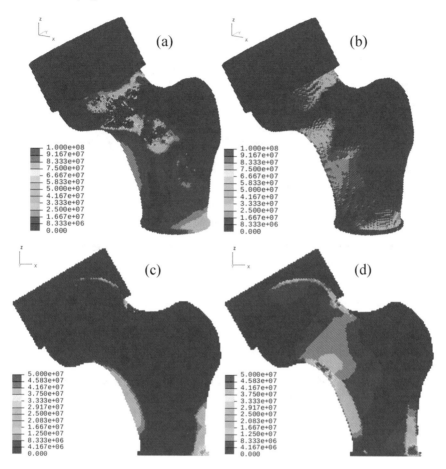

Fig. 20 Distribution of pore pressure in the model for the test with chosen poroelastic parameters (**a, c**) and undrained test (**b, d**), **c** and **d** are corresponding cross-sections

supporting plate. But the most dangerous seems the tensile mean stress in the upper part of the femoral neck (Fig. 19b).

Figure 20 shows fields of the pore pressure in the cases with interstitial fluid. It can be seen that ability to flow results in filtration of the fluid to the regions of large tensile and shear stresses of the cortical bone, but not the maximum tensile mean stress (upper part of the femoral neck). At the same time, one can see negligible pore pressure in the cancellous bone in the case of fluid filtration (Fig. 20c).

5.3.5 Modeling the Bone-Endoprosthesis System

In the last sub-section, we consider the femur–endoprosthesis system, where the bone is described using poroelastic approach and the implant is made of the TiN-coated

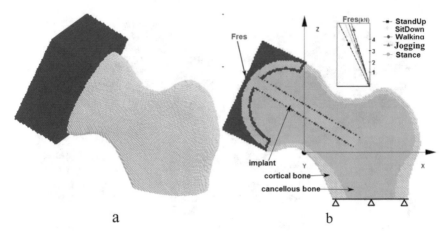

Fig. 21 Model system "bone-endoprosthesis". **a** 3D view; **b** cross-section view with scheme of loading

titanium alloy Ti6Al4V. The values for the main physico-mechanical properties of the titanium alloy and TiN coating produced by the PIRAC deposition mode 2 are taken from Table 1. The properties for the bone tissues are taken from Table 3. A model of the bone-endoprosthesis system was constructed by analogy with Sect. 5.2 and is shown in Fig. 21a.

The main feature of this section is that here we vary the loading similar to the different types of human activity. Thus, dynamic loading F_{res}, which is equivalent to the physiological one for a person weighing 75 kg, was applied to the upper part of the implant (Fig. 21b). According to [78], this force lies in the medial plane ZX and is inclined under different angles relative to the bone axis Z for different kinds of activity. Standing up load is characterized by the total force of 3.6 kN and applied at the angle of 24°; sitting down load of 3 kN is applied at the angle of 20°; a load of walking is 3 kN and applied at the angle of 17°; jogging is characterized by 4.5 kN and applied under angles of 15°; stance position is characterized by 3.2 kN and applied under angle of 16°. Here, the loading is simulated through the setting constant velocity to the automata of the external surface of the loading block marked by the blue color in Fig. 21 up to the moment when the required loading value of the resistance force is reached, similarly as it was done in [78, 79]. The value of the loading velocity is 1 m/s for walking, sitting, standing up and position while standing, for jogging the loading velocity is 2 m/s [79].

Typical patterns of the mean stress fields obtained by simulations are shown in Fig. 22. According to the presented results, the highest tensile and compressive stresses are concentrated in the upper and lower parts of the femur neck, respectively.

Analysis of the images also shows that the angle of the load application for the same value of the resulting force significantly affects the distribution of compressive stresses in the proximal femur (Fig. 22a–c). Thus, an increase in the angle of the load application by 4° with a simultaneous increase in the force by 20% leads to an increase

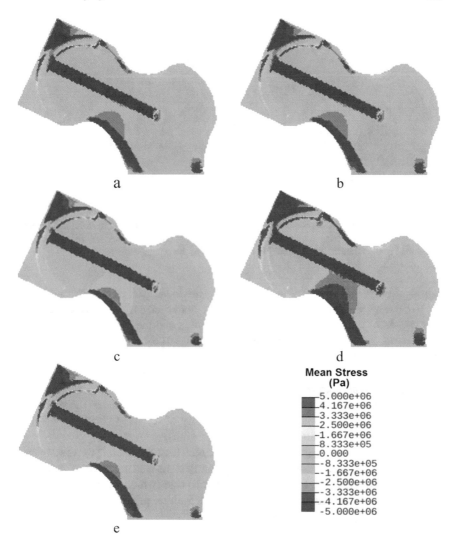

Fig. 22 Fields of mean stress in the system "bone-endoprosthesis" under different physiological loading: **a** standing up, **b** sitting down, **c** walking, **d** jogging and **e** stance position

in the area of compressive stresses by 20% (Fig. 22b, a, respectively). A decrease in the angle of the load application by 3° at the same value of the force (the cases of sitting down and walking) causes an increase in the localization area of maximum compressive stresses by about 15% (Fig. 22b, c, respectively). At the same time, a decrease in the angle of the load application by 2° relative to the nominal direction with an increase in the loading force by 7% (the cases of walking and standing position) leads to a decrease in the localization area of maximum compressive stresses by 5% (Fig. 22c, e, respectively). The highest stress concentration is observed for

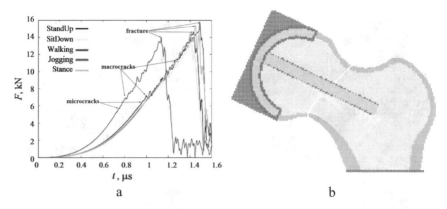

a   b

Fig. 23 **a** Plots of the loading force versus time for the considered types of physiological loading; **b** fracture pattern of the bone–endoprosthesis system (the lines connect bonded automata, the cracks correspond to the white zigzag "curves" in the network of the bonded automata)

jogging: compressive stresses are observed in the lower part of the neck under the implant pin, as well as in the femoral head area under the casing cap (Fig. 22d).

Analysis of the loading plots shown in Fig. 23a and the corresponding fracture patterns (one example is depicted in Fig. 23b) allows us to conclude that for all types of physiological loads microcracks arise at the loading of 5–6 kN, while formation of a macrocrack is observed at the force above 6 kN. When the force value reaches 14–16 kN the complete failure of the femoral neck takes place; according to [80] the resulted pattern of crack depicted in (Fig. 23b) may be defined as the subcapital fracture. It worth noting that the least probability of fractures occurs at walking, while the greatest when sitting down.

One more interesting peculiarity that can be seen from the loading curves depicted in Fig. 23a, is a lower failure force at jogging (red curve). It can be explained by the minimum angle of the load inclination for this case, which means the maximum arm of the force for breaking the femoral neck. Failure, in this case, occurs earlier because of the higher loading velocity.

Thus, we finally considered the mechanical behavior of the femur bone with hip resurfacing under conditions of real physiological loads. The simulation data obtained, indicate that with a decrease in the angle of the force application relative to the femur axis, an increase in the area of compressive stresses is observed, as well as the appearance of such stresses in the femur head in the region closed to the prosthesis cap. With a further increase in loading force, corresponding to an increase in the body weight or the performance of physical exercises with weighting, a tendency to fracture is observed. Furthermore, the accumulation of micro and macro-cracks in the bone tissue under the loading force values, slightly exceeding the natural physiological levels, can also under certain circumstances lead to failure. A critical physiological load, in which the destruction of bone tissue starts to appear in the case of resurfacing endoprosthesis, corresponds to sitting down in excess weight conditions or using additional weights.

6 Conclusions and Future Work

This chapter presents the multiscale numerical models that can be used as a simulation tool for the virtual analysis of the femur–endoprosthesis system for the case of hip resurfacing arthroplasty. The movable cellular automaton method, which is used for simulation of the mechanical behavior of the materials, allows explicit accounting for contact loading of rough surfaces of the TiN coatings used in the friction pair of the endoprosthesis as well as fracture of the bone tissues at different scales of the model. The implementation of the Boit theory of poroelasticity within the movable cellular automaton method allows revealing the role of the interstitial biological fluid in the mechanical behavior of the bone tissues.

Based on the analysis of the obtained simulation results, the following plan for future works is proposed. First, it is necessary to use real geometry of the bone and endoprosthesis, which are more complicated than was considered here. This means that we need to consider not only the proximal part of the femur, but at least a whole bone, and maybe with the knee joint. Real resurfacing caps have many geometrical features that also have to be included in future models. All these requirements lead to the use of the small size of the automata and hence the huge number of them and large computational costs. Second, the loads have to be also refined in order to be more realistic. The third important problem that has to be considered in the future is the detained description of the interface between an implant and bone, including osseointegration.

Acknowledgements The research presented in this chapter was supported by the Russian Foundation for Basic Research (Grant No. 20-08-00818, simulation results) and the Government research assignment for ISPMS SB RAS (in-house software development).

References

1. Bitar D, Parvizi J (2015) Biological response to prosthetic debris. World J Orthop 6(2):172–189. https://doi.org/10.5312/wjo.v6.i2.172
2. Sovak G, Weiss A, Gotman I (2000) Osseointegration of Ti6A14V alloy implants coated with titanium nitride by a new method. J Bone Joint Surg 82(2):290–296. https://doi.org/10.1302/0301-620X.82B2.0820290
3. Gotman I, Gutmanas EY, Hunter G (2017) In: Ducheyne P (ed) Comprehensive biomaterials II. Elsevier, Amsterdam, pp 165–203. https://doi.org/10.1016/B978-0-12-803581-8.09795-2
4. Wu SJ, Li H, Wu SY, Guo Q, Guo B (2014) Preparation of titanium carbide–titanium boride coatings on Ti6Al4V by PIRAC. Surf Eng 30(9):693–696. https://doi.org/10.1179/1743294414Y.0000000304
5. Wu Si, Ma S, Wu Sh, Zhang G, Dong N (2018) Composition, microstructure, and friction behavior of PIRAC chromium carbide coatings prepared on Q235 and T/P 24. Int J Appl Ceram Technol 15:501–507. https://doi.org/10.1111/ijac.12803
6. Kot M, Rakowski W, Lackner JM, Major T (2013) Analysis of spherical indentations of coating-substrate systems: experiments and finite element modeling. Mater Des 43:99–111. https://doi.org/10.1016/j.matdes.2012.06.040

7. Lofaj F, Németh D (2017) The effects of tip sharpness and coating thickness on nanoindentation measurements in hard coatings on softer substrates by FEM. Thin Solid Films 644:173–181. https://doi.org/10.1016/j.tsf.2017.09.051

8. Marchiori G, Lopomo N, Boi M, Berni M, Bianchi M, Gambardella A, Visani A, Russo A (2016) Optimizing thickness of ceramic coatings on plastic components for orthopedic applications: a finite element analysis. Mater Sci Eng C 58:381–388. https://doi.org/10.1016/j.msec.2015.08.067

9. Bouzakis KD, Michailidis N, Hadjiyiannis S, Skordaris G, Erkens G (2002) The effect of specimen roughness and indenter tip geometry on the determination accuracy of thin hard coatings stress–strain laws by nanoindentation. Mater Charact 49(2):149–156. https://doi.org/10.1016/S1044-5803(02)00361-3

10. Jiang WG, Su JJ, Feng XQ (2008) Effect of surface roughness on nanoindentation test of thin films. Eng Fract Mech 75(17):4965–4972. https://doi.org/10.1016/j.engfracmech.2008.06.016

11. Sliwa A, Mikuła J, Gołombek K, Tanski T, Kwasny W, Bonek M, Brytan Z (2016) Prediction of the properties of PVD/CVD coatings with the use of FEM analysis. Appl Surf Sci 388(PartA):281–287. https://doi.org/10.1016/j.apsusc.2016.01.090

12. Skordaris G, Bouzakis KD, Kotsanis T, Charalampous P, Bouzakis E, Breidenstein B, Bergmann B, Denkena B (2017) Effect of PVD film's residual stresses on their mechanical properties, brittleness, adhesion and cutting performance of coated tools. CIRP J Manuf Sci Technol 18:145–150. https://doi.org/10.1016/j.cirpj.2016.11.003

13. Zlotnikov I, Dorogoy A, Shilo D, Gotman I, Gutmanas E (2010) Nanoindentation, modeling, and toughening effects of zirconia/organic nanolaminates. Adv Eng Mater 12(9):935–941. https://doi.org/10.1002/adem.201000143

14. Li J, Beres W (2006) Three-dimensional finite element modelling of the scratch test for a TiN coated titanium alloy substrate. Wear 260:1232–1242. https://doi.org/10.1016/j.wear.2005.08.008

15. Holmberg K, Laukkanen A, Ronkainen H, Wallin K, Varjus S (2003) A model for stresses, crack generation and fracture toughness calculation in scratched TiN-coated steel surfaces. Wear 254(3–4):278–291. https://doi.org/10.1016/S0043-1648(02)00297-1

16. Pandure PS, Jatti V, Singh TP (2014) Three dimensional FE modeling and simulation of nanoindentation and scratch test for TiN coated high speed steel substrate. Int J Appl Eng Res 9(15):2771–2777

17. Toparlj M, Sasaki S (2002) Evaluation of the adhesion of TiN films using nanoindentation and scratch testing. Philos Mag A 82(10):2191–2197. https://doi.org/10.1080/014186102082 35729

18. Zhang T, Harrison NM, McDonnell PF, McHugh PE, Leen SB (2013) A finite element methodology for wear–fatigue analysis for modular hip implants. Tribol Int 65:113–127. https://doi.org/10.1016/j.triboint.2013.02.016

19. Kruger KM, Tikekar NM, Heiner AD, Baer TE, Lannutti JJ, Callaghan JJ, Brown TD (2014) A novel formulation for scratch-based wear modelling in total hip arthroplasty. Comput Methods Biomech Biomed Eng 17(11):1227–1236. https://doi.org/10.1080/10255842.2012.739168

20. Donati D, Colangeli M, Colangeli S, Di Bella C, Mercuri M (2008) Allograft-prosthetic composite in the proximal tibia after bone tumor resection. Clin Orthop Relat Res 466(2):459–465. https://dx.doi.org/10.1007%2Fs11999-007-0055-9

21. Koukal M, Fuis V, Florian Z, Janíček P (2011) A numerical study of effects of the manufacture perturbations to contacts of the total hip replacement. Eng Mech 18(1):33–42

22. Ashkanfar A, Langton DJ, Joyce TJ (2017) A large taper mismatch is one of the key factors behind high wear rates and failure at the taper junction of total hip replacements: a finite element wear analysis. J Mech Behav Biomed Mater 69:257–266. https://doi.org/10.1016/j.jmbbm.2017.01.018

23. Askari E, Flores P, Dabirrahmani D, Appleyard R (2016) A review of squeaking in ceramic total hip prostheses. Tribol Int 93(A):239–256. https://doi.org/10.1016/j.triboint.2015.09.019

24. Kuhl E, Balle F (2005) Computational modeling of hip replacement surgery: total hip replacement vs. hip resurfacing. Technische Mechanik 25(2):107–114

25. Dickinson A, Taylor A, Browne M (2012) Implant–bone interface healing and adaptation in resurfacing hip replacement. Comput Methods Biomech Biomed Eng 15(9):935–947. https://doi.org/10.1080/10255842.2011.567269
26. Dickinson AS, Brown M, Roques AC, Taylor AC (2014) A fatigue assessment technique for modular and pre-stressed orthopaedic implants. Med Eng Phys 36(1):72–80. https://doi.org/10.1016/j.medengphy.2013.09.009
27. Cowin SC (1999) Bone poroelasticity. J Biomech 32(3):217–238. https://doi.org/10.1016/S0021-9290(98)00161-4
28. Zhang D, Cowin SC (1996) Mechanics of poroelastic media. Springer, Dordrecht (NL), pp 273–298. https://doi.org/10.1007/978-94-015-8698-6_16
29. Manfredini P, Cocchetti G, Maier G, Redaelli A, Montevecchi FM (1999) Poroelastic finite element analysis of a bone specimen under cyclic loading. J Biomech 32(2):135–144. https://doi.org/10.1016/S0021-9290(98)00162-6
30. Abousleiman Y, Cui L (1998) Poroelastic solutions in transversely isotropic media for wellbore and cylinder. Int J Solids Struct 35(34–35):4905–4929. https://doi.org/10.1016/S0020-7683(98)00101-2
31. Psakhie SG, Horie Y, Korostelev SY, Smolin AY, Dmitriev AI, Shilko EV, Alekseev SV (1995) Method of movable cellular automata as a tool for simulation within the framework of physical mesomechanics. Russ Phys J 38(11):1157–1168. https://doi.org/10.1007/BF00559396
32. Shilko EV, Psakhie SG, Schmauder S, Popov VL, Astafurov SV, Smolin AYu (2015) Overcoming the limitations of distinct element method for multiscale modeling of materials with multimodal internal structure. Comput Mater Sci 102:267–285. https://doi.org/10.1016/j.commatsci.2015.02.026
33. Smolin AYu, Shilko EV, Astafurov SV, Kolubaev EA, Eremina GM, Psakhie SG (2018) Understanding the mechanisms of friction stir welding based on computer simulation using particles. Defence Technol 14(6):643–656. https://doi.org/10.1016/j.dt.2018.09.003
34. Smolin AY, Smolin IY, Shilko EV, Stefanov YP, Psakhie SG (2019) Handbook of mechanics of materials. Springer, Singapore, pp 1675–1714. https://doi.org/10.1007/978-981-10-6884-3_35
35. Psakhie SG, Smolin AY, Shilko EV, Dimaki AV (2019) Handbook of mechanics of materials. Springer, Singapore, pp 1311–1345. https://doi.org/10.1007/978-981-10-6884-3_79
36. Datasheet M (2000) Properties and processing of TIMETAL 6–4s. Titanium Metals Corporation, Dallas
37. Bonello T, Avelar-Batista Wilson JC, Housden J, Gutmanas EY, Gotman I, Matthews A, Leyland A, Cassar G (2014) Evaluating the effects of PIRAC nitrogen-diffusion treatments on the mechanical performance of Ti–6Al–4V alloy. Mater Sci Eng A 619:300–311. https://doi.org/10.1016/j.msea.2014.09.055
38. Giannakpoulos AE, Suresh S (1999) Determination of elastoplasic properties by instrumented sharp indentation. Scripta Mater 40(10):1191–1198. https://doi.org/10.1016/S1359-6462(99)00011-1
39. Eremina GM, Smolin AY (2019) Multilevel numerical model of hip joint accounting for friction in the hip resurfacing endoprosthesis. Facta Universitas Ser Mech Eng 17(1):29–38. https://doi.org/10.22190/FUME190122014E
40. Oliver W, Pharr GM (1992) An improved technique for determining hardness and elastic modulus using load and displacement sensing indentation experiments. J Mater Res 7(6):1564–1583. https://doi.org/10.1557/JMR.1992.1564
41. Avelar-Batista Wilson JC, Wu S, Gotman I, Housden J, Gutmanas EY (2015) Duplex coatings with enhanced adhesion to Ti alloy substrate prepared by powder immersion nitriding and TiN/Ti multilayer deposition. Mater Lett 157:45–49. https://doi.org/10.1016/j.matlet.2015.05.054
42. Eremina GM, Smolin AYu (2019) Numerical modeling of wearing two rough surfaces of a biocompatible ceramic coating. AIP Conf Proc 2167:020089. https://doi.org/10.1063/1.5131956
43. Attard B, Leyland A, Matthews A, Gutmanas EY, Gotman I, Cassar G (2018) Improving the surface characteristics of Ti-6Al-4V and Ti metal 834 using PIRAC nitriding treatments. Surf Coat Technol 339:208–223. https://doi.org/10.1016/j.surfcoat.2018.01.051

44. Kang J, Wang M, Yue W, Fu Z, Zhu L, She D, Wang C (2019) Tribological behavior of titanium alloy treated by nitriding and surface texturing composite technology. Materials 12(2):301. https://doi.org/10.3390/ma12020301

45. Kao WH, Su YL, Horng JH, Hsieh YT (2017) Improved tribological properties, electrochemical resistance and biocompatibility of AISI 316L stainless steel through duplex plasma nitriding and TiN coating treatment. J Biomater Appl 32(1):12–27. https://doi.org/10.1177/088532821 7712109

46. Eremina GM, Smolin AY (2019) Numerical model of the mechanical behavior of coated materials in the friction pair of hip resurfacing endoprosthesis. In: Oñate E, Wriggers P, Zohdi T, Bischoff M, Owen DRJ (eds) VI international conference on particle-based methods. Fundamentals and applications. PARTICLES 2019, 28–30 Oct 2019, CIMNE, Barcelona, pp 197–203

47. Eremina GM, Smolin AYu (2019) Numerical modeling of the mechanical behavior of hip resurfacing endoprosthesis and healthy bone. AIP Conf Proc 2167:020087. https://doi.org/10.1063/1.5131954

48. Le Pense S, Chen Y (2017) Contribution of fluid in bone extravascular matrix to strain-rate dependent stiffening of bone tissue—a poroelastic study. J Mech Behav Biomed Mater 65:90–101. https://doi.org/10.1016/j.jmbbm.2016.08.016

49. Xia Z, Kwon YM, Mehmood S, Downing C, Jurkschat K, Murray DW (2011) Characterization of metal-wear nanoparticles in pseudotumor following metal-on-metal hip resurfacing. Nanomed Nanotechnol Biol Med 7(6):674–681. https://doi.org/10.1016/j.nano.2011.08.002

50. Cheung G, Zalzal P, Bhandari M, Spelt JK, Papini M (2004) Finite element analysis of a femoral retrograde intramedullary nail subject to gait loading. Med Eng Phys 26(2):93–108. https://doi.org/10.1016/j.medengphy.2003.10.006

51. Todo M (2018) Biomechanical analysis of hip joint arthroplasties using CT-image based finite element method. J Surg Res 1:34–41. https://doi.org/10.26502/jsr.1002005

52. Gerhardt LC, Boccaccini AR (2010) Bioactive glass and glass-ceramic scaffolds for bone tissue engineering. Materials 3(7):3867–3910. https://doi.org/10.3390/ma3073867

53. Biot MA (1957) The elastic coefficients of the theory of consolidation. J Appl Mech 24:594–601

54. Detournay E, Cheng AH-D (1993) Comprehensive rock engineering: principles, practice & projects, vol 2. Elsevier, Amsterdam, pp 113–171. https://doi.org/10.1016/B978-0-08-040615-2.50011-3

55. Basniev KS, Dmitriev NM, Chilingar GV, Gorfunkle M, Nejad AGM (2012) Mechanics of fluid flow. Wiley, Hoboken. https://doi.org/10.1002/9781118533628.ch3

56. Smita TH, Huygheb JM, Cowin SC (2002) Estimation of the poroelastic parameters of bone. J Biomech 35(6):829–836. https://doi.org/10.1016/S0021-9290(02)00021-0

57. Lim TH, Hong JH (2000) Poroelastic properties of bovine vertebral trabecular bone. J Orthop Res 18(4):671–677. https://doi.org/10.1002/jor.1100180421

58. Kohles SS, Roberts JB (2002) Linear poroelastic cancellous bone anisotropy: trabecular solid elastic and fluid transport properties. J Biomech Eng 124(5):521–526. https://doi.org/10.1115/1.1503374

59. Cardoso L, Schaffler MB (2015) Changes of elastic constants and anisotropy patterns in trabecular bone during disuse-induced bone loss assessed by poroelastic ultrasound. J Biomech Eng 137(1):011008. https://doi.org/10.1115/1.4029179

60. Sandino C, McErlain DD, Schipilow J, Boyd SK (2015) The poro-viscoelastic properties of trabecular bone: a micro computed tomography-based finite element study. J Mech Behav Biomed Mater 44:1–9. https://doi.org/10.1016/j.jmbbm.2014.12.018

61. Smolin AYu, Eremina GM, Dimaki AV, Shilko EV (2019) Simulation of mechanical behaviour of the proximal femur as a poroelastic solid using particles. J Phys Conf Ser 1391:012005. https://doi.org/10.1088/1742-6596/1391/1/012005

62. Shilko EV, Dimaki AV, Smolin AYu, Psakhie SG (2018) The determining influence of the competition between pore volume change and fluid filtration on the strength of permeable brittle solids. Procedia Struct Integrity 13:1508–1513. https://doi.org/10.1016/j.prostr.2018.12.309

63. Shilko EV, Dimaki AV, Psakhie SG (2018) Strength of shear bands in fluid-saturated rocks: a nonlinear effect of competition between dilation and fluid flow. Sci Rep 8:1428. https://doi.org/10.1038/s41598-018-19843-8
64. Seeman E, Delmas PD (2006) Bone quality-the material and structural basis of bone strength and fragility. New Engl J Med 354(21):2250–2261. https://doi.org/10.1056/NEJMra053077
65. Johnell O, Kanis JA (2006) An estimate of the worldwide prevalence and disability associated with osteoporotic fractures. Osteoporos Int 17(12):1726–1733. https://doi.org/10.1007/s00198-006-0172-4
66. Judex S, Boyd S, Qin YX, Miller L, Müller R, Rubin C (2003) Combining high-resolution micro-computed tomography with material composition to define the quality of bone tissue. Curr Osteoporos Rep 1(1):11–19. https://doi.org/10.1007/s11914-003-0003-x
67. Diez-Perez A, Güerri R, Nogues X, Cáceres E, Peña MJ, Mellibovsky L, Randall C, Bridges D, Weaver JC, Proctor A, Brimer D, Koester KJ, Ritchie RO, Hansma PK (2010) Microindentation for in vivo measurement of bone tissue mechanical properties in humans. J Bone Miner Res 25(8):1877–1885. https://doi.org/10.1002/jbmr.73
68. Tomanik M, Nikodem A, Filipiak J (2016) Microhardness of human cancellous bone tissue in progressive hip osteoarthritis. J Mech Behav Biomed Mater 64:86–93. https://doi.org/10.1016/j.jmbbm.2016.07.022
69. Link TM (2012) Osteoporosis imaging: state of the art and advanced imaging. Radiology 263(1):3–17. https://doi.org/10.1148/radiol.12110462
70. Eremina GM, Smolin AYu, Shilko EV (2019) Numerical modeling of the indentation of cancellous. AIP Conf Proc 2167(1):020090. https://doi.org/10.1063/1.5131957
71. Wahlquist JA, DelRio FW, Randolph MA, Aziz AH, Heveran CM, Bryant SJ, Neu CP, Ferguson VL (2017) Indentation mapping revealed poroelastic, but not viscoelastic, properties spanning native zonal articular cartilage. Acta Biomater 64:41–49. https://doi.org/10.1016/j.actbio.2017.10.003
72. Makuch AM, Skalski KR (2018) Human cancellous bone mechanical properties and penetrator geometry in nanoindentation tests. Acta Bioeng Biomech 20(3):153–164. https://doi.org/10.5277/ABB-01176-2018-02
73. Bembey AK, Oyen ML, Bushby AJ, Boyde A (2006) Viscoelastic properties of bone as a function of hydration state determined by nanoindentation. Phil Mag 86(33–35):5691–5703. https://doi.org/10.1080/14786430600660864
74. Wang B, Chen R, Chen F, Dong J, Wu Z, Wang H, Yang Z, Wang F, Wang J, Yang X, Feng Y, Huang Z, Lei W, Liu H (2018) Effects of moisture content and loading profile on changing properties of bone micro-biomechanical characteristics. Med Sci Monit 24:2252–2258. https://dx.doi.org/10.12659%2FMSM.906910
75. Marcián P, Florian Z, Horáčková L, Kaiser J, Borák L (2017) Microstructural finite-element analysis of influence of bone density and histomorphometric parameters on mechanical behavior of mandibular cancellous bone structure. Solid State Phenom 258:362–365. https://doi.org/10.4028/www.scientific.net/SSP.258.362
76. Fan Z, Rho JY (2003) Effects of viscoelasticity and time-dependent plasticity on nanoindenttion measurements of human cortical bone. J Biomed Mater Res Part A 67:208–214. https://doi.org/10.1002/jbm.a.10027
77. Chittibabu V, Rao KS, Rao PG (2016) Factors affecting the mechanical properties of compact bone and miniature specimen test techniques: a review. Adv Sci Technol Res J 10(32):169–183. https://doi.org/10.12913/22998624/65117
78. Stansfield BW, Nicol AC, Paul JP, Kelly IG, Graichen F, Bergmann G (2003) Comparison of calculated hip joint contact forces with those measured using instrumented implants. An evaluation of a three-dimensional mathematical model of the lower limb. J Biomech 36(7):929–936. https://doi.org/10.1016/s0021-9290(03)00072-1
79. Stansfield BW, Nicol AC (2002) Hip joint contact forces in normal subjects and subjects with total hip prostheses: walking and stair and ramp negotiation. Clin Biomech 17(2):130–139. https://doi.org/10.1016/s0268-0033(01)00119-x

80. Fabbri D, Orsini R, Moroni A (2018) Stress fracture of proximal femur after hip resurfacing treated with cannulated screw. Joints 6(2):128–130. https://dx.doi.org/10.1055%2Fs-0038-166 0815

Abstract Methods on Mesoscopic Scales of Friction

Georg-Peter Ostermeyer and Andreas Krumm

Abstract In recent years, research has increasingly focused on the complex processes involved in friction contacts. Especially in tribological high-loaded contacts, characterized by the presence of contact modifying wear particles, macroscopic friction shows a surprisingly high dynamic complexity on many temporal and local scales. There are dominant effects on mesoscopic scales such as the geometric self-organization structures of the wear dust in the contact, which can significantly change the local contact surfaces. For the description and simulation of these phenomena, abstract methods have shown their effectiveness. One class of methods are cellular automata, both volume- and particle-based. The latter are in particular the Movable Cellular Automata developed by Sergey Psakhie. The scales of these discrete methods are freely selectable in wide ranges between the macro world and the atomic scale. Nevertheless, they provide reliable information on mesoscopic balances in the boundary layer and thus also on the macroscopic behavior of the tribocontact. The success of these methods is shown by the example of an automotive brake. The question of the relative insensitivity of the scales of these mesoscopic methods is examined in detail.

Keywords Dynamic friction · Dynamic equilibrium · Mesoscopic scales · Cellular automata · Particle methods · Dissipation of information

1 Introduction

Normally, the introduction to works on friction provides a quick overview or excerpt from 5000 years of history on friction. Starting with the fire generated by friction in the Stone Age to modern theories of friction. Although friction is omnipresent, it is still far from being understood well.

G.-P. Ostermeyer (✉) · A. Krumm
Institut für Dynamik und Schwingungen, Technische Universität Braunschweig, Schleinitzstraße 20, 38106 Braunschweig, Germany
e-mail: gp.ostermeyer@tu-braunschweig.de

© The Author(s) 2021
G.-P. Ostermeyer et al. (eds.), *Multiscale Biomechanics and Tribology of Inorganic and Organic Systems*, Springer Tracts in Mechanical Engineering, https://doi.org/10.1007/978-3-030-60124-9_6

This paper provides not a historical but an "engineering" view to friction as a "friction machine". Imagine you have to design a machine that generates the following:

- forces,
- heat,
- vibration,
- noise,
- particles and dust

  and in addition includes:

- surface roughening,
- surface smoothing,
- welding processes,
- modifications of the crystalline surface structure and
- chemical oxidation processes.

It is very simple to realize this kind of machine: Two solid bodies in contact, which are shifted tangentially against each other, can show all the phenomena mentioned above (see Fig. 1).

The paradox between the geometric simplicity of the machine and the complexity of the physical phenomena observed is the reason for the difficulty in describing friction comprehensively. Only in very few cases is it possible to adjust some of the processes mentioned above a priori by designing this "friction machine" or "boundary layer machine" [1].

It is interesting that people have frequently tried to describe these machines with a single scalar quantity μ [2]. This quantity is defined as the quotient of frictional and normal force.

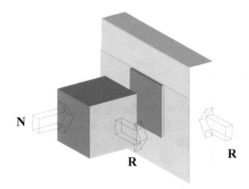

Fig. 1 Design of a friction machine

$$\mu = \frac{\left\| \vec{R} \right\|}{\left\| \vec{N} \right\|} \tag{1}$$

Measurements repeatedly confirmed that the coefficient of sliding friction depends on the speed, surface roughness, the materials involved, the ambient medium and many other parameters. It may even depend on quantities that can be explained by the measuring principle of the friction force rather than by the physics of friction. Recent investigations suggest that this coefficient is dynamic, i.e. it is only indistinctly represented by stationary measurements [3].

The following interpretation may explain why the friction coefficient is nevertheless useful. If the numerator and denominator in (1) are multiplied by the tangential velocity vector,

$$\mu = \frac{\vec{R} \cdot \vec{v}}{\left\| \vec{N} \right\| \cdot \| \vec{v} \|} \tag{2}$$

the friction coefficient can be interpreted as friction power related to a characteristic system performance [1]. Power is certainly a good measure to describe the effect of a machine. With this interpretation energetic connections to the above mentioned phenomena of a friction machine can be presented and explained. This applies in particular to heat regeneration or the averaged wear volume.

If you want to take a closer look at the physics of the machine, you have to move from the symbolic point contact (1) to the consideration of the area of the frictional contact. This is connected with a change of scale of the considered lengths and times. The scales are very different for the processes of the boundary layer machine mentioned above. They often lie between the macroscopic world of experience and the atomic scale. Consideration of mesoscopic scales have proven to be very useful for establishing connections both with macro and micro scales. To approach the meso scale, one can start from the atomic scale ("Bottom Up") or the macroscopic scale ("Top Down").

2 Bottom-Up View

On the atomic scale, with a characteristic length of 10^{-10} m, energy distribution processes can be traced by phonons and atomic movements. Today's non-equilibrium molecular dynamics programs are able to describe atomic or molecular assemblies in a thermodynamically and quantum-mechanically correct way. In principle, this method yields approximately 10^{25} equations per cm^3 of matter. The time constants of molecular dynamics are on the order of femtoseconds. With today's computers, therefore, no general statements about friction on the macroscopic scale are possible.

The transition from the atomic scale to larger length scales is combined with a drastic reduction in the number of descriptive equations of motion. Classical mechanics offers a perspective on this matter. There are different ways to transfer atomic information into the macroscopic world.

On the one hand, atomic or molecular assemblies can be made ever larger by reducing the number of particles but increasing their mass. This results in an ensemble of particles which are the core of the so-called Discrete Element Methods (DEM). If these atomic substitutes are further coarsened, only a few rigid bodies may remain. These are objects of the Multibody Systems Methods (MBS).

Another very successful way of mechanics to transform information from the atomic level to the macroscopic world is the smoothing out of the atomic structure to a massy continuum. Like the other two methods, this can be treated mathematically with very small effort. Numerical methods like finite element methods or boundary element methods (FEM, BEM) are being used.

In this way, mass-equivalent models can be created that are easy to handle mathematically. Interestingly, the force influences that are essential for the equations of motion are determined by measurement from the macroscopic world of experience alone and integrated into the models. Any phenomena for the material or motion behaviour can be approximated with arbitrary accuracy using a few generic force elements. This works quite perfectly in may situations, but fails completely in the case of friction forces.

One method for a better description of friction could be to transfer force information from the atomic scale to the macroscopic experiential using the mesoscopic particle method [4, 5].

In fact, as early as 1890, Boltzmann made the first proposals to connect the microscopic particle world with the macroscopic mechanical world [6]. Pioneering papers on this subject have been published by Greenspan [7, 8]. DEM provides excellent results for macroscopic motion and deformation dynamics, but it shows its weakness in thermodynamics. Although the thermodynamic processes can be calculated well with the MD simulation by statistical and stochastic methods, these methods cannot describe dynamic effects in the macroscopic world due to their time constants.

The approach of mesoscopic particles is to apply an atomic substitute world on a mesoscopic scale, which is below but close to the macroscopic scale (see Fig. 2). This mesoscopic world makes it possible to couple the particles to the mechanical bodies by "hidden" degrees of freedom. These additional degrees of freedom allow a correct thermodynamic description of mechanical systems even with particle worlds.

The idea is to describe the frictional system on the macroscopic time scale, but the friction boundary layer on the microscopic scale. The dynamics on the microscopic scale are hidden for the macroscopic scale, just as the atomic dynamics are hidden for the macroscopic world. It only serves to detect macroscopically observable phenomena, for example heat or wear. Thus, it is possible to describe the friction process and the associated phenomena (vibrations, heat, wear, etc.) with the mesoscopic particles with arbitrarily adjustable accuracy (see Fig. 3).

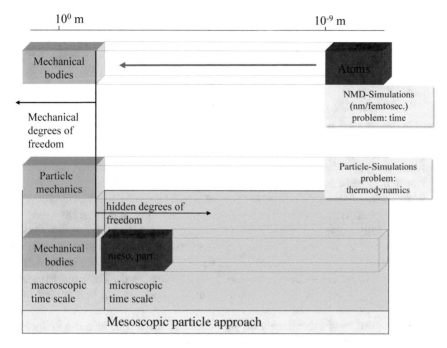

Fig. 2 Classification of the mesoscopic scale

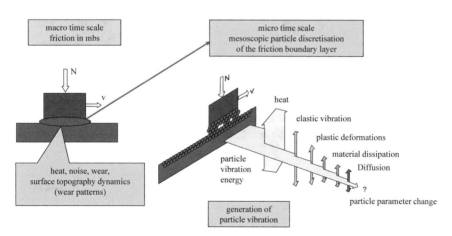

Fig. 3 Combination of macroscopic and microscopic scales in a friction system

To illustrate how this method works, a simple rod as in Fig. 4a is considered. If a hammer beats on the rod, the rod heats up. In this thought experiment the mechanical degrees of freedom of the rod are not considered, the elastic waves in the rod are not important. Therefore, the rod is only modelled with a sub-mechanical degree of

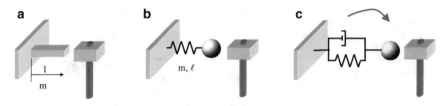

Fig. 4 **a** Heating a rod; **b** particle discretisation; **c** mesoscopic particle discretization [9]

freedom to detect heat in the rod. This sub-mechanical degree of freedom can be represented by a simple spring-mass system (see Fig. 4b).

Beaten with a hammer, the mass starts to vibrate. The system has absorbed energy from this impact. This energy can be interpreted as heat energy, the vibration itself is macroscopically irrelevant. If a second beat is given to this system, it usually leads to a violation of the thermodynamic laws, because with the second beat, the system depends essentially on the point in time at which the beat occurs, that is in which phase relation the beat and the already existing vibration stands. Thus, the vibration can be completely eliminated or the energy of the vibration can be doubled (see Fig. 5).

However, this means that the energy introduced by the second beat can no longer be interpreted as a heat gain. This energy should be added to the existing heat energy. This is exactly where the thermodynamic laws are violated.

If this mass point is, in addition to its mass property, given an internal dynamic variable, the temperature, then ultimately the sub-mechanical vibration dynamics induced by the beat can be transformed into heat. For this purpose, the vibration is dissipated by the damper, but the dissipated energy is stored in the inner variable T. In Fig. 6 this procedure is shown by means of the potential curve (blue) of the system. By the first beat a certain energy ΔE is entered into the system (red). Through the damper integrated in the system, the energy introduced as an vibration is now stored in the system as heat and the potential curve is raised to this energy level. Then, a second beat can be introduced and the sub-mechanical vibration of the system induced by this

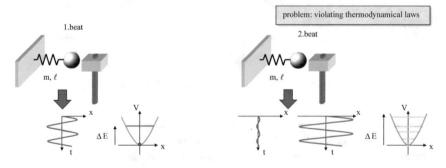

Fig. 5 Violation of the thermodynamic laws in conventional particle methods

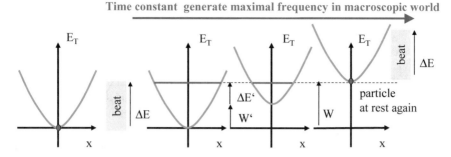

Fig. 6 Storage of thermal energy in mesoscopic particles

can be added to the existing heat energy. It should be noted that this method defines a minimum constant in terms of time, which defines the minimum distance between macroscopic impacts in order to correctly represent the thermodynamic behaviour of the system via the sub-mechanical vibrations. Such time grids can be found in all discrete systems, the grid width is a function of the typical frequencies in the grid. They limit the maximum possible frequency in the model. Every discretization method induces such grid constants. In numerical mathematics, this property is taken for granted as an essential element for the interpretation of results. However, these grid sizes can be made as small as desired by refining the sub-mechanical degrees of freedom.

The quantity of the time constant is determined by the time needed to adjust the potential curve to the correct energy level.

Mathematically this can be described by the following Lagrangian equations. The Lagrange function and the dissipation function of the system have the form

$$L = E_{kin} - V = \frac{1}{2}m\dot{x}^2 - \frac{1}{2}cx^2 \text{ and } D = \frac{1}{2}b\dot{x}^2, \tag{3}$$

where m is the mass, c the stiffness of the spring, and b the damping coefficient.

The equations of motion are as follows:

$$\frac{d}{dt}\left(\frac{\partial}{\partial \dot{x}}L\right) - \frac{\partial}{\partial x}L = -\frac{\partial}{\partial \dot{x}}D \tag{4}$$

and the temperature change is calculated as

$$\dot{T} = \frac{1}{m \cdot c_v}\dot{x}\frac{\partial}{\partial \dot{x}}D, \tag{5}$$

where c_v is the specific heat capacity of the material (per unit mass) and T is the temperature.

The inner variable T gives an extended energy integral:

$$\frac{1}{2}m\dot{x}^2 - \frac{1}{2}cx^2 + mc_vT = const. \tag{6}$$

This approach can be transferred to all possible mechanical systems. In addition to temperature, other internal variables can also be used. For example, different energy fields with different time constants are superimposed with each other. The different levels provide their macroscopically observable phenomena, and the sub-mechanical texture of the model allows the thermodynamically correct transition from one form of energy to another.

Macroscopic forces either have a zero range, for example contact or impact forces, or are described by some interaction law with a specific coordinate dependence. Forces of finite range are characteristic of atomic forces. They can be described by Lennard Jones-like potentials. In many cases, the macroscopic forces can be approximated very accurately [7].

Such Lennard–Jones potentials can also be designed temperature-dependent in such a way that transitions of the mesoscopic particles from solid to fluid or gaseous phases can be described dynamically. Figure 7 shows a metal block that is melted on a heat plate.

At the beginning of the simulation the heating plate is switched off and the particles have a defined starting temperature. After the plate is switched on, it can be observed that the temperature in the block slowly increases. As expected, the temperature of the particles in contact with the heating plate increases first. Through heat transmission, the temperature of the particles in contact with the heating plate increases. After some time it can be observed that the melting temperature of the material is reached and particles are released from the block.

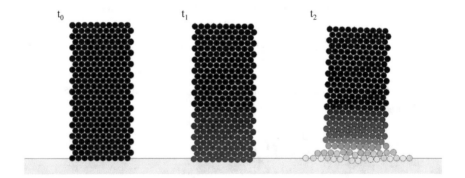

Fig. 7 A melting metal block

3 Top-Down View

Friction in the macroscopic world can also be investigated more intensively using a top-down method. This has been used intensively especially for tribological high-load contacts such as a vehicle brake. In Fig. 8, such a technical brake is shown.

Brakes show a highly dynamic friction coefficient. This is correlated with an extremely rich topographical and chemical dynamic in the boundary layer during the braking process.

Due to the complex material composition in brake pads (often more than 25 different materials), the friction process is supplied through a true a warehouse of materials in the boundary layer. Depending on the load and its duration, this process uses various chemical and physical processes which are technically designed to guarantee the quality and comfort of the brake pad. These processes are only understood to a small extend, and often appeared in a trial and error process to become the know-how of a company. In addition to chemical processes, the boundary layer shows a characteristic topography. The creation of this topography can be explained with Fig. 9.

The main chemical structure of the brake material is a relatively soft polymer matrix with embedded small and very hard particles (e.g. SiO_2 particles). All other components seem to only modify the process, which is described below and they will be ignored in this study [10]. A small section of the simplified coating with only one hard inhomogeneity is shown below.

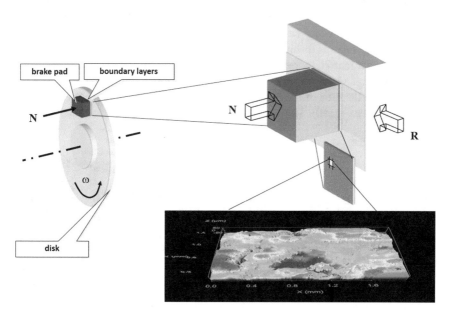

Fig. 8 Technical brake and presentation of the topography of the friction boundary layer

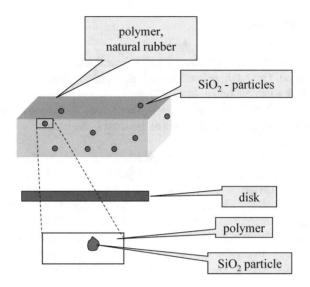

Fig. 9 Composition of a brake pad [3]

When the brake disc is pressed against the brake pad, abrasion occurs on the soft polymer matrix. The wear particles are transported along the contact zone. Some of the wear particles stay on the brake disc and after the disc has rotated, they return to the contact zone, while other particles are released into the environment [10].

This wear causes the hard particles to come to the surface of the pad. In the following, two modes may be observed. On the one hand, the flow of wear particles of the boundary layer is disturbed. On the other hand, the hard particles are pressed into the polymer matrix, as the soft matrix around the particle wears much more than the particle itself. As a result, an increase in the normal and tangential stress can be observed in the area of the hard particle. This results in an increase of the local temperature (see Fig. 10). The increase in stress and heat results in a process similar

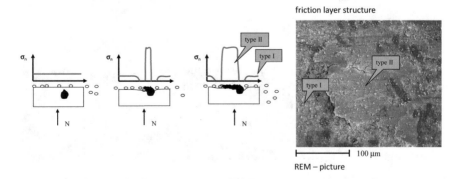

Fig. 10 Contact areas of the pad surface with normal stress distribution [10]

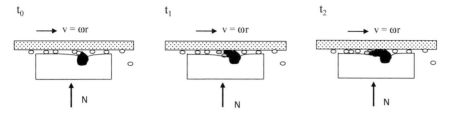

Fig. 11 Growth of contact patches [10]

to melting or sintering, whereby the wear particles together with the hard particle form an agglomeration modified by alloying processes and thin hard contact patches are formed (see Fig. 11) [10].

This process will divide the boundary layer surface into two contact areas. Type 1 is a contact area that consists of a rather soft polymer matrix that is relatively rough, heavily worn and has little frictional power. The second contact area is represented by the very hard and more wear-resistant patches (Type 2), which have a smooth surface and are mainly responsible for the main part of the friction power. They are ultimately responsible for the grey cast iron disc of the brake to wear and for iron entering the boundary layer chemistry.

These contact areas are dynamic. The patches grow due to friction power. However, their size is limited. If the total frictional power of a contact patch becomes too large, the patch cracks (see Fig. 12). The worn patch fragments represent an essential part of the wear particles of the braking process.

In the friction boundary layer, heat and wear are significant for the described dynamic process of topography change. Wear produces particles, and together with the heat the friction-intensive contact patches are born. Wear itself also ensures that the service life of these patches is limited. The total area of the patches is correlated with the current frictional power.

Friction is therefore a dynamic equilibrium. Not the friction coefficient itself, but its time derivative is defined by this process (see Fig. 13).

Dynamic friction laws can be derived from this idea [10]. Coulomb's friction law can be represented as the simplest dynamic equilibrium possible

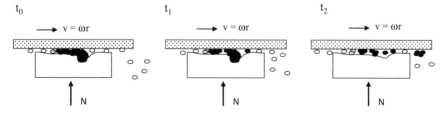

Fig. 12 Wear of contact patches [10]

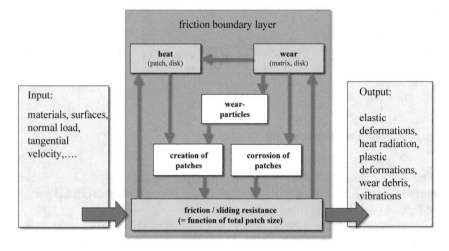

Fig. 13 Interactions in the boundary layer during the friction process [10]

$$\dot{\mu} = 0.$$

Volume-based Cellular Automata (CA) are a useful tool to simulate patch dynamics in the boundary layer (see Fig. 14). For the implementation, two areas are separated, as shown in Fig. 13. The relatively soft polymer matrix and the hard patches are discretized by cells with different normal elasticity.

In the first step, the normal stresses in the cells are calculated with a specified normal force. The contact forces in the patch areas are significantly higher. The

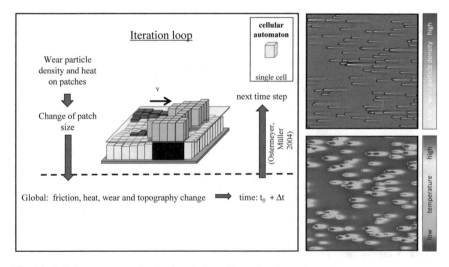

Fig. 14 Cellular automaton for the description of boundary layer dynamics

tangential speed of the brake disc above the cells is also given as well as different friction coefficients on the polymer cells and contact patch cells.

This allows a wear volume or a wear particle density to be calculated in the contact area of the polymer matrix (see Fig. 14, right). Calculating the frictional power on the patch cells the temperature can be obtained on the respective patches. This temperature uses the density of the wear particles from the matrix to determine the growth of the patches. The total dissipation over time, the temperature level, and the size control the time of patch destruction.

This simulation model creates an iterative process that maps the patch dynamics and allows a statement to be made about global friction, heat, wear and topography changes. This allows the CA to discretize the local geometry of the brake pad, to introduce local material data, to map the local chemical status of the boundary layer and to formulate local balance equations. The balance equations serve to formulate the thermodynamic, physical and chemical rules of the CA [11, 12].

The simulation model confirmed the dynamics of the surface structure and the heterogeneous heat distribution of the pad. However, the simulation also shows that this boundary layer dynamics is a very general effect of a stable and robust self-organisation process on highly loaded friction surfaces. These simulations confirmed the structure of the transient friction coefficient of Ostermeyer in [13, 10], that the temperature T and the wear as a function of the friction power according to (2) are considered in the flow balance.

$$\dot{\mu} = -\alpha\big((\beta + |v_r \cdot N|) \cdot \mu - \gamma \cdot T_p\big)$$
$$\dot{T} = -\delta\big(T_p - T_0 - \varepsilon \cdot |v_r \cdot N|\big) \tag{7}$$

4 Natural Principles of Dissipation of Information

Mesoscopic scales between the macroscopic and atomic world seem to be particularly suitable for observing locally resolved dynamics in the friction boundary layer. The core of these considerations, however, is the reduction of information that would theoretically make an exact calculation of friction on the atomic scale possible.

The problem with friction is that its characteristic dissipation on the macroscopic level does not exist on the atomic scale. It is neither the dissipation of energy nor the dissipation of material that is the core of friction, but the dissipation of information that should explain the macroscopic manifestations.

In search for general mechanisms of information reduction, one quickly finds that the mechanisms that are being used in physics are a natural way.

One of them is heat. Heat is a scalar field that collects all high-frequency effects on a molecular basis in a single parameter. Another mechanism that can be interpreted as dissipation of information is wear. Wear constantly changes the topography on an atomic scale. Therefore, it does not make sense to describe the geometric details at any time. Here, dynamic topography change rates as a scalar field of wear quantities make more sense.

Heat and wear are very good candidates to dissipate dynamic and geometric micro information, see Fig. 13, which is the basis for the dynamic friction models (7).

Certainly there are other elementary dissipation mechanisms of information which can be applied here in a useful way. Another mechanism is probably the uncertainty with which our experiences and measurements in the macroscopic world are recorded. A mathematically promising approach is offered by the Polymorphic Uncertainty Analysis.

Frictional systems can perform self-excited vibrations. During braking, they occur as an unpleasant noise, the so-called squealing. These self-excited states can be analyzed considering an eigenvalue problem. The uncertainty in the friction prediction induces an uncertainty in the stability limits of the model. Minimizing these uncertainties is necessary to enable more reliable brake designs. In Fig. 15 an uncertainty calculation has been carried out in the complex eigenvalue analysis of a simple brake model [14].

Here, Coulomb's friction law is compared to the dynamic friction law with respect to the confidence interval for stability limits in such an eigenvalue analysis. With the confidence interval, it is possible to quantify uncertainties that are present in the friction laws. If the parameters of Coulomb's friction law and the parameters of Ostermeyer's friction law are determined from concrete measured values, and the uncertainty of the friction coefficient is determined from these values, the model of the brake and the technique of complex eigenvalue analysis result in some uncertainty of the stability limits.

It is shown that the uncertainty interval for Coulomb's description is 15 times greater than for the dynamic friction law. This large difference is a consequence of the dynamic a-priori knowledge, which is represented in the flow balance conditions.

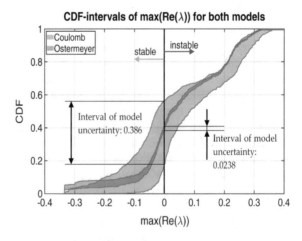

Fig. 15 Comparison of the Coulomb friction model and the Ostermeyer friction model with respect to polymorphic uncertainty and stability analysis [14]

5 Conclusion and Discussion

Friction is a highly complex and dynamic phenomenon that is still not fully understood. One of the challenges in describing friction are the manifold interactions of dynamic effects on different time and length scales. Mesoscopic methods are an excellent tool to cope with the complexity of friction.

In this paper, two different approaches are discussed, which can describe the friction process in technical brakes with higher precision. One is specific particle methods, the other is volume-based cellular automata. Both methods use the mobility on the temporal and local scales between the atomic and macroscopic scale.

More general is the method of Movable Cellular Automata, which combines the advantages of the particle world with the advantages of the world of Cellular Automata. The creator of this method, Sergey Psakhie was a visionary. His methods have found the way not only into friction, but into many other areas [15].

References

1. Ostermeyer G-P (2003) Thesen zur Modellierung von Reibung und Verschleiß. Tribol Schmierungstech 4:18–22
2. Coulomb CA (1785) Die Theorie einfacher Schwingungen. Memoires de mathematique et de physique de l'Academie des Sciences 10:161–331
3. Ostermeyer G-P (2010) Dynamic friction laws and their impact on friction induced vibrations. SAE technical paper 2010-01-1717, pp 1–27
4. Ostermeyer G-P (1999) A mesoscopic particle method for description of thermomechanical and friction processes. Phys Mesomech 6:25–32
5. Ostermeyer G-P (2007) The mesoscopic particle approach. Tribol Int 40(6):953–959
6. Boltzmann L (1897) Vorlesungen über die Prinzipe der Mechanik. Verlag von Johann Ambrosius Barth, Leipzig
7. Greenspan D (1972) A discrete numerical approach to fluid dynamics. Inf Process 71:1297–1304
8. Greenspan D (1988) Particle modeling in science and technology. In: Numerical methods, Proceedings 4th conference, Miskolc/Hung, 1986, Colloquia Mathematica Societatis Janos Bolyai 50, pp 51–66
9. Ostermeyer G-P (1996) Many particle systems. Dynamical problems in mechanical systems. In: Proceedings of the 4th Polish-German workshop, July 30–Aug 05, 1995, Polska Akademia Nauk, IPPT Warszawa, Warsaw (PL), pp 249–259
10. Ostermeyer G-P (2003) On the dynamics of the friction coefficient. Wear 254(9):852–858
11. Ostermeyer G-P, Müller M (2005) Dynamic Interaction of friction and surface topography in brake systems. Tribol Int 39(5):370–380
12. Ostermeyer G-P, Müller M (2008) New insights into the tribology of brake systems. Proc Inst Mech Eng Part D J Autom Eng 222(7):1167–1200
13. Ostermeyer G-P (2001) Friction and wear of brake systems. Forsch Ingenieurwes 66:267–272
14. Ostermeyer G-P, Müller M, Brumme S, Srisupattarawanit T (2019) Stability analysis with an NVH minimal model for brakes under consideration of polymorphic uncertainty of friction. Vibration 2:135–156
15. Psakhie SG, Horie Y, Ostermeyer G-P, Smolin AYu, Korostelev SYu, Shilko EV, Dmitriev AI, Blatnik S, Špegel M, Zavšek S (2001) Movable cellular automata method for simulating materials with mesostructured. Theoret Appl Fract Mech 37(1–3):311–334

Study of Dynamics of Block-Media in the Framework of Minimalistic Numerical Models

Alexander E. Filippov and Valentin L. Popov

Abstract One of the principal methods of preventing large earthquakes is stimulation of a large series of small events. The result is a transfer of the rapid tectonic dynamics in a creep mode. In this chapter, we discuss possibilities for such a transfer in the framework of simplified models of a subduction zone. The proposed model describes well the basic characteristic features of geo-medium behavior, in particular, statistics of earthquakes (Gutenberg Richter and Omori laws). Its analysis shows that local relatively low-energy impacts can switch block dynamics from stick–slip to creep mode. Thus, it is possible to change the statistics of seismic energy release by means of a series of local, periodic, and relatively low energy impacts. This means a principal possibility of "suppressing" strong earthquakes. Additionally, a modified version of the Burridge-Knopoff model including a simple model for state dependent friction force is derived and studied. The friction model describes a velocity weakening of friction between moving blocks and an increase of static friction during stick periods. It provides a simplified but qualitatively correct stability diagram for the transition from smooth sliding to a stick–slip behavior as observed in various tribological systems. Attractor properties of the model dynamic equations were studied under a broad range of parameters for one- and two-dimensional systems.

Keywords Earthquakes · Block-media motion · Numerical simulation · Burridge-knopoff model · Stick–slip · Phase diagram · Seismic shocks · Phase transition · Time dependent friction

A. E. Filippov (✉)
Donetsk Institute for Physics and Engineering, NASU, Donetsk 83114, Ukraine
e-mail: filippov_ae@yahoo.com

V. L. Popov
Technische Universität Berlin, 10623 Berlin, Germany
e-mail: v.popov@tu-berlin.de

© The Author(s) 2021
G.-P. Ostermeyer et al. (eds.), *Multiscale Biomechanics and Tribology of Inorganic and Organic Systems*, Springer Tracts in Mechanical Engineering, https://doi.org/10.1007/978-3-030-60124-9_7

1 Introduction

An important and interesting application of studies of spatial–temporal pattern formation of mechanical systems is the formation of geological faults and their dynamics.
Of particular interest is the study of the possibility of changing the mode of the
fault dynamics into a slow one, thus preventing strong seismic shocks. The fact that
the statistics of earthquake magnitude and their time correlations meet the laws of
Gutenberg Richter [1, 2] and Omori [1, 3], typical for self-organized critical systems
[4, 5], is often used for the conclusion that it principally occurs on all spatial scales
ranging from microscopic to continental plate scale. Therefore, it is impossible to
exert a targeted influence on the dynamics of earthquakes by local effects of limited
energy. However, the works [6–10], based both on modeling by movable cellular
automaton and full-scale experiments (on one of the active faults of the Baikal rift
zone) suggest the principal possibility of releasing the accumulated elastic energy
due to controlled low energy actions (vibration load and watering).

   Based on this idea, we develop and study a model of the behavior of contact zones
of block media and analyze the possibility of controlling the mode of displacement
as was found experimentally.

2 Mechanical Model

Minimalistic mechanical model demonstrating the time correlations typical for
systems showing self-organized criticality has been suggested by the authors. In
the conceptual form, it is shown in Fig. 1. The plate is moved by the external force of
F_{ext}. The plate is inclined by an angle that determines the ratio between the vertical
component of the force $F_{ext} \cos(\alpha)$ acting against the force of Archimedes, which
supports the plate "in magma", and the horizontal component, which results in the

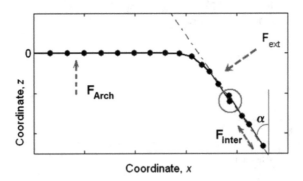

Fig. 1 Schematic diagram of the mechanical model. The force exerted by the mainland, F_{ext}, the
force of Archimedes F_{Arch} and the force of elastic interaction F_{inter} are indicated. The circle shows
the event of "bark fracture"

displacement of the entire system along the x-axis and deformation of the elastic plate.

In a numerical simulation, the plate is transformed into a set of discrete elements connected by a (nonlinear) elastic force that tries to maintain a fixed distance between them. Let us first consider the simplest, two-dimensional version of the problem. In this case, the plate is transformed into an elastic chain, and the model equations are reduced to the following form:

$$
\begin{aligned}
\partial x/\partial t &= F_{\text{inter}, x} + F_x + \xi_x; \\
\partial z/\partial t &= F_{\text{inter}, z} + F_{\text{Arh}, z} + F_z + \varepsilon_z; \\
\partial X/\partial t &= \sum_j F_x(j) + F_{\text{ext}, x} + \varepsilon.
\end{aligned}
\tag{1}
$$

Here, $F_x = F_{\text{ext}} \sin(\alpha)$ and $F_z = F_{\text{ext}} \cos(\alpha)$ are the projections of the force of the chain's interaction with the "mainland plate" (we neglect the vertical movement of the heavy mainland plate here), and the summation is made over all elements of the chain. The external force is assumed to be constant (acting on the drifting continent from the magma side). The levitation force of Archimedes $F_{\text{Arch}, z}$ is given by a condition:

$$
F_{\text{Arch}, z} = const \cdot U_0 > 0 \text{ at } z < 0 \text{ and } F_{\text{Arch}, z} = 0 \text{ at } z > 0
\tag{2}
$$

and $F_{\text{inter}, x}$ and $F_{\text{inter}, z}$, components of nonlinear elasticity between the segments of the chain, are equal:

$$
F_{\text{inter}, x} = -\partial U_{\text{inter}}/\partial x \text{ and } F_{\text{inter}, z} = -\partial U_{\text{inter}}/\partial z,
\tag{3}
$$

where the distance-dependent effective interaction potential U_{inter} looks like:

$$
U_{\text{inter}}(r) = Kr^2(1 - r^2/r_0^2).
\tag{4}
$$

Here, K is the elastic constant. To fix the distance between the elements, in the simplest, most widespread approach [11–14], the potential of the fourth order (4) is used, for which the components of the forces between the elements contain cubic nonlinearity, providing the required rigidity. At the same time, in 2D (or 3D) space, the chain (surface) can bend under the influence of the force F_{ext} and its elements move in the vertical (and/or horizontal) direction.

In numerical modeling, this leads to the following fracture condition. If the x- or z-projection of the vector connecting two consecutive segments of the chain is negative and its absolute value exceeds some threshold value, then the "fracture" occurs in this place. The fragment of the chain from its beginning to this point is subsequently removed. In the absence of resistance on the part of the removed fragment, the speed of the continental plate sharply increases, up to its deceleration by further segments of the chain. Random influences from the surrounding subsystems are included through the δ-correlated noise source:

$$\langle \xi_{xz}(t, x, z)\xi_{xz}(t', x', z')\rangle = D_1\delta_{xz}\delta(t - t')\delta(x - x')\delta(z - z');$$
$$\langle \xi(t, x, z)\xi(t', x', z')\rangle = D_2\delta(t - t')\delta(x - x')\delta(z - z'). \qquad (5)$$

Here $\delta(\ldots)$ is the impulse function of Dirac, δ_{xz} is the symbol of Kronnecker, and in each case some effective temperature can be assigned to the "diffusion coefficient" $D_{1,2} = 2k_B T_{1,2}$. The dissipative constant can be selected arbitrarily. It sets a characteristic time scale, and should be fitted a posteriori by experimental data. Random influences on the heavy mainland plate can be neglected, which leads to the assumption $D_1 \gg D_2$.

3   Statistical Properties of the Model

At a constant external force F_{ext}, the movement occurs with a constant—on the large time scales—average velocity $\langle V \rangle = dX/dt$. If we subtract Vt from the $X(t)$ curve, the fine structure of the derivative dX/dt becomes clearly visible when the chain breaks (see Fig. 2). The right side of Fig. 2 shows the distribution of the lengths of "jumps" and intervals between them.

Let us compare these distributions at different noise intensities. With exception of the expected reduction in the average interval between the jumps, the results do weakly dependend on the noise intensity up to the values comparable to the fracture threshold. Moderate noise (which does not exceed the dynamic chaos intensity of this

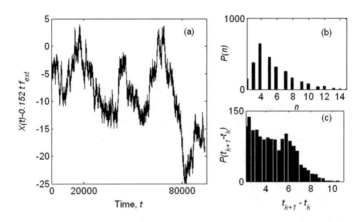

Fig. 2 The fine structure of the jumps **a** obtained by subtracting Vt and typical distributions of lengths of the chain fragments removed after fracture events (**b**) intervals between the jumps (**c**) obtained for parameter values $\alpha = \pi/6$, $F_{ext} = 10$, $U_0 = 0.1$, $K = 0.1 + 0.9\xi$

nonlinear system) only increases standard deviations. Therefore, we present below the results obtained at negligibly low noise.

Of interest is an analysis of the influence of regular (periodic) spatially localized actions on the system. We simulated the influence of sinusoidal and impulse actions of different intensity. Note that exactly the latter type of action is used in field experiments.

In the presence of noise, such actions reduce the average time between jumps. However, being spatially localized, they fix quite precisely both the time and length of each jump. This can be achieved by selecting resonant frequencies, amplitudes and force application points. Weak impact plays here the role of the trigger mechanism, provoking its own, more powerful processes in the system.

Formally, the model is able to achieve an accurate resonance optimum such that the random components are practically suppressed. In field conditions, the parameters are not as controlled as in the numerical experiment, and the histogram of jump distribution acquires additional lines near the resonant one.

4 Three-Dimensional System and Reduced Frontal Motion Model

Real tectonic systems are three-dimensional. Therefore, even for a minimalistic model, in addition to coordinates (x, z) also the y-coordinate along the edge of the fault has to be considered. In a self-consistent approximation, we can assume that the $3D$ system is composed of many equivalent $2D$ systems, which interact only through a common front of contact with the "continent". All system Eq. (1) acquire an additional index [layer number (x, z)] along the y-axis

$$\partial x(k)/\partial t = F_{inter, x}(k) + F_x(k) + \xi_x(k),$$
$$\partial z(k)/\partial t = F_{inter, z}(k) + F_{Arh, z}(k) + F_z(k) + \xi_z(k). \qquad (6)$$

The last equation of the system (1) is modified into equation:

$$\partial X(k)/\partial t = \sum_j F_x(k)(j) + F_{ext, x}(k) + K[X(k+1) + X(k-1) - 2X(k)]. \qquad (7)$$

At small deviations between the neighboring layers, they behave quasi-independently according to the $2D$ model described above. If the deviations between $X(k+1) + X(k-1)$ and $2X(k)$ increase, they are suppressed by the elastic bond $K[X(k+1) + X(k-1) - 2X(k)]$ determined by the constant K.

In the following, we will shift each following action along the front by some value so that at the present moment the small jump will be (with certain probability) provoked in another place of the front. Figure 3 illustrates the resulting propagation

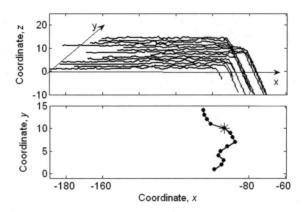

Fig. 3 The movement of the frontal area of $X(k)$ caused by periodic impacts along it

of the front on example fraction of the front. Moving the impact point along the front (asterisk in Fig. 3) generates multiple front jumps (instead of one), thus facilitating small jumps. The corresponding histograms contain some contribution of jumps, large in amplitude and time intervals between them, which, however, is much smaller than for an unperturbed system.

Taking into account the practical importance of such a problem, as well as the general scientific interest, it is useful to construct a simplified minimalist model of frontal motion, in which the connection between the layers would be taken into account in the rules of advancement of its fragments. For this purpose, let us consider a $2D$ front line in the plane (x, y), each segment of which moves forward under the action of a constant external force.

Such a minimalistic model successfully reproduces the basic properties of a more general $3D$-model, and is compact enough for large parameter studies. It is stable against varying model parameters in a very wide range. First, we checked that the change of the elastic constant by three orders of magnitude did not led to any substantial change in the distribution function $P[y(k)]$.

The method of inducing local surges described above works with the reduced model in the same manner as described above in a 3D model. In other words, it is possible to select such periodicity and distance between the local impacts that they provoke a wave of small jumps, which leads to an almost regular movement of the entire front line. The instantaneous state of this process is shown in Fig. 4. For the sake of clarity, the planar front has been chosen here as the initial condition. The area near the artificial influence is marked with a grey circle. The instantaneous position of the impact zone is shown by a dark solid circle; the spontaneous jumps caused by it are marked with bold dots inside the circle.

The numerical model above, in its minimalistic variant, was published by the authors [12] in 2006 and was used to study the dynamics of subduction zone dynamics. Later on, in 2008, the detected correlations were used by us [13] to select the optimal scenarios of other "weak and cheap" local energy effects. As a result, the

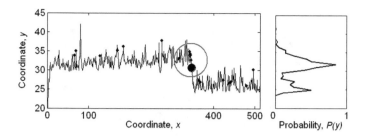

Fig. 4 The initial stage of the formation of an equilibrium distribution of probability $P[y(k)]$ in the minimalist model at the 2D front

efficiency of such impacts was significantly increased making it possible to switch the movement of blocks between "stick–slip" and creep modes. The average energy of single seismic shocks was significantly reduced and it became possible to "suppress" strong earthquakes. The model adequately described the laws of deformation for a block system and the temporal correlations typical of systems with self-organized criticality. The proposed model differs in principle from those studied previously (see, e.g., [2–11]) by taking into account the real topology of the creep of a continental platform on a thinner oceanic platform.

5 Correlation Functions

The simplicity and numerical efficiency of the model described in the previous Sections can be used to accumulate the statistics of correlations of the motion of separate blocks at the front. These correlations, in turn, can be used for selecting an optimum scheme of external action leading to the transformation of the system dynamics into a creep regime. Consider a $2D$ front in the (x, y) plane, each segment of which is moved forward by an over damped external force. Let us assume that every subsequent "tectonic" jump of each segment takes place upon its displacement by a certain distance, which is generated by a random number generator $\langle \xi(t, x, z)\xi(t', x', z')\rangle = D_2\delta(t - t')\delta(x - x')\delta(z - z')$ with zero mean value $\langle \xi(t, x, z)\rangle = 0$.

The jump magnitude ψ is also assumed to be a random quantity with zero mean value $\langle \psi(k, t)\rangle = 0$ and is set by the average weak noise intensity $D_\psi \ll D_\xi$ as follows:

$$\langle \psi(k, t)\psi(k', t')\rangle = D_\psi \delta_{kk'}\delta(t - t'). \tag{8}$$

The particular noise intensity D_ψ has to be reconstructed a posteriori from experimental data. Following article [12], we assumed that the coordinate $X(k)$ depends on the number of layers k and that different layers are coupled by the elastic force $F_{elastic} = K[X(k + 1) + X(k - 1) - 2X(k)]$ already used in Eq. (7). Other details

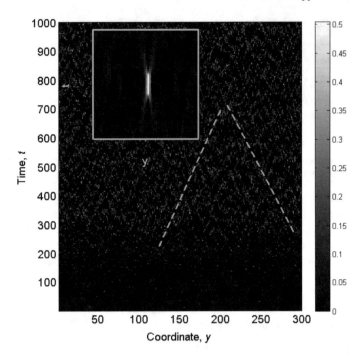

Fig. 5 Spatial–temporal map of jump distribution represented by the $M(t, x)$ matrix, with the intensities indicated by grayscale on the right. The initial stage corresponds to the establishment of a stationary process. Dashed lines correspond to the characteristic velocities of wave propagation (pre- and aftershocks). The inset shows the corresponding $G(t, x)$ correlation function

of the model are the same as described in [12] and above in this Chapter. The initial condition for the further study was selected in the form of a planar front. The resulting spatio-temporal map of jump length distribution along the front is depicted in Fig. 5.

At every step of the numerical procedure, system (1)–(3) is solved and a set of tectonic displacements $\delta X(k)$ (including zero shifts) distributed along the y-coordinate is obtained, which represents a row of the $M(t, y)$ matrix at the given time. This procedure is repeated and the entire geological history of the system is recorded in the form of the $M(t, y)$ matrix, which is represented in Fig. 5 by a gray scale map. This spatial–temporal map exhibits a clear initial transition period. A particular scenario is determined by the initial configuration (here, a planar front). As can be seen from Fig. 5, the stationary regime reveals a well-pronounced correlated character. The neighboring regions at the front interact by means of the elastic force $F_{elastic} = K[X(k + 1) + X(k - 1) - 2X(k)]$, so that large jumps of one segment induce several jumps in the neighboring segment that propagate as decaying waves in both directions from a strong local "earthquake." Arriving at a certain "weak" segment, i.e., a segment potentially close to a spontaneous break, such waves can initiate this break, inducing a new "tectonic shear" with accompanying waves and so on. In other words, each significant event in the system is surrounded by a set of pre- and aftershocks that lead to correlations in the $M(t, y)$ matrix.

The dashed lines in Fig. 5 correspond to the characteristic velocities of wave propagation. Although the velocity of these waves is not universal and varies depending on the constants in Eqs. (1)–(3) their existence is a consequence of the structure of the system under consideration. Physically, the velocity of correlation spreading depends on the composition and strength of rocks, the level of friction forces between blocks, etc.; therefore, each geo-logical region has a certain characteristic velocity, which can be determined from experimental statistics of local secondary shocks accompanying earthquakes. The correlation can be quantitatively described by a spatial–temporal correlation function $G(t - t', y - y') = \langle M(t, y)M(t', y') \rangle$ similar to that depicted in the inset to Fig. 5.

This correlation function was calculated for a particular realization of the $M(t, y)$ matrix over the $t - t_0$ time interval (beginning with the time t_0 found from the termination of the transient process). The gray scale reflects the absolute values of correlations between jumps at the front. The slopped ridges of density $G(t - t', y - y')$ on both sides of the central maximum correspond to the averaged (typical) velocities of propagation of the interacting events in the given system.

The aim of our investigations is the practical usage of a theoretically justified effect of weak controlled spatially localized impacts on a given system. Previously, we modeled the effect of periodic pulses of variable intensity and preset on/off ratios. If the action is localized in a single (x, z)-layer, this layer gradually proceeds forward and pulls the neighboring layers behind to form a protrusion on the $X(k)$ front. Then, increasing deviations $X(k+1) + X(k-1) - 2X(k)$ are sup-pressed by the elastic coupling with neighboring layers. Nevertheless, we succeeded in suggesting a strategy [12] that retained the applicability of the proposed method in a distributed system. For this purpose, each subsequent point of action was shifted over several (x, z) layers along the front so that a small jump would be initiated at a different site of the front, stimulating new neighboring regions. This shift was selected in both the 3D-model and its reduced variant.

Figure 6a illustrates such an artificially stimulated process in the same system (and same notation) as in Fig. 5. Here, the clearly distinguished straight lines correspond to periodic impacts regularly shifted along the front, which virtually completely suppress the spontaneous jumps in the systems. Unfortunately, this scenario requires large-scale preliminary works irrespective of whether the probable earthquakes will actually take place. At the same time, the correlations of spontaneous events suggest a constructive idea; it is possible to apply the artificial impacts at the sites of statistically anticipated aftershocks rather than over the entire front, thus only producing a controlled initiation of small jumps at the sites where these jumps are stimulated by intrinsic correlations.

Figure 6b shows the distribution of events caused by such a self-consistent action. This pattern appears as more densely filled with jumps as compared to that in Fig. 6a. However, the scales of jump lengths in Fig. 6a, b are also substantially different. The main consequence of this procedure is a sharp drop in the fraction of spontaneous events taking place when the system reaches the level of critical stresses.

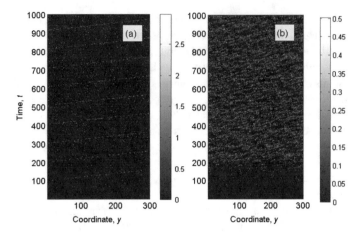

Fig. 6 Spatial–temporal map as in Fig. 5, but in the presence of artificial impacts initiating local jumps of the front segments in the case of **a** regular shift of the impact site along the entire front and **b** adaptive reaction to events selected using the $G(t, x)$ correlation function

Figure 7 shows the temporal variation of the number of such events that were not prevented by the economic adaptive scenario mentioned previously. As can be seen, the relative fraction of critical events is formally large (even reaching unity) only in the initial transient stage, where the events are not yet correlated. However, this stage is an evident artifact of the numerical procedure with a planar initial front (since the initial configuration was a priori not known). In a stationary stage, where the system attains a self-consistent regime, the fraction of spontaneous jumps not prevented by the economic adaptive action falls within 0.1–0.2. In other words, the

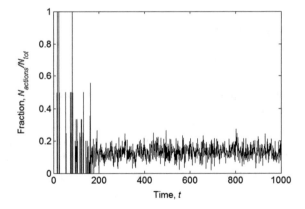

Fig. 7 Temporal variation of the fraction of spontaneous jumps not prevented by the proposed adaptive scenario (mapped in Fig. 6b). In the initial transient stage, the events are not yet correlated and the fraction of spontaneous jumps can be large (reaching unity), while in a stationary stage this fraction falls within 0.1–0.2

economic scenario allows 80–90% of spontaneous earthquakes in the system to be prevented.

6 Burridge-Knopoff (BK) Model

In addition to the above model, we also studied in [14] the well known Burridge-Knopoff (BK) model [15] initially proposed to investigate statistical properties of earthquakes. Numerical studies by Carlson et al. [16, 17] have demonstrated that the BK model can reproduce characteristic empirical features of tectonic processes such as the Gutenberg-Richter law for the magnitude distribution of earthquakes, or the Omori law for statistics of aftershocks [16–19], both properties stemming from the so called "self-organized criticality" of this system. It has been intensively used to simulate different aspects of the problem [18–38] and to discuss general properties of earthquakes statistics as well as predictability of earthquakes.

Numerical simulations give evidences that the self-organized criticality and the corresponding fractal attractor of the system is closely related to dynamic structures with "traveling waves" [21], their ordering and specific "phase transitions" [22] controlled by a number of parameters (external driving velocity, springs stiffness, number of blocks, their mutual interaction and so on). It is in particular the dependence of the dynamic properties of the BK model on the spring stiffness, which makes it necessary to introduce the generalization of the friction law proposed in this paper.

The physical reason for the stick–slip instability in the Burridge-Knopoff model is the assumed decrease of the friction force with the sliding velocity [16, 17]. Motion of a single block with this friction law is always unstable which does not correspond to properties of real tribological systems. The real law of rock friction is more complicated [26–31]. In this article, we proceed from more realistic friction laws described in [31, 32]. The main qualitative picture of a realistic law is: (a) approximately logarithmic increase of the static friction force as a function of contact time—the property, found already by Coulomb [39], and (b) a logarithmic dependence of the sliding friction on the sliding velocity. Both properties can be described in the framework of "state dependent" friction laws by introducing additional internal variables describing the state of the contact. Up to now, there were no attempts to study the dynamic and statistic properties of the Burridge-Knopoff model with a state dependent friction law.

In the paper [14], we proposed a modified version of the BK model with a state dependent friction force, reproducing in the simplest way both the velocity weakening friction and the increase of static friction with time when the block is not moving. To validate the model, we studied the stability diagram. It qualitatively reproduces typical diagrams found for almost all tribological systems. The state dependent friction law was also used in [40, 41] where an extensive numerical simulation of the one-dimensional spring-block model with such a friction law has been performed and the magnitude distribution and the recurrence-time distribution were studied.

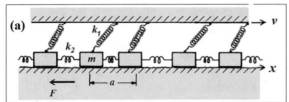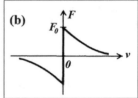

Fig. 8 Burridge-Knopoff (BK) model. Subplot **a** presents a conceptual scheme of the model. In the simulations we use $m = 1$ for the mass and $a = 1$ for the distance between the blocks. Subplot **b** reproduces the original velocity weakening stick–slip friction law used in the Carlson, Langer version of the BK model

Let us start from the original BK model, the conceptual structure of which is depicted in Fig. 8. Blocks of mass m are attached to a moving surface by springs with stiffness k_1 and are coupled to each other by springs with stiffness k_2. The moving surface has a velocity v, and the blocks are in contact with a rough substrate. The friction force F between the blocks and the rough surface is assumed to depend only on the velocity. Subplot (b) of Fig. 8 reproduces the original velocity dependence of the friction force used in Carlson and Langer version of the BK model. The equations of the BK model can be written in the following form:

$$m\frac{\partial^2 u_j}{\partial t^2} = k_2\left(u_{j+1} + u_{j-1} - 2u_j\right) + k_1\left(vt - u_j\right) - F(v_j) \qquad (9)$$

where $v = const$ is the external driving velocity and $v_j \equiv \frac{\partial u_j}{\partial t}$ is an array of individual block velocities $(j = 1, \ldots, N)$.

In the BK model, the sliding friction force is supposed to decreases monotonously from a constant initial value F_0. It is further supposed that the static friction $F(v_j \to 0)$ can possess any necessary negative value to prevent back sliding:

$$F(v_j) = \begin{cases} \frac{F_0}{1+2\alpha v_j/(1-\sigma)}, & F_0 = 1 - \sigma; \quad \partial u_j/\partial t > 0 \\ (-\infty, 0] & \partial u_j/\partial t = 0 \end{cases} \qquad (10)$$

Here, according to the original works [16, 17], the parameter α defines a rate of friction decrease when block starts to slide, and σ is the acceleration of a block at the instant when slipping begins.

7 Modified BK Model

The friction Eq. (10) is a drastic oversimplification of real properties of static and kinetic friction. This equation does not reproduce the correct stability diagram for sliding; with this friction equation the system is always unstable. Experiments show that, in most cases of dry friction, sliding stabilizes for either sufficiently large velocities or a sufficiently large stiffness of the system. These friction properties are now well understood and explained in details in the books [31, 32]. Based on friction experiments with rocks, Dieterich [26, 27] has proposed friction equations with internal variables. In his approach, the friction force depends on an additional variable that describes the state of the contact zone. This variable is, in a sense, a "melting parameter." The friction force at non-zero velocity drops down from its initial value due to a "shear melting" effect which may have various physical origins [14–17]. When the motion stops, the surfaces start to form new bonds and the static friction increases with time. These observations become especially important if the model is to be used for describing phenomena with geological characteristic times like earthquakes. Below we follow the ideology of "shear melting effect" and use additional kinetic equations for the friction force.

The dynamics of systems with state dependent friction has been investigated in a number of papers [26–34]. All these studies have been devoted to the simple one-particle version of the model. In the present paper, we investigate dynamics of the many-body BK model with a state dependent friction. The basic dynamics equations of the model are the same.

$$m\frac{\partial^2 u_j}{\partial t^2} = k_2\left(u_{j+1} + u_{j-1} - 2u_j\right) - \eta\frac{\partial u_j}{\partial t} + k_1\left(vt - u_j\right) - F_j[v_j(t)]. \quad (11)$$

However, the friction force is not a function of velocity, but is defined by the additional kinetic equation:

$$\frac{\partial F_j[v_j(t)]}{\partial t} = \beta_1\left(F_0 - F_j\right) + \beta_2 v_j, \quad \text{with } \beta_2 < 0\, v_j > 0, \quad (12)$$

$$F(v_j) = -\infty \quad v_j \leq 0. \quad (13)$$

The parameters in first of these equations β_1 and β_2 have the following physical (and geophysical) meaning. When a block starts to slide with $v_j > 0$, its friction force monotonously decreases from an initial value F_0. The general time scale of this process [in relation to other time-scales of the problem, defined by the terms of Eq. (11)] is determined by the first parameter β_1 and an effectiveness of the melting is given by a relation between the absolute value of the negative parameter $\beta_2 < 0$ and β_1.

Static friction $F(v_j \to 0)$ in second line of Eq. (12) can possess any necessary negative value to prevent back sliding, as in Eq. (10). It is the only nonlinear part of the system, which is found to be enough to create all nontrivial properties of the model. For general applicability to tribological problems Eq. (11) contains also a viscous term $\eta\partial u_j/\partial t$. Our calculations show that attractor properties are weakly influenced by the viscous term. Moreover, this influence exists only at sufficiently high velocities v_j. For generality, in the results presented below, we keep small nonzero value $\eta = 0.05$.

It should be noted that other forms of the kinetic equations for $\partial F_j/\partial t = \beta_1(F_0 - F_j) + \beta_2 v_j$ have been proposed [14, 15] leading to qualitatively similar decreasing behavior of the dynamic friction at $v_j > 0$. Equation (12) is the simplest form. It is also linear. Another reason for this choice is the following. A realistic model must reproduce the correct stick–slip and sliding behavior in appropriate parameter regions. In particular, it must reproduce the correct stability diagram on the plane $\{v, k_1\}$, which typically has an unstable region at small system velocities and small system stiffness [14, 15]. Note that the original BK model in the form of Eqs. (9) and (10) does not reproduce a stable sliding regime at small sliding velocities and large stiffness. The modified model qualitatively reproduces a typical known stability diagram for sliding friction under various conditions. We have studied the stability diagram by direct dynamic simulation of the BK-system. For this sake, we control the behavior of the complete velocity arrays $\{v_j(t)\}$ for all blocks $j = 1, \ldots, N$. To visualize the dynamics of the system, it is convenient to calculate the mean velocity:

$$\langle v \rangle \equiv \langle v_j(t) \rangle = \frac{\sum_{j=1}^{N} v_j(t)}{N}. \tag{14}$$

Unless otherwise specified, we keep the number of the blocks equal to $N = 512$. The main control parameters of the system are the driving velocity v and the stiffness k_1 of the "springs" connecting the block with the external subsystem. In order to study general properties of the model, let us vary parameters v and k_1 around the values of order of unity. Typical time dependencies of the mean velocity $\langle v \rangle$ found at free boundary conditions for three representative regions of driving velocity and stiffness constant $v = 0.7$ and $k_1 = 0.2$; $v = 0.7$ and $k_1 = 2$; $v = 2$ and $k_1 = 0.2$ are shown in Fig. 9 subplots (a–c), respectively.

In regions (b) and (c), the fluctuations of the velocity disappear after a transient period, whereas in the stick–slip region (a) the velocity fluctuates continuously. The dynamics of the friction force

$$\langle F_{fric} \rangle = \langle F_j(t) + \eta v_j(t) \rangle \tag{15}$$

is shown in the same plot.

Performing the simulation for various combinations of parameters, we have found the desired stability diagram, shown in Fig. 10, for the modified BK model in the plane $\{v, k_1\}$. The line separates the regions of sliding and stick–slip motion. Higher intensity of gray color corresponds to bigger oscillations of friction force. As discussed above, this characterizes the amplitude of the force variation at different spring stiffness k_1 and driving velocity v. If the interaction between the blocks is relatively weak, these strong variations of friction can lead, in principle, to a state in which different blocks of the chain can simultaneously be found in moving and stacked states. This observation is important for studying of the model further. Below we

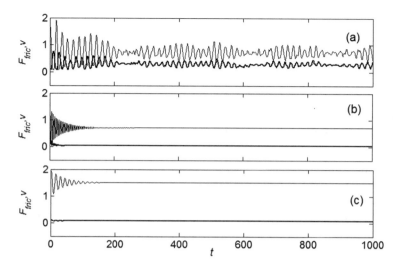

Fig. 9 Time dependencies of mean velocity $\langle v \rangle$ and friction force $\langle F_{fric} \rangle = \langle F + \eta v \rangle$ found for three representative regions of the driving velocity and stiffness constant $v = 0.7$ and $k_1 = 0.2$; $v = 0.7$ and $k_1 = 2$; $v = 2$ and $k_1 = 0.2$ shown in (**a**)–(**c**) subplots, respectively. In the sliding regions **b** and **c** the fluctuations of the velocity disappear after a transient period of initial oscillations, whereas in stick–slip region (**a**) both mean velocity and a correlated friction force remain perpetually variant

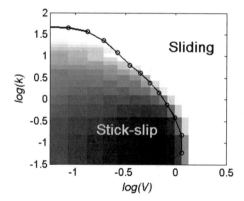

Fig. 10 Stability diagram for the modified BK model (MBK). The line separates the region of stable sliding from that of variable stick–slip motion. The grayscale map represents the standard deviation value of the complete friction force $\langle F_{fric} \rangle = \langle F_j[v_j(t)] + \eta v_j(t) \rangle$ in stationary stick–slip regime. It characterizes the amplitude of the force variation at different spring constants k and driving velocities v. Higher gray color saturation corresponds to bigger friction force oscillations (with black color corresponding to maximum of standard deviation equal to 0.72)

study the dynamic and statistical properties of the new model with the previously described realistic state dependent friction law. We will refer to the set of Eqs. (12), (13) as the modified BK model (MBK).

8 Attractor Properties, Wave State and Phase Transition in a 1-Dimensional Model

The equations of motion in (12) are a discrete representation of a nonlinear wave equation. During the last few decades various nonlinear wave equations have been widely studied, starting from the very early implementation of the numerical simulations [35]. In particular, in the context of dynamic "thermalization" it has been shown that N interacting segments of the nonlinear chain form a collective attractor with energy transfer performed by nonlinear excitations [36, 37, 42, 43]. Depending on the total energy and/or on the strength of the interaction between the blocks k_2, the chain can form (nearly) uniform or strongly non-uniform structures and phase patterns.

Analogous behavior should be expected in the MBK model as well. It is obvious that instantaneously moving blocks must be involved with stationary ones in the overall motion. This causes a "detachment" wave, propagating along the chain. Such solitary waves were found and studied in the original BK model [15], and the analogous process exists in the MBK model. There are two possible types of traveling waves: (a) after a transition time all the blocks are provoked to move simultaneously, (b) in steady-state, some of the blocks can be found instantaneously motionless while others move. One would expect that the response of the chain will depend on the strength of the interaction between the blocks. This has been studied for the original BK model in [17], and a specific "phase transition" between correlated and uncorrelated behavior has been found.

Let us study this problem for the MBK model. Figure 11 presents typical waves of local block velocities in a "stationary" regime for parameter values $k_2 = 4$, $k_1 = 1$ and $v = 0.2$. Non-zero velocities are shown here by the spatiotemporal mesh surface. Traveling and mutually scattering waves are clearly visible. Let us draw the attention to the areas of intensive local spikes appearing as result of mutually scattering of waves. These spikes have displacement amplitudes $|u_j| \gg \langle u_j \rangle$ and velocities $|v_j| \gg \langle v_j \rangle$ much higher than the mean values. They are relatively rare (the area occupied by these events is much smaller than the total space–time area), and in applications for geodynamics they should be treated as "earthquake events".

To characterize the difference of the correlated and uncorrelated behavior, an order parameter has been introduced in Ref. [22]. If we define a value h_j in block number j by the condition:

$$h_j = \begin{cases} 1 & \partial u_j/\partial t \neq 0 \text{ (the block is moving)}; \\ 0 & \partial u_j/\partial t = 0 \end{cases} \tag{16}$$

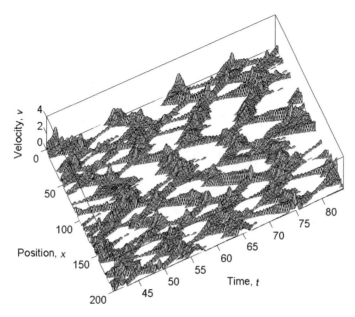

Fig. 11 Typical waves of the local velocity (of each discrete block of the system) for the developed regime in a one-dimensional system. The non-zero velocities are shown by the spatiotemporal mesh surface. The areas of intensive local events due to mutual scattering of waves are clearly visible. Parameter values: $v = 0.7$, $k_1 = 0.2$, $k_2 = 4$, $N = 512$, $\beta_1 = 1$, $\beta_2 = 25$, $\eta = 0.05$, $m = 1$.

then the local density of the order parameter H_j^* can be written as $H_j^* = h_j(h_{j+1} + h_{j-1})$. This function takes the unit value $H_j^* = h_j(h_{j+1} + h_{j-1}) = 1$ if block j is moving and exactly one of its nearest neighbors is also moving. Further, $H_j^* = h_j(h_{j+1} + h_{j-1}) = 2$ when both nearest neighbors of the moving block are in motion. All other cases yield $H_j^* = 0$.

Our observations with the MBK model show that even for blocks at vanishing inter-block interaction $k_2 \to 0$ (when the motion of neighboring blocks is almost uncorrelated), the fraction of configurations with moving sets of neighboring blocks is still relatively high. The combination $H_j^* = h_j(h_{j+1} + h_{j-1})$ does not vanish in such an uncorrelated system. It is therefore convenient to construct another simple combination:

$$H_j = h_j h_{j+1} h_{j-1} = \begin{cases} 1 & \text{all 3 blocks } j, j+1, j-1 \text{ in contact are moving} \\ 0 & \text{otherwise} \end{cases} . \quad (17)$$

This yields $H_j = h_j h_{j+1} h_{j-1} = 0$ zero in all cases except when both neighboring blocks of a central sliding block are also in motion. This combination can be used as an "order parameter".

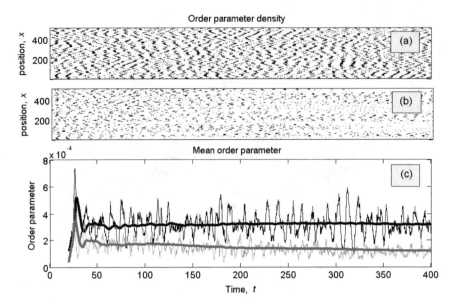

Fig. 12 Order parameter density and its time dependence in two different regimes. Subplots **a** and **b** show the grayscale maps of the order parameter density depending on time and space for strong and weak coupling between the neighboring blocks ($k_2 = 4$ and $k_2 = 1.2$ respectively). Time evolution of the ensemble-averaged order parameter for (**a**) and (**b**) cases are presented in the subplot (**c**) by black and gray tick lines respectively. The time averages for these values are shown by the bold lines of corresponding color. Parameter values: $v = 0.7$, $k_1 = 0.2$, $k_2 = 4$, $N = 512$, $\beta_1 = 1$, $\beta_2 = 25$, $\eta = 0.05$, $m = 1$.

Figure 12 shows the order parameter density and time dependence of its ensemble average $\langle H_j(t) \rangle$ in two different regimes. Subplots (a) and (b) show grayscale maps of the order parameter density depending on time and space for strong and weak coupling between neighboring blocks ($k_2 = 4$ and $k_2 = 1.2$ respectively). Time dependence of the mean value $\langle H_j(t) \rangle$ is presented in the subplot (c) with black and gray thick lines for the (a) and (b) cases, respectively.

To extract integral quantitative information about the steady ordered and disordered states let us calculate the time evolution of the ensemble-averaged order parameter:

$$H(t) \equiv \langle H_j(t) \rangle = \frac{1}{t} \int_0^t \langle H_j(t) \rangle dt. \qquad (18)$$

These time-averaged values for the cases (a) and (b) are shown in the subplot (c) by the bold lines overlapping respective thick curves $\langle H_j(t) \rangle$. The long-time stationary asymptote $\langle H_j(t) \rangle \to const$ can characterize the behavior of the system in an integral manner.

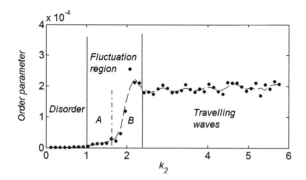

Fig. 13 Phase transition from correlated to uncorrelated motion of blocks in a one-dimensional system. The order parameter tends to a constant asymptote at high mutual interaction $k_2 \gg 1$ and vanishes below transition point $k_2 = k_2^{critical} \simeq 1$. Other parameters are the same as in Fig. 5. The ordered state corresponds to the nearly regular waves (seen clearly in Fig. 11). Two intermediate fluctuation regions A and B correspond to states with short-range and long-range correlated nonlinear excitations, respectively

The dependence of the integrated order parameter on the stiffness k_2, showing a transition from correlated to uncorrelated block motion in a one-dimensional system, is shown in Fig. 13. Two limiting cases can be identified: (a) the order parameter tends to a constant non-zero asymptote $\langle H_j(t) \rangle \to const \neq 0$ at strong interaction $k_2 \gg 1$, and (b) it vanishes $\langle H_j(t) \rangle \to 0$ below the transition point $k_2 \approx 1$.

The ordered state corresponds to the nearly regular waves presented in Fig. 12. We distinguish two fluctuation regions A and B at intermediate interaction. These regions can be characterized by two clearly different order parameter mean values in Fig. 13. One can verify further that they also differ dynamically and correspond to states with short range- and long-range correlated nonlinear excitations, respectively. This intermediate behavior may characterize the physically important features of the model under consideration.

9   Study of the 2-Dimensional Model

The real contact of two surfaces is two-dimensional. Let us generalize the MBK model for the 2D case. The generalized model is very similar to the 1-dimensional model, but incorporates a 2D array of blocks connected by elastic springs in both directions. All other components of the MBK remain unchanged.

The system of equations of motion takes the form:

$$m\frac{\partial^2 u_{j,n}}{\partial t^2} = k_2\left(u_{j+1,n} + u_{j-1,n} + u_{j,n+1} + u_{j,n-1} - 4u_{j,n}\right)$$

$$- \eta\frac{\partial u_{j,n}}{\partial t} + k_1\left(vt - u_{j,n}\right) + F(v_{j,n}); \tag{19}$$

$$\frac{\partial F_{j,n}(v_{j,n}(t))}{\partial t} = \beta_1\left(F_0 - F_{j,n}\right) + \beta_2 v_{j,n}; \text{ with } \beta_2 < 0, v_{j,n} > 0, \qquad (20)$$

$$F(v_{j,n}) = -\infty \quad v_{j,n} \leq 0$$

where: $j = 1, \ldots, N_x$ and $n = 1, \ldots, N_y$. N_x and N_y are the numbers of elements in the x- and y-directions. It is possible to repeat all the simulations of the previous section, reproduce all the results presented in Figs. 9, 10, 11, 12 and 13, and show that these properties are quite common between the 1D and 2D MBK models.

In particular, one can obtain a wave state in two dimensions. The only difficulty appears in a visualization of the results, depending on 3 coordinates $\{x, y; t\}$. As an example, Fig. 14 presents the mentioned wave state in the two-dimensional model. In contrast to the 2-dimensional $\{x; t\}$ space–time maps with a complete history of events in Figs. 11 and 12 the subplots (a) and (b) now represent instantaneous

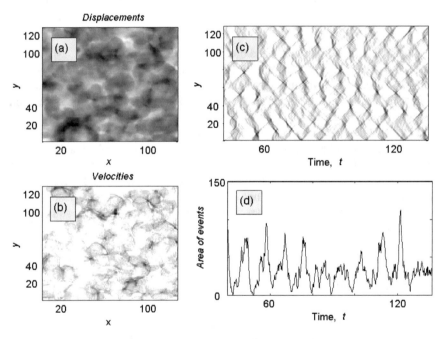

Fig. 14 Wave state in the two-dimensional model. Subplots **a** and **b** represent snapshots of the instantaneous densities of the local displacements $u = u(x, y; t)$ and velocities $v = v(x, y; t) \equiv \partial u/\partial t$, respectively. The darkest color corresponds to the value 3 indimensionless units and white corresponds to zero. Mutual scattering manifests itself in high sharp peaks of the "events" reproduced here by the dark gray spots of the corresponding densities. A time–space representation of this process is shown in (**c**) by the cross-section of the $\{x, y; t\}$-space along one of the $x = const$ planes. Ensemble averaged area of the events corresponding to the same process is plotted in subplot (**d**). The number of blocks is equal to $N_x \times N_y = 128 \times 128$ and other parameters are the same as in Fig. 12

snapshots of the density distribution (for the local displacements $u = u(x, y; t)$ and velocities $v = v(x, y; t) \equiv \partial u / \partial t$ respectively). However, direct observation of the time-dependent numerical simulations shows that the waves (clearly visible in the Fig. 14) are moving $2D$ fronts of the excitations. These fronts conserve their shape for relatively long periods of time. Their mutual scattering manifests itself in high and sharp peaks of the "events". Corresponding spikes are well reproduced by dark gray spots of the density distributions in subplots (a) and (b) of Fig. 14.

Some record of the process which has led to the presented instantaneous distributions is shown in subplot (c) by means of a cross-section of the $\{x, y; t\}$-space along one of the planes where $x = const$. One can also calculate an ensemble averaged area of the events corresponding to the same process. This is plotted in Fig. 14 subplot (d). There is obvious correlation between the subplots (c) and (d); however, the correlation is not complete. In reality, the total area of events includes a summation over all the planes $j = 1, \ldots, N_x$ and involves plenty of impacts from other $x = const$ planes which are invisible in the subplot (c).

Nevertheless, the correlation is easily seen and looks much stronger than one would expect in such case. The traveling waves influence the motion of blocks neighboring in both directions $\{x, y\}$. Therefore, there is a certain correlation between the densities along all 3 time–space coordinates $\{x, y; t\}$. In order to reproduce this in a static picture, we present a 3-dimensional density distribution for the exact same process in Fig. 15. This figure combines the density of "events" (black volumes) with the grayscale maps discreetly depicted for certain sub-planes: $t = const, x = const$ and $y = const$.

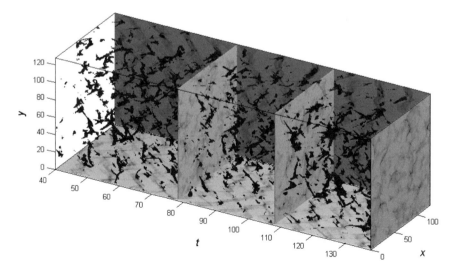

Fig. 15 Density of events (black volumes) in a two-dimensional system $N_x \times N_y = 128 \times 128$ at the same parameters as in Figs. 12 and 13. Grayscale maps (with the same gradations as in previous figure) for some representative planes $t = const, x = const$ and $y = const$ are added to compare with Fig. 13

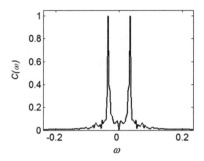

Fig. 16 Fourier transform $C(\omega)$ of the two-time correlation function $G(t_2 - t_1)$

There is a noticeable periodicity in the total area of events in Figs. 14 and 15. To illustrate this, we have calculated a two-time correlation function $G(t_2 - t_1)$ for the total area and taken its Fourier transform G_ω. The resulting Fourier transform is presented in Fig. 16. It smoothes random impacts from the time-fluctuations and possesses obvious maxima corresponding to a characteristic frequency of the total area oscillation. The frequency of these large-scale collective oscillations is determined by the parameters of the problem. According to our numerical experiments, the characteristic frequency can be varied mainly by changing the driving velocity, external springs and constants $\beta_{1,2}$ in the equation $\partial F_{j,n}[v_{j,n}(t)]/\partial t = \beta_1(F_0 - F_{j,n}) + \beta_2 v_{j,n}$. It is important to stress here that the observed behavior corresponds to a global attractor of the dynamic system (19) and (20). This means that it corresponds to the stationary asymptotic behavior of the system, independent of initial conditions.

In all the cases presented in previously mentioned figures, we omitted initial time intervals corresponding to the transient period. This part of the evolution can be different and depends on the initial conditions. We have checked this by starting the simulations from almost uniform distribution of the low velocities, from small displacements, from intensive random noise, or by changing open boundary conditions (normally used here) to the periodic ones, and so on. In all cases, the system quickly suppresses unfavorable fluctuations, vents to an attracting "large river" common for all the transient scenarios, and slowly attracts along the "river" to the stationary scenario. This kind of evolution is mathematically typical for many nonlinear systems [38] and the MBK model is no exception.

The attractor manifests itself in a stationary distribution $\rho = \rho\{u, F_{fric}, v, \ldots\}$ of the dynamic variables in a phase space. Figure 17a, b present its projections in two different sub-planes of the phase-space: planes $\{u, v\}$ and $\{F_{fric}, v\}$, respectively. By accumulating the density $\rho = \rho\{u, \ldots, v, \ldots\}$ onto a grayscale map in sub-space $\{u, v\}$, one can see the correlation between the time-depending fluctuations of displacements $u(x, y; t)$ and velocities $v(x, y; t)$.

It can be shown that the dense central part of the distribution is mainly due to multiple but weak oscillations of small amplitude ("phonons") and basal areas of the traveling waves. The widely extended depopulated gentle slopes with low density

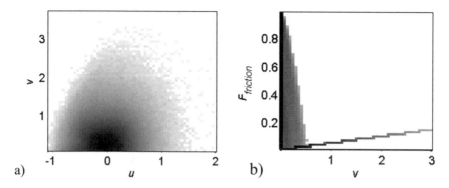

Fig. 17 a Grayscale map of the density of dissipative attractor states projected onto the phase sub-space $\{u, v\}$. **b** The same attractor as in Fig. 17a projected onto the sub-space $\{v, F_{friction}\}$, where $\langle F_{friction} \rangle = \langle F_j[v_j(t)] + \eta v_j(t) \rangle$

$\rho = \rho\{u, \ldots, v, \ldots\}$ in peripheral regions of the $\{u, v\}$ surface are produced by rare intensive "events" [which cause high spikes of displacements $u(x, y; t)$ and velocities $v(x, y; t)$]. In other words, statistical study of the rare "earthquake events" in the frame of the MBK model is equivalent to the study of the outer periphery of its dissipative attractor.

Another projection of the attractor onto the $\{F_{fric}, v\}$ subspace shown in the Fig. 17b can be used to control correct correspondence between statistically preferable behavior of the dynamically complete friction force $F_{fric} = F[v_{n,j}(t)] + \eta v_{n,j}(t)$ with the "naive", physically expected dependence $F = F_{fric}(v)$. Finally, let us return again to the discussion of Fig. 17a. The inherent structure of the attractor with extended gentle slopes of the density $\rho = \rho\{u, \ldots, v, \ldots\}$ corresponding to rare intensive "events" gives a simple and clear image for the origination of scaling asymptotic distributions. To obtain these, one must cut off the outer areas along both the displacement and velocity coordinates. Corresponding asymptotic distributions obtained after such a cut-off are reproduced in subplots (a) and (b) of Fig. 18, respectively. The inserts to the figures illustrate the power-law nature of both distributions.

Comparing the models one can conclude that the standard BK model utilizes a velocity weakening friction force to reproduce the correct statistical behavior of "events". In contrast, the MBK model includes an additional phenomenological equation, subsequently providing a self-consistent dynamic description of the velocity depending friction force. This modification has at least two advantages: it realistically generates the velocity weakening friction force of the moving blocks and provides growth of static friction for the locked blocks. The model was studied for different driving velocities and driving springs elastic constants. It was possible to build a stability diagram for the transition between smooth sliding and stick–slip behavior, which was in good qualitative agreement with what is expected experimentally. Further numerical study under a broad range of parameters proved that the MBK model reproduces all important features of the standard BK model (traveling

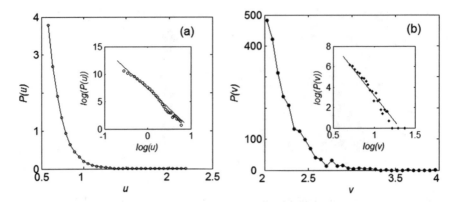

Fig. 18 Scaling relations in the model. To get scaling behavior one must cut-off the external light-gray regions with power-law density of states from the attractor in Fig. 17. Physically it corresponds to a selection of the rare but intensive "events", which is compatible with the ideology of the empirical Gutenberg-Richter law

waves, attractor properties of dynamic equations in one- and two-dimensional cases, and so on).

References

1. Gutenberg B, Richter CF (1944) Frequency of earthquakes in California. Bull Seismol Soc Am 34(4):185–188
2. Bak P, Christensen K, Danon L, Scanlon T (2002) Unified scaling law for earthquakes. Phys Rev Lett 88:178501
3. Omori F, Coll J (1895) On the aftershocks of earthquakes. J Coll Sci Imperial Univ Tokyo 7:111–120
4. Bak P, Tang Ch, Wiesenfeld K (1987) Self-organized criticality: an explanation of the 1/f noise. Phys Rev Lett 59(4):381–384
5. Carlson JM, Langer JS (1989) Properties of earthquakes generated by fault dynamics. Phys Rev Lett 62:2632
6. Ruzhich VV, Smekalin OP, Shilko EV, Psakhie SG (2002) About nature of "slow waves" and initiation of displacements at fault regions. In: Proceedings of international conference "new challenges in mesomechanics", Aalborg University, Denmark, Aug 26–30, 2002, vol 1, pp 311–318
7. Ruzhich VV, Truskov VA, Chernykh EN, Smekalin OP (1999) Russ Geol Geophys 40:356
8. Psakhie SG, Shilko EV, Astafurov SV (2004) Peculiarities of the mechanical response of heterogeneous materials with highly deformable interfaces. Tech Phys Lett 30:237–239
9. Psakhie SG, Ruzhich VV, Shilko EV, Popov VL, Dimaki AV, Astafurov SV, Lopatin VV (2005). Influence of the state of interfaces on the character of local displacements in fault-block and interfacial media. Tech Phys Lett 31(8):712–715. https://link.springer.com/article/10.1134/1.2035374
10. Ruzhich VV, Psakhie SG, Bornyakov SA, Smekalin OP, Shilko EV, Chernykh EN, Chechelnitsky VV, Astafurov SV (2002) Investigation of influence of vibroimpulse excitations on regime of displacements in seismically active fault regions. Phys Mesomech 5(5–6):85

11. Fermi E, Pasta J, Ulam S, Tsingou M (1955) Studies of non linear problems. Los Alamos Report Laboratory of the University of California, LA-1940
12. Filippov AÉ, Popov VL, Psakhie SG, Ruzhich VV, Shilko EV (2006) Converting displacement dynamics into creep in block media. Tech Phys Lett 32:545–549
13. Filippov AÉ, Popov VL, Psakhie SG (2008) Correlated impacts optimizing the transformation of block medium dynamics into creep regime. Tech Phys Lett 34(8):689–692
14. Filippov AÉ, Popov VL (2010) Modified Burridge-Knopoff model with state dependent friction. Tribol Int 43:1392–1399
15. Burridge R, Knopoff L (1967) Model and theoretical seismicity. Bull Seismol Soc Am 57(3):341–371
16. Bruce SE, Carlson JM, Langer JS (1992) Patterns of seismic activity preceding large earthquakes. J Geophys Res Solid Earth 97(B1):479
17. Carlson JM, Langer JS, Shaw BE (1994) Dynamics of earthquake faults. Rev Mod Phys 66(2):657–670
18. Saito T, Matsukawa H (2007) Size dependence of the Burridge-Knopoff model. In: Journal of physics: conference series, International conference on science of friction, 9–13 Sept 2007, Irago, Aichi, Japan, vol 89, p 012016
19. Mandelbrot BB (1982) The fractal geometry of nature. Freeman and Co, San Francisco
20. Gutenberg B, Richter CF (1954) Seismicity of the earth and associated phenomena. Princeton University Press, Princeton (New Jersey)
21. Muratov CB (1999) Traveling wave solutions in the Burridge-Knopoff model. Phys Rev E 59:3847
22. Huisman BAH, Fasolino A (2005) Transition to strictly solitary motion in the Burridge-Knopoff model of multicontact friction. Phys Rev E 72(1Pt2):016107
23. Clancy I, Corcoran D (2005) Criticality in the Burridge-Knopoff model. Phys Rev E 71(4Pt2):046124
24. Clancy I, Corcoran D (2006) Burridge-Knopoff model: exploration of dynamic phases. Phys Rev E 73(4Pt2):046115
25. Mori T, Kawamura H (2008) Simulation study of earthquakes based on the two-dimensional Burridge-Knopoff model with long-range interactions. Phys Rev E 77(5Pt1):051123
26. Dieterich JH (1974) Earthquake mechanisms and modeling. Annu Rev Earth Planet Sci 2:275–301
27. Dieterich JH (1978) Time-dependent friction and the mechanics of stick-slip. Pure Appl Geophys 116:790–806
28. Rice JR (1983) Earthquake aftereffects and triggered seismic phenomena. Pure Appl Geophys 121:187–219
29. Gu JC, Rice JR, Ruina AL, Tse ST (1984) Slip motion and stability of a single degree of freedom elastic system with rate and state dependent friction. J Mech Phys Solids 32(3):167–196
30. Marone C (1998) Laboratory-derived friction laws and their application to seismic faulting. Annu Rev Earth Planet Sci 26:643–696
31. Persson BNJ (2000) Sliding friction, physical properties and applications. Springer, Berlin
32. Popov VL (2010) Contact mechanics and friction. Foundations and applications. Springer, Berlin
33. Popov VL (2000) A theory of the transition from static to kinetic friction in boundary lubrication layers. Solid State Commun 115(7):369–373
34. Filippov AE, Klafter J, Urbakh M (2004) Friction through dynamical formation and rupture of molecular bonds. Phys Rev Lett 92(13):135503
35. Fermi E, Pasta J, Ulam S, Tsingou M (1993) In: Mattis DC (ed) The many-body problem: an encyclopedia of exactly solved models in one dimension. World Scientific Publishing Co Pte Ltd., Singapore
36. Braun OM, Hu B, Filippov A, Zeltser A (1998) Traffic jams and hysteresis in driven one-dimensional systems. Phys Rev E Stat Phys Plasmas Fluids Relat Interdisc Top 58(2):1311
37. Fillipov A, Hu B, Li B, Zeltser A (1998) Energy transport between two attractors connected by a Fermi-Pasta-Ulam chain. J Phys A Gen Phys 31(38):7719

38. Filippov AE (1994) Mimicry of phase transitions and the large-river effect. JETP Lett 60(2):141
39. Popova E, Popov VL (2015) The research works of Coulomb and Amontons and generalized laws of friction. Friction 3(2):183–190
40. Persson BNJ, Popov VL (2000) On the origin of the transition from slip to stick. Solid State Commun 114(5):261–266
41. Ohmura A, Kawamura H (2007) Rate-and state-dependent friction law and statistical properties of earthquakes. EPL (Europhys Lett) 77(6):69001
42. Ford J (1992) The Fermi-Pasta-Ulam problem: paradox turns discovery. Phys Rep 213(5):271–310
43. Lichtenberg AJ, Lieberman MA (1992) Regular and chaotic dynamics. Springer, New York

Material Transfer by Friction Stir Processing

Alexander A. Eliseev, Tatiana A. Kalashnikova, Andrey V. Filippov, and Evgeny A. Kolubaev

Abstract Mechanical surface hardening processes have long been of interest to science and technology. Today, surface modification technologies have reached a new level. One of them is friction stir processing that refines the grain structure of the material to a submicrocrystalline state. Previously, the severe plastic deformation occurring during processing was mainly described from the standpoint of temperature and deformation, because the process is primarily thermomechanical. Modeling of friction stir welding and processing predicted well the heat generation in a quasi-liquid medium. However, the friction stir process takes place in the solid phase, and therefore the mass transfer issues remained unresolved. The present work develops the concept of adhesive-cohesive mass transfer during which the rotating tool entrains the material due to adhesion, builds up a transfer layer due to cohesion, and then leaves it behind. Thus, the transfer layer thickness is a clear criterion for the mass transfer effectiveness. Here we investigate the effect of the load on the transfer layer and analyze it from the viewpoint of the friction coefficient and heat generation. It is shown that the transfer layer thickness increases with increasing load, reaches a maximum, and then decreases. In so doing, the average moment on the tool and the temperature constantly grow, while the friction coefficient decreases. This means that the mass transfer cannot be fully described in terms of temperature and strain. The given load dependence of the transfer layer thickness is explained by an increase in the cohesion forces with increasing load, and then by a decrease in cohesion due to material overheating. The maximum transfer layer thickness is equal to the feed to rotation rate ratio and is observed at the axial load that causes a stress close to the yield point of the material. Additional plasticization of the material resulting from the acoustoplastic effect induced by ultrasonic treatment slightly reduces the transfer layer thickness, but has almost no effect on the moment, friction coefficient, and temperature. The surface roughness of the processed material is found to have a similar load dependence.

A. A. Eliseev · T. A. Kalashnikova · A. V. Filippov · E. A. Kolubaev (✉)
Institute of Strength Physics and Materials Science SB RAS, 634055 Tomsk, Russia
e-mail: eak@ispms.tsc.ru

© The Author(s) 2021
G.-P. Ostermeyer et al. (eds.), *Multiscale Biomechanics and Tribology of Inorganic and Organic Systems*, Springer Tracts in Mechanical Engineering, https://doi.org/10.1007/978-3-030-60124-9_8

Keywords Friction stir welding · Ultrasound · Aluminum alloy · Transfer layer ·
Roughness · Friction coefficient

1 Introduction

Friction stir welding is a solid-state process of permanent joining, which is based on
mass transfer. A nonconsumable tool, rotating during welding, heats the two pieces of
a material to a plastic state ($\approx 0.8 \cdot T_m$) due to friction force and mixes them as it moves
along the joint line (Fig. 1). Being a nonmelting process, friction stir welding allows
joining unweldable and dissimilar materials. Weld material moves in a complex way,
undergoing not only circular but also upward and downward flow. In this case, the
material can be taken as a quasi-viscous fluid, while still remaining a solid phase.
Insufficient or excessive heat generation causes macrodefects to form in the seam,
such as voids, wormholes, lack, etc. About 80% of heat is produced due to friction
of the tool shoulders against the workpiece surface [1], and mass transfer in the bulk
material is mainly implemented by the tool pin. The combination of elevated temperatures and severe plastic deformation during welding induces a series of processes
at various stages: recovery, annealing, static and dynamic recrystallization, etc. [2].

These processes can significantly alter a structure and properties of the material.
For example, the grain structure of rolled 2024 aluminum alloy becomes equiaxial
and refines by an order of magnitude (Fig. 2). Restructuring occurs at all hierarchical levels of the material: grains, subgrains, intermetallic compounds, coherent
and semi-coherent particles of secondary phases, and lattice curvature. This feature
makes friction stir welding applicable to surface modification. In particular, friction stir processing is capable of hardening the surface due to plastic deformation,
forming composite materials by mixing particles of dissimilar materials, removing
surface defects, for example, pores in products of powder additive manufacturing,
etc., which offers promise for many industries. However, this comparatively recent
technology (developed in 1991 at the Welding Institute of the United Kingdom) has
its limitations and is still not clearly understood. In particular, deformation and heat
are combined in a complex way during welding, which makes it difficult to predict

Fig. 1 Scheme of friction
stir welding process

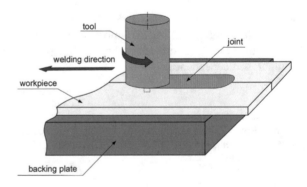

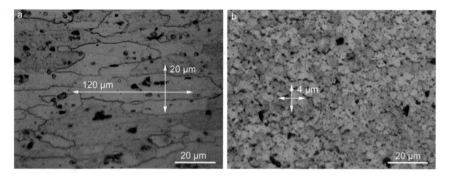

Fig. 2 Metallographic images of initial rolled sheets AA2024 (**a**) and after friction stir welding (**b**) with grain size

their effect on structure and properties. It is known from general materials science that heating coarsens grains of a solid solution and particles of secondary phases (an example is thermally hardened aluminum alloys), while deformation, due to dynamic recrystallization and strain-induced dissolution of phases, leads to grain refinement and dissolution of intermetallides in the solid solution. The influence of welding process parameters on heat generation and deformation is however nonlinear. Moreover, deformation and heat generation prove to be interrelated processes. A part of deformation energy goes into moving material macrovolumes, and a part is expended in increasing internal energy, i.e. temperature, due to internal friction. Heating, in its turn, changes viscosity of the material. Changed viscosity affects strength of the material and conditions of adhesion to the tool, i.e. deformation. To ensure the necessary temperature and deformation in practice, an optimal welding/processing mode is chosen using methods of parameter optimization, fuzzy logic, and neural networks. However, advances in the production of materials and compounds do not help to elucidate fundamental processes of friction and deformation. This work is aimed at summarizing the scientific results in this field and synthesizing them with the original experimental data.

2 Influence of Process Parameters

The main friction stir process parameters are tool rotation rate, tool feed rate, and axial load. Without adaptive adjustment during the process, the first two parameters are kept constant. As for load, the "soft" and "hard" modes are distinguished. The soft mode is characterized by constant axial load acting on the tool during its plunging and further processing/welding. In the hard mode, the tool plunge rate is set, and further processing occurs without significant axial load. The soft mode takes account of natural flow of the material around the tool. In the hard mode, the load varies greatly, making the process unstable as heat generation and deformation change at each process stage. However, the hard mode is more convenient in mass production and therefore the most common. From considerations of intensity of the impact on a material and quality of products, processing under constant axial load is favored. It also provides another way to control the process. As tool pressure, the load enters one of the classical equations of heat generation due to friction stir welding [3]:

$$Q_S = \frac{4}{3}\pi^2 \mu P \omega R_S^3 \tag{1}$$

where Q_S is the friction heat generated by the tool shoulders, μ is the friction coefficient, P is the pressure under the shoulders, ω is the tool rotation rate, and R_S is the tool shoulder radius.

This equation can be conveniently used at the specified constant friction coefficient, which is often the case. However, the friction coefficient is impossible to directly measure during processing; tribological model experiments also fail to provide accurate values as the coefficient depends on the quality of surfaces, materials, rotation rate, and temperature. Thus, the friction coefficient will change not only at different process parameters but also during welding and even along the tool pin height. The coefficient can be determined analytically and verified indirectly. In one of the first equations proposed in [4], the friction coefficient for the shoulder was expressed via the moment:

$$\mu = \frac{3M}{2F_z R_S} \tag{2}$$

where M is the measured moment, and F_z is the axial force.

The calculated friction coefficients for aluminum alloys 5182 and F-357 are very much different at the same moment (0.35–0.55 and 0.6–1.3, respectively), which is explained by different axial load. The applied load, in turn, depends on the material properties. Kumar et al. [5] used the same approach to determine the friction coefficient but with moment in place of tangential force. He found that a higher load, under parameters being equal, increases the friction coefficient (in the range 500–1500 N and 200–1400 rev/min for alloy AA7020-T6). The friction coefficient rises to 1.4 at the maximum load and rotation rate. The measured temperature correlates with the friction coefficient and amounts to 450 °C at the maximum parameters. Thus, the temperature and friction coefficient exhibit synergy. Based on this, modeling of welding/processing should take account of the temperature dependence of the friction coefficient.

The last successful attempt was described in [6]:

$$\mu = \frac{\tau_0 - \tau_1}{\left(1 - \frac{\tau_1 - \tau_0 \sin \alpha}{(1 - \sin \alpha)\tau_y}\right) P_0 (1 - \sin \alpha)} \tag{3}$$

where τ_0 are the shear stresses under the pin and shoulders, τ_1 are the shear stresses on the lateral surface of the pin, P_0 is the axial contact pressure, α is the tool pin cone angle, τ_y are the shear stresses determined using the von Mises yield criterion and Johnson–Cook material model with regard to temperature and strain rate. The experimentally determined and modeled welding temperatures in various modes differed by no more than 3%, which confirms the adequacy of the approach.

Thus, a variation in load affects heat generation, and the related temperature variation can change the friction coefficient. These processes can influence the material adhesion to the tool and cohesion in the material. According to the adhesion-cohesion mechanism of friction stir processing, changes in mass transfer are possible, which, however, have been still disregarded. Analytical evaluation of strain is usually implemented via geometry, feed and rotation rates of the tool. For example, the classical equation of strain due to friction stir welding is [7]

$$\varepsilon = \ln\left(\frac{l}{APR}\right) + \left|\ln\left(\frac{APR}{l}\right)\right| \tag{4}$$

where APR is advance per revolution, and l is the maximum deformed length. In fact, by APR is meant a material volume deformed per tool revolution without consideration for its adhesive transfer. This evaluation method may be adequate if the tool does not slip in the material, i.e., at optimal adhesion. However, a deformed volume does not conform to APR in every instance and in full measure. For a better understanding of deformation during friction stir processing, it is necessary to study the features of mass transfer.

3   Adhesion-Cohesion Concept of Mass Transfer

Mass transfer is characterized by the thickness of a transfer layer. The transfer layer is used to mean a certain material volume that sticks to the tool surface due to adhesion forces and is transferred in the rotation direction. This layer, when moving, grows in mass by capturing the surrounding material due to cohesion forces. When the critical mass is reached, the transfer layer breaks away from the tool on the trailing edge where the driving shear force of the tool and shear stresses due to the surrounding material are differently directed. A continuous material is thus formed layer by layer behind the tool. In some materials, for example, aluminum alloys, the thickness of this layer can be precisely determined at the front surface where the tool passes and in an etched longitudinal section within the material (Fig. 3).

In most works (for example in [8]), the transfer layer thickness h is expressed as the ratio of the parameters:

$$h = \frac{V}{\omega} \tag{5}$$

where V is the feed rate. Similar results are derived when introducing a marking material into the joint between the workpieces or placing it at the front surface. In

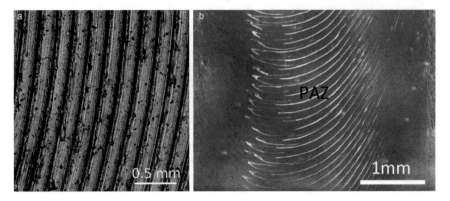

Fig. 3 Transfer layers on the front surface (**a**) and on the planar cross section in [9] (**b**)

this case, planar and cross sections demonstrate fragmented lamellae of the marking material at the boundaries of transfer layers [9]. Appearance of the marking material at the transfer layer boundary is evidence for the process of separation of materials with different densities, i.e. the vortical motion of a quasi-liquid material during processing.

However, the transfer layer thickness can be inconsistent with the given formula or can depend on conditions being neglected in it. For example, in [10] we revealed a variation in the transfer layer thickness when friction stir welding of 2024 aluminum alloy is accompanied by ultrasonic vibrations transmitted into the workpiece. Ultrasonic vibrations activate the acoustoplastic effect, which, in this context, means deformation intensification, making changes in adhesive-cohesive transfer possible. In [11] described an adhesion model based on the number of valence electrons and interatomic distance. Severe plastic deformation causes curvature of the crystal lattice and motion of electron gas [12]. Thus, any slight variation in temperature or load during friction stir welding can change transfer conditions. Under significant loads, strains, and strain rates, the material behaves differently. An example is cyclic phase transformation with intermediate amorphization [13], which can also affect the adhesive-cohesive transfer.

In [14] we performed a model experiment on ball-on-disk dry sliding friction at the ambient temperatures 25, 100, and 200 °C. Test specimens were balls made of bearing structural 100Cr6 steel 6 mm in diameter and a disk made of 5056 aluminum alloy 50 mm in diameter. The test conditions were the load acting on the ball 1 N, sliding velocity 0.5 m/min, and sliding distance 4 km. The friction force was observed to oscillate and vary in testing, which can be explained by the sticking and detaching cycles of the aluminum alloy as well as by local temperature increase. The higher was the test temperature, the greater was the friction coefficient variation. The friction coefficient decreased with sliding distance but increased on average with temperature (Fig. 4a).

The analysis of the wear surface of the steel balls showed the presence of a transfer layer in the form of separate islands of transferred material. In the longitudinal

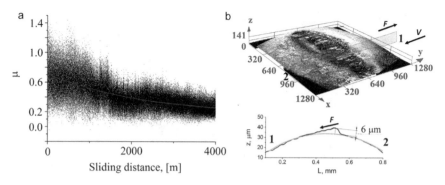

Fig. 4 Friction coefficient at sliding (**a**), and the wear scar and transfer layer on the ball bearing steel surface after sliding (**b**) [14]

section the islands had a wedge-shaped form, and the layer thickness was maximum at the trailing edge of the ball relative to its central axis (Fig. 4b). The transfer layer parameters were average thickness, maximum thickness, width of islands, and contact spot area. All the layer parameters increased with temperature. The most intensive transfer was on the specimen tested at 200 °C. The analysis of friction tracks shows the presence of aluminum particles transferred back and smoothed above the tribological layer. The mechanically stirred tribological layer had marks of plastic deformation in the form of sliding bands and curved grain boundaries. In so doing, the layer depth increased at higher test temperature. The presence of such a layer is typical for sliding friction, but bulges smoothed above it testify reverse transfer of the material from ball to disc. The analysis of the transfer layer thickness shows that the reverse process is most intensive at the ambient temperature 100 °C. At the temperature 25 °C, normal wearing with weak transfer of aluminum to the steel ball prevails, and at the temperature 200 °C direct transfer to the ball prevails. Reverse transfer is weakly pronounced in this case as in overheating the transfer layer serves as a lubricant and is uniformly distributed over the disc. In fact, these three modes take place between the tool and workpiece during friction stir processing, but underheating or overheating (with the adhesion-cohesion balance being broken) causes defects to form.

Technically, detachment of the transfer layer form the tool is as a rule incomplete during welding and should be so. After welding, there is always a certain layer of the workpiece material adherent to the tool. This occurs in welding of any materials: aluminum, titanium, copper alloys, etc. Therefore, the tool used to weld one alloy must not treat another material, unless required by the experimental condition. Moreover, a tool first time in use never demonstrates optimal processing results, even at the specified process parameters, until it is covered by a layer of the material being processed. This can be evidenced by the friction coefficient variation depending on the distance passed by the steel ball. Thus, a natural surface is formed on the tool, which provides an effective transfer of the material during friction stir welding. This layer is adherent due to diffusion, as was previously shown [15], and acts as a

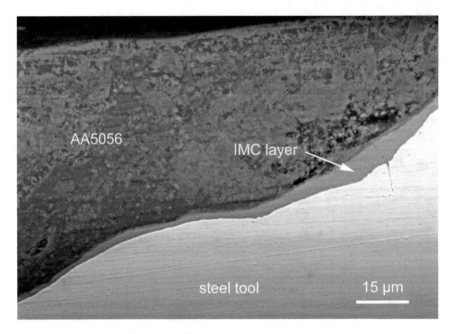

Fig. 5 Transfer layer and intermetallic compound on tool [15]

protector of the tool. In welding/processing of the aluminum alloy by the steel tool, a thin intermetallic Fe–Al layer up to 10 μm in thickness is formed at the interface due to diffusion, which is harder than the tool material (Fig. 5). The intermetallic layer is coated by the processed material, with the transfer layer sticking to it due to cohesion forces. In overheating and durable operation of the tool, diffusion of the material into the tool increases, grain-boundary diffusion occurs in the form of specific intermetallide protrusions and the tool surface fractures. Thus, the adhesive-cohesive transfer in friction stir processing is a two-step process: formation of a natural protective coating on the tool and cohesive transfer of the material by this coating.

4   Influence of Load on the Transfer Layer

In the present work, we study the dependence of the transfer layer on the axial load. Since the relation of the tool feed and rotation rates to the layer thickness is indisputable, of particular interest is load, which is rarely studied, although it can affect the heat generation and friction coefficient. The study is performed on 2024 aluminum alloy sheets 8 mm in thickness. The material is milled to remove 0.5 mm of its clad surface and then processed by the friction stir tool with the 5 mm long pin. The pin length is chosen so as not to touch the platform if the tool would sink at high loads. The processing scheme is shown in Fig. 6.

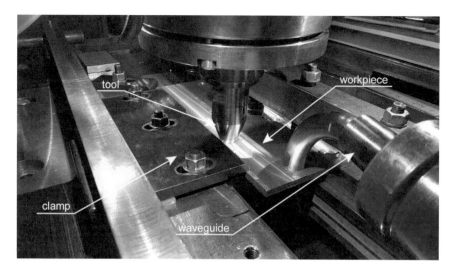

Fig. 6 Friction stir processing set up

Since the divergence of the measured transfer layer thickness from the previously calculated one is also observed at the acoustoplastic effect, friction stir processing is performed with ultrasonic assistance at the same parameters. Processing modes are given in Table 1. The ultrasonic power is 0.6 kW. Ultrasound is transmitted into the sheet at the rigid fixation at the free end. The ultrasound generator works in an adaptive mode and provides a resonant input of vibrations without attenuation throughout the workpiece length. This method is described and investigated elsewhere [16].

Polished and etched longitudinal sections cut from the center of the processed material show a typical pattern of transfer layers (Fig. 6). Due to variation of the material etchability across the thickness, the layers are of different contrast. From the front surface to the root, the layers rotate in the direction of the processing pass. Since the tool plunge depth is less than the material thickness, a layer of untreated material is observed at the root surface, with a structure corresponding to the thermomechanically affected zone and the base material. At high loads, the untreated layer becomes narrower due to a deeper penetration of the tool shoulders.

The longitudinal section clearly demonstrates that rings on the front surface correspond to the etched transfer layers, i.e. they are of the same nature, despite the fact that they are formed by the shoulders on the front surface and by the pin in the bulk. Since transfer layers on the front surface are more pronounced, they are viewed in an Olympus LEXT-OLS4100 laser scanning microscope. 3D images of friction surfaces are used to construct 2D profiles of transfer layers (Fig. 7). From the profiles it is seen that the layers are asymmetric, often with two—high and low—peaks. Asymmetry of the peaks is apparently associated with material extrusion from under the tool shoulders in the opposite direction. However, where the layer starts and how the second peak is formed are nontrivial questions that require further research. Although peak asymmetry was observed earlier, for example, by Zuo et al. [17], the morphology

Table 1 Friction stir processing modes

#	Feed rate, (mm/min)	Rotation rate (RPM)	Axial load (kg)	US	#	Feed rate (mm/min)	Rotation rate (RPM)	Axial load (kg)	US
1	90	450	2400	–	11	90	450	2400	+
2	90	450	2450	–	12	90	450	2450	+
3	90	450	2500	–	13	90	450	2500	+
4	90	450	2550	–	14	90	450	2550	+
5	90	450	2600	–	15	90	450	2600	+
6	90	450	2650	–	16	90	450	2650	+
7	90	450	2700	–	17	90	450	2700	+
8	90	450	2800	–	18	90	450	2800	+
9	90	450	2900	–	19	90	450	2900	+
10	90	450	3000	–	20	90	450	3000	+

Fig. 7 Longitudinal sections
of friction stir processed
AA2024

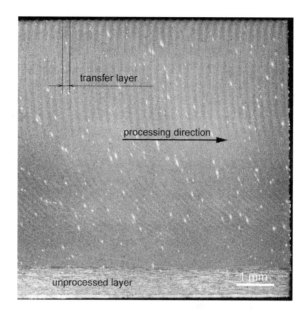

remains unexplained. Evidently, the transfer layer formation is a gradual process. A layer possibly starts at the beginning of the first peak, and ends at the subsequent valley. Assuming that a layer is formed per one revolution of the tool, the layer thickness will vary from 0 to 2π ($0°$–$360°$). Thus, the shoulders gradually build up material layers with the peak formation, and the load causes them to rotate and extend. Each transfer layer actually consists of smaller layers of the material that is spread over the entire surface of the shoulders.

Pronounced peaks in the 2D profile allow an accurate construction of the load dependence of the transfer layer thickness. The measurement results are plotted in Fig. 8. The results show that the deviations of the transfer layer thickness are quite small and are within 6 µm. However, there is a clear dependence of these

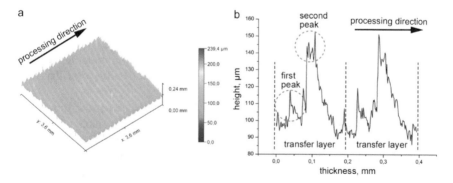

Fig. 8 Typical 3D images of the friction surface (**a**) and 2D profile of transfer layers (**b**)

deviations on the load, which was previously predicted [10]. The diagram resembles a Gaussian or parabola, and its maximum values correspond to the feed to rotation rate ratio during processing at the load 2650–2700 kg. The application of ultrasonic vibrations alters slightly the diagram, though the general pattern remains. The transfer layer thickness decreases slightly in this case (up to 3%). A decrease in the transfer layer thickness with lower or higher load relative to the maximum is associated with insufficient adhesion/cohesion force during mass transfer. At insufficient load, the adhesive-cohesive bond is weaker, and, if the load is too high, the material overheats, working as a lubricant, which reduces the reverse transfer from the tool. In terms of technology and material quality, such a small change in the transfer layer thickness will probably be of little importance, but this allows a better understanding of the fundamental processes that occur during processing/welding.

The maximum thickness of the transfer layer is observed at the load 2700 kg with and without ultrasonic treatment, which is obviously associated with the material properties. During processing, the longitudinal force is measured along the tool path. Since aluminum alloys have high thermal conductivity, this force decreases during processing due to heating of the material in front of the tool. The longitudinal force is however independent of the load and ultrasonic vibrations and amounts to 409 $\pm$ 35 kg. By relating this value to the pin projection area, we derive the longitudinal welding stress, which is equal to 99 MPa. This stress corresponds to the yield point of 2024 alloy in compression at a temperature of 450 °C [18]. A close axial stress can be obtained at the load 3000 kg related to the projected pin area (101 MPa). As mentioned above, at this load the tool shoulders become heated and sink into the workpiece. The axial stress at the load 2700 kg is 91 MPa, which is somewhat lower but still close. Thus, the largest mass transfer is observed in conditions close to the yield point of the material. A further increase in the load can lead to overheating of the material, a drop in its yield point and decrease in mass transfer.

During friction stir processing, the tool moment is also measured. The moment increases with load and ranges 15–21 Nm. Consequently, the friction force also increases. The experimentally measured moment is used to calculate the coefficient of friction of the pin by a formula similar to (2):

$$\mu = \frac{M}{F_z \overline{R_p} \sin(\alpha/2)} \tag{6}$$

where $\overline{R_p}$ is the average pin radius (4 mm), and α is the pin cone angle (30°). Geometrically, the sine of the half-cone angle is equal to the cosine of the angle between axial force and support force.

The dependence of the calculated friction coefficient on the processing time shows (Fig. 9) that an increase in the load leads to an insignificant but stable decrease in the friction coefficient, despite a rise in the moment and friction force. This means that the moment and load growth is disproportional due to changes in the friction behavior. Ultrasound impact exerts no effect on the moment and friction coefficient at any load.

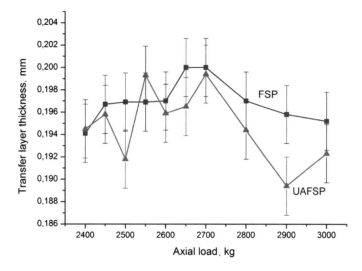

Fig. 9 The relationship between the transfer layer thickness and the axial load at friction stir processing and ultrasonic-assisted friction stir processing

This method of evaluating the friction coefficient or friction force is not quite adequate since the measured moment is a characteristic of the entire system. The evaluated value is an averaged one, but each tool region has its own moment, friction force, and friction coefficient on account of different sliding velocities at each radius. As is known, the friction coefficient depends on the speed not only in the case of viscous friction, but also during sliding. For example, in friction of a steel ball against lead or indium, the friction coefficient increases with speed, achieves a plateau, and falls again in the speed range 10^{-10} to 10 cm/s [19]. Thus, the real friction coefficient at the average pin radius can differ strongly from the evaluated one. However, the above evaluation can be used to compare technological modes at the same radius (Fig. 10).

The process temperature is measured using a FLIR 655 SC thermal imaging camera. Thermal images are taken only of the material near the tool contour, which quickly cools, so the measured temperature will be less than the real one. However, the measurement results allow an estimation of the influence of the technological mode on the heat generation. The average process temperature, from the tool penetration to its removal, is used as a factor (Table 2). An increase in the load is expected to rise the average process temperature, which is associated with a more intense heat generation. An increase in the temperature during deformation additionally plasticizes the material, resulting in a decrease in the friction coefficient.

The application of ultrasonic vibrations, on the contrary, leads to a slight decrease in temperature, which is less obvious since the ultrasonic treatment itself causes heating of the material. During friction stir welding at the slip contact of the waveguide, the material is usually heated by 10–15 °C [20]. At the rigid fixation of the

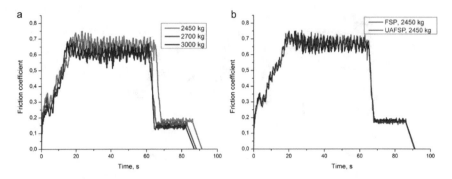

Fig. 10 The dependence of the calculated friction coefficient on the load (**a**); influence of ultrasonic impact on the calculated friction coefficient (**b**)

Table 2 The average friction stir processing temperature

#	Load (kg)	US	$\bar{t}$ (°C)
2	2450	–	216
12	2450	+	213
7	2700	–	221
17	2700	+	212
10	3000	–	227
20	3000	+	222

waveguide, the activation of the acoustoplastic effect apparently enhances the dissipation of thermal energy into the surrounding material. Nevertheless, this hardly affects the friction coefficient and the general moment.

Thus, the most effective mass transfer during friction stir processing occurs at loads that induce stresses close to the yield point of the material at the process temperature. Under these conditions, the transfer layer thickness is equal to the feed to rotation rate ratio. A reduction in the load decreases adhesion/cohesion and consequently mass transfer. An increase in the load also decreases mass transfer due to overheating of the material and reduces the friction coefficient. Intensification of deformation resulting from the acoustoplastic effect activated by ultrasound affects insignificantly the mass transfer characteristics.

5 Surface Topography and Roughness

The quality of structural components produced by friction stir welding/processing involves not only the strength characteristics of the material. An important criterion is also the surface quality of the components. The characteristic surface relief in the form of "onion rings", which is formed behind the advancing tool and is associated with the mass transfer process, can lead to undesirable consequences. The given part of the structure may be more prone to contamination, oxidation, corrosion, wear, etc. However, this issue has not yet been adequately addressed. The surface quality after welding/processing is evaluated only visually to identify macrodefects, such as shrinkage, tunnels, holes, oxidation, etc., because the fracture of an operating structure begins at these defects, if there are no other larger scale internal macrodefects [21]. Another defect is burr formation on both sides of the advancing tool, which reduces the cross-sectional area of the material in the stir zone. Burrs usually indicate that the welding/processing parameters were not properly selected, which leads, e.g., to overheating. However, the surface quality does not always imply a good quality of the joint in terms of strength. For example, as was shown in [22], an increase in the rotation rate during welding of AA5052 alloy led to a visually smoother surface, as well as to the formation of a tunnel defect. Thus, when selecting the optimal parameters, one should be guided by the quality criteria that are closest to the performance requirements.

The performance characteristics are affected not only by the presence of visible defects, but also by roughness. Visible defects on the surface are only a first approximation. With optimal parameters in terms of strength, these defects are usually absent, but the surface roughness is still pronounced. The roughness influences the fatigue characteristics and resistance to corrosion and wear [23]. In order to reduce the surface roughness, the process parameters can be further optimized within the range of previously selected optimal parameters or the surface can be post-processed. From a fundamental point of view, the surface topography and roughness can explain the mass transfer processes occurring in friction stir processing.

Within a certain range of process parameters, the roughness is significantly reduced with increasing rotation rate and decreasing feed rate, which is explained by a change in the transfer layer thickness according to Eq. (5). This was shown for 7075 aluminum alloy [17] and in dissimilar welding of A5052P-O aluminum and AZ31B-O magnesium alloys [24]. In [17] it was shown that the topography of the front surface is self-similar, and its fractal dimension linearly correlates with the roughness. However, the given regularity is not observed for all materials. For example, for friction stir processed 7075 aluminum alloy/CBA, WFA, CSA PKSA or CFA matrix composites, the dependence of the roughness on the rotation rate is unstable, up to directly proportional one, i.e., the larger the rotation rate, the higher the roughness [25]. With somewhat higher or lower parameters, the roughness dependence is also nonlinear.

The search for optimal processing parameters in terms of roughness was made for 2017 aluminum alloy [26]. The authors clearly showed that the dependence of the roughness on process parameters is not always linear, even for homogeneous materials. In particular, an increase in the rotation rate at a low feed rate can lead to material overheating and numerous overlaps, which increases the roughness. The axial load, which has not been previously investigated in the given context, also has a nonlinear effect on the roughness. For example, with a rotation rate of 900 rpm and a feed rate of 50 mm/min, an increase in the load from 500 to 1500 N led to an increase in the roughness, which obviously resulted from overheating due to large heat input.

However, with other rotation rate and feed rate values, the roughness decreased with increasing load. These results indicate that various mechanisms are involved in the surface relief formation, and a linear dependence is observed only in special cases. In this regard, an important factor in addition to the process parameters is the thermal conductivity of the material and the amount of heat generated by friction.

Note that not all materials exhibit a well-defined morphology in the form of the onion ring structure. For example, this kind of structure was not observed in friction stir processed commercially pure titanium [27]. Judging by the relief topography, mass transfer on the surface was extremely unstable. The roughness was reduced with increasing load and the surface was generally smoothed, but unlike more ductile materials the surface demonstrated overlaps and tear.

In the present work, the surface roughness of friction stir processed 2024 aluminum alloy was examined using a laser scanning microscope Olympus LEXT-OLS4100. The processing parameters are given in Table 1. Since the front surface is undulated and consists of rings (transfer layers), the roughness was analyzed for two cases: with and without subtracting the undulation (Fig. 11). In the case of subtracted undulation, the roughness is obviously lower, because it is measured with respect to a curved surface. In both cases, the load dependence of the roughness approximately resembles the load dependence of the transfer layer thickness with a maximum at 2700 kg. In general, this is consistent with the results obtained in [26]. An increase in the axial load enhances the extrusion of the material behind the tool and hence the roughness increases, but above a threshold load value it decreases. As noted earlier, the yield point of the material is reached at a 3000 kg load, which may explain surface smoothing. Without subtracting the undulation, the roughness values changed drastically within the range of 7–21 μm. The roughness may also increase due to increasing plunge depth of the tool.

The application of ultrasonic vibrations destabilized the load-roughness dependence, which points to a less uniform dependence of the transfer layer thickness. This behavior is observed only when the undulation is subtracted. Without subtracting the undulation, the roughness increases with increasing load. The above behavior of the

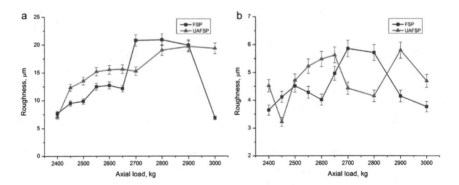

Fig. 11 Roughness of front surface without (**a**) and with subtracting the undulation (**b**)

curve generally indicates that it is inappropriate to control the roughness through the load, because its maximum is achieved at an optimal load from the viewpoint of adhesive-cohesive mass transfer in friction stir processing. But it is precisely at this load that the surface roughness decreases as a result of ultrasound application. It is currently unclear whether the result is occasional or regular, but ultrasonic assistance can be considered as a way to reduce roughness.

At present, if necessary, the surface of the component in the stir zone is smoothed by mechanical post-processing, which is considered technologically inefficient in modern production. The use of more advanced post-processing techniques does not always provide the desired result. For example, as shown in [23], laser peening and shot peening did not significantly reduce the surface roughness of the joint. The friction characteristics of various samples differed slightly, although laser peening made the weld surface less stepped. Obviously, it is not always possible to achieve a good surface quality by varying the process parameters within the admissible parameter range; the quality often depends on the material. That is why the friction stir processing technology is being improved. The surface quality can be improved by using a stationary shoulder that allows for a smoother surface [28]. Drawbacks of this approach are that when the surface is smoothed immediately behind the tool it is not always possible to visually inspect surface defects.

6 Conclusion

The friction stir welding and processing technology has greatly evolved over almost 30 years of its development. Nevertheless, the fundamental explanation of the processes can be found in the works of 15 years ago. The then proposed scientific foundations in terms of thermomechanical processes generally well explain changes in the mechanical properties and microstructure, but they do not fully predict the behavior of materials. This is due to the nonlinear dynamics of the friction stir processes. In particular, deformation and heat generation are interdependent, which is often not taken into account in analytical solutions. Moreover, the material behaves differently at such large strains and strain rates in comparison with ordinary deformation. The melting temperature rises under high loading conditions, and other material properties, e.g., fluidity, change. Thus, the process dynamics implicitly depends on process parameters. These problems are approximately solved by modeling methods, without complete understanding of the friction stir processes as they are based on adhesive-cohesive mass transfer.

The complex mass transfer pattern is due to a combination of temperature, load, friction coefficient, material properties and sliding velocity, which are also largely interdependent. Here we showed that the load increase at a constant feed rate and tool rotation rate first causes an increase in the transfer layer thickness and then a decrease. This cannot be fully explained by the change in the coefficient of friction and temperature, because the average moment and temperature increase while the friction coefficient decreases. With increasing load, the adhesive-cohesive force between the tool and the surrounding material increases, resulting in larger material transfer. The maximum thickness of the transfer layer is observed at a stress close to the yield point of the material. A further increase in the load overheats the material and impairs mass transfer. The application of ultrasonic vibrations during friction stir processing, which causes the acoustoplastic effect and amplifies the deformation but does not increase the temperature and does not change the friction coefficient, leads to a decrease in the transfer layer thickness. This is evidently due to a change in the path length of the transferred material. Thus, the mass transfer in friction stir processing depends on the fundamental physical processes of bond and structure formation in the material at a fine level. The same effects influence the surface roughness of the processed material. In general, the load dependence of the roughness resembles that of the transfer layer thickness. An increase in the load enhances the extrusion of the transfer layer material from under the tool, but in overheating the shoulders smooth the surface. The roughness changes only slightly with ultrasound application, but it was found that at the optimal load in terms of the transfer layer thickness the ultrasonic treatment reduces the roughness.

These issues concerning friction stir processing/welding are rarely discussed as they are not of acute practical interest. However, their investigation can provide better fundamental understanding of mass transfer mechanisms. Presumably, the answers can be obtained from the studies of the transfer layer structure. Of particular interest is the asymmetry of the material asperities on the front surface as well as the different etchability of aluminum alloys over the layer thickness, which will be discussed in subsequent papers.

Acknowledgements The reported study was funded by RFBR and Tomsk Oblast according to Research Project No. 19-42-700002 (the results are reported in Sections 4–5) and performed within the frame of the Fundamental Research Program of the State Academies of Sciences for 2013–2020, line of research III.23.2.4 (the results in Sections 1–3).

References

1. Schmidt HNB, Hattel JH, Wert J (2004) An analytical model for the heat generation in friction stir welding. Modell Simul Mater Sci Eng 12(1):143–157
2. Mishra RS, De PS, Kumar N (2014) Friction stir welding and processing: science and engineering. Springer, Basel
3. Frigaard Ø, Grong Ø, Midling OT (2001) A process model for friction stir welding of age hardening aluminum alloys. Metall Mater Trans A 32(5):1189–1200
4. Colligan KJ, Mishra RS (2008) A conceptual model for the process variables related to heat generation in friction stir welding of aluminum. Scripta Mater 58:327–331
5. Kumar K, Kalyan C, Kailas SV, Srivatsan TS (2009) An investigation of friction during friction stir welding of metallic materials. Mater Manuf Process 24(4):438–445
6. Meyghani B, Awang MB, Poshteh RGM, Momeni M, Kakooei S, Hamdi Z (2019) The effect of friction coefficient in thermal analysis of friction stir welding (FSW). In: IOP Conference series: materials science and engineering, vol 495, Art. 012102. Science and engineering (CUTSE) international conference, 26–28 Nov 2018, Sarawak, Malaysia

7. Long T, Tang W, Reynolds AP (2007) Process response parameter relationships in aluminium alloy friction stir welds. J Sci Technol Weld Joining 12(4):311–317
8. Schneider JA, Nunes AC Jr (2004) Characterization of plastic flow and resulting microtextures in a friction stir weld. Metall Mater Trans B 35(4):777–783
9. Liu XC, Wu CS (2015) Material flow in ultrasonic vibration enhanced friction stir welding. J Mater Process Technol 225:32–44
10. Eliseev AA, Kalashnikova TA, Gurianov DA, Rubtsov VE, Ivanov AN, Kolubaev EA (2019) Ultrasonic assisted second phase transformations under severe plastic deformation in friction stir welding of AA2024. Mater Commun 21:100660. https://doi.org/10.1016/j.mtcomm.2019.100660
11. Wills JM, Harrison WA (1984) Further studies on interionic interactions in simple metals and transition metals. Phys Rev B 29:5486–5490
12. Panin VE, Surikova NS, Lider AM, Bordulev YuS, Ovechkin BB, Khayrullin RR, Vlasov IV (2018) Multiscale mechanism of fatigue fracture of Ti-6A1-4V titanium alloy within the mesomechanical space-time-energy approach. Phys Mesomech 21(5):452–463
13. Glezer M, Metlov LS (2010) Physics of megaplastic (severe) deformation in solids. Phys Solid State 52(6):1162–1169
14. Tarasov SY, Filippov AV, Kolubaev EA, Kalashnikova TA (2017) Adhesion transfer in sliding a steel ball against an aluminum alloy. Tribol Int 115:191–198. https://doi.org/10.1016/j.triboint.2017.05.039
15. Tarasov SY, Kalashnikova TA, Kalashnikov KN, Rubtsov VE, Eliseev AA, Kolubaev EA (2015) Diffusion-controlled wear of steel friction stir welding tools used on aluminum alloys. AIP Conf Proc 1683:020228. https://doi.org/10.1063/1.4932918
16. Tarasov SY, Rubtsov VE, Fortuna SV, Eliseev AA, Chumaevsky AV, Kalashnikova TA, Kolubaev EA (2017) Ultrasonic-assisted aging in friction stir welding on Al-Cu-Li-Mg aluminum alloy. Weld World 61(4):679–690
17. Zuo L, Zuo D, Zhu Y, Wang H (2018) Effect of process parameters on surface topography of friction stir welding. Int J Adv Manuf Technol 98:1807–1816
18. Seidt JD, Gilat A (2013) Plastic deformation of 2024–T351 aluminum plate over a wide range of loading conditions. Int J Solids Struct 50:1781–1790. https://doi.org/10.1016/j.ijsolstr.2013.02.006
19. Burwell JT, Rabinowicz E (1953) The nature of the coefficient of friction. J Appl Phys 24:136–139. https://doi.org/10.1063/1.1721227
20. Shi L, Wu CS, Sun Z (2018) An integrated model for analysing the effects of ultrasonic vibration on tool torque and thermal processes in friction stir welding. Sci Technol Weld Joining 23(5):365–379. https://doi.org/10.1080/13621718.2017.1399545
21. Derazkola HA, Aval HJ, Elyasi M (2015) Analysis of process parameters effects on dissimilar friction stir welding of AA1100 and A441 AISI steel. Sci Technol Weld Joining 20(7):553–562. https://doi.org/10.1179/1362171815Y.0000000038
22. Moshwan R, Yusof F, Hassan MA, Rahmat SM (2015) Effect of tool rotational speed on force generation, microstructure and mechanical properties of friction stir welded Al–Mg–Cr–Mn (AA 5052-O) alloy. Mater Des 66:118–128. https://doi.org/10.1016/j.matdes.2014.10.043
23. Hatamleh O, Smith J, Cohen D, Bradley R (2009) Surface roughness and friction coefficient in peened friction stir welded 2195 aluminum alloy. Appl Surf Sci 255(16):7414–7426. https://doi.org/10.1016/j.apsusc.2009.04.011
24. Shigematsu I, Kwon YJ, Saito N (2009) Dissimilar friction stir welding for tailor-welded blanks of aluminum and magnesium alloys. Mater Trans 50(1):197–203. https://doi.org/10.2320/matertrans.MER2008326
25. Ikumapayi OM, Akinlabi ET (2019) Experimental data on surface roughness and force feedback analysis in friction stir processed AA7075 – T651 aluminium metal composites. Data Brief 23:103710
26. Langlade C, Roman A, Schlegel D, Gete E, Noel P, Folea M (2017) Influence of friction stir process parameters on surface quality of aluminum alloy A2017. In: MATEC web of conferences, vol 94, Art. 02006. https://doi.org/10.1051/matecconf/20179402006

27. Eliseev AA, Amirov AI, Filippov AV (2019) Influence of axial force on the pure titanium surface relief during friction stir processing. AIP Conf Proc 2167(1):020077
28. Zhou Z, Yue Y, Ji S, Li Z, Zhang L (2017) Effect of rotating speed on joint morphology and lap shear properties of stationary shoulder friction stir lap welded 6061–T6 aluminum alloy. Int J Adv Manuf Technol 88:2135–2141

Nanomaterials Interaction with Cell Membranes: Computer Simulation Studies

Alexey A. Tsukanov and Olga Vasiljeva

Abstract This chapter provides a brief review of computer simulation studies on the interaction of nanomaterials with biomembranes. The interest in this area is governed by the variety of possible biomedical applications of nanoparticles and nanomaterials as well as by the importance of understanding their possible cytotoxicity. Molecular dynamics is a flexible and versatile computer simulation tool, which allows us to research the molecular level mechanisms of nanomaterials interaction with cell or bacterial membrane, predicting in silico their behavior and estimating physicochemical properties. In particular, based on the molecular dynamics simulations, a bio-action mechanism of two-dimensional aluminum hydroxide nanostructures, termed aloohene, was discovered by the research team led by Professor S. G. Psakhie, accounting for its anticancer and antimicrobial properties. Here we review three groups of nanomaterials (NMs) based on their structure: nanoparticles (globular, non-elongated), (quasi)one-dimensional NMs (nanotube, nanofiber, nanorod) and two-dimensional NMs (nanosheet, nanolayer, nanocoated substrate). Analysis of the available in silico studies, thus can enable us a better understanding of how the geometry and surface properties of NMs govern the mechanisms of their interaction with cell or bacterial membranes.

Keywords Nanoparticle · Nanotube · Nanosheet · Lipid membrane · Molecular dynamics

A. A. Tsukanov
Skolkovo Institute of Science and Technology, Moscow 121205, Russia

A. A. Tsukanov (✉) · O. Vasiljeva
Institute of Strength Physics and Materials Science SB RAS, Tomsk 634055, Russia
e-mail: a.a.tsukanov@yandex.ru

O. Vasiljeva
e-mail: olga.vasiljeva@ijs.si

O. Vasiljeva
Jozef Stefan Institute, 1000 Ljubljana, Slovenia

© The Author(s) 2021
G.-P. Ostermeyer et al. (eds.), *Multiscale Biomechanics and Tribology of Inorganic and Organic Systems*, Springer Tracts in Mechanical Engineering, https://doi.org/10.1007/978-3-030-60124-9_9

1 Introduction

Over the past decade, the variety of different nanoparticles (NPs) and nanomaterials (NMs) were considered an important contributors to multiple medical diagnostics and the therapy applications [1, 2]. Nanomaterials can be used as devices [3], contrast drug carriers, drug delivery systems [4, 5], adjuvant, therapeutic and theranostic agents [6–8]. All these applications need an understanding of how the nanomaterial interacts with the membranes of the cell and intracellular organelles. Before the consideration of a NM-biomembrane interaction, it is important to describe what the typical cell membrane is.

Amphiphilic organic molecules, lipids, are the basic building units of the typical cell membrane [9]. Lipid consists of two parts: long hydrophobic tails and a comparatively compact hydrophilic head. Due to their composition, lipids form stable structures in water solution: e.g. liposomes, spherical and cylindrical micelles, bilayer membrane, etc. Lipid bilayer is stabilized by hydrophobic interaction between lipid tails in the inner part of the membrane, as well as by the interaction of head groups with water and with each other at the surface regions [10]. Lipid bilayer is a basis of the cell membrane.

The rapid development of multiple techniques, as well as growth and availability of high-performance computers, contribute to an increasing number of computer simulation studies of complex molecular systems, which may include sophisticated biological objects and complex nanomaterials. There are several computational approaches to study the NM-biomembrane interaction: a self-consistent field and density functional theory approach [11, 12], stochastic-elastic modeling [13, 14], all-atom [15] and coarse-grained [16] molecular dynamics, including classical unbiased and constrained or steered molecular dynamics (SMD) [17], Monte Carlo methods [18], etc.

It is worth mentioning several previously published reviews concerning the interaction of molecules and NMs with biomembrane models. Small compounds, drug molecules, biomolecules, and fullerenes interaction with the cell membranes was considered in [19, 20]. In particular, cases of fullerenes, their aggregates and derivatives impact to the membranes were summarized as, pure fullerenes C_{60} and their clusters tend to penetrate inside the lipid bilayer accumulating in the hydrophobic interior of the membrane (membrane width is several times larger than the diameter of C_{60} fullerene) [21–23]. It was noted that larger fullerenes or fullerenes in high concentrations can cause significant disturbances in the membrane structure [24–26]. However, accordingly to [27] the presence of C_{60}–C_{180} fullerenes inside the membrane with fullerene-to-lipid ratio about ~1:1 results in a mechanical strengthening of the lipid bilayer. It was also reported, that the pure C_{60}, C_{70} fullerenes can bind with ion channels, embedded into the lipid membrane, thereby affecting their structure or function [28, 29]. Fullerenes with a functionalized surface exhibit tendency to anchor their polar or charged groups either in a lipid–water interface [21, 23] or in hydrophilic-hydrophobic interface between lipid heads and tails parts [26, 30]. The review of molecular dynamics studies of small compounds permeation through the lipid bilayer as well as the interaction of proteins with the membranes can be found in [31]. Simulations of the impact of different carbon NMs as fullerenes, nanotubes, and combustion-generated carbon NPs, having an arbitrary structure, on the cell membrane are reviewed in [32]. In-depth review describing the computer modeling studies of nanomaterials interaction with the cell membranes and other different biological nano-objects can be also found in a work [33].

2 Nanoparticles

In this section, we consider papers focused on the numerical simulations of the interaction of lipid membranes with organic and inorganic nanoparticles such as, dendrimers, functionalized gold NPs, bimetallic NPs of immiscible metals, Janus NPs and abstractive nanoparticles with a surface charge.

2.1 Dendrimers and Dendritic Nanostructures

Dendrimers and dendritic nanostructures attract much attention in biochemistry, nanotechnology, and pharmaceutical sciences, due to the possibility to precise control of its size, shape, and location of functional groups [34, 35]. Dendrimers play an important role in biomedicine, as contrast agents, gene-transfection agents, and antibacterial substrates [36]. The number of branches and the size of dendrimer of a certain kind depends on its generation number G. A schematic of poly(amidoamine) dendrimers (PAMAM) G2 and a fragment of G11 dendrimers in the vicinity of the cell membrane is shown in Fig. 1. The 3D model of PAMAM G11 was built using the structural data from [37].

Lee and Larson using MD simulations showed that charged PAMAM, interacting with a cell, can induce pore formation in the lipid membrane. Herewith, this was observed at a temperature of 310 K, while at lower T = 277 K the described phenomenon did not occur, which is explained by transition of lipids into a condensed phase [38]. Moreover, it was also found that at a high ion concentration in water (about 0.5 M of NaCl) large charged dendrimers demonstrate no tendency to penetrate into the bilayer because of the screening of electrostatic interaction between dendrimer surface and lipid head groups. A simulation of PAMAM dendrimers and several copies of a peptide—poly-L-lysine (PLL) near the membrane showed that the ability to disturb the membrane is dependent on the generation number of the dendrimer, and hence on its size, as well as on the dendrimers concentrations near the membrane surface [39, 40]. Large dendrimers in a high concentration are capable to induce significant distortion of the bilayer structure, as well as the formation and stabilization of pores in the membrane. Kelly and co-workers performed MD simulations of neutral, positive, and negative dendrimers with the membrane in implicit water model [41]. It was shown that charged PAMAM dendrimers of third generation (G3)

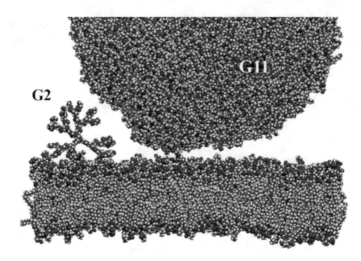

Fig. 1 Schematic of PAMAM dendrimers with generation numbers G2 and G11 near a lipid bilayer. Colors: carbon—grey, hydrogen—white, oxygen—red, nitrogen—blue, phosphorus—brown

more intensively interact with the membrane surface than neutral dendrimers. The interaction of charged dendrimers with the membrane in a liquid-phase state may also be accompanied by a hydrophobic interaction between the dendrimer interior and the lipid tail groups [42]. The permeation of the charged dendrimers through the lipid membrane under an elastic tension was studied using the coarse-grained MD simulations in [43]. The obtained results showed that the elastic tension enhances the permeability of the membrane for the charged PAMAM dendrimers of G3, G5 and G7 generation. Ting and Wang, using a self-consistent field theory, showed that in the absence of the lateral tension membrane prefer to wrap the charged dendrimer partially [44]. An increase in the dendrimer charge density increases the extent and stability of the wrapped state. But in a case of slightly tensioned membrane large charged nanoparticles as G5 (or higher) dendrimers can induce the formation of metastable pores in a lipid bilayer whereas for G3 dendrimers the pores are unstable.

2.2 Abstractive Nanoparticles

An abstractive NP means a model particle, which is not representing a certain material with a defined chemical composition, but mimics the NP with defined properties such as shape, size, charge, hydrophilic-hydrophobic balance etc.

In order to find out how the size and charge of nanoparticles affect their interaction with the cell membrane, Ginzburg and Balijepalli [45] used the self-consistent field/density functional theory of block copolymer/nanoparticle mixtures proposed in [11, 12]. The results revealed that a neutral nanoparticle of a diameter 16–32 Å penetrates into the bilayer, forming a single-layer hybrid micelle in which lipids orient

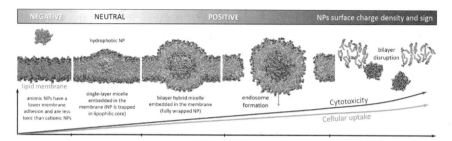

Fig. 2 A generalized scheme of the interaction of charged and neutral NPs with a cell membrane, combining the results from [45] and [49]. Anionic NP is less likely to translocate across the cell membrane than cationic NP. Hydrophobic NPs can integrate into the membrane, forming a single-layer embedded micelle. Cationic NPs having high enough surface charge density intensively participate in endosomes formation and cellular uptake. Strongly charged cationic NPs may generate pores in the bilayer, causing the overt membrane disruption

their hydrophobic tails toward the nanoparticle (see Fig. 2). If the NP charge density increases, a hybrid bilayer micelle is formed in the membrane. Further increase in the charge density or/and size of NP causes separation of the endosome, containing the nanoparticle, which may be followed by the loss of membrane integrity (Fig. 2).

The interaction of hydrophobic and semihydrophilic NPs with the membranes was studied in [46]. Using the coarse-grained MD simulations it was shown that a hydrophobic nanoparticle with a radius of about 5 nm can easily penetrate into a lipophilic bilayer interior, while a semihydrophilic NPs remain on the lipid membrane surface. An amphiphilic nanoparticle with Janus structure, consisting of both a hydrophilic and a hydrophobic part, can exhibit more complex behavior. A coarse-grained MD study of the interaction of a Janus nanoparticles with a cell membrane were conducted by Alexeev et al. [47]. The considered Janus NPs have a size comparable to the bilayer thickness. The capability of such nanoparticles to stabilize pores in the lipid membrane was found. Stabilized nanopores can be opened at relatively low elastic stress applied to the membrane, thereby providing permeability of membranes to water, ions and other compounds. Thus, using Janus NPs, it is possible to control membrane permeability altering its tension, which may be caused by changing local environmental conditions such as temperature or pH.

Besides the physicochemical properties of the NP surface, the shape and local curvature of the surface play an important role in NP interaction with a cell membrane. Yang and Ma investigated the translocation of NP with different shapes and sizes using coarse-grained model [48]. It was found that the shape anisotropy and the initial orientation of a particle are decisive for the character of its interaction with the lipid membrane. The penetrability of NP through a lipid bilayer is determined by the contact area between the particle and bilayer, and by the local surface curvature of the particle at the point of contact. The increase in curvature facilitates the translocation. It was also reported that NP volume indirectly affects penetration to a lesser extent.

The influence of the elastic lateral tension on the character of membrane interaction with abstractive dendrimer-like soft nanoparticles, having different generation number is investigated in [50]. It was found that the way, in which the membrane interacts with soft particles, depends on both the value and the sign of membrane surface tension. The researchers defined three typical phases of interaction of a soft NP and the membrane: penetration at high positive tension, penetration and partial wrapping at low positive tension, and full wrapping at low negative tension.

In addition, the interaction of multi-molecular complex comprising a membrane-soluble outer shell and nanoporous core with the cell membrane can be considered in this subsection. Studying the membrane-NP interaction, Carr et al. proposed a possible route for forming a synthetic ion channel in the cell membrane by embedding a supramolecular complex of dimethyldioctadecylammonium (DODA) surfactant capsule with a porous polyoxomolybdate (POM) particle into the lipid bilayer [51]. The POM nanoparticle has a strong negative charge (-72 e) that makes their embedment in the membrane impossible (Fig. 2). Carr and colleagues using coarse-grained MD showed that the mixed capsule of amphiphilic cations DODA and POPC lipids, formed around the POM nanoparticle, facilitates the NP embedment into the membrane. The positive charge of the detergent liposomal structure partially screens the negative POM charge, and the further liposome fusion with the lipid bilayer leads to the embedding and stabilization of NP in the membrane center.

2.3 Metallic Nanoparticles

Gold nanoparticles (AuNP) have a wide range of possible biomedical applications [52] that explains the great attention paid to studies of the interaction of AuNP with the membrane, including in silico approaches. The interaction of functionalized gold nanoparticles with electroneutral and negatively charged lipid membranes was investigated in [49]. The considered AuNPs were functionalized with charged and/or hydrophobic ligands. Cationic ammonium groups and anionic carboxylate groups were used to provide AuNP with a positive or negative surface charge, respectively. Different ratios of charged/hydrophobic ligands coated gold core to make different surface charge densities of the NPs. The results of coarse-grained computer simulations showed that the AuNP may spontaneously adhere to the membrane surface or penetrate into the bilayer. The way they interact with membrane depends on both sign and density of NPs surface charge (Fig. 2). Using SMD simulations it was found that the approach of anionic AuNP to the negatively charged membrane is complicated by electrostatic repulsion. In both cases of the neutral and negative membranes, anionic NPs have free energy minima near bilayer surface in the adsorbed state. The free energy profile for gold NPs with hydrophobic ligands only reach minimal values inside a lipophilic bilayer region. In case of neutral lipid membrane, the energetically more preferable configuration for cationic AuNP (with 70% of charged ligands) as well as for anionic one is the adsorbed state without any significant bilayer distortions. Interaction of AuNP, having cationic ligands, with the negatively charged membrane is more intensive, strong binding, and immersion of NP were observed, which cause a large deformation of bilayer as well as a formation of hydrated region within the membrane. The free energy profile has a minimum value corresponding to NP position inside the lipid head groups region. Increasing the surface charge density of positively charged AuNP enhances the ability of such particles for cellular uptake; however, beginning with a certain surface charge, severe deformation and disintegration of the cell membrane take place (Fig. 2). It was pointed out that there is a range of surface charges of cationic NPs in which a balance between cellular uptake and cytotoxicity may be attained [49].

The influence of the shape and surface functionalization of gold nanoparticles on the character of its interaction with a negatively charged membrane was investigated in [53]. Based on the coarse-grained simulations, the estimates of the free energy barriers and translocation rate constants were obtained depending on the nanoparticle shape and charge density. It was shown that anionic NPs were electrostatically repelled from the membrane surface and their translocation through the bilayer is less probable in comparison with cationic NPs. Furthermore, shape anisotropy may result in the reorientation rotations of the charged NPs in the contact region with membrane, thus distorting the lipids self-assembly and possibly causing cytotoxic effect. For the studied cases, it was also found that translocation rate constants may differ 60 orders of magnitude, in spite of equal sizes of NPs [53].

It is important that both the sign and the density of NP surface charge can be tuned by grafting of the ligands with certain charge, polarity and hydrophilic-hydrophobic balance. Recent theoretical research of Professor S. G. Psakhie's scientific group, devoted to bimetallic nanoparticles of immiscible or partially miscible metals and their interaction with bacterial and cell membranes, showed that the electrostatic properties of NP surface can be tuned without grafting of the charged ligands [54]. Using the embedded atom method (EAM) [55, 56] and MD simulations it was shown that the surface of bimetallic Ag–Cu NP is formed by silver atoms independent on nanoparticle composition ($Ag_{70}Cu_{30}$ or $Cu_{70}Ag_{30}$) due to lower surface energy of Ag (Fig. 3a). Furthermore, due to different electronegativity of Ag and Cu atoms, silver and copper, in average, will have different partial charges. Using the density functional theory (DFT) [57, 58], the mean values of partial atomic charges of Ag and Cu metals in small Ag–Cu clusters were estimated as a function of Ag-to-Cu ratio in the nanocluster. Based on this result it was concluded that magnitude of the surface charge density of considered bimetallic Ag–Cu NP is adjustable in the range from -2.9 to -7.3 e/nm^2, by the changing of Ag-to-Cu ratio from 7/3 to 3/7. Moreover, choosing metals by taking into account both the difference in their electronegativity and the difference in their surface energies, it is possible to synthesize bimetallic NP having a given permanent surface charge density of a certain sign [54]. In addition, using the force field based SMD simulations, in which metal atoms are uncharged, the free energy of interaction of $Ag_{70}Cu_{30}$, $Cu_{70}Ag_{30}$ and $Cu_{70}Ag_{30}O_4$ nanoparticles of a diameter about 40 Å with bacterial and cell membranes were estimated. It was found, that lipopolysaccharide (LPS-DPPE) membrane adsorbs the bimetallic NPs on the membrane-water interface (Fig. 3b), whereas lipid bilayer membrane demonstrates the tendency to wrap (or partially wrap) pure $Ag_{70}Cu_{30}$, $Cu_{70}Ag_{30}$ NPs (Fig. 3c). In the case of $Cu_{70}Ag_{30}O_4$ NP, the presence of surface oxide groups prevent the wrapping of the NP by the lipid membrane.

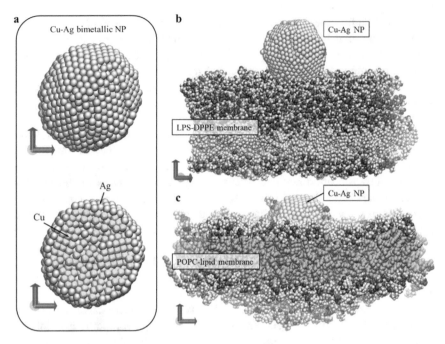

Fig. 3 Bimetallic NPs of immiscible or partially miscible metals: **a** structure of $Cu_{70}Ag_{30}$ NP, silver as a metal with lower surface energy forms a surface of NP; **b, c** interaction of Cu–Ag NP with lipopolysaccharide (LPS-DPPE) membrane and POPC-bilayer, respectively. The images were available using the 3D data from work [54]. Colors: Ag—grey, Cu—yellow, C—cyan, O—red, H—white, N—blue (water is not shown)

3 One-Dimensional Nanomaterials

3.1 Carbon Nanotubes

Nanomaterials with a structure as nanotube (NT), nanofiber, nanowire, nanorod, etc. belong to the class of (quasi)one-dimensional NMs (1D-NMs). The carbon nanotubes (CNT) occupies a special place among 1D-NMs and remains for a long time in the focus of researchers attention. CNTs are promising structures for the biotechnology and biomedicine, due to their biocompatibility, high stability, mechanical elasticity, thermal, electrical, and optic properties as well as the adaptability for chemical modification with bioactive compounds [59]. The single-walled carbon nanotubes (SWNTs) structure can be open or can be terminated with caps (hemispheres of a fullerene). In the former case, NT has a nanocapsule-like structure with the internal cavity isolated from the environment. Ligands in functionalized CNTs can be chemically bound with the cylindrical surface or ends. Functional groups can also be bound with CNT by non-covalent bonds. Nanotubes can be wrapped with surfactants, proteins, lipids, DNA, etc., forming the supramolecular complexes due to the Van der Waals forces, hydrophobic interaction, and electrostatic polarization. Depending on chirality, SWNTs can exhibit metallic or semiconductor properties. Electronic properties of the SWNTs provide a formation of stable DNA-CNT nanohybrids [60, 61]. It is important to note, that typically in the force field-based MD simulation the polarizability of CNT is not taken into account, and carbon atoms are modeled as Lennard-Jones particles with zero partial charges.

One of the pioneering MD studies of the interaction of CNT with the membrane is reported in [62]. The results showed that open carbon SWNT of 13 Å in a diameter with hydrophilic groups at the ends, due to the formation of salt bridges with lipid head groups, is capable of the spontaneous embedment into the membrane, forming a transmembrane channel permeable for water molecules. Two phases of a transmembrane channel formation were distinguished: (1) location and embedment of the NT into the membrane surface, and (2) turn to a transmembrane position. As a hydrophilic NT end moved through the membrane, lipids attached from the nearest membrane monolayer were transferred to the opposite one. The example of a transmembrane position of a zigzag-type SWNT with chirality indices (25, 0) and with ends functionalized by OH-groups is shown schematically in Fig. 4 (left CNT).

The formation and transport properties of CNT-based transmembrane channel were investigated by Zimmerli and Koumoutsakos, using the all-atom MD simulations [63]. The results showed a possibility of electrophoretic transport of short RNA segments (20 adenosine nucleotides) through a synthetic channel based on transmembrane SWNT with a diameter of 18.7 Å. It was found that an electrostatic potential difference of about 1–2 V maintains the RNA fragment translocation with a velocity of about 1–30 nucleotides per nanosecond.

The study of interaction of non-functionalized NTs, having different length, with the membrane showed that short NTs, being inside the lipid bilayer, prefer to be oriented parallel to lipid molecules [64]. An increase in the length of NT change the preferred orientation to parallel to the membrane plane. The example of such

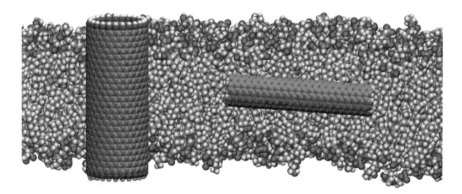

Fig. 4 Schematic of a transmembrane SWNT with functionalized faces (at the left) and hydrophobic thin CNT inside the bilayer. Color legend: carbon of CNT—orange, C (lipids)—grey, H—white, O—red, N—blue

an orientation of armchair-type SWNT with chirality (6, 6) in the bilayer center is shown schematically in Fig. 4 (right CNT).

Based on a coarse-grained MD model, two ways of CNT penetration through the membrane were revealed: by wrapping with lipids and by direct piercing [65]. It is reported that the key factor, which defines the way of penetration, is the diameter of NT. Nanotubes of a rather small diameter pierce membranes, whereas those of larger diameter penetrate through the bilayer via wrapping. The wrapping of a thinner NT is accompanied by considerable bending of the membrane, which is less energetically favorable than the separation of lipids to a comparatively small distance. For a NT of diameter comparable with the membrane thickness, the distance to which lipids need to be displaced for piercing is significant, however, the membrane curvature would be low if membrane wraps the NT. This explains why penetration via wrapping become more energetically preferable than piercing.

According to the computation results [66] based on a single chain mean field theory (SCMF) [67], the energy required for penetration of a perpendicularly oriented NT with a hydrophilic surface is about hundred of $k_B T$. It was noted, that an orthogonal penetration produces the least damages to the membrane and has the least energy barrier. A nanotube with a hydrophobic surface is "attracted" by the hydrophobic core of the bilayer, which prevents its movement and separation from the membrane due to thermal motion. The SCMF theory was also applied for estimating the penetration parameters of a NT decorated with alternate hydrophilic and hydrophobic bands on its surface [68]. It was shown that a specific pattern on the surface of a NT can facilitate its penetration through a lipid bilayer.

The penetration of NTs through the membrane is dependent on a composition of the membrane, e.g. on cholesterol content, as it was shown in a comparative study of a POPC-lipids bilayer and POPC/cholesterol membrane [69]. Based on a constant velocity SMD simulation of NT penetration, it was demonstrated that the presence of 30% of cholesterol molecules stabilizes the membrane and increases its rigidity. Moreover, MD modeling of lipid membranes reinforced with SWNTs showed that intercalated nanotubes limit the degrees of freedom of neighboring lipids, strengthen membranes, and make them more stable and resistant to temperature increase [70]. The decrease in lipid mobility, particularly, the diminution of self-diffusion coefficient of lipids in the presence of NTs intercalated in the bilayer was also reported [71].

Kraszewski and co-workers, investigated the interaction of pristine (non-functionalized) and amino derivative-functionalized SWNTs with the membrane [72]. The results of MD simulations showed that a closed pristine SWNT freely penetrates into the bilayer through three phases of passive diffusion: landing and floating on the bilayer surface (60 ns), fast penetration through the zone of head groups (20–40 ns), and finally, sliding through the lipid tails region. The penetration of functionalized SWNTs is similar, except that with a comparatively large number of functional groups where the phase of landing is in fact absent. During the functionalized SWNT translocation through the bilayer, amine groups were deprotonated in the simulation and then, when NT approached the opposite membrane side, the charges were recovered. It is pointed out that the presence of functional groups slows down the penetration of NT through the bilayer. Open SWNT can cause sticking of lipids at their open ends, thereby violating the local structure of the membrane. This effect increases the free energy barrier for translocation of open SWNTs through the membrane in comparison with capped ones. The simulation results [73] lead to the same conclusions. The penetration of open and closed SWNTs of the armchair type through a cholesterol-containing lipid membrane was studied using the all-atom model. It was shown that the penetration of a closed NT causes smaller membrane disturbances than an open one does. A closed SWNT has a lower free energy barrier to penetration through the membrane, which makes it a successful choice for the drug delivery, serving as nano-carriers or nano-containers.

As an example of the use CNT as nano-carrier, the delivery of paclitaxelum (PTX) encapsulated into an open nanotube (PTX@SWNT) through a lipid membrane was considered in [74]. Paclitaxelum is a mitotic inhibitor used in cancer chemotherapy, which molecule has a polar core. Four SMD calculations were performed for the penetration of a PTX@SWNT complex through a lipid bilayer at different velocities. The PTX molecule was located in the far region of the SWNT. In the simulation, external forces were applied both to the PTX and to the SWNT. The highest resistance to penetration was observed when the complex passed through the region of lipid tails which was due to the formation of hydrophobic bonds between the lipophilic groups and the outer SWNT surface. Entry of both the water molecules and the lipids into the inner volume of the nanotube was observed during the simulation. Moreover, the simulations revealed the formation of hydrogen bonds between the PTX and water molecules penetrated into the SWNT. It was noted that a stabilization of the PTX molecule inside the SWNT occurs due to both the Van der Waals forces between the PTX and inner surface of SWNT wall, and the hydrogen bonding between the PTX and water molecules penetrated inside the nanotube cavity.

In order to investigate the features of cell membrane interaction with pristine and functionalized SWNTs differing in diameter, length, chemical modification, and location of functional groups, the simulations with seven types of SWNTs with closed ends and aggregates of SWNTs have been conducted in [75]. The nanotubes had different positions of hydrophilic groups: fully hydrophilic SWNTs, hydrophilic end groups, and fully hydrophobic SWNTs. It was shown that small nanotubes spontaneously penetrate into the membrane. The most stable position of closed hydrophobic SWNTs is the membrane center with an orientation parallel to the bilayer plane. However, in the case of NT with a length smaller than the membrane thickness, the energy minimum for perpendicular (or transmembrane) orientation in the bilayer is somewhat lower. For SWCNTs with functionalized ends and a length slightly greater than the membrane thickness, it is advantageous to have a transmembrane position with some angle to the bilayer at which the difference between the nanotube length and the membrane thickness is compensated. Fully functionalized hydrophilic SWNTs are preferably adsorbed by the membrane at the water-lipid interface in an orientation parallel to the membrane plane. The probability of crossing the membrane or embedding in the bilayer for such SWNTs is extremely low. It is also shown that bundles of several nanotubes self-aggregate in a water solution. These aggregates entering the bilayer significantly disrupt the membrane structure and the perturbation increases with increasing cluster size.

3.2 Boron Nitride Nanotubes

Despite the similar to CNT tubular geometry, analogous crystal structure, and unique physicochemical properties, the nanoscale mechanisms of the boron nitride nanotubes (BNNT) interaction with the biomembranes and other bio-object are mostly unknown and very poorly covered by MD-based studies. To date, there are only two MD studies of BNNT-lipid bilayer interaction with the atomic description of the system [76, 77].

The pioneering MD study of the BNNT interaction with the lipid membrane was conducted by Hilder and co-workers [76]. The BNNTs with chirality (10, 0) and (10, 10), having a length of 2 nm, were considered. The partial atomic charges of B and N atoms were 0.4 e and –0.4 e, respectively. It was shown that the mechanism of BNNT insertion is similar to the mechanism of insertion of amino-functionalized carbon nanotubes, previously considered by Kraszewski et al. [72], into the lipid bilayer, except for the final stage of realignment. The following four stages were determined for the BNNT insertion into the membrane: (1) formation of BNNT-bilayer contact, (2) BNNT reorientation to become almost parallel to the membrane plane, (3) an interaction of one end of BNNT with the lipid head groups, and (4) an insertion of BNNT, allowing it to slide into the bilayer at an angle of about 45°. As a result the BNNT is partially inserted and is stable inside the bilayer. In general, the behavior of BNNTs is similar to the behavior of functionalized carbon nanotubes, which is less cytotoxic than pure CNTs. Although Hilder et al. performed both the unbiased and steered MD simulations, the energy characteristics of BNNT-membrane interaction were not obtained.

In order to estimate the free energy profile, including the energy barrier, and the depth of energy well, the SMD simulations with a potential of mean force (PMF) analysis of short BNNT and analogous CNT insertion into the lipid bilayer were conducted by the Professor S. G. Psakhie scientific group [77]. Previously, using DFT calculation, it has been found that the partial atomic charges of B and N are strong environmentally dependent [78]. In particular, in case of BNNT with chirality (5, 5) in a vacuum, boron atoms have partial charge of $+0.4$ e, nitrogen -0.4 e, but for the BNNT containing 5–7 water molecules inside, the absolute values of partial charges are significantly higher ± 1.05 e. For this reason, two models of BNNT were examined in the study [77]: BNNT (± 0.4) and BNNT (± 1.05). A diameter of BNNT was 6.9 Å, length ~11.3 Å.

The result of SMD simulations with the PMF analysis showed that: (1) both BNNT have local minimum of energy about -10 kJ/mol at the lipid-water interface, (2) the BNNT (± 0.4) has a global minimum of -72 kJ/mol inside the lipophilic core of the bilayer, the depth is about 30 kJ/mol smaller than that in case of analogous CNT, (3) the CNT was staying empty inside the lipid membrane whereas BNNT

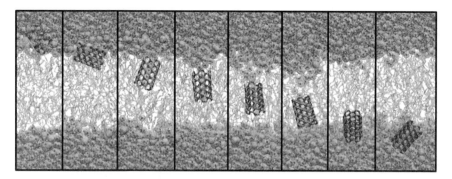

Fig. 5 Boron nitride nanotube (BNNT) translocation through the POPC membrane in steered MD simulation. The orientation of BNNT is transmembrane when it is crossing the tail region of the membrane. BNNT tilts orientation when overcoming the head group—water interface. Water is observed inside BNNT. Images use the XYZ-data from work [77]. Colors: B—pink, N—blue, lipids—light-green, water—light-blue

brings in several water molecules, (4) the penetration of the BNNT into the bilayer is accompanied by a formation of "water defect"—a several-water-molecule long tail behind the BNNT (Fig. 5), (5) the difference in orientation of short BNNT and CNT translocating into the bilayer is that the CNT tends to be in a perpendicular orientation to the membrane plane during whole time of insertion, whereas BNNT ($\pm$0.4) has a tilt from 0 (parallel) to 45° being in the head-group region (Fig. 5), (6) the insertion of the BNNT ($\pm$1.05) is not energetically favorable. The former means an absence of impact on the cell membrane, which indirectly indicates a less cytotoxicity of this nanomaterial.

Besides the two abovementioned researches there is a study of BNNT-membrane interaction, in which, using the mixed model—the Lennard-Jones potential together with the continuum approximation, the relationship between the location of energy minimum, the radius of a cylindrical hole in the membrane, and the perpendicular distance of the BNNT from the hole was determined [79].

4 Two-Dimensional Nanomaterials

4.1 Graphene

One of the pioneering MD studies on the interaction of graphene nanosheet (GNS) with a cell membrane was reported in [80]. Using a coarse-grained MD model, it was shown that GNS can form composite sandwich-like structures in the internal hydrophobic region of a lipid bilayer. Such hybrid graphene–membrane structure can be obtained by forming hydrated micelles of individual GNS coated with a phospholipid layer which can then be absorbed by the membrane. The GNS dimensions in the simulation were about 60×60 Å^2. It was found that the absorbed graphene nanosheet residing at the bilayer center parallel to the membrane plane is a stable system. Slow diffusion motion of the frustrated graphene along the membrane was observed. It is pointed out that the nanosheet does not change the bilayer thickness. The absorption of a lipid micelle containing the GNS by the membrane was also investigated. It was shown that during the absorption of the micelle, the graphene sheet is fixed in the hydrophobic core of the bilayer and lipids from the shell transfer into the nearest membrane leaflet. Full absorption and stabilization of the nanosheet in the bilayer center required about 500 ns. The dependence of displacement of graphene balanced inside the bilayer on the force applied to its edge perpendicular to the system plane was also estimated. Further integration of the obtained data gave an estimate of the energy required for the extraction of graphene from the membrane. In the case of three parallel nanosheets are absorbed, the membrane increases in thickness by about 1 nm, the absorption of eight-GNS sandwich leads to an increase of thickness more than 1.5 times. If the size of a nanosheet on one of the sides coincides with the bilayer thickness and its opposite faces are functionalized by hydrophilic groups, the graphene fragment takes a stable transmembrane position inside the bilayer perpendicular the membrane plane.

The penetration of single square and circular fragments of pristine graphene GNSes of different sizes into the membrane was studied in [81]. It was shown that as the nanosheet size is increased, the bilayer tends to form semi-spherical vesicle, and thus the membrane experiences substantial deformation. During internalization, the angle between the membrane plane and GNS was estimated. The behavior of this angle revealed three stages: (1) spontaneous orientation before the nanosheet touches the membrane surface, (2) embedment of the nanosheet into the bilayer mainly at an angle of 47°, and (3) rotation of the nanosheet into membrane-parallel position with the formation a sandwich-like superstructure. The MD simulation was also performed to study how the thickness of single- and multilayered graphene nanosheets and the degree of their surface oxidation influence the interaction with a lipid membrane [82]. The penetration of a GNS coated with lipids was also considered. It was shown that for the pristine GNS, the position in the bilayer center parallel to the membrane plane is stable, whereas a nanosheet with oxide at the boundary (10% degree of oxidation) is attached at an angle to the membrane plane. The degree of boundary oxidation strongly affects the final GNS orientation in the bilayer. At oxidation values of less than 5%, the GNS behaves as a pristine one. It was noted that there are two appreciable energy drops in the system during penetration: (1) when the GNS is embedded in the bilayer, and (2) when it rotates inside the bilayer to take its final position. The second energy drop is lower than the first one. The simulation of coated GNS was performed for different amounts of coating lipids. At a low surface absorption density, the nanosheet edge zones were open to water. The penetration was no different from the internalization of pristine graphene sheets. At a relatively high lipid concentration on the GNS surface, no penetration within the rather limited time of MD simulation was observed. For penetration of a multilayered graphene into the bilayer, piercing of the membrane by its corner and parallel orientation inside the membrane, like for single GNS, was detected.

Using coarse-grained MD and all-atom SMD simulations, the mechanism of spontaneous penetration of multilayered (few-layer) GNS into a cell though piercing the membrane by their edge or corner was studied in [83]. The model took into account the roughness (asperities) of the GNS edge the texture of which was borrowed from an image of the real structure taken with a transmission electron microscope. The results of MD calculations showed that monolayer graphene in the form of a rhomb (with an edge length of 64 Å and most acute angle of 30° and 60°) that turns due to thermal motion resulting in one of its acute corner sites to be directed to the bilayer. When the region of head groups is pierced, complete GN absorption by the membrane can be initiated through hydrophobic interaction of graphene with lipid tail groups. Additionally, a series of calculations with triangular, square, and hexahedral fragments of both pristine GNS and GNS functionalized at the corners or along the perimeter were performed. The results of MD calculations showed that absorption occurs with the non-functionalized corner forward, and if no such corners are present, the GNS remains on the membrane surface. Using the SMD method, the free energy barrier for graphene penetration into the membrane was estimated. It was shown that the penetration of the GNS with an ideal boundary into the membrane is almost impossible at room temperature, whereas inhomogeneous topography of its boundary can greatly decrease the free energy barrier. It was noted that when absorbing the GNS, the membrane structure is hardly disturbed, except that the lipid tail groups are straitened along the nanosheet surface due to high adhesion.

A series of coarse-grained calculations were performed to study the interaction of pristine graphene and graphene oxide with a lipid membrane for different nanosheet sizes and degrees of oxidation [84]. The graphene sheets were square flakes with side lengths of 35, 70, and 105 Å; the density of oxide groups was varied from 0 to 40% of the total number of carbon atoms. Nanosheets oxidized only along the perimeter were also considered. In total, four variants were simulated: pristine GNS, graphene oxidized along the perimeter (epitaxial graphene oxide, EGO), graphene with oxide along the perimeter and 20% on the surface (GO20), and graphene with oxide along the perimeter and 40% on the surface (GO40). Each calculation took about 45 μs. It was found that the graphene interaction with the membrane provides one of the four configurations: a sandwich-like structure with a nanosheets located in the bilayer center and parallel to it (for GNS with sides $l = 35, 70$ Å); a graphene-containing semi-spherical vesicle immersed into the bilayer (for GNS and EGO with $l = 105$ Å); a nanosheet lying on the membrane surface and parallel to it (GO20 and GO40 with $l = 35$ Å); and crosswise oriented graphene inside the bilayer. It was found that the scale of observed membrane distortions depends on the degree of graphene oxidation: the more this degree, the higher the distortion of the bilayer structure. Increasing the side of square graphene fragments increases these disturbances up to the point of membrane continuity disruption.

The results of experimental and numerical studies of the interaction of pristine graphene and graphene oxide with a bacterial membrane were reported in [85]. According to them, GNS can embed and cut the external and internal cell membranes of *E. coli*. This involves degradation of the membrane with possible destruction of the cell wall and partial cytoplasm loss. Using the all-atom MD model, it was shown that there exist three phases of graphene and graphene oxide interaction with lipid membranes. The first phase is diffusion motion of the nanosheet till it touches the membrane; the second phase is its comparatively fast insertion into the bilayer; the third one is an extraction of lipids from the bilayer due to Van der Waals forces and hydrophobic interaction on the GNS surface. This involves substantial depression and deformation of the membrane, facilitating disruption of its continuity. Such an action decreases the bacterial cell viability and allows consideration of graphene as an antibacterial nanomaterial. Schematic molecular model illustrating the extraction of the lipids on GNS surface is presented at Fig. 6.

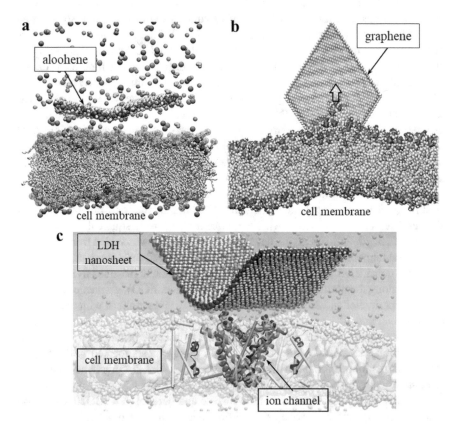

Fig. 6 Two-dimensional nanomaterials and cell membrane interaction: **a** aloohene doesn't penetrate to the bilayer and doesn't extract lipids, its impact to a cell is a dysregulation of ionic balance (the result of long unbiased molecular dynamics simulation). Reproduced with permission from ACS [6]. **b** Schematic molecular model illustrating the extraction of the lipids onto GNS surface (orange) from the cell membrane. **c** Nanosheet of layered double hydroxide interacting with outer loops of bacterial sodium channel $Na_V Ab$, embedded into the membrane (the model was built using XYZ-data from [86])

4.2 Metal (Oxy)hydroxides

Unlike GNS, the layered metal hydroxides and oxyhydroxides are hydrophilic compounds with large amount of polar OH-groups on the surface. Layered metal hydroxides and oxyhydroxides as a rule are chemically inert, non-toxic, biocompatible, have a large specific surface area with non-zero surface charge, can form the hybrid complexes with organic molecules, and be used as selective ion exchangers with a high proton and/or hydroxyl buffer capacity. These unique physicochemical properties make the layered metal (oxy)hydroxides an important material for a wide range of applications in different areas, including biomedicine.

One of the first MD research of biomembrane interaction with metal (oxy)hydroxide was performed by Shroll and Straatsma [87]. They considered an adhesion of the gram-negative bacteria outer membrane on FeOOH (goethite). Although the goethite FeOOH is not a layered oxyhydroxide but the oxyhydroxide with bulk crystal structure, we will consider this case, since the modeled fragment has a thickness of several atomic layers. The modelled outer membrane of *Pseudomonas aeruginosa* bacteria has bilayer structure and consist of lipopolysaccharide (LPS) monolayer (outer leaflet) and the monolayer of typical phospholipids (inner leaflet). The LPS has several carboxyl and phosphate groups, which are the acceptors of proton. Each LPS molecule brings strong negative charge of -13 e. The strong adhesion within LPS-NM interface was observed. It was found that outermost LPS groups must compete with water molecules resided near the FeOOH surface and may replace them forming multiple hydrogen bonds with hydroxyl groups of the mineral surface. Due to the formation of these bonds the significant changes in the structure of outermost saccharide groups of the membrane were observed. No obvious tendency of FeOOH-induced membrane disruption is reported.

A similar result was obtained in the unbiased MD simulation of the layered double hydroxide (LDH) near the outer surface of the membrane with bacterial ion channel embedded [87]. The hydrogen bonding of outer carboxyl-contained groups of the $Na_V Ab$ voltage-gated sodium-selective ion channel with the hydroxide groups on the LDH surface was observed, without any tendency to disrupt the membrane (Fig. 6).

The layered metal oxyhydroxides nanomaterial based on the AlOOH (boehmite), termed "aloohene", was developed by the scientific group of Professor S. G. Psakhie [6] and showed promise for multiple biomedical applications. The all-atom MD simulations of the aloohene fragment in the vicinity of the cell membrane demonstrate that aloohene doesn't tend to internalize into the lipid bilayer. Moreover, due to the electrostatic and amphoteric properties of its surface, the aloohene dysregulates the composition and the ion balance of the tumor cell microenvironment.

It is supposed that the electrostatic action on the membrane and perturbation of the ion concentration near its surface can greatly affect the functioning of membrane proteins sensitive to changes in membrane polarization. This effect can be critical for some metabolic and intracellular processes, including those which support viability and proliferation of tumor cells.

5 Summary

Computer simulations of interactions on the molecular level provide a unique opportunity to explore the processes and mechanisms in such complex and heterogeneous systems as biomembrane-nanomaterial interfaces at picosecond-microsecond timescales. The understanding of the basic rules, which govern adsorption, insertion, accumulation, wrapping, uptake, and disruption in these systems allows one to determine the key factors to control the cell-nanomaterial or cell-nanodevice behavior. This knowledge is of fundamental importance for the further progress in nanomedicine, theranostics, oncology, and related fields.

Acknowledgements The part of work related to the carbon nanomaterials was performed in the framework of the research project supported from the Russian Foundation for Basic Research (RFBR), Grant No. 18-29-19198.

References

1. Shi J, Kantoff PW, Wooster R, Farokhzad OC (2017) Cancer nanomedicine: progress, challenges and opportunities. Nat Rev Cancer 17(1):20–37
2. Lee H, Lee Y, Song C, Cho HR, Ghaffari R, Choi TK, Kim KH, Lee YB, Ling D, Lee H, Yu SJ, Choi SH, Hyeon T, Kim D-H (2015) An endoscope with integrated transparent bioelectronics and theranostic nanoparticles for colon cancer treatment. Nat Commun 6:10059
3. Song S, Faleo G, Yeung R, Kant R, Posselt AM, Desai TA, Tang Q, Roy S (2016) Silicon nanopore membrane (SNM) for islet encapsulation and immunoisolation under convective transport. Sci Rep 6(1):1–9
4. Mikhaylov G, Mikac U, Magaeva AA, Itin VI, Naiden EP, Psakhye I, Babes L, Reinheckel T, Peters C, Zeiser R, Bogyo M, Turk V, Psakhye SG, Turk B, Vasiljeva O (2011) Ferri-liposomes as an MRI-visible drug-delivery system for targeting tumours and their microenvironment. Nat Nanotechnol 6(9):594–602
5. Li D, Zhang Y-T, Yu M, Guo J, Chaudhary D, Wang C-C (2013) Cancer therapy and fluorescence imaging using the active release of doxorubicin from MSPs/Ni-LDH folate targeting nanoparticles. Biomaterials 34(32):7913–7922
6. Lerner MI, Mikhaylov G, Tsukanov AA, Lozhkomoev AS, Gutmanas E, Gotman I, Bratovs A, Turk V, Turk B, Psakhye SG, Vasiljeva O (2018) Crumpled aluminum hydroxide nanostructures as a microenvironment dysregulation agent for cancer treatment. Nano Lett 18(9):5401–5410
7. Gupta A, Landis RF, Li C-H, Schnurr M, Das R, Lee Y-W, Yazdani M, Liu Y, Kozlova A, Rotello VM (2018) Engineered polymer nanoparticles with unprecedented antimicrobial efficacy and therapeutic indices against multidrug-resistant bacteria and biofilms. J Am Chem Soc 140(38):12137–12143
8. Zhou W, Pan T, Cui H, Zhao Z, Chu PK, Yu X-F (2019) Black phosphorus: bioactive nanomaterials with inherent and selective chemotherapeutic effects. Angew Chem (International ed. English) 131(3):779–784
9. Gennis RB (2013) Biomembranes: molecular structure and function. Springer Science & Business Media, New York
10. Mitaku S (1993) The role of hydrophobic interaction in phase transition and structure formation of lipid membranes and proteins. Phase Transit Multinational J 45(2–3):137–155
11. Thompson RB, Ginzburg VV, Matsen MW, Balazs AC (2001) Predicting the mesophases of copolymer-nanoparticle composites. Science 292(5526):2469–2472

12. Thompson RB, Ginzburg VV, Matsen MW, Balazs AC (2002) Block copolymer-directed assembly of nanoparticles: forming mesoscopically ordered hybrid materials. Macromolecules 35(3):1060–1071
13. Qian J, Gao H (2010) Soft matrices suppress cooperative behaviors among receptor-ligand bonds in cell adhesion. PLoS ONE 5(8):e12342
14. Gao H, Qian J, Chen B (2011) Probing mechanical principles of focal contacts in cell–matrix adhesion with a coupled stochastic–elastic modelling framework. J R Soc Interface 8(62):1217–1232. https://doi.org/10.1098/rsif.2011.0157
15. MacKerell AD Jr, Bashford D, Bellott M, Dunbrack RL Jr, Evanseck JD, Field MJ, Fischer S, Gao J, Guo H, Ha S, Joseph-McCarthy D, Kuchnir L, Kuczera K, Lau FTK, Mattos C, Michnick S, Ngo T, Nguyen DT, Prodhom B, Reiher WE, Roux B, Schlenkrich M, Smith JC, Stote R, Straub J, Watanabe M, Wiórkiewicz-Kuczera J, Yin D, Karplus M (1998) All-atom empirical potential for molecular modeling and dynamics studies of proteins. J Phys Chem B 102:3586–3616
16. Monticelli L, Kandasamy SK, Periole X, Larson RG, Tieleman DP, Marrink S-J (2008) The MARTINI coarse-grained force field: extension to proteins. J Chem Theory Comput 4(5):819–834
17. Izrailev S, Stepaniants S, Isralewitz B et al (1997) Computational molecular dynamics: challenges, methods, ideas. Springer, Berlin, pp 39–65 (Steered molecular dynamics)
18. Pogodin S, Werner M, Sommer JU, Baulin VA (2012) Nanoparticle-induced permeability of lipid membranes. ACS Nano 6(12):10555–10561
19. Orsi M, Essex JW (2010) Molecular simulations and biomembranes: from biophysics to function. Royal Society of Chemistry, London, pp 76–90 (Passive permeation across lipid bilayers: a literature review)
20. Tsukanov AA, Psakhie SG (2015) A review of computer simulation studies of cell membrane interaction with neutral and charged nano-objects. Quasi-zero-dimensional nanoparticles, drugs and fullerenes. Adv Biomater Dev Med 2(1):44–53
21. Qiao R, Roberts AP, Mount AS, Klaine SJ, Ke PC (2007) Translocation of C_{60} and its derivatives across a lipid bilayer. Nano Lett 7(3):614–619
22. Wong-Ekkabut J, Baoukina S, Triampo W, Tang I-M, Tieleman DP, Monticelli L (2008) Computer simulation study of fullerene translocation through lipid membrane. Nat Nanotechnol 3:363–368
23. D'Rozario RSG, Wee CL, Wallace EJ, Sansom MSP (2009) The interaction of C_{60} and its derivatives with a lipid bilayer via molecular dynamics simulations. Nanotechnology 20:115102
24. Jusufi A, DeVane RH, Shinoda W, Klein ML (2011) Nanoscale carbon particles and the stability of lipid bilayers. Soft Matter 7(3):1139–1146
25. Zhang S, Mu Y, Zhang JZH, Xu W (2013) Effect of self-assembly of fullerene nano-particles on lipid membrane. PLoS ONE 8(10):e77436
26. Bozdaganyan ME, Orekhov PS, Shaytan AK, Shaitan K (2014) Comparative computational study of interaction of C_{60}-fullerene and tris-malonyl-C_{60}-fullerene isomers with lipid bilayer: relation to their antioxidant effect. PLoS ONE 9(7):e102487
27. Lai K, Wang B, Zhang Y, Zheng Y (2013) Computer simulation study of nanoparticle interaction with a lipid membrane under mechanical stress. Phys Chem Chem Phys 15(1):270–278
28. Kraszewski S, Tarek M, Treptow W, Ramseyer C (2010) Affinity of C_{60} neat fullerenes with membrane proteins: a computational study on potassium channels. ACS Nano 4:4158–4164
29. Monticelli L, Barnoud J, Orlowski A, Vattulainen I (2012) Interaction of C_{70} fullerene with the Kv1.2 potassium channel. Phys Chem Chem Phys 14(36):12526–12533
30. Kraszewski S, Tarek M, Ramseyer C (2011) Uptake and translocation mechanisms of cationic amino derivatives functionalized on pristine C_{60} by lipid membranes: a molecular dynamics simulation study. ACS Nano 5:8571–8578
31. Tieleman DP (2006) Computer simulations of transport through membranes: passive diffusion, pores, channels and transporters. Proc Aust Physiol Soc 37:15–27

32. Monticelli L, Salonen E, Ke PC, Vattulainen I (2009) Effects of carbon nanoparticles on lipid membranes: a molecular simulation perspective. Soft Matter 5(22):4433–4445

33. Makarucha AJ, Todorova N, Yarovsky I (2011) Nanomaterials in biological environment: a review of computer modelling studies. Eur Biophys J 40:103–115

34. Esfand R, Tomalia DA (2001) Poly(amidoamine) (PAMAM) dendrimers: from biomimicry to drug delivery and biomedical applications. Drug Discov Today 6(8):427–436

35. Kolotylo M, Holovatiuk V, Bondareva J, Lukin O, Rozhkov V (2019) Synthesis of sulfonimide-based dendrimers and dendrons possessing mixed $1 \to 2$ and $1 \to 4$ branching motifs. Tetrahedron Lett 60(4):352–354

36. Rozhkov VV, Kolotylo MV, Onys'ko PP, Lukin O (2016) Synthesis of sulfonimide-based branched arylsulfonyl chlorides. Tetrahedron Lett 57(3):308–309

37. Maiti PK, Çağın T, Wang G, Goddard WA (2004) Structure of PAMAM dendrimers: generations 1 through 11. Macromolecules 37(16):6236–6254

38. Lee H, Larson RG (2006) Molecular dynamics simulations of PAMAM dendrimer-induced pore formation in DPPC bilayers with a coarse-grained model. J Phys Chem B 110:18204–18211

39. Lee H, Larson RG (2008) Coarse-grained molecular dynamics studies of the concentration and size dependence of fifth- and seventh-generation PAMAM dendrimers on pore formation in DMPC bilayer. J Phys Chem B 112:7778–7784

40. Lee H, Larson RG (2008) Lipid bilayer curvature and pore formation induced by charged linear polymers and dendrimers: the effect of molecular shape. J Phys Chem B 112:12279–12285

41. Kelly CV, Leroueil PR, Nett EK, Wereszczynski JM, Baker JR, Orr BG, Banaszak Holl MM, Andricioaei I (2008) Poly(amidoamine) dendrimers on lipid bilayers I: free energy and conformation of binding. J Phys Chem B 112(31):9337–9345

42. Kelly CV, Leroueil PR, Orr BG, Banaszak Holl MM, Andricioaei I (2008) Poly(amidoamine) dendrimers on lipid bilayers II: effects of bilayer phase and dendrimer termination. J Phys Chem B 112:9346–9353

43. Yan LT, Yu X (2009) Enhanced permeability of charged dendrimers across tense lipid bilayer membranes. ACS Nano 3(8):2171–2176

44. Ting CL, Wang ZG (2011) Interactions of a charged nanoparticle with a lipid membrane: implications for gene delivery. Biophys J 100(5):1288–1297

45. Ginzburg VV, Balijepalli S (2007) Modeling the thermodynamics of the interaction of nanoparticles with cell membranes. Nano Lett 7(12):3716–3722

46. Li Y, Chen X, Gu N (2008) Computational investigation of interaction between nanoparticles and membranes: hydrophobic/hydrophilic effect. J Phys Chem B 112(51):16647–16653

47. Alexeev A, Uspal WE, Balazs AC (2008) Harnessing Janus nanoparticles to create controllable pores in membranes. ACS Nano 2(6):1117–1122

48. Yang K, Ma YQ (2010) Computer simulation of the translocation of nanoparticles with different shapes across a lipid bilayer. Nat Nanotechnol 5(8):579–583

49. Lin J, Zhang H, Chen Z, Zheng Y (2010) Penetration of lipid membranes by gold nanoparticles: insights into cellular uptake, cytotoxicity, and their relationship. ACS Nano 4(9):5421–5429

50. Guo R, Mao J, Yan LT (2013) Unique dynamical approach of fully wrapping dendrimer-like soft nanoparticles by lipid bilayer membrane. ACS Nano 7(12):10646–10653

51. Carr R, Weinstock IA, Sivaprasadarao A, Müller A, Aksimentiev A (2008) Synthetic ion channels via self-assembly: a route for embedding porous polyoxometalate nanocapsules in lipid bilayer membranes. Nano Lett 8(11):3916–3921

52. Dykman L, Khlebtsov N (2012) Gold nanoparticles in biomedical applications: recent advances and perspectives. Chem Soc Rev 41:2256–2282

53. Nangia S, Sureshkumar R (2012) Effects of nanoparticle charge and shape anisotropy on translocation through cell membranes. Langmuir 28(51):17666–17671

54. Tsukanov AA, Pervikov AV, Lozhkomoev AS (2020) Bimetallic Ag–Cu nanoparticles interaction with lipid and lipopolysaccharide membranes. Comput Mater Sci 173:109396

55. Daw MS, Baskes MI (1984) Embedded-atom method: derivation and application to impurities, surfaces, and other defects in metals. Phys Rev B 29(12):6443

56. Williams PL, Mishin Y, Hamilton JC (2006) An embedded-atom potential for the Cu–Ag system. Modell Simul Mater Sci Eng 14(5):817
57. Hohenberg P, Kohn W (1964) Inhomogeneous electron gas. Phys Rev 136:B864
58. Kohn W, Sham LJ (1965) Self-consistent equations including exchange and correlation effects. Phys Rev 140(4A):A1133
59. Bekyarova E, Ni Y, Malarkey EB, Montana V, McWilliams JL, Haddon RC, Parpura V (2005) Applications of carbon nanotubes in biotechnology and biomedicine. J Biomed Nanotechnol 1(1):3–17
60. Rotkin SV (2010) Electronic properties of nonideal nanotube materials: helical symmetry breaking in DNA hybrids. Annu Rev Phys Chem 61:241–261
61. Tsukanov AA, Grachev EA, Rotkin SV (2007) Modeling of the SWNT-DNA complexes in the water solution. APS March meeting, American Physical Society, 5–9 Mar 2007, abstract id. V28.002
62. Lopez CF, Nielsen SO, Moore PB, Klein ML (2004) Understanding nature's design for a nanosyringe. PNAS Proc Nat Acad Sci U S A 101:4431–4434
63. Zimmerli U, Koumoutsakos P (2008) Simulations of electrophoretic RNA transport through transmembrane carbon nanotubes. Biophys J 94(7):2546–2557
64. Höfinger S, Melle-Franco M, Gallo T, Cantelli A, Calvaresi M, Gomes JANF, Zerbetto F (2011) A computational analysis of the insertion of carbon nanotubes into cellular membranes. Biomaterials 32:7079–7085
65. Shi XH, Kong Y, Gao HJ (2008) Coarse grained molecular dynamics and theoretical studies of carbon nanotubes entering cell membrane. Acta Mech Sin 24:161–169
66. Pogodin S, Baulin VA (2010) Can a carbon nanotube pierce through a phospholipid bilayer? ACS Nano 4:5293–5300
67. Ben-Shaul A, Szleifer I, Gelbart WM (1985) Chain organization and thermodynamics in micelles and bilayers. AIP J Chem Phys 83(7):3597–3611
68. Pogodin S, Slater NKH, Baulin VA (2011) Surface patterning of carbon nanotubes can enhance their penetration through a phospholipid bilayer. ACS Nano 5:1141–1146
69. Gangupomu VK, Capaldi FM (2011) Interactions of carbon nanotube with lipid bilayer membranes. J Nanomater 2011:830436
70. Shityakov S, Dandekar T (2011) Molecular dynamics simulation of POPC and POPE lipid membrane bilayers enforced by an intercalated single-wall carbon nanotube. NANO 6(01):19–29
71. Parthasarathi R, Tummala NR, Striolo A (2012) Embedded single-walled carbon nanotubes locally perturb DOPC phospholipid bilayers. J Phys Chem B 116:12769–12782
72. Kraszewski S, Bianco A, Tarek M, Remaseyer C (2012) Insertion of short amino-functionalized single-walled carbon nanotubes into phospholipid bilayer occurs by passive diffusion. PLoS ONE 7(7):e40703
73. Raczyński P, Górny K, Pabiszczak M, Gburski Z (2013) Nanoindentation of biomembrane by carbon nanotubes—MD simulation. Comput Mater Sci 70:13–18
74. Mousavi SZ, Amjad-Iranagh S, Nademi Y, Modarress H (2013) Carbon nanotube-encapsulated drug penetration through the cell membrane: an investigation based on steered molecular dynamics simulation. J Membr Biol 246(9):697–704
75. Baoukina S, Monticelli L, Tieleman DP (2013) Interaction of pristine and functionalized carbon nanotubes with lipid membranes. J Phys Chem B 117:12113–12123
76. Thomas M, Enciso M, Hilder TA (2015) Insertion mechanism and stability of boron nitride nanotubes in lipid bilayers. J Phys Chem B 119(15):4929–4936
77. Tsukanov AA, Psakhie SG (2016) Potential of mean force analysis of short boron nitride and carbon nanotubes insertion into cell membranes. Adv Biomater Dev Med 3(1):1–9
78. Won CY, Aluru NR (2008) Structure and dynamics of water confined in a boron nitride nanotube. J Phys Chem C 112(6):1812–1818
79. Alshehri MH (2018) Interactions of boron nitride nanotubes with lipid bilayer membranes. J Comput Theor Nanosci 15(1):311–316

80. Titov AV, Kral P, Pearson R (2010) Sandwiched graphene–membrane superstructures. ACS Nano 4:229–234
81. Guo R, Mao J, Yan L-T (2013) Computer simulation of cell entry of graphene nanosheet. Biomaterials 34(17):4296–4301
82. Wang J, Wei Y, Shi X, Gao H (2013) Cellular entry of graphene nanosheets: the role of thickness, oxidation and surface adsorption. RSC Adv 3(36):15776–15782
83. Li Y, Yuan H, von dem Bussche A, Creighton M, Hurt RH, Kane AB, Gao H (2013) Graphene microsheets enter cells through spontaneous membrane penetration at edge asperities and corner sites. PNAS Proc Nat Acad Sci U S A 110(30):12295–12300
84. Mao J, Guo R, Yan L-T (2014) Simulation and analysis of cellular internalization pathways and membrane perturbation for graphene nanosheets. Biomaterials 35(23):6069–6077
85. Tu Y, Lv M, Xiu P, Huynh T, Zhang M, Castelli M, Liu Z, Huang Q, Fan C, Fang H, Zhou R (2013) Destructive extraction of phospholipids from *Escherichia coli* membranes by graphene nanosheets. Nat Nanotechnol 8(8):594–601
86. Tsukanov AA, Psakhie SG (2016) Adsorption of charged protein residues on an inorganic nanosheet: computer simulation of LDH interaction with ion channel. AIP Conf Proc 1760(1):020066
87. Shroll RM, Straatsma TP (2003) Molecular basis for microbial adhesion to geochemical surfaces: computer simulation of *Pseudomonas aeruginosa* adhesion to goethite. Biophys J 84(3):1765–1772

Application of Crumpled Aluminum Hydroxide Nanostructures for Cancer Treatment

Aleksandr S. Lozhkomoev, Georgy Mikhaylov, Vito Turk, Boris Turk, and Olga Vasiljeva

Abstract The tumor microenvironment regulates tumor progression and the spread of cancer in the body. Applications of nanomaterials that can dysregulate tumor-microenvironment are emerging as a promising anti-cancer approaches, which can improve the efficacy of existing cancer treatments. We have reported that agglomerates of radially assembled Al hydroxide crumpled nanosheets with the disordered defective surface structure have a large positive charge and therefore can lead to ion imbalance at the cell perimembranous space through the selective adsorption of extra-cellular anionic species. This effect was demonstrated in vitro by reduced viability and proliferation of tumor cells, and further validated in a murine melanoma cancer model. Furthermore, crumpled Al hydroxide nanostructures showed a much stronger suppressive effect on tumor growth in combination with a minimally effective dose of doxorubicin. Taken together, the described approach of tumor microenvironment dysregulation through selective adsorption properties of folded crumpled nanostructures opened a new avenue for development of innovative anticancer therapy strategies.

Keywords Aluminum hydroxide · Nanosheets · Folded crumpled nanostructures · Tumor microenvironment · Anti-cancer therapy · Aloohene · Ion imbalance

G. Mikhaylov · V. Turk · B. Turk · O. Vasiljeva (✉)
Jozef Stefan Institute, 1000 Ljubljana, Slovenia
e-mail: olga.vasiljeva@ijs.si

A. S. Lozhkomoev
Institute of Strength Physics and Materials Science, Tomsk 634055, Russia

G.-P. Ostermeyer et al. (eds.), *Multiscale Biomechanics and Tribology of Inorganic and Organic Systems*, Springer Tracts in Mechanical Engineering, https://doi.org/10.1007/978-3-030-60124-9_10

1 Introduction to Low-Dimensional Aluminum (Hydro)oxides

Low-dimensional (2D) nanostructures have been extensively studied in the last 10–15 years [1–5]. The interest can be explained by the fact that two-dimensional nano-objects exhibit physicochemical properties significantly different from those of three-dimensional (bulk) materials. This is due to both the dominant role of the free surface and the morphological instability of the planar shape of 2D systems [6–9]. It should be noted that the current trend in the development of nanotechnology in the field of synthesis of low-dimensional nanostructures, possessing the necessary physicochemical and biomedical properties, is characterized by a broader use of the methods based on controlled self-organization (or self-assembling) of the structure directly in the synthesis process [10, 11]. In case of ultrathin 2D-nanostructures, the self-assembly of such ensembles is of high importance to fundamental and applied sciences, since it gives an opportunity not only to investigate a substance behavior in the two-dimensional state, but also to produce complex hybrid (organic-inorganic) nanosystems for a wide range of applications ranging from elements for sensor devices to nanostructures for biomedical applications [10, 12]. One of the promising directions in the formation of 2D nanostructures with a controlled structure is self-assembly during the oxidation of nanoparticles [10, 13].

The perspectives of new drugs development for the diagnostics and treatment of various diseases, including cancer, are associated with studies of low-dimensional nanostructures interaction with living cells [14–16]. The effects of colloidal gold particles [17], nanofibers or nanowires of zinc oxide [18, 19], titanium oxide [20], carbon nanotubes [21] and graphene oxide [22] have been the most studied as yet. Low-dimensional structures via self-assembling can form more complex ensembles [23]; however, the interaction of cells with such ensembles can be also studied at the level of individual low-dimensional components of the complex.

From the point of view of medical applications and targeted action on biological objects the low-dimensional aluminum (hydro)oxides should be considered as a promising basis for development of nanostructures with bio-activity. This is due to both their unique surface properties [12, 24], which determine their biological activity [25, 26], as well as their low toxicity [27] and the possibility to generate nanostructures with certain morphology during their synthesis [28].

2 Synthesis of Aluminum Oxyhydroxide Low-Dimensional Nanostructures

A simple and fast method for the synthesis of nanostructures based on the aluminum oxide and hydroxide phases is oxidation of aluminum nanopowder by water obtained via electrical dispersion of an aluminum wire [29, 30]. The synthesis proceeds in a single stage at atmospheric pressure and a temperature of 50–70 °C. The reaction mixture contains only water and aluminum. Therefore, the reaction products do not contain organic and ionic contaminants. This is a major advancement from the previous studies, where oxidation of aluminum nanoparticles by water was studied only as a method for a hydrogen production, and much less attention was paid to the reaction products [31–36]. Nanoparticles of the aluminum nitride composition (Al/AlN) provide a better starting material for the reaction due to the presence of a thin oxide film on the surface of the particles and the release of ammonia during hydrolysis, which increases the pH of the reaction medium [37]. The oxidation process of Al/AlN nanoparticles proceeds without an induction period, due to a thinner oxide film, and is accompanied by complete oxidation of aluminum as a result of ammonia excretion during the reaction. In order to obtain the aluminum oxyhydroxide low-dimensional nanostructures, the AlN/Al nanopowder produced by electric explosion of an aluminum wire in a nitrogen atmosphere, was used in our research.

Nanopowder is represented by a mixture of spherical and faceted particles. According to the results of the elemental analysis, nitrogen is found in all particles, however it is distributed in spherical particles non-uniformly (Fig. 1a), indicating

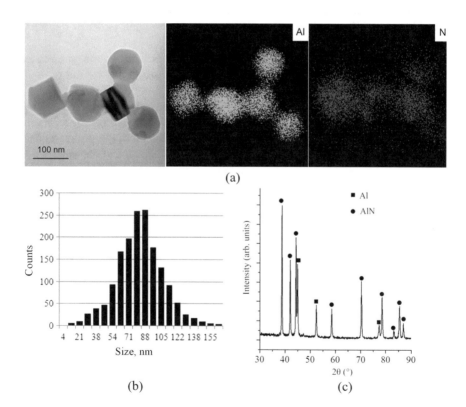

Fig. 1 TEM image and EDS mapping (**a**), particle size distribution and XRD pattern (**c**) of AlN/Al nanoparticles

(a)   (b)

Fig. 2 TEM (**a**) and SEM (**b**) images of agglomerates of crumpled nanosheets. Reproduced with permission from ACS [38]

presence of metallic aluminum in them. The average particle size is 85 nm (Fig. 1b). According to the X-ray diffraction (XRD) analysis, the main peaks in the XRD pattern correspond to the Al and AlN phases (Fig. 1c).

As the result of AlN/Al nanoparticles oxidation in water at 60 °C the agglomerates of size up to 2 μm, consisting of nanosheets of size up to 200 nm and thickness of about 2–5 nm, are formed (Fig. 2).

According to the X-ray phase analysis (Fig. 3), the main peaks of the AlN/Al oxidation products correspond to crystalline boehmite. Both the shift of the reflection in the (020) plane toward smaller angles and the broadening of the peaks indicate the absence of long-range order in the arrangement of atoms and the amorphous structure, which is characteristic for pseudo-boehmite.

Fig. 3 XRD pattern of the agglomerates of AlOOH crumpled nanosheets

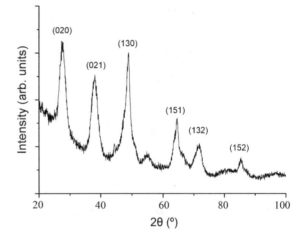

Nitrogen N_2 adsorption-desorption isotherms obtained for AlOOH nanosheet agglomerates have a hysteresis loop in the region of capillary condensation at a relative pressure of $P/Po \geq 0.4$. Such isotherms are typical for mesoporous materials with slit-shaped pores. The maximum of pore size distribution, determined by the method of Barrett, Joyner and Halenda, is within the region of 4 nm. The specific surface area, determined by Brunauer-Emmett-Teller method was 286 m^2/g. The slit-like structure of the mesopores and the large specific surface area of the AlOOH nanosheets agglomerates provide them with high adsorption activity (Fig. 4).

Figure 5 shows synthesized nanostructures zeta-potential dependency on pH level of the medium. The nanostructures have a large positive zeta-potential of about 30 mV at pH ~6.5–8.0 and 37 °C. The pH of the zero charge point pH_{IEP} of AlOOH nanostructures was estimated as 9.53.

The above results suggest nonequilibrium morphology and structural state of AlOOH nanosheets agglomerates. The crumpled structure causes local deformations and multiple micro-stresses, which contribute to the formation of a larger number of surficial active centers due to surface defect zones. It should be noted that the resulting folded crumpled nanostructures due to the presence of a large number of (active) hydroxyl groups per unit surface area can acquire a significant pH-dependent electric charge. Such aluminum nanomaterial is an amphoteric compound, which in the case of an acidic and even neutral medium relatively easily releases the OH-groups into the solution. Dissociation of OH^- groups is facilitated in the areas of defects of the nanomaterial, similarly to the case of hydroxyl groups deprotonation, observed in cationic clays [39]. Therefore, it can be assumed that AlOOH nanosheets with a

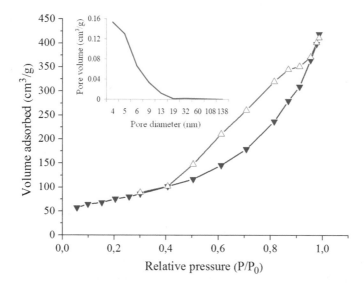

Fig. 4 The N_2 adsorption/desorption isotherms and pore size distributions of AlOOH crumpled nanosheets

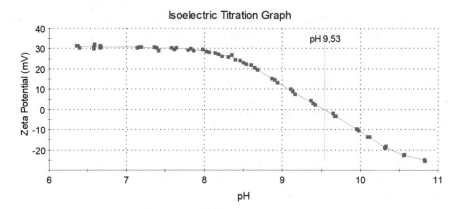

Fig. 5 Dependency of zeta potential of AlOOH crumpled nanosheets on pH of the medium at 37 °C

disordered defective surface structure will have a larger positive charge in comparison with regular aluminum oxyhydroxide.

3 Anticancer Activity of Radially Assembled Al Hydroxide Crumpled Nanosheets

Applied to tumor microenvironment, interaction of the positively charged crumpled nanosheets with extracellular anionic species can lead to disturbance of extracellular ion concentration, and thus result in anti-cancer activity. To investigate the anticancer activity of the agglomerates of radially assembled crumpled AlOOH nanosheets, termed Alohene ("Aloohene") referring to its chemical formula AlOOH, we have therefore performed a series of in vitro and in vivo studies using different cancerous cell lines and a mouse melanoma tumor model, respectively.

3.1 Effect of Aloohene on Tumor Cells Viability and Proliferation in Vitro

First, we have investigated the effect of the synthesized nanomaterial on the viability of three human cancer cell lines (MCF-7, UM-SCC-14C and Hela) cells. After 24 h incubation of Aloohene, a significant decrease in tumor cell viability was measured for all tested cells compared to the non-treated controls (Fig. 6a). Furthermore, a 30–37% decrease in proliferation of the tested cells was detected using the BrdU (5-bromo-2′-deoxyuridine) assay (Fig. 6b). Taken together, these data indicate that Aloohene affects tumor cells viability, most likely through inhibition of their proliferation.

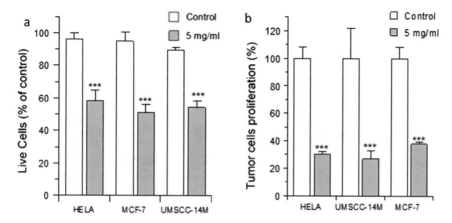

Fig. 6 Analysis of cellular toxicity of Aloohene. **a** FACS analysis of Annexin V and PI staining of Hela, MCF-7 and UM-SCC-14C cells with or without co-culture with 5 mg/ml of Aloohene at 37 °C for 24 h. Fluorescence intensity was measured by flow cytometry and data were analyzed by the Cell Quest software. **b** Proliferation of Hela, UM-SCC-14C and MCF-7 cells, as measured by BrDU assay. Cell cultures were incubated with Aloohene (5 mg/ml) at 37 °C for 24 h. Fluorescence intensity was measured 48 h after BrDU labeling at excitation and emission wavelengths of 370 nm and 470 nm, respectively. Results are means of 3 independent experiments. ***$p < 0.001$. Reproduced with permission from ACS [38]

3.2 Evaluation of Antitumor Activity of Aloohene in Mouse Model of Cancer

We next investigated whether the in vitro cancer cell growth inhibition effect of Aloohene would translate into antitumor efficacy in vivo in a melanoma mouse cancer model based on intradermal administration of 5×10^4 B16F10 cells into the C56BL/6J mice. When the tumor volume reached 70 mm^3, mice were treated with Aloohene, doxorubicin and a combination of both reagents. Aloohene was administered at a dose of 10 mg/ml weekly via two intratumoral injections and doxorubicin was injected intraperitoneally at a single dose of 10 mg/kg. The horizontal and vertical tumor diameters were measured by a digital calliper every second day until the end of treatment and volume was calculated using the formula $V = \frac{\pi}{6}ab^2$ where a and b are the longer and shorter diameter of the tumour, respectively. After two weeks of treatment, Aloohene treatment resulted in a significant decrease of tumor growth as compared to the vehicle treated control animals (Fig. 7). A combination therapy with doxorubicin, the standard-of-care anticancer drug, and Aloohene demonstrated a much stronger suppressive effect on tumor growth, also surpassing the anti-tumor effect of doxorubicin alone (Fig. 7).

To address the mechanism of tumor growth inhibition by Aloohene, we have measured the markers of tumor cells proliferation, cell death and vascularization. Notably, a significant decrease in the proliferation rate for all treatment regimens as compared to the control group was measured by immunohistochemical (IHC)

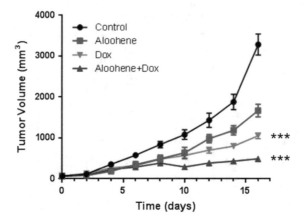

Fig. 7 Antitumor effect of Aloohene in vivo in a mouse melanoma model. **a** Effect of treatment with Aloohene alone and in combination with doxorubicin. Mice were treated with 10 mg/ml of Aloohene, 10 mg/kg of doxorubicin and their combination, and tumor volumes were measured twice a week. Data are presented as mean tumor volume $\pm$ standard errors of mean ($n = 10$ per treatment group). The statistical significance of differences between the groups was assessed by Student's t-test. ***$p < 0.001$ compared with the control group. Reproduced with permission from ACS [38]

staining for the proliferation marker Ki67 (Fig. 8), with the most profound effect detected after the combinatorial treatment of doxorubicin and Aloohene.

Further, the level of tumor necrosis, which could be indicative of treatment effect, was assessed by TUNEL (terminal deoxynucleotidyl transferase dUTP nick end labelling) staining. Large areas of dead cells were detected by this method in tumors treated with Aloohene, doxorubicin and their combination unlike in the control group ($p < 0.05$) (Fig. 9).

No difference in distribution of the endothelial cell marker CD31 was detected in the analyzed tumor sections, suggesting that tested treatments had no effect on tumor vascularization (Fig. 10). Taken together, these results demonstrate a significant in vivo antitumor effect of Aloohene and its potential to evolve into a novel strategy of cancer treatment, possibly in combination with established chemotherapy drugs, such as doxorubicin.

4 Summary

Collectively, our data demonstrates that AlOOH nanosheets with a disordered defective surface structure that have a large positive charge are capable of disturbing tumor-microenvironment extracellular ion balance, therefore representing a novel class of inorganic materials exhibiting strong antitumor effect. Here we report that Al hydroxide nanostructures in the form of agglomerates of crumpled and radially assembled nanosheets (Aloohene) trigger cancer cell death in vitro and inhibit

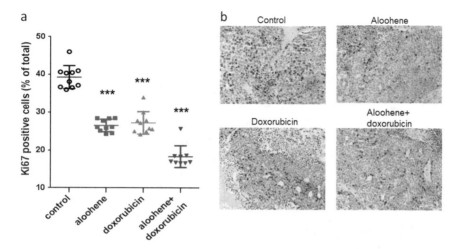

Fig. 8 Effect of Aloohene on tumor proliferation in vivo in mouse melanoma model. **a** Cell proliferation in primary melanomas determined by immunodetection of Ki67 in non-treated mice and mice treated with doxorubicin, Aloohene or their combination. Ki67-positive cells calculated as percentage of total cells. **b** Corresponding illustrative images of Ki67 staining in tumors of all tested groups. ***$p < 0.001$ compared with the control group. Reproduced with permission from ACS [38]

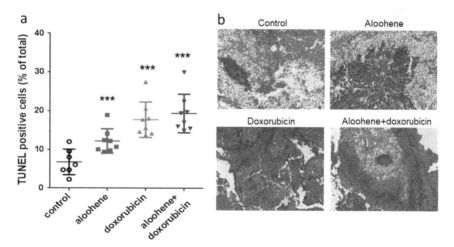

Fig. 9 Effect of Aloohene on tumor necrosis in vivo in mouse melanoma model. **a** The percentage of areas of dead cells was determined on high-power fields of primary melanomas of non-treated mice and mice treated with doxorubicin, Aloohene or their combination, based on immunohistochemical detection by terminal dUTP nick-end labeling staining (brown areas). The percentage of proliferative, necrotic and apoptotic cells are presented as means and standard errors, n = 10. **b** Representative images are shown for control mice and mice treated with Aloohene, doxorubicin, and their combination. The statistical significance of differences between the groups was assessed by Student's t-test. ***$p < 0.001$ compared with the control group. Reproduced with permission from ACS [38]

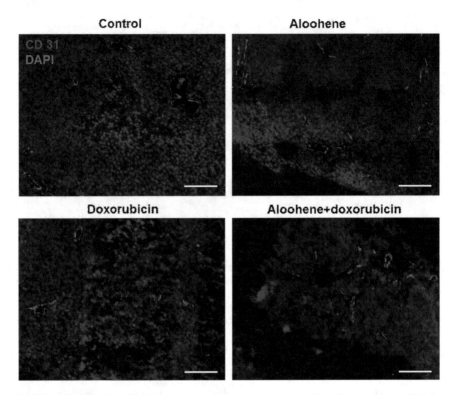

Fig. 10 Vascularization of skin melanoma tumors after treatment with Aloohene, doxorubicin or their combination. Representative images of immunofluorescence staining of the endothelial cell specific marker CD31 (red staining) in cryopreserved tumor sections. The scale bar corresponds to 200 μm. Reproduced with permission from ACS [38]

tumor growth in vivo. The disturbing effect of Aloohene on ion balance in the tumor-microenvironment is supported by our direct molecular dynamics simulation [38]. Furthermore, our results show that Aloohene nanostructures can potentiate the anticancer action of the cytotoxic agent doxorubicin and thus could improve the efficacy of state-of-the-art chemotherapy when used in combination. The findings of the present research highlight the important role of Aloohene, a novel class of tumor-microenvironment dysregulating nanomaterial based on AlOOH nanosheets agglomerates, in the development of more effective anticancer strategies.

Acknowledgements This work was inspired and motivated by Professor S. Psakhie. His visionary ideas for innovative scientific approaches layered foundation for multiple projects we have been collaborating over the years.

Author Contributions AL investigated the patterns of synthesis and the physicochemical characteristics of crumpled aluminum hydroxide nanostructures. G. M. and O. V. evaluated the effects of Aloohene on tumor cell growth inhibition in vitro and in vivo.

References

1. Yu MF, Files BS, Arepalli S, Ruoff RS (2000) Tensile loading of ropes of single wall carbon nanotubes and their mechanical properties. Phys Rev Lett 84(24):5552–5555
2. Marie X, Urbaszek B (2015) 2D materials: ultrafast exciton dynamics. Nat Mater 14(9):860–861
3. Hayamizu Y, So CR, Dag S, Page TS, Starkebaum D, Sarikaya M (2016) Bioelectronic interfaces by spontaneously organized peptides on 2D atomic single layer materials. Sci Rep 6:33778
4. Wang ZL (2004) Nanostructures of zinc oxide. Materialstoday 7(6):26–33
5. Mamalis AG (2007) Recent advances in nanotechnology. J Mater Process Technol 181(1):52–58
6. Tallinen T, Åström JA, Timonen J (2009) The effect of plasticity in crumpling of thin sheets. Nat Mater 8(1):25–29
7. Zang J, Ryu S, Pugno N, Wang Q, Tu Q, Buehler MJ, Zhao X (2013) Multifunctionality and control of the crumpling and unfolding of large-area graphene. Nat Mater 12(4):321–325
8. Wang Q, Zhao X (2015) A three-dimensional phase diagram of growth-induced surface instabilities. Sci Rep 5:8887
9. Holmes DP, Ursiny M, Crosby AJ (2008) Crumpled surface structures. Soft Matter 4(1):82–85
10. Lozhkomoev AS, Glazkova EA, Bakina OV, Lerner MI, Gotman I, Gutmanas EY, Kazantsev SO, Psakhie SG (2016) Synthesis of core–shell AlOOH hollow nanospheres by reacting Al nanoparticles with water. Nanotechnology 27(20):205603
11. Cai W, Chen S, Yu J, Hu Y, Dang C, Ma S (2013) Template-free solvothermal synthesis of hierarchical boehmite hollow microspheres with strong affinity toward organic pollutants in water. Mater Chem Phys 138(1):167–173
12. Tsukanov AA, Psakhie SG (2016) Energy and structure of bonds in the interaction of organic anions with layered double hydroxide nanosheets: a molecular dynamics study. Sci Rep 6:19986
13. Bakina OV, Svarovskaya NV, Glazkova EA, Lozhkomoev AS (2015) Flower-shaped ALOOH nanostructures synthesized by the reaction of an AlN/Al composite nanopowder in water. Adv Powder Technol 26(6):1512–1519
14. Mikhaylov G, Mikac U, Magaeva AA, Itin VI, Naiden EP, Psakhye I, Babes L, Reinheckel T, Peters C, Zeiser R, Bogyo M, Turk V, Psakhye SG, Turk B, Vasiljeva O (2011) Ferri-liposomes as an MRI-visible drug-delivery system for targeting tumours and their microenvironment. Nat Nanotechnol 6(9):594–602
15. Thakor AS, Gambhir SS (2013) Nanooncology: the future of cancer diagnosis and therapy. CA Cancer J Clin 63(6):395–418
16. Allegra A, Penna G, Alonci A, Rizzo V, Russo S, Musolino C (2011) Nanoparticles in oncology: the new theragnostic molecules. Anti-Cancer Agents Med Chem (Formerly Curr Med Chem-Anti-Cancer Agents) 11(7):669–686
17. Joseph D, Tyagi N, Geckeler C, Geckeler KE (2014) Protein-coated pH-responsive gold nanoparticles: microwave-assisted synthesis and surface charge-dependent anticancer activity. Beilstein J Nanotechnol 5(1):1452–1462
18. Akhtar MJ, Ahamed M, Kumar S, Khan MM, Ahmad J, Alrokayan SA (2012) Zinc oxide nanoparticles selectively induce apoptosis in human cancer cells through reactive oxygen species. Int J Nanomed 7:845–857
19. Bisht G, Rayamajhi S (2016) ZnO nanoparticles: a promising anticancer agent. Nanobiomedicine (Rij) 3:9
20. Kulkarni M, Mazare A, Gongadze E, Perutkova Š, Kralj-Iglič V, Milošev I, Schmuki P, Iglič A, Mozetič M (2015) Titanium nanostructures for biomedical applications. Nanotechnology 26(6):062002
21. Elhissi AMA, Ahmed W, Hassan IU, Dhanak VR, D'Emanuele A (2011) Carbon nanotubes in cancer therapy and drug delivery. J Drug Deliv 2012, Article ID 837327
22. Theodosopoulos GV, Bilalis P, Sakellariou G (2015) Polymer functionalized graphene oxide: a versatile nanoplatform for drug/gene delivery. Curr Org Chem 19(18):1828–1837

23. Kumar VB, Kumar K, Gedanken A, Paik P (2014) Facile synthesis of self-assembled spherical and mesoporous dandelion capsules of ZnO: efficient carrier for DNA and anti-cancer drugs. J Mater Chem B 2(5):3956–3964
24. Lin N, Sun JM, Hsiao J-H, Hwu Y (2009) Spontaneous emergence of ordered phases in crumpled sheets. Phys Rev Lett 103(26):263902
25. Wang T, Wan Y, Zhanqiang L (2016) Fabrication of hierarchical micro/nanotopography on bio-titanium alloy surface for cytocompatibility improvement. J Mater Sci 51(21):9551–9561
26. Ulanova M, Tarkowski A, Hahn-Zoric M, Hanson LÅ (2001) The common vaccine adjuvant aluminum hydroxide up-regulates accessory properties of human monocytes via an interleukin-4-dependent mechanism. Infect Immun 69(2):1151–1159
27. Dong E, Wang Y, Yang S-T, Yuan Y, Nie H, Chang Y, Wang L, Liu Y, Wang H (2011) Toxicity of nano gamma alumina to neural stem cells. J Nanosci Nanotechnol 11(9):7848–7856
28. Xie Y, Kocaefe D, Kocaefe Y, Cheng J, Liu W (2016) The effect of novel synthetic methods and parameters control on morphology of nano-alumina particles. Nanoscale Res Lett 11(1):1–11
29. Lerner MI, Svarovskaya NV, Psakhie SG, Bakina OV (2009) Production technology, characteristics, and some applications of electric-explosion nanopowders of metals. Nanotechnol Russ 4(11):741–757
30. Svarovskaya NV, Bakina OV, Glazkova EA, Lerner MI, Psakh'e SG (2010) The formation of nanosheets of aluminum oxyhydroxides from electroexposive nanopowders. Russ J Phys Chem A 84(9):1566–1569
31. Yang Y, Gai W-Z, Deng Z-Y, Zhou J-G (2014) Hydrogen generation by the reaction of Al with water promoted by an ultrasonically prepared Al(OH)₃ suspension. Int J Hydrogen Energy 39(33):18734–18742
32. Gai W-Z, Deng Z-Y (2014) Effect of initial gas pressure on the reaction of Al with water. Int J Hydrogen Energy 39(25):13491–13497
33. Rosenband V, Gany A (2010) Application of activated aluminum powder for generation of hydrogen from water. Int J Hydrogen Energy 35(20):10898–10904
34. Deng Z-Y, Ferreiraw JMF, Tanaka Y, Ye JH (2007) Physicochemical mechanism for the continuous reaction of gamma-Al₂O₃-modified aluminum powder with water. J Am Ceram Soc 90(5):1521–1526
35. Sarathi R, Sankar B, Chakravarthy SR (2010) Influence of nano aluminium powder produced by wire explosion process at different ambience on hydrogen generation. J Electr Eng-Elektrotechnicky Cas 61(4):215–221
36. Ivanov VG, Safronov MN, Gavrilyuk OV (2001) Macrokinetics of oxidation of ultradisperse aluminum by water in the liquid phase. Combust Explosion Shock Waves 37(2):173–177
37. Bakina OV, Svarovskaya NV, Glazkova EA, Lozhkomoev AS, Khorobraya EG, Lerner MI (2015) Flower-shaped ALOOH nanostructures synthesized by the reaction of an AlN/Al composite nanopowder in water. Adv Powder Technol 26(6):1512–1519
38. Lerner MI, Mikhaylov G, Tsukanov AA, Lozhkomoev AS, Gutmanas E, Gotman I, Bratovs A, Turk V, Turk B, Psakhye SG, Vasiljeva O (2018) Crumpled aluminum hydroxide nanostructures as a microenvironment dysregulation agent for cancer treatment. Nano Lett 18(9):5401–5410
39. Choy J-H, Park M (2004) Clay surfaces: fundamentals and applications. Academic Press, Cambridge, pp 403–424 (Cationic and anionic clays for biological applications)

Influence of Lattice Curvature and Nanoscale Mesoscopic Structural States on the Wear Resistance and Fatigue Life of Austenitic Steel

Viktor E. Panin, Valery E. Egorushkin, and Natalya S. Surikova

Abstract The gauge dynamic theory of defects in a heterogeneous medium predicts the nonlinearity of plastic flow at low lattice curvature and structural turbulence with the formation of individual dynamic rotations at high curvature of the deformed medium. The present work is devoted to the experimental verification of the theoretical predictions. Experimentally studied are the influence of high-temperature radial shear rolling and subsequent cold rolling on the internal structure of metastable Fe–Cr–Mn austenitic stainless steel, formation of nonequilibrium ε- and α′-martensite phases, appearance of dynamic rotations on fracture surfaces, fatigue life in alternating bending, and wear resistance of the material. Scratch testing reveals a strong increase in the damping effect in the formed hierarchical mesosubstructure. The latter is responsible for a nanocrystalline grain structure in the material, hcp ε martensite and bcc α′ martensite in grains, a vortical filamentary substructure on the fracture surface as well as for improved high-cycle fatigue and wear resistance of the material. This is related to a high concentration of nanoscale mesoscopic structural states, which arise in lattice curvature zones during high-temperature radial shear rolling combined with smooth-roll cold rolling. These effects are explained by the self-consistent mechanical behavior of hcp ε-martensite laths in fcc austenite grains and bcc α′-martensite laths that form during cold rolling of the steel subjected to high-temperature radial shear rolling.

Keywords Gauge dynamic theory of defects · Nanoscale mesoscopic structural states · Lattice curvature · Damping effect · Dynamic rotations · Fatigue failure · Wear resistance

V. E. Panin (✉) · V. E. Egorushkin · N. S. Surikova
Institute of Strength Physics and Materials Science SB RAS, Tomsk 634021, Russia
e-mail: paninve@ispms.tsc.ru

N. S. Surikova
e-mail: surikova@ispms.ru

V. E. Panin
National Research Tomsk Polytechnic University, Tomsk 634050, Russia

National Research Tomsk State University, Tomsk 634050, Russia

G.-P. Ostermeyer et al. (eds.), *Multiscale Biomechanics and Tribology of Inorganic and Organic Systems*, Springer Tracts in Mechanical Engineering,
https://doi.org/10.1007/978-3-030-60124-9_11

1 Introduction

Mechanical behavior of metals were subject of intensive theoretical and experimental studies over many decades. Conventionally, mechanisms of mechanical behavior under various loading conditions are associated with various-scale strain-induced defects, including cracks, in a translation-invariant crystal lattice. However, in most cases, the translational invariance of the crystal lattice in a deformed solid is strongly violated, which does not allow for a correct description of plastic deformation and fracture within the linear approach of Newtonian mechanics. In this work, we thoroughly study the mechanical behavior of Fe–Cr–Mn austenitic steel, in which complex structural phase transformations of the initial fcc lattice with the formation of hcp and bcc martensite phases occur in uniaxial tension. According to [1, 2], crystal lattice transformations are related to the appearance of nanoscale mesoscopic structural states at the interstices of lattice curvature zones. Thus, the account for lattice curvature provides a basis for nonlinear solid mesomechanics.

In this work, crystal lattice curvature throughout the Fe–Cr–Mn austenitic steel specimen is produced using the complex treatment by high-temperature radial shear rolling + cold rolling at room temperature. This complex treatment reveals important new regularities in the mechanical behavior of austenitic steel.

2 Gauge Dynamic Theory of Defects in the Heterogeneous Medium

2.1 Basic Equations of the Gauge Theory

A deformable solid as a multilevel hierarchically organized system requires a self-consistent description at the nano-, micro-, meso- and macroscopic levels. Gauge dynamic theory of defects in the heterogeneous medium provides a framework for such description. In this theory, fluxes of strain-induced defects J and their density α are described by the equations

$$\frac{\partial}{\partial x_\alpha} J_\mu^\alpha = -\frac{\partial \ln u_\alpha(x, t)}{\partial t}, \tag{1}$$

$$\varepsilon_{\mu\chi\delta} \frac{\partial J_\delta^\alpha}{\partial x_\chi} = -\frac{\partial \alpha_\mu^\alpha}{\partial t}, \tag{2}$$

$$\frac{\partial \alpha_\mu^\alpha}{\partial x_\alpha} = 0, \tag{3}$$

$$\varepsilon_{\mu\chi\delta} \frac{\partial \alpha_\delta^\alpha}{\partial x_\chi} = \frac{1}{c^2} \frac{\partial J_\mu^\alpha}{\partial t} + \frac{\partial \ln u_\beta(x, t)}{\partial x_\nu} \frac{C_{\alpha\beta}^{\mu\nu}}{E} - P_\nu^\beta \frac{C_{\alpha\beta}^{\mu\nu}}{E}, \tag{4}$$

$$\frac{1}{c^2}\frac{\partial^2 \ln u_\alpha(x,t)}{\partial t^2} + \frac{\partial^2 \ln u_\beta(x,t)}{\partial x_\mu \partial x_\nu}\frac{C_{\alpha\beta}^{\mu\nu}}{E} = -P_\nu^\beta \frac{C_{\alpha\beta}^{\mu\nu}}{E} \tag{5}$$

Equations (1)–(5) have the following sense and use the following notations:

(1) is the continuity equation for a defected medium, which indicates that the source of plastic flow causes the defect flow rate;

(2) is the condition of plastic-strain compatibility, the time variation of the medium density is determined in this case by the flux rotor, i.e. its heterogeneity, rather than by the divergence;

(3) is the condition of continuity of defects, which corresponds to the absence of charges of the rotational component of the plastic strain field $\left(\alpha_\chi^\beta = \varepsilon_{\chi\mu\nu}\partial_\mu P_\nu^\beta\right)$;

(4) is the governing equation for the medium with plastic flow;

(5) is the equation of quasi-elastic equilibrium, which presents the known continuum mechanics equation. Along with the elastic strain, it contains plastic distortions on the right-hand side. In fact, this summand corresponds to the nucleation of strain-induced defects in local zones of hydrostatic tension due to stress concentrators.

Expression (4) is applicable only to the medium with plastic flow. It relates the time variation in plastic flow to the anisotropic spatial variation in the defect density $\varepsilon_{\mu\chi\delta}\partial\alpha_\delta^\alpha/dx$ and sources $\left(\sigma_\mu^\alpha - P_\nu^\beta C_{\alpha\beta}^{\mu\nu}/E\right)$. Equations (4) and (5) differ from the corresponding elastic equations in that the time variation in plastic strain rate is determined by the stresses themselves rather than by $\partial\sigma_\mu^\alpha/\partial x$, as in the elastic case. In addition, the right-hand side of (4) includes plastic distortion $P_\nu^\beta(x,t)$ taken as sources, which indicates the duality of defects as field sources.

From the system of Eqs. (1)–(5), wave equations can be found for the dimensionless quantities of defect flux J and density α:

$$\frac{1}{c^2}\frac{\partial^2 J_\alpha^\mu}{\partial t^2} - \frac{\partial^2 J_\alpha^\mu}{\partial x^2} = \frac{\partial}{\partial t}\left\{\frac{\partial \ln u_\alpha(x,t)}{\partial x_\mu} - \frac{1}{E}\frac{\partial \ln u_\beta}{\partial x_\nu}C_{\alpha\beta}^{\mu\nu} - \frac{1}{E}P_\nu^\beta C_{\alpha\beta}^{\mu\nu}\right\} \tag{6}$$

$$\frac{1}{c^2}\frac{\partial^2 \alpha_\alpha^\mu}{\partial t^2} - \frac{\partial^2 \alpha_\alpha^\mu}{\partial x_\nu^2} = \varepsilon_{\mu\chi\sigma}\left\{\frac{\partial^2 \ln u_\beta(x,t)}{\partial x_\chi \partial x_\nu}C_{\alpha\beta}^{\mu\nu} - \frac{\partial P_\nu^\beta}{\partial x_\chi}C_{\alpha\beta}^{\mu\nu}\right\}\frac{1}{E}, \tag{7}$$

subject to the source compatibility condition

$$\frac{\partial N_\mu}{\partial t} + \varepsilon_{\mu\ell m}\frac{\partial M_\mu}{\partial x_\ell} = 0, \tag{8}$$

where M is the right-hand side of expression (6), N is the right-hand side of expression (7), and $u(x,t)$ is the inelastic displacements in the wave of inelastic localized strain.

The right-hand side of Eq. (6) characterizes defect flow sources. They are determined by the rate of quasi-elastic strain $\frac{\partial}{\partial t}\left(E_\mu^\alpha E - E_\nu^\beta C_{\alpha\beta}^{\mu\nu}\right)\frac{1}{E}$. Parenthesized is the difference between internal compressive (tensile) stresses and shear stresses

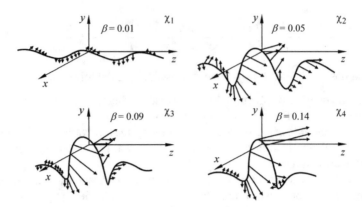

Fig. 1 Shape and velocity of plastic deformation depending on the curvature of the deformed region

associated with the stress distribution in the stress concentration zone. Relaxation processes of defect rearrangement (such as various atomic configurations or their conglomerates) are presented in (6) by the term $P_\nu^\beta C_{\alpha\beta}^{\mu\nu}/E$.

The right-hand side of Eq. (7) expresses a source of the strain-induced defect density. This is vorticity $\varepsilon_{\mu\chi\delta}\frac{\partial}{\partial x}\left(E_\nu^\beta - P_\nu^\beta\right)\frac{C_{\alpha\beta}^{\mu\nu}}{E}$ of shear strain induced by the shear stress relaxation in local zones of hydrostatic tension during the defect formation.

The wave pattern of strain-induced defect flows is defined by the right-hand side of Eqs. (6) and (7). Plastic distortion $P_\nu^\beta(x, t)$ plays a major role in the wave pattern of localized plastic flow.

Equation (6) for the strain-induced defect flux along the direction L (at $r < L$) is solved as

$$\overrightarrow{J} = \frac{b_1 - b_2}{4\pi}\chi(s, t)\overrightarrow{b}(s, t)\left(\ln\frac{2L}{r} - 1\right) - \nabla f, \tag{9}$$

where $\overrightarrow{b}$ is the binormal vector in the local coordinate system, $\vec{n}$ is the normal, $\vec{t}$ is the tangent, χ is the curvature variation in a deformed region due to external loading, s is the current length of the region, b_1 and b_2 are the Burgers vector moduli of the bulk translational and subsurface rotational inconsistency, respectively, ∇f is the gradient part of the flux due to third-party sources.

The expression for the defect flux includes curvature χ of the deformed medium. Let us analyse the role of this factor in the behavior of the strain-induced defect flow.

2.2   Structural Turbulence at Severe Lattice Curvature

The wave pattern of localized plastic flow in Eqs. (6) and (7) is generally shown in Fig. 1. It demonstrates the dependence of the wave profile of localized plastic flow on the curvature of the deformed region $\chi(x, t) = 4\beta \text{sech}[2\beta(x + 4\upsilon t)]$.

In Fig. 1, it can be seen that an increase in the curvature of the deformed region greatly changes the shape and velocity of plastic deformation. The experiment confirms the theoretical prediction [1].

At a high curvature of the deformed medium, plastic distortion $P_\upsilon^\beta(x, t)$ strongly grows, too. This means that the right-hand sides of Eqs. (6) and (7) increase significantly. The wave pattern of plastic flow cannot be preserved in this case, and plastic deformation becomes turbulent, breaking up into individual dynamic rotations. Such a theoretical prediction of structural turbulence in a deformed solid was impossible under the conditions of translational invariance of the crystal. The violation of translational invariance and account of curvature in a deformed medium are fundamentally new. Their experimental confirmation was earlier obtained [2].

Severe curvature of plastic flow was previously achieved by introducing titanium carbonitride nanoparticles into a deformed solid [2]. Unlike carbides, which have a spherical configuration, carbonitrides have a cubic structure [3, 4]. Due to plastic flow in the presence of titanium carbonitride nanoparticles, a severe curvature is formed in the deformed medium, especially under shock loading. Figure 2 shows the fracture surface of low-carbon steel 09Mn2Si when measuring impact toughness with 0.15% titanium carbonitride nanoparticles introduced into the steel. The New View profilometer discovers pronounced dynamic rotations of turbulent flow on the surface in the form of individual vortical protrusions. The scratch test shows a significant decrease in the groove size in the presence of nanoparticles (Fig. 3b), in contrast to

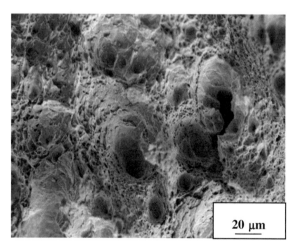

Fig. 2 Structural turbulence on the fracture surface of low-carbon steel 09Mn2Si, with 0.15% TiCN nanoparticles introduced into it

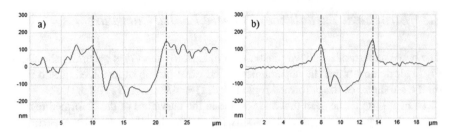

Fig. 3 Cross profiles of the scratch groove in the 09Mn2Si specimens: initial (**a**) and with 0.15% TiCN nanoparticles (**b**)

a deep groove forming in the absence of nanoparticles (Fig. 3a).

Structural turbulence in a solid is characterized by relay-race transfer of momentum from particle to particle. This is what makes it different from classical turbulence characterized by the Avogadro number [5, 6].

Structural turbulence is even more pronounced on the fracture surface of a chevron-notched specimen shown in Fig. 4. Counter shear loads from the specimen notches induce couple stresses that form dynamic rotations (Fig. 5). A filamentary structure appears around the notches, which bears witness not to translational invariance but to the formation of special structural states in lattice curvature zones. These states also arise during the formation of dynamic rotations (Fig. 6). In other words, at the crystal structure curvature the electronic subsystem forms special structural states that are responsible for dynamic rotations by the mechanism of plastic distortion.

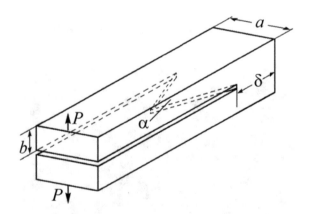

Fig. 4 Geometry of a chevron-notched specimen

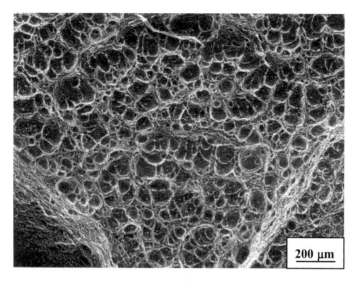

Fig. 5 Formation of a filamentary structure in the vicinity of the edge-cut notches in the chevron-notched 09Mn2Si steel specimen in fracture

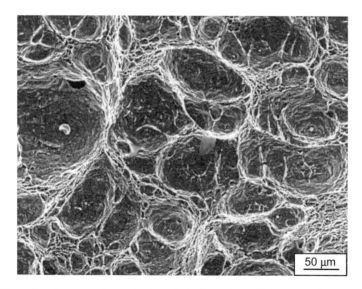

Fig. 6 Formation of noncrystallographic boundaries due to interstitial mesoscopic structural states in turbulent rotations on the fracture surface of a chevron-notched 09Mn2Si steel specimen

3 Role of Lattice Curvature in the Mechanical Behavior of Austenitic Steel

3.1 Influence of High-Temperature Radial Shear Rolling and Subsequent Smooth-Roll Cold Longitudinal Rolling on the Austenitic Steel Microstructure

High-temperature radial shear rolling exerts no effect on the phase composition of austenitic steel. It forms a layered structure with varying degrees of the crystal lattice curvature-torsion and refines the grain structure. In the near-surface layers, the material has a globular structure with the average grain size $d \approx 0.57$ µm; grains in the near-axial zone are elongated along the rolling direction. A small-angle substructure is well developed within the grains. The fraction of special twin boundaries with a misorientation of ~60° is significantly reduced. The yield stress $\sigma_{0.2}$ and ultimate strength σ_B increase, respectively, from 400 to 620 MPa and from 850 to 1050 MPa. The ductility decreases from 90 to 55% (the true strain $\varepsilon = 0.7$).

Radial shear and cold rolling to the cold strain $\varepsilon = 1.8$ changes significantly the austenitic steel microstructure. In austenite grains, a two-phase (ε martensite $+ \alpha'$ martensite) nanocrystalline mesosubstructure is formed. Grains containing α' martensite with a distorted bcc structure prevail. According to X-ray structural estimates, the volume fraction of α martensite after rolling to $\varepsilon_{true} \sim 1.8$ comprises $\approx 85.6\%$, and the size of coherent scattering regions is 40 nm. Cold rolling forms a heterogeneous structure. In the specimen areas with a high plastic strain, the grain size is 40–100 nm. The specimen areas with a lower strain (Fig. 7) contain austenite grains 200–400 nm in size (Fig. 7b) and grains with ferrite zones 30–400 nm in size (Fig. 7d).

Fragmentation processes during cold rolling begin with the formation of fine stacking faults and ε-martensite nuclei (Fig. 8a) or α'-martensite plates (Fig. 8b) in various grains of the material.

The results presented in Figs. 7 and 8 indicate that ε-martensite and α'-martensite laths are important intermediate phases for the structural transformation of fcc austenite to bcc ferrite. Of special note is a strong texture of the intragranular structure. In austenite crystals, the direction and plane of rolling are close to $\langle 111 \rangle_\gamma$ and $\{110\}_\gamma$, respectively; in ferrite zones, to $\langle 110 \rangle_\alpha$ and $\{111\}\alpha_\gamma$.

Thus, within a nanostructured austenitic steel subjected to the complex treatment by radial shear rolling + smooth-roll cold rolling a multiscale hierarchically organized structure is formed that allows translational-rotational modes of plastic deformation from macro to nanoscale levels. Let us consider the mechanisms of such plastic deformation in fracture by uniaxial tension.

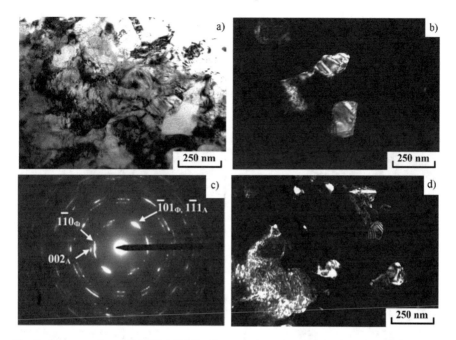

Fig. 7 Structure of steel 12Cr15Mn9NiCu after stepwise radial shear hot rolling to ε ~ 65% and subsequent cold rolling to ε ~ 80%. **a** The bright field; **c** the selected-area diffraction image, shown are the two zones of reflections with azimuthal misorientations: [110] austenite zone and [111] ferrite zone; **b** the dark-field image in the 002$_A$ austenite reflection; **d** the dark-field image in the $1\bar{1}0_\Phi$ ferrite reflection

3.2 Fracture Surface in Uniaxial Tension of Austenitic Steel Specimens After Various Treatments

The fracture mode in tension of austenitic steel in the initial state and after the complex treatment is schematized in Fig. 9. In the initial steel, fracture starts with the propagation of an opening mode crack (Fig. 9a) and ends with the propagation of a tearing mode crack. Specimens subjected to radial shear rolling are fractured only by the propagation of a tearing mode crack (Fig. 9b).

The fracture surfaces of the steel specimens in the initial state (Figs. 10, 11 and 12) bears witness to the spatial stress state in fracture. After the treatment by radial shear rolling + cold rolling, the specimens under tension are fractured in the plane stress state (Fig. 13).

The fracture pattern of the initial steel specimens in the scale hierarchy of the grain structure is shown in Figs. 10, 11 and 12. Grain-boundary sliding of grain conglomerates (Fig. 10) is accompanied by stochastic microcracking and initiates individual accommodation rotations of individual grains with the formation of micropores in them (Fig. 11). A pronounced dimple relief on the opening mode surface (Fig. 12a) accompanied by the formation of microporosity is a sign of ductile fracture [7, 8].

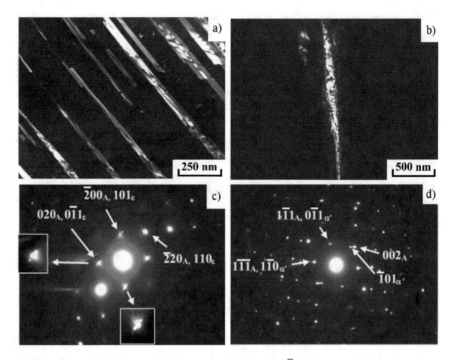

Fig. 8 **a** The dark-field image of ε-martensite laths in the $0\bar{1}1_\varepsilon$ reflection. **c** The selected-area diffraction image for (**a**), shown are reflections in the two zones: [001] austenite zone and [$\bar{1}11$] ε-martensite zone; the arrangement and form of the reflections are given at higher magnification on the left and at the bottom of the image. **b** The dark-field of α′-martensite laths in the $10\bar{1}_\alpha$ reflection. **d** The selected-area diffraction image for (**b**), shown are reflections in the two parallel zones: [110] austenite zone and [111] α′-martensite zone

The micropore formation is associated with the coalescence of vacant lattice sites under the conditions of plastic distortion at the lattice curvature interstices [9, 10].

Quasi-elastic cleavages in the zones of tearing mode cracks (Figs. 11b and 12b) also exhibit microporosity. This shows the important role of accommodation processes of plastic distortion at the nanoscale level at various fracture mechanisms of the initial austenitic steel. In other words, the self-consistency of rotational deformation modes in a wide range of scales, from macro to nano, causes the high ductility $\delta = 90\%$ in uniaxial tension of the initial austenitic steel.

A high concentration of transition martensite phases in fcc austenite grains with ε martensite and in grains with bcc ferrite zones in α′ martensite, which is due to radial shear plus cold rolling of the steel, provides a means for dynamic rotations as a mechanism of structural turbulence (Fig. 13). The phenomenon of structural turbulence in a deformed solid is detailed elsewhere [5, 6]. The presence of nanoscale mesoscopic structural states in the lattice curvature interstices makes it possible to synthesize vortical nanofilaments of a material under the structural turbulence condition. A hierarchy of dynamic rotations on fracture surfaces was numerously

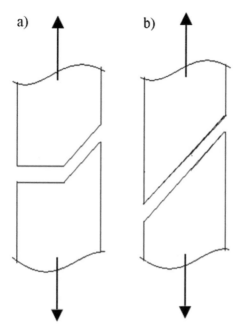

Fig. 9 Schematic of the propagation of the main crack in fracture by uniaxial tension of the specimens of the initial austenite steel (**a**) and after the complex treatment by radial shear rolling + cold rolling (**b**)

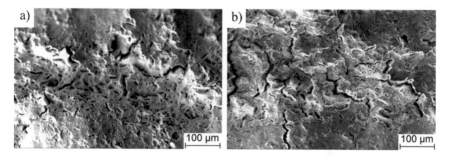

Fig. 10 Stochastic cracks on the fracture surface in tension of the specimens made of the initial austenitic steel: **a** opening mode zone; **b** tearing mode zone

observed [9, 11–13]. However, the mechanism of their formation and their relation to structural turbulence of plastic flow of atoms in lattice curvature zones have not been discussed yet.

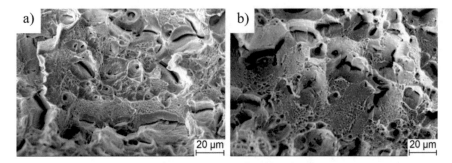

Fig. 11 Rotational mode of a grain conglomerate in the initial steel in the opening mode zone accommodated by rotations of individual grains with the formation of micropores in them (**a**); cleavages of grain conglomerates with the formation of micropores in the tearing mode zone (**b**)

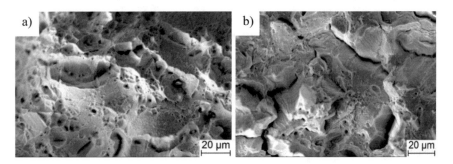

Fig. 12 Ductile dimple fracture of the initial steel in the opening mode zone (**a**); quasi-elastic cleavages of grain conglomerates of the initial steel in the tearing mode zone (**b**)

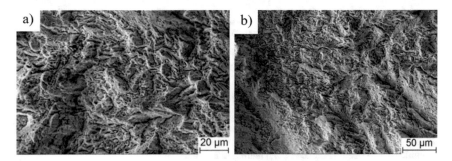

Fig. 13 Dynamic rotations with the formation of a vortical filamentary structure on the fracture surface of the steel specimens after radial shear and cold rolling: initial (**a**) and final (**b**) zones of the main tearing mode crack

3.3 Damping Effect in the Structure of Austenitic Steel After the Treatment by Radial Shear Rolling + Cold Rolling

The damping effect in the heterogeneous internal structure of austenitic steel is studied by scratch testing at the indenter loads 50, 100, and 200 mN. The damping effect is distinct at all the studied loads. Figures 14 and 15 show groove profiles at the initial and treated surfaces of the steel specimens at the indenter load 100 mN.

As can be seen from Fig. 14, the highly plastic initial steel forms a pronounced smooth groove ~170 nm in depth. Bands of plastic shear are visible at the groove surface. A qualitatively different type of grooves is formed at the specimen surface treated by radial shear rolling + cold rolling (Fig. 15). The groove depth is only 15 nm, and its central zone is extruded ~15 nm above the initial surface. No traces of

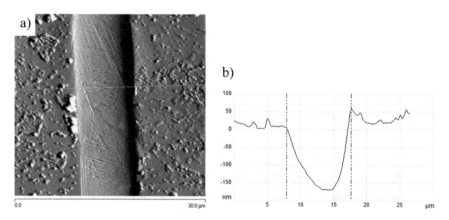

Fig. 14 Smooth surface of a deep groove (**a**) and its profile (**b**) for the steel in the initial state

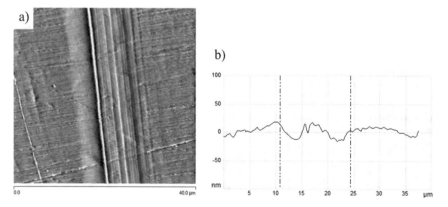

Fig. 15 Corrugated surface of a groove with the damping effect (**a**) and its profile (**b**) for the steel treated by radial shear rolling with subsequent cold rolling

Table 1 Nanohardness H, elastic modulus E^*, and shape recovery factor R at different states of the steel

State	Characteristics		
	H (GPa)	E^* (GPa)	R (%)
Initial	3.7 ± 0.2	229 ± 31	7
After radial shear and cold rolling	4.9 ± 0.3	190 ± 12	14

plastic shear are observed at the groove surface in the treated steel (Fig. 15a). Plastic extrusion is due to nanostructural transformations. The damping effect is very strong, while it is associated with a nonuniform stress distribution at the interface between the groove bottom and the substrate material.

Mechanical characteristics of the austenitic steel in different states are presented in Table 1.

From the tabulated data it is seen that complex radial shear and cold rolling increases the nanohardness H, decreases the elastic modulus E^*, and doubles the shape recovery factor R. The ductile mode of the material extrusion in scratch testing appears not only within the groove, but also in the material on its left and right. All these zones have no traces of plastic shear, but the stress distribution heterogeneity is clearly manifested geometrically at the interface between the surface layer plastically deformed by the moving indenter and the elastic substrate.

3.4 Influence of the Treatment by Radial Shear Rolling + Cold Rolling on the Development of Gigacycle Fatigue and Wear Resistance of Austenitic Steel

Nanostructuring of the austenitic steel and the strong damping effect of the material in scratch testing should increase the fatigue life of the crystal lattice [14, 15]. Moreover, it is known that nanostructuring of a material promotes the development of gigacycle fatigue. This is fully confirmed for the austenitic steel processed by radial shear rolling with subsequent smooth-roll cold rolling. The investigation results are shown in Table 2.

Table 2 Influence of the complex treatment of austenitic steel by radial shear rolling (RShR) + cold rolling (CR) on the characteristics of its fatigue life and wear resistance

Characteristics	State		
	Initial	After RShR	After RShR + CR
The number of cycles to fatigue failure	3 mln		65 mln
Wear rate coefficient, 10^5 mm^3/N m	8.58	8.25	1.30

Table 2 presents two fundamentally important results. First, the obtained data confirm high-cycle fatigue failure (3×10^6 cycles to failure) in the initial austenitic steel and gigacycle fatigue failure (more than 65×10^6 cycles to failure) after radial shear and cold rolling of this steel.[1] Gigacycle fatigue is usually realized at a significant reduction in external applied stresses [15, 16]; however, the tabulated results are derived at a high external stress. In other words, at high external stresses the transition from high-cycle to gigacycle fatigue can occur due to a specific internal substructure formed at the nanoscale structural level. Secondly, the wear resistance of austenitic steel does not vary after high-temperature radial shear rolling. To achieve this requires additional cold rolling that forms bcc ferrite zones in austenite grains of the steel. The mechanism of formation of wear particles is also associated with fatigue failure of the tribocontact. When the counterbody moves along the flat surface of the material, each mesovolume is first compressed and then stretched. Such cyclic deformation causes fatigue fragmentation of the material and the formation of wear particles.

A heterogeneous hierarchical structure formed in austenitic steel during radial shear and cold rolling effectively functions in tribological conditions. When the counterbody compresses mesovolumes of the heterogeneous austenitic steel, bcc ferrite grains are elastically compressed and hcp ε-martensite laths are embedded into the close-packed fcc austenite structure by the mechanism of forward + reverse martensitic transformation. In subsequent tension of this mesovolume, the hcp ε-martensite laths are recovered at the interstitial nanoscale structural states, and local stresses in the bcc ferrite grains relax. These processes are reversible and significantly retard plastic deformation, cracking, and the formation of wear particles. We emphasize that this effect is also associated with reversible structural transformations at the nanoscale structural level, where nanoscale mesoscopic structural states can exist at the lattice curvature interstices.

4   Structural Turbulence and Gigacycle-Fatigue Processes in a Solid with Lattice Curvature

4.1   Structural Turbulence of Plastic Flow at Lattice Curvature and in the Presence of Nanoscale Mesoscopic Structural States at Its Curvature Interstices

No turbulent plastic flow can exist in a translation-invariant crystal. However, the appearance of lattice curvature zones and of nanoscale mesoscopic structural states at the lattice curvature interstices radically changes the mechanisms of plastic deformation and fracture of solids. This concerns the effect of plastic distortion [11],

---

[1] At the time of publication the cyclic loading experiment for specimens after radial shear and cold rolling is being continued.

formation of a vortical filamentary mesosubstructure [9, 12], structural turbulence, and dynamic rotations [2, 10, 13].

Structural turbulence of plastic flow was predicted when modeling grain-boundary sliding by the excitable cellular automaton method for grain boundaries that lack translational invariance [17]. In this case, consideration was given to lattice curvature at grain boundaries and in near-boundary regions.

Clusters of excess vacancies, that number in ~500 at the cluster size ~3.5 nm, in localized strain bands were viewed by Matsukawa and Zinkle [18] in the transmission electron microscope column in tension of a gold foil. Tetrahedra of stacking faults form in such vacancy clusters, which can move in the $\langle 110 \rangle$ direction at the migration energy $E_t = 0.19$ eV. Recall that the migration energy of a single vacancy in gold is $E_V = 0.85$ eV. This means that the migration of vacancy tetrahedra is not a diffusion process, but it is associated with structural transformations of nanostructured tetrahedra in the $\langle 110 \rangle$ direction.

As noted above, coherent scattering regions of the size ~40 nm appear in the structure of the martensite phase, whose volume amounts to as much as 85.6%. This is a very important nanostructural element, which contributes to the formation of nanostructured grains in austenite and ferrite grains, as demonstrated above in Figs. 1 and 2. Obviously, both in localized strain bands and in nonequilibrium martensite laths in Figs. 7 and 8, a variety of structural configurations can form: highly mobile stacking fault tetrahedra, slow-moving stacking fault octahedra, misoriented nanofragments, nanograins of various composition, including nano carbides, nanocarbonitrides, and others. The presence of the 85.6% nonequilibrium martensite phase in the metastable austenitic steel with coherent scattering regions of the size ~40 nm makes possible reversible structural-phase transformations in the steel specimens under cyclic loading. The mechanism of such transformations is discussed below.

4.2 Influence of the Mechanism of Reversible Structural-Phase Transformations on Gigacycle Fatigue and Wear Resistance Increase in Austenitic Steel After Radial Shear and Cold Rolling

Since individual volumes of a specimen periodically undergo alternating tension-compression under cyclic loading, this process can be reversible without cracking only subject to the condition of reversible structural-phase transformations. Nanostructured fcc austenite grains and bcc ferrite zones have different yield stresses and are surrounded by the martensite phase, which arises on the basis of interstitial nanoscale structural states in lattice curvature zones that lack translational invariance. Laths of the hcp ε martensite in compression can be embedded into the fcc austenite structure, transforming into its close-packed configuration. This governs inelastic compression deformation of the specimen. Laths of the bcc α′ martensite in compression will elastically change the spatial orientation of covalent d bonds and generate local stresses. In tension under cyclic loading, elastic stresses in the bcc ferrite zone will relax and cause a recovery of the hcp ε martensite in the austenite grains, implementing inelastic tensile deformation. Such processes of structural transformations are reversible in nanostructured materials [19–21], which determines the damping effect in their structure under cyclic loading, an increase in the gigacycle fatigue life and wear resistance.

A similar damping effect is revealed during scratch testing (Fig. 15). When the indenter moves during scratch testing, the martensite phase is first compressed. Laths of the hcp ε martensite transform their structure into the fcc lattice of close-packed austenite. Spatially oriented along the cube diagonals, structural elements of the bcc α' martensite associated with d electrons undergo quasi-elastic compression. After the indenter passes, the ε-martensite recovers its hcp structure, and the groove in Fig. 15 exhibits a damping effect. Relaxation of high local elastic stresses in the α'-martensite initiates the groove recovery after the indenter passes. Thus, structural transformations in austenitic steel after high-temperature radial shear and cold rolling are indeed reversible under cyclic external influences.

An increase in the fatigue life of austenitic steel, when loaded below the yield stress of a translation-invariant material, is explained by the nonequilibrium nanostructured martensite structure produced by radial shear and cold rolling and associated with lattice curvature. An important functional role is played by the spatial distribution of ε- and α'-martensite laths [22]. This distribution governs a complex spatial distribution of lattice curvature, different nanoscale mesoscopic structural states, the appearance of high local internal stresses, and the possibility of gigacycle fatigue without fatigue cracking at sufficiently high external stresses.

5 Conclusions

The description of a deformable solid as a multiscale hierarchically organized system is usually limited in the literature to a microscale structural level, where strain-induced defects of a translation-invariant crystal lattice are considered. An important role in the problem of the mechanical behavior of materials is played by curvature of the crystal lattice, in whose interstices nanoscale mesoscopic structural states arise [9, 10, 22]. In the present study, such nanoscale mesoscopic structural states were formed in Fe–Cr–Mn austenitic stainless steel using the complex treatment by multistage high-temperature radial shear rolling with subsequent smooth-roll cold rolling to the resulting plastic strain 1.8–2.0.

Such complex treatment causes the formation of nanostructured fcc austenite grains in the steel, bcc ferrite zones, and lattice curvature of nonequilibrium ε- and α'-martensite phases in the interstitial space based on nanoscale mesoscopic structural states. Under mechanical loading, the nonequilibrium heterogeneous martensitic structure of the specimens undergoes reversible structural-phase transformations, which are responsible for a nanocrystalline structure of the material, a vortical filamentary structure and dynamic rotations on the fracture surface, an increased wear resistance, and the transition of high-cycle fatigue life of the initial material to gigacycle fatigue without reducing external applied stress.

This work was performed within the State contract for the Program of Fundamental Research of the State Academies of Sciences for 2013–2020 (project III.23.1.1), RFBR projects (No. 18-08-00221 and 17-01-00691), and Integration Project of the SB RAS No. II.1.

References

1. Panin VE, Egorushkin VE, Panin AV (2012) Nonlinear wave processes in a deformable solids as a multiscale hierarchically organized system. Phys Usp 55(12):1260–1267
2. Panin VE, Egorushkin VE, Kuznetsov PV, Galchenko NK, Shugurov AR, Vlasov IV, Deryugin YY (2019) Structural turbulence of plastic flow and ductile fracture in low alloy steel under lattice curvature conditions. Phys Mesomech 22(4):16–28
3. Turchanin AG, Turchanin MA (1991) Thermodynamics of refractory carbides and carbonitrides. Metallurgia, Moscow
4. Averin VV, Revyakin AV, Fedorchenko VI, Kozina LN (1976) Nitrogen in metals. Metallurgia, Moscow
5. Mukhamedov AM (2015) Deindividuation phenomenon: links between mesodynamics and macroscopic phenomenology of turbulence. Phys Mesomech 18(1):24–32
6. Mukhamedov AM (2018) Geometrodynamical models of the mesomechanics of a continuum: dynamic degrees of freedom with a non-Eulerian space-time evolution. Fiz Mezomekh 21(4):13–21
7. Trefilov VI, Milman YV, Firstov AS (1975) Physical foundations of the strength of refractory metals. Naukova Dumka, Kiev
8. Rybin VV (1986) Severe plastic deformation and fracture of metals. Metallurgia, Moscow
9. Panin VE, Egorushkin VE, Elsukova TF, Surikova NS, Pochivalov YI, Panin AV (2018) Multiscale translation-rotation plastic flow in polycrystals. Handbook of mechanics of materials. Springer Nature, Singapore. https://doi.org/10.1007/978-981-10-6855-3_77-1
10. Panin VE, Derevyagina LS, Panin SV, Shugurov AR, Gordienko AI (2019) The role of nanoscale strain-induced defects in the sharp increase of low-temperature toughness in low-carbon and low-alloy steels. Mater Sci Eng A 768:138491
11. Panin VE, Egorushkin VE (2013) Curvature solitons as generalized structural wave carriers of plastic deformation and fracture. Phys Mesomech 16(4):267
12. Panin VE, Egorushkin VE, Derevyagina LS, Deryugin EE (2013) Nonlinear wave processes of crack propagation in brittle and brittle-ductile fracture. Phys Mesomech 16(3):183–190
13. Surikova NS, Panin VE, Derevyagina LS, Lutfullin RY, Manzhina EV, Kruglov AA, Sarkeeva AA (2015) Micromechanisms of deformation and fracture in a VT6 titanium laminate under impact load. Phys Mesomech 18(3):250–260
14. Shanyavsky AA (2007) Modeling of fatigue fracture of metals. Synergetics in aviation. Monograph, Ufa
15. Shanyavsky AA (2015) Scales of metal fatigue cracking. Phys Mesomech 18(2):163–173
16. Mughrabi H (2006) Specific features and mechanisms of fatigue in the ultrahigh-cycle regime. Int J Fatigue 28:1501–1508
17. Panin VE, Moiseenko DD, Elsukova TF (2014) Multiscale model of deformed polycrystals. Hall–Petch problem. Phys Mesomech 17(1):1–14
18. Matsukawa Y, Zinkle SJ (2007) One-dimensional fast migration of vacancy clusters in metals. Science 318:959–962
19. Steed JV, Atwood JL (2009) Supramolecular chemistry. Wiley, New York

20. Ragulya AV, Skorohod VV (2007) Consolidated nanostructured materials. Naukova Dumka, Kiev
21. Noskova NI, Mulyukov RR (2003) Submicrocrystalline and nanocrystalline metals and alloys. Ural. Otd. Ross. Akad. Nauk., Yekaterinburg
22. Panin VE, Panin AV, Perevalova OB, Shugurov AR (2019) Mesoscopic structural states at the nanoscale in surface layers of titanium and its alloy Ti–6Al–4V in ultrasonic and electron beam treatment. Phys Mesomech 22(5):345–354

Autowave Mechanics of Plastic Flow

Lev B. Zuev

Abstract The notions of plastic flow localization are reviewed here. It have been shown that each type of localized plasticity pattern corresponds to a given stage of deformation hardening. In the course of plastic flow development a changeover in the types of localization patterns occurs. The types of localization patterns are limited to a total of four pattern types. A correspondence has been set up between the emergent localization pattern and the respective flow stage. It is found that the localization patterns are manifestations of the autowave nature of plastic flow localization process, with each pattern type corresponding to a definite type of autowave. Propagation velocity, dispersion and grain size dependence of wavelength have been determined experimentally for the phase autowave. An elastic-plastic strain invariant has also been introduced to relate the elastic and plastic properties of the deforming medium. It is found that the autowave's characteristics follow directly from the latter invariant. A hypothetic quasi-particle has been introduced which correlates with the localized plasticity autowave; the probable properties of the quasi-particle have been estimated. Taking the quasi-particle approach, the characteristics of the plastic flow localization process are considered herein.

Keywords Elasticity · Plasticity · Localization · Crystal lattice · Self-organization · Autowaves · Quasi-particle

1 Introduction. General Consideration

In past few decades, the nature and salient features of plastic deformation in solids were investigated. A wealth of new experimental data has been collected, which add strong support to our understanding of plasticity problem. Naturally, we can do no more than mention a few experimental and theoretical studies related to dislocation physics and solids mechanics, e.g. [1–14]. We have recently made a significant

L. B. Zuev (✉)
Institute of Strength Physics and Materials Science, Siberian Branch of Russian Academy of Sciences, 634055 Tomsk, Russia
e-mail: lbz@ispms.tsc.ru

© The Author(s) 2021
G.-P. Ostermeyer et al. (eds.), *Multiscale Biomechanics and Tribology of Inorganic and Organic Systems*, Springer Tracts in Mechanical Engineering, https://doi.org/10.1007/978-3-030-60124-9_12

discovery that the deforming medium is a self-organizing system, which is in a state far from thermodynamic equilibrium; such media are addressed by [15–24]. Moreover, the plastic flow in solids is found to have a space-time periodic nature, which is discussed at length by [25–28]. On the macro-scale level the plastic deformation exhibits an inhomogeneous localization behavior from yield point to failure. Hence, localization is a general feature of the plastic flow process, which should be taken properly into account to markedly advance our understanding of the deforming medium stratification into alternating deforming and non-deforming layers about 10^{-2} m thick. Similar layers form localized plasticity pattern.

In this line of research, considerable experimental study has been given to the problem of plastic deformation macrolocalization; the investigation results were summarized in a monograph by [29].[1] In what follows, we discuss new approaches to the same problem.

1.1 Experimental Technique

The experimental procedure was as follows. The flat samples having work part 50 × 6 × 2 mm were tested in tension along the axis x at a rate of 3.5×10^{-5} s^{-1} in a test machine at 300 K. The non-metallic materials were subjected to compression. Traditional 'stress–strain', $\sigma(\varepsilon)$, diagrams recording were completed by double-exposure speckle photography [30] for reconstruction of the displacement vector field $r(x, y)$. Special device for these purposes has field of vision ~100 mm, real-time mode of operation and spatial resolution ~1...2 μm. According to [31], the plastic distortion tensor for plane stressed state is

$$\beta^{(p)} = \nabla r(x, y) = \begin{bmatrix} \varepsilon_{xx} & \varepsilon_{xy} \\ \varepsilon_{yx} & \varepsilon_{yy} \end{bmatrix} + \omega_z. \tag{1}$$

Longitudinal, ε_{xx}, transverse, ε_{yy}, shear $\varepsilon_{xy} = \varepsilon_{yx}$ components and rotation ones, ω_z can be calculated for different points of the test sample.

This developed method enables visualization of the deformation inhomogeneities. Thus, Fig. 1a demonstrates a macro-photograph of the deforming material structure and $\varepsilon_{xx}(x, y)$ distribution for this case. The diagram $X(t)$ is shown in Fig. 1b (here X is the x-coordinate of deformation nucleus and t is time); it illustrates the procedure employed for measuring the spatial and temporal periods of the deformation process, i.e. the values λ, T and rate of the strain nucleus, $V_{aw} = \lambda/T$.

---

[1]Professor S. G. Psakhie has promoted this book as the Editor.

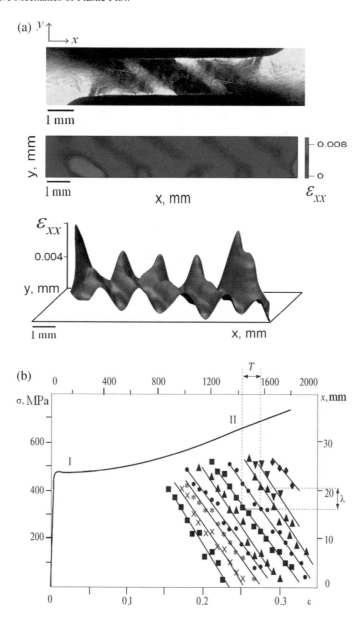

Fig. 1 Localized plastic flow pattern observed for the test sample of Fe-3 wt% Si alloy: **a** macropho-tograph and distribution of strains ε_{xx} (halftone photograph) **b** illustration of the method for measuring λ and T values

1.2   Studied Materials

We have studied by now nearly forty single-crystal and polycrystalline metals, alloys and other materials, which differ in chemical composition, crystal lattice type (FCC, BCC or HCP), grain size and deformation mechanism, i.e. dislocation glide [32], twinning [33] or phase transformation induced plasticity [34]. Later our study has been extended to ceramics, single alkali halide crystals and rocks.

1.3   Preliminary Results

The localization behavior of plastic deformation is its most salient feature. Thus, space-time periodic structures, so-called deformation patterns, emerge in the deforming sample from the yield limit to failure by constant-rate tensile loading. The following features are common to all the localization patterns observed thus far:

- *localization structures will occur spontaneously in the sample by constant-rate loading in the absence of any specific action from the outside;*
- *in the course of plastic deformation a changeover in the types of localized plasticity patterns is observed;*
- *due to work hardening, the deforming medium's defect structure undergoes irreversible changes, which are suggestive of its non-linearity and are reflected in the emergent patterns.*

Recent independent evidence supports the validity of the present conception about the macroscopic localization by deformation [35–37]. Acharia et al. [38] observed a localization nucleus traveling along the extension axis in the single Cu crystal at the linear work hardening stage; a stationary localization pattern at the parabolic work hardening stage in the samples of Fe-Mn alloy was described by [39].

2   Deformation Pattern. Localized Plastic Flow Viewed as Autowaves

The plastic flow has an attribute, which is common to all deforming solids. On the macro-scale level, the deformation is found to exhibit a localization behavior from the yield point to failure. In the cause of plastic flow, plastic flow curve would occur by stages; each work hardening stage involves a certain dislocation mechanism [40–42]. To gain an insight about the nature of plasticity, the existence of explicit connection between the above two attributes must be discovered.

2.1   Plastic Flow Stages and Localized Plasticity Patterns

We focused our efforts on proving this connection. To provide a proof for this asser-
tion, a localized plasticity pattern was matched against the respective work hardening
stage. These can be readily distinguished on the stress-strain curve $\sigma(\varepsilon)$, describing
by the Lüdwick equation [43]

$$\sigma(\varepsilon) = \sigma_y + K\varepsilon^n, \tag{2}$$

where σ_y is the yield point and K is hardening coefficient. The convenient charac-
teristic of the deformation process is the exponent $n = (\ln \varepsilon)^{-1} \cdot \ln\left[(\sigma - \sigma_y)/K\right]$.

According to [43] and [44], the exponent n varies with the stages of plastic
deformation process as

- for the yield plateau or easy glide stage, $n \approx 0$;
- for the linear stage of work hardening, $n = 1$;
- for the parabolic stage of work hardening, $n = \frac{1}{2}$,
- for the prefailure stage, $\frac{1}{2} > n > 0$.

A set of kinetic diagrams $X(t)$ was obtained simultaneously for the above stages;
it is schematized in Fig. 2. Similar sets were plotted for all deforming materials,
no matter what their microstructure or deformation mechanism. The localization
patterns arise in a consecutive order that is governed by the work hardening law $\theta(\varepsilon)$
alone. The emergent localization patterns can be distinguished from the dependencies
$X(t)$ obtained for the work hardening stages. The distinctive features of localization
patterns remain the same; differences are quantitative ones. It should be pointed out
that plastic deformation inhomogeneities have to be correlated for the entire material
volume. The localized plasticity nuclei are distributed periodically; in all studied
materials the distance between nuclei $\lambda \approx 10^{-2}$ m.

A qualitative analysis of experimental data rests on the observation that these
regularities serve to provide a unified explanation of the plastic flow behavior. A

Fig. 2 Plastic flow
development. Schematic
representation of autowave
evolution process

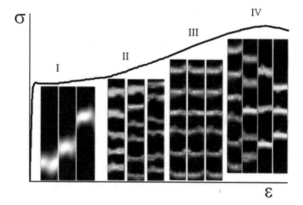

total of four types of localized plasticity patterns have been observed experimentally for all studied materials. A definite type of localization pattern would emerge at each flow stage. It is thus asserted that

- yield plateau is identified with a single mobile plastic flow nucleus;
- linear work hardening stage, with a set of equidistant moving nuclei;
- parabolic work hardening stage, with a set of equidistant stationary nuclei;
- prefailure stage, with a set of moving nuclei, which are forerunners of failure.

Taken together, these results provide a reliable proof that one-to-one correspondence exists between the localization patterns on the one hand, and the respective plastic flow stages on the other.

2.2 Localized Plastic Flow Autowaves

A key aspect of this many-faceted problem to be dealt with is the nature of localized plasticity. Our basic viewpoint is that there are striking parallels between the localized plasticity patterns described herein and the dissipative structures synergetics deals with. We suggest that the dissipative structures are *autowaves* [45], *self-excited waves* [46] or *pseudo-waves* [15]. Special modes of these structures, which are also known as switching autowaves, phase autowaves or stationary dissipative structures, have been studied in detail for a number of chemical and biological systems.

The difference between the wave and autowave processes needs careful explanation. The majority of well-known wave processes are described by functions $\sin(\omega t - kx)$, which are solutions of hyperbolic differential equations of the type $\ddot{Y} = c^2 Y''$. Here the value c is wave propagation rate, which is a finite quantity determined by material characteristics. The second derivative with respect to time is applicable to reversible physical processes alone, e.g. elastic deformation.

The autowaves have long been recognized as solutions to parabolic differential equations of the type $\dot{Y} = \varphi(x, y) + \kappa Y''$ [47]. These equations can be derived formally by adding to the right part of the equation $\dot{Y} = \kappa Y''$ the nonlinear function $\varphi(x, y)$; the value κ is a transport coefficient having the dimensionality $L^2 \cdot T^{-1}$. The availability of the first derivative with respect to time implies that the above equations are suitable for addressing irreversible processes similar to those involved in the plastic deformation.

The speculation that the localization patterns in question are equivalent to autowaves was originally prompted by a formal resemblance between the two kinds of phenomena. In what follows, we provide unequivocal physical evidence for the validity of this viewpoint by focusing our attention on the autowave nature of processes of interest. Thus, the well-known Lüders front can be regarded as a boundary between the elastically and plastically deforming material volumes. As the Lüders front propagates along the tensile sample, it leaves behind an ever increasing volume of deformed material [48]. Due to the structural changes, the deforming material volume acquires a new state, which is characterized by increasing density of defects; its deformation occurs via dislocation glide mechanism. With growing total deformation, the plastic flow will exhibit an intermittent behavior on the macro-scale level. Therefore, the Lüders band propagation is regarded as *a switching autowave*. A different scenario is realized at the linear work hardening stage where a set of mobile nuclei is observed. The nuclei move at a constant rate along the test sample (see Fig. 2). In this case, the phase constancy condition is fulfilled, i.e. $\omega t - kx = const$. This pattern will be designated *a phase autowave*. The linear stage over, the parabolic stage of work hardening begins for $n = \frac{1}{2}$ and $V_{aw} = 0$ (see Fig. 2). At this stage the emergent localization pattern fits the definition of *a stationary dissipative structure* [15, 16]. At the prefailure stage the deformation development is nearing completion. For $n < \frac{1}{2}$, *collapse of the autowave* would be observed [49, 50]. Thus, the types of localized plasticity patterns have been unambiguously identified with the respective modes of autowave processes.

It is therefore concluded that

- a solitary localized deformation nucleus traveling at the yield plateau is a switching autowave;
- a set of equidistant localization nuclei propagating at a constant rate along the sample at the linear work hardening stage is a phase autowave;
- a set of stationary equidistant localization nuclei emergent at Taylor's stage corresponds to a stationary dissipative structure;
- a pattern of synchronously moving nuclei, which finally merge at the prefailure stage, fits neatly the definition of collapse of localized plasticity autowave.

Taken together, these regularities are *Correspondence rule*. Accordingly, it can thus be asserted that the plastic flow process occurring in the deforming can be addressed as continuous evolution of localized plastic flow autowaves. Hence, it can be claimed with confidence that the transition from one flow stage to the next involves a changeover in the types of autowaves generated by the deformation. With due regard to the correspondence rule, it is maintained that the plastic flow stages and the respective autowaves modes are closely related. This is favorable ground for inferring that the process of localized plastic flow is evolution of autowave patterns. Due to a changeover in the flow stages, the autowaves will emerge from a random strain distribution in an orderly sequence (Fig. 2): elastic wave → switching autowave → phase autowave → stationary dissipative structure → collapse of autowave. In some materials, however, individual stages might be missing and this sequence can be broken.

It is necessary to remind here that a special-purpose reaction cell has to be designed for carrying out experimental investigations of autowaves in chemistry or biology. Such cells differ widely in type and size, depending on the kind of studied system and its chemical composition as well as the kinetics of chemical reactions involved, temperatures employed, etc. However, it is found for plastic deformation that the autowaves will be generated spontaneously in the tensile sample practically at any temperature. From this point of view, the deforming solid can be regarded as a universal reaction cell [51], which can be conveniently used for modelling and studying the generation and evolution of various autowave modes.

2.3 Autowaves Observed for the Linear Work Hardening Stage

The plastic flow exhibit generally a regular localization behavior, which is markedly pronounced at the linear work hardening stage. In this case, the localization nuclei move in a concerted manner at a constant rate along the sample, forming a phase autowave. The experimental data on propagation rate, dispersion and material structure response obtained for these autowaves are demonstrated in Fig. 3.

The propagation rates of localized plasticity autowaves in all studied materials are in the range $10^{-5} \leq V_{aw} \leq 10^{-4}$; they depend solely upon the work hardening coefficient, $\theta = E^{-1} \cdot d\sigma/d\varepsilon$, and are given as (Fig. 3a)

$$V_{aw}(\theta) = V_0 + \Xi/\theta \sim \theta^{-1}, \tag{3}$$

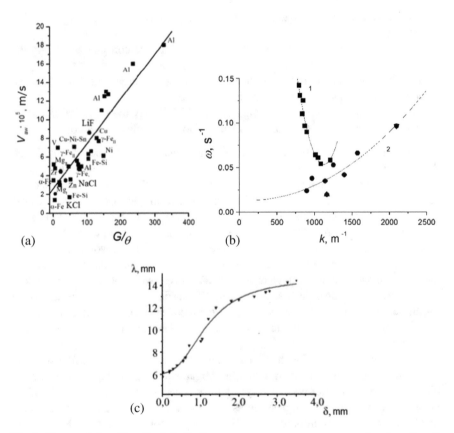

(a) (b) (c)

Fig. 3 Characteristics of localized plastic flow autowaves: **a** autowave rate as a function of work hardening coefficient for all studied materials; **b** dispersion observed for γ-Fe single crystals (2) and polycrystalline Al (1); **c** autowave length as a function of grain size plotted for polycrystalline Al

where $V_0 <$ and Ξ are empirical constants, having the dimensionality of rate.

We have also obtained dispersion relation, $\omega(k)$ (here $\omega = 2\pi/T$ is frequency and $k = 2\pi/\lambda$ is wave number) for localized plasticity autowaves [29, 52–55]. This relation has quadratic form (Fig. 3b)

$$\omega(k) = \omega_0 + \alpha(k - k_0)^2. \tag{4}$$

Using the values $\omega = \omega_0 \cdot \tilde{\omega}$ and $k = k_0 + \frac{\tilde{k}}{\sqrt{\alpha/\omega_0}}$ (here $\tilde{\omega}$ and $\tilde{k}$ are dimensionless frequency and wave number), Eq. (4) can reduce to the form $\tilde{\omega} = 1 + \tilde{k}^2$.

Finally, the grain size dependence of autowave length illustrated in Fig. 3c has the form of logistic curve

$$\lambda(\delta) = \lambda_0 + \frac{a_1/a_2}{1 + C \cdot \exp(-a_1\delta)}, \tag{5}$$

where a_1, and a_2 are empirical coefficients, $\lambda_0 \approx 5$ mm and $C \approx 2.25$. The inflection point of Eq. (5) is found from the condition $d^2\lambda/d\delta^2 = 0$; this corresponds to the boundary value of grain size $\delta = \delta_0 \approx 0.15 \ldots 0.2$ mm. The dependence $\lambda(\delta)$ has two limiting cases, i.e. $\lambda \sim \exp(\delta/\delta_0)$ for $\delta < \delta_0$ and $\lambda \sim \ln(\delta/\delta_0)$ for $\delta > \delta_0$. The quantity λ generally depends only weakly on the structural characteristics of the deforming medium. Thus variation in the grain size of polycrystalline Al from 5 μm to 5 mm corresponds to a 2.5-fold increase in the value λ [56].

Thus, the most significant features of localized plastic flow at the linear stage of work hardening are specified herein by Eqs. (3) through (5). By way of summing up our findings, we contend that in the course of plastic flow a large-scale deformation structure would form. Its elements are characterized by the nontrivial dependence $V_{aw}(\theta) \sim \theta^{-1}$, the quadratic dispersion law $\tilde{\omega} = 1 + \tilde{k}^2$ and the logistic dependence of autowave length on material structure, $\lambda(\delta)$.

2.4 Plastic Flow Viewed as Self-organization of the Deforming Medium

We hypothesize herein that the localization of plastic flow might be regarded as a process of self-organization occurring spontaneously in an open thermodynamic system. The validity of our hypothesis can be objectively confirmed by the observations of localized plasticity patterns emergent at the linear stage of work hardening, which provide strong indications that the medium is separating into deforming and undeforming layers. The fruitful concept of self-organization has been proposed by [17], which states that a self-organizing system can attain spatial, temporal or functional inhomogeneity in the absence of any specific action from the outside. Note that the definition is used in a restricted, phenomenological, meaning; it implies no concrete underlying mechanism responsible for the realization of self-organization process.

The concept of self-organization is frequently and successfully used for explanation of the formation of structure in active media studied in physics, chemistry, materials science or biology. The deforming medium can be similar to an active medium far from thermodynamic equilibrium in which the sources of energy are distributed over material volume.

We furnish strong evidence that the plastic flow also involves self-organization phenomena. According to [57], the generation of localized plastic flow autowaves causes a decrease in the entropy of the deforming system, which is the principal attribute of self-organization processes.

2.5 Autowave Equations

To offer adequate tools for describing autowave processes, a set of two equations has to be produced [45, 58] to describe the rate of change in the catalytic and damping factors. The justification of the choice of these two factors is far from to be trivial. By addressing plasticity, it is convenient to introduce plastic deformation, ε, and stress, σ, as catalytic and damping factors, respectively. Hence, equations for the rates $\dot{\varepsilon}$ and $\dot{\sigma}$ have to be derived on the base of general principles. The equation for $\dot{\varepsilon}$ is deduced from the condition of deformation flow continuity [43] as

$$\dot{\varepsilon} = \nabla \cdot (D_\varepsilon \nabla \varepsilon), \tag{6}$$

where the value D_ε is a transport coefficient and the term $D_\varepsilon \nabla \varepsilon$ is the deformation flow in the deformation gradient field $\nabla \varepsilon$; the coefficient D_ε depends on coordinates. By restricting our analysis to the case of uniaxial deformation along the axis x, we obtain

$$\partial \varepsilon / \partial t = \partial \varepsilon / \partial x \cdot \partial D_\varepsilon / \partial x + D_\varepsilon \partial^2 \varepsilon / \partial x^2 = f(\varepsilon) + D_\varepsilon \partial^2 \varepsilon / \partial x^2, \tag{7}$$

where $f(\varepsilon) = \partial \varepsilon / \partial x \cdot \partial D_\varepsilon / \partial x$ is a non-linear strain function.

The equation for $\dot{\sigma}$ can be derived from Euler's equation for hydrodynamic flow [59] as

$$\frac{\partial}{\partial t} \rho V_i = -\frac{\partial \Pi_{ik}}{\partial x_k}. \tag{8}$$

In the case of viscous medium, the momentum flux density tensor is given as $\Pi_{ik} = p\delta_{ik} + \rho V_i V_k - \sigma_{vis} = \sigma_{ik} - \rho V_i V_k$; the value δ_{ik} is unit tensor and V_i and V_k are flow rate components. The stress tensor $\sigma_{ik} = -p\delta_{ik} + \sigma_{vis}$ includes viscous stresses. By plastic deformation, $-p\delta_{ik} \equiv \sigma_{el}$; hence, $\sigma = \sigma_{el} + \sigma_{vis}$ or $\dot{\sigma} = \dot{\sigma}_{el} + \dot{\sigma}_{vis} = g(\sigma) + \dot{\sigma}_{vis}$.

The origination of viscous stresses, σ_{vis}, is due to plastic deformation inhomogeneity; the value σ_{vis} is related to variation in the rate of elastic waves propagating in the deforming medium, i.e. $\sigma_{vis} = \hat{\eta} \nabla V_t$. Here $\hat{\eta}$ is the dynamic viscosity of the medium and V_t is the propagation rate of transverse ultrasound waves. The equation $\sigma_{vis} = \hat{\eta} \nabla V_t$ can be written as $\partial \sigma_{vis}/\partial t = V_t \nabla \cdot \left(\hat{\eta} \nabla V_t \right) = \hat{\eta} V_t \partial^2 V_t / \partial x^2$. The value V_t depends on the acting stresses as $V_t = V_* + \varsigma \sigma$ [60]. Hence, equation for describing the rate of stress change may have the form analogous to that of Eq. (7), i.e. $\partial \sigma_{vis}/\partial t = \hat{\eta} V_t \partial^2 V_t / \partial x^2 = \hat{\eta} \varsigma V_t \partial^2 \sigma / \partial x^2$. Thus, we obtain

$$\dot{\sigma} = g(\sigma) + D_\sigma \partial^2 \sigma / \partial x^2, \tag{9}$$

where $D_\sigma = \hat{\eta} \varsigma V_t$ is the transport coefficient. It was shown in [60] that Eqs. (7) and (9) can be used to adequately describe plastic flow regimes.

2.6   On the Relation of Autowave Equations to Dislocation Theory

The problem of plasticity can be addressed in the frame of two different approaches, i.e. the autowave model proposed herein and dislocation theory. Of particular importance is the possible interrelation between the two approaches. Almost all the dislocation theories of plasticity are based on the Taylor-Orowan equation, which is used to describe the dislocation mechanism of plasticity [32], i.e.

$$\dot{\varepsilon} = b \rho_m V_{disl}, \tag{10}$$

where b is the Bürgers vector and ρ_m is the density of dislocations moving at rate $V_{disl}(\sigma)$ under the action of applied stress. The first term in the right side of Eq. (7) is transformed by assuming that dislocation distribution is homogeneous and $d\varepsilon/dx \approx 1/s \cdot b/s \approx b\rho_m$. Here s is the distance between dislocations; the quantity b/s has the meaning of shear strain for dislocation path s and $s^{-2} \approx \rho_m$. For the case of plastic flow, it may be written $\eta_\varepsilon \approx L_{disl} \cdot V_{disl}$ (here $L_{disl} \approx \zeta x$ and $V_{disl} = const$ are, respectively, dislocation path and rate). Hence,

$$\dot{\varepsilon} = \zeta b \rho_m V_{disl} + D_\varepsilon \partial^2 \varepsilon / \partial x^2. \tag{11}$$

The right part of Eq. (11) accounts for two deformation flows, i.e. $\zeta b \rho_m V_{disl} \sim V_{disl}$ and $D_\varepsilon \partial^2 \varepsilon / \partial x^2$. The former flow is 'hydrodynamic' in character [59] and the latter is a 'diffusion' one.

Evidently, elimination of the second term will transform Eq. (11) into Eq. (10); hence, the Taylor-Orowan equation might be regarded as a special case of Eq. (11). Clearly, Eq. (10) finds limited use for plastic flow description, since it describes chaotic distributions of dislocations, which form no complex ensembles; hence,

Eq. (10) corresponds to work hardening due to long-range stress fields alone. Dislocation theory based on long-range stress fields might be called a linear one, while theory describing both flows from Eq. (11) might be regarded as an extended version of the dislocation theory.

A thorough analysis of Eq. (11) suggests that an appropriate dislocation model can be developed with the proviso that both terms in the right side of Eq. (11) are taken into account. In case the term $D_\varepsilon \partial^2 \varepsilon / \partial x^2$ is not eliminated from Eq. (11), it will initiate non-linear corrections of dislocation theory equations and thus expand significantly the area of application of dislocation theory. Such corrections might be of significance for crystals having high dislocation density.

Thus, the application of new experimental technique for plastic flow investigation enabled discovery of a new class of deformation phenomena—localized plastic deformation autowaves. These phenomena are addressed above, taking different but complementary approaches, i.e. autowave plasticity theory and dislocation theory. There is good reason to believe that compelling evidence has been provided for the existence of localized plasticity autowaves.

3   Elastic-Plastic Strain Invariant

The uniformity of localized plasticity phenomena observed for a wide range of materials suggests the existence of a general law for the localized plastic flow autowaves. This section focuses on the search for a quantitative relationship between the characteristics of elastic waves and autowaves.

3.1   Introduction of Elastic-Plastic Strain Invariant

We suggest a link between plastic flow macro-parameters and crystal lattice characteristics. For this purpose, two products are matched, i.e. λV_{aw} and χV_t, which characterize plastic flow and elastic deformation, respectively. The quantities χ and V_t are interplanar spacing of crystal lattice and transverse ultrasound wave velocity, respectively. Numerical analysis was performed using experimentally obtained values λ and V_{aw} as well as hand-book values χ and V_t. The data listed in Table 1 allow one to write the equality

$$\left\langle \frac{\lambda V_{aw}}{\chi V_t} \right\rangle = \widehat{Z} \approx const. \tag{12}$$

It can be seen from Fig. 4 that Eq. (12) holds true for all studied materials. The averaging of the value α was performed for seventeen materials to give $\langle \widehat{Z} \rangle = 0.49 \pm 0.05 \approx 1/2 < 1$. This result constitutes both formal and physical proofs

Table 1 Data for verification of Eq. (12) for the elastic-plastic invariant of strain (data for λV_{aw} and χV_t are multiplied by 10^7 m²/s)

	Metals											
	Cu	Zn	Al	Zr	Ti	V	Nb	γ-Fe	α-Fe	Ni	Co	Mo
λV_{aw}	3.6	3.7	7.9	3.7	2.5	2.8	1.8	2.55	2.2	2.1	3.0	1.2
χV_t	4.8	11.9	7.5	11.9	7.9	6.2	5.3	4.7	6.5	6.0	6.0	7.4
$\lambda V_{aw}/\chi V_t$	0.75	0.3	1.1	0.3	0.3	0.45	0.33	0.54	0.34	0.35	0.5	0.2

	Metals					
	Sn	Mg	Cd	In	Pb	Ta
$\lambda \cdot V_{aw}$	2.4	9.9	0.9	2.6	3.2	1.1
$\chi \cdot V_t$	5.3	15.8	3.5	2.2	2.0	4.7
$\lambda V_{aw}/\chi V_t$	0.65	0.63	0.2	1.2	1.6	0.2

	Alkali-halide crystals			Rocks	
	KCl	NaCl	LiF	Marble	Sandstone
$\lambda \cdot V_{aw}$	3.0	3.1	4.3	1.75	0.6
$\chi \cdot V_t$	7.0	7.5	8.8	3.7	1.5
$\lambda V_{aw}/\chi V_t$	0.43	0.4	0.5	0.5	0.4

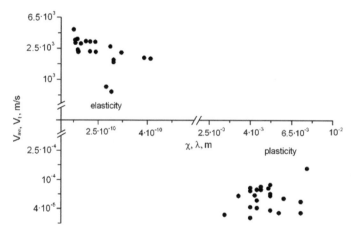

Fig. 4 Verification of invariant (12). The quantities χ, V_t (left) and λ, V_{aw} (right) are grouped in logarithmic coordinates in the neighborhood of average values

that the elastic and plastic processes involved in the deformation are closely related. Therefore, Eq. (12) has been labeled as *The elastic-plastic strain invariant*.

Apparently, $V_t \approx \chi \omega_D$ (here ω_D is the Debye frequency); hence, we write

$$\lambda V_{aw} \approx \widehat{Z}\frac{V_t^2}{\omega_D} \approx \widehat{Z}\frac{G}{\omega_D \rho} \approx \widehat{Z}\frac{\partial^2 W/\partial \upsilon^2}{(\omega_D \chi)\rho} \approx \widehat{Z}\frac{\partial^2 W/\partial \upsilon^2}{\xi_1}, \tag{13}$$

where $\upsilon \ll \chi$ is atomic displacement near interparticle potential minimum (W); the elastic modulus is expressed in terms of interparticle potential as $G = \chi^{-1}\partial^2 W/\partial \upsilon^2$ [61] and the value $\xi_1 = (\omega_D \chi)\rho = V_t \rho$ is specific acoustic resistance of the medium, which is shown to be related to crystal lattice perturbation due to dislocation motion [28]. The interparticle potential from Eq. (13) is

$$W(\upsilon) \approx \frac{1}{2} \cdot \left(\partial^2 W/\partial \upsilon^2\right)\upsilon^2 + \frac{1}{6}\left(\partial^3 W/\partial \upsilon^3\right)\upsilon^3 = \frac{1}{2}p\upsilon^2 - \frac{1}{3}q\upsilon^3, \qquad (14)$$

where p is the coefficient of quasi-elastic coupling and q is anharmonicity coefficient. With the proviso that $\frac{1}{2}p \cdot \upsilon^2 \gg \left|-\frac{1}{3}q\upsilon^3\right|$, Eq. (13) assumes the form

$$\lambda V_{aw} \approx p/\xi_1 \approx p/V_t \rho, \qquad (15)$$

where λV_{aw} can be taken as a criterion of plasticity [29].

3.2 Generalization of Elastic-Plastic Strain Invariant

The criterion λV_{aw} from Eq. (15) also holds good for deformation initiated by chaotically distributed dislocations. Let mobile dislocation density be ρ_m; then the average distance between dislocations, which is equal to the dislocation path, is given as $\langle s \rangle = \rho_m^{-1/2}$. According to [32], $\sigma \approx (Gb/2\pi)\rho_m^{1/2}$; hence, we can write $\rho_m^{-1/2} = \langle s \rangle = Gb/2\pi\sigma \sim \sigma^{-1}$. The rate of quasi-viscous motion of dislocations is $V_{disl} = (b/B) \cdot \sigma$. Here B is the coefficient of dislocation drag by the phonon and electron gases [62]. Hence,

$$l \cdot V_{disl} \approx const = \frac{Gb^2}{2\pi \cdot B}. \qquad (16)$$

Using the values $G \approx 40$ GPa and $B \approx 10^{-4}$ Pa s, which are conventionally employed for dislocation motion descriptions, we obtain $lV_{disl} \approx 10^{-6}$ m^2/s. The latter value is close to the calculated value of the product $\hat{Z}\chi V_t$ obtained for studied materials (see Table 1).

The above suggests that we have established a reliable quantitative criterion for analyzing the interaction between the elastic deformation, which occurs on the micro-scale level, and the macro-scale plastic deformation. This criterion, in its universal form, applies to autowaves in question as well as to elastic and plastic deformation via dislocation motion. Therefore, this criterion is considered as a more general form of the elastic-plastic strain invariant:

$$\lambda V_{aw} = lV_{disl} = \hat{Z}\chi V_t \approx 10^{-6} \text{ m}^2/\text{s}. \qquad (17)$$

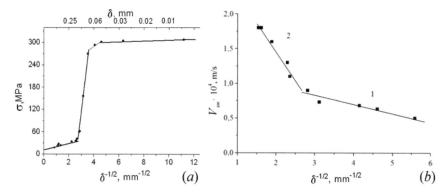

Fig. 5 Characteristics of autowaves obtained for polycrystalline Al: **a** strength limit as a function of grain size; **b** autowave rate as a function of grain size (sections 1 and 2 correspond, respectively, to the ranges $0.005 \leq \delta \leq 0.15$ mm and $0.15 \leq \delta \leq 5$ mm)

To provide a framework for validating the proposed strain invariant, a special-purpose series of experiments were carried on for polycrystalline aluminum samples having grain sizes in the range $5 \times 10^{-3} \leq \delta \leq 5$ mm. We obtained grain size dependencies of strength limit, $\sigma_B(\delta)$. The Hall-Petch relation $\sigma_B(\delta) = \sigma_0 + k_B \delta^{-1/2}$ [63] has been plotted; it is illustrated in Fig. 5a. It can be seen that for $\delta_0 \approx 0.1 \ldots 0.2$ mm, a jump-wise variation occurs in the value $\sigma_B(\delta)$. The dependence $V_{aw}(\delta)$ demonstrated in Fig. 5b has a similar form. The value V_{aw} would also vary over the entire range of grain sizes; however, the surprising thing is that the ratio $\lambda V_{aw}/\chi V_t \approx 1/2$ would remain constant for both $\delta > \delta_0$ and $\delta < \delta_0$.

In view of the above, invariant (12) shows promise for gaining an insight into the nature of localized plastic flow. All the basic regularities of localized plastic flow autowaves can be deduced from Eq. (12). Therefore, the invariant is expected to play an important role in the development of new notions of plasticity.

3.3 On the Strain Invariant and Autowave Equations

Now certain theoretical considerations concerning the origin of invariant (12) will be explored. Particular emphasis is placed upon the fact that the quantities λV_{aw} and χV_t from Eq. (12) have evidently the dimension $L^2 \, T^{-1}$. Note that in the right sides of Eqs. (7) and (9) the coefficients D_ε and D_σ for the terms containing second-order derivatives $\partial^2/\partial x^2$ have the same dimension.

In view of the above, we assume that

$$\lambda V_{aw} \equiv D_\varepsilon \qquad (18)$$

and

$$\chi V_t \equiv D_\sigma. \tag{19}$$

The above identification is valid, considering the dimensions of the quantities λV_{aw} and χV_t; moreover, Eq. (9) contains expression for the coefficient D_σ in which velocity V_t appears. According to Eq. (12), $\widehat{Z} < 1$, i.e. $D_\sigma > D_\varepsilon$. Hence, the condition required for the autowave generation is satisfied [45].

Given their dimensions, these quantities may be either diffusion coefficients or kinematic viscosities of a media. In what follows, the reasoning behind the latter suggestion is formulated. Using the dimensionality analysis, we write $\lambda \approx \sqrt{2\eta \cdot t}$; here the value $\eta \approx 10^{-6}$ m^2/s and the value $t \approx T_{aw} \approx 10^2$ s. Apparently, $\lambda \approx 10^{-2}$ m, which is identical to the autowave length.

3.4 Some Consequences of the Strain Invariant

Implicitly, it is generally assumed that the total deformation is a sum of elastic and plastic strains, i.e. $\varepsilon_{tot} = \varepsilon_e + \varepsilon_p$. By virtue of $\varepsilon_e \ll \varepsilon_p$ and $\varepsilon_{tot} \approx \varepsilon_p$, the contribution of elastic strain is frequently neglected altogether. However, invariant (12) implies that the quantities ε_e and ε_p are closely related; hence, these quantities should be taken properly into account by addressing plastic flow localization. A model for simulation of plastic deformation should be based on the fundamental assumption that plastic form changing involves both elastic and plastic deformation mechanisms that are interdependent in principle.

To provide arguments in favor of the strain invariant, its consequences have been analyzed. It is found that the regularities of plastic flow localization, which are given by Eqs. (3) through (5), can be derived from Eq. (12) as well. Consider the respective procedures step by step.

It follows from Eq. (3) that the propagation rate of localized plasticity autowave is inversely proportional to the work hardening coefficient. To prove it, we differentiate Eq. (12) with respect to deformation ε

$$\lambda \frac{dV_{aw}}{d\varepsilon} + V_{aw} \frac{d\lambda}{d\varepsilon} = \widehat{Z}\chi \frac{dV_t}{d\varepsilon} + \widehat{Z}V_t \frac{d\chi}{d\varepsilon}. \tag{20}$$

Hence,

$$V_{aw} = \left(\frac{d\lambda}{d\varepsilon}\right)^{-1} \left(\widehat{Z}\chi \frac{dV_t}{d\varepsilon} + \widehat{Z}V_t \frac{d\chi}{d\varepsilon} - \lambda \frac{dV_{aw}}{d\varepsilon}\right). \tag{21}$$

Since the interplanar spacing of crystal is independent of plastic deformation, in Eq. (21) $\widehat{Z}V_t \frac{d\chi}{d\varepsilon} \approx 0$. Hence,

$$V_{aw} = \widehat{Z}\chi \frac{dV_t}{d\lambda} - \lambda \frac{dV_{aw}}{d\lambda}. \tag{22}$$

Further we shall refer to [32] who reasoned that work hardening coefficient can be expressed as a ratio of two parameters of the deforming medium structure, e.g. $\chi \ll \lambda$, i.e. $\theta \approx \chi/\lambda$ and $dV_{aw}/d\lambda < 0$. Thus, rearrangement of Eq. (22) yields the same result as the experiment does, i.e.

$$V_{aw} = \widehat{Z}\chi\frac{dV_t}{d\lambda} - \chi\frac{dV_{aw}}{d\lambda}\frac{\lambda}{\chi} \approx V_0 + \frac{\Xi}{\theta}. \tag{23}$$

Equation (4) also follows from Eq. (12), which can be rewritten as

$$V_{aw} = \frac{\Theta}{\lambda} = \frac{\Theta}{2\pi}k, \tag{24}$$

where $\Theta = \chi V_t/2$. If $V_{aw} = d\omega/dk$, then $d\omega = (\Theta/2\pi) \cdot k \cdot dk$. Integration of the latter equality is performed:

$$\int_{\omega_0}^{\omega} d\omega = \frac{\Theta}{2\pi}\int_{0}^{k-k_0} kdk \tag{25}$$

to yield dispersion law of quadratic form

$$\omega = \omega_0 + \frac{\Theta}{4\pi}(k - k_0)^2, \tag{26}$$

which is equivalent to Eq. (4) with the proviso that $\alpha = \Theta/4\pi$.

The grain size dependence of autowave length, $\lambda(\delta)$, given by Eq. (5) also follows from invariant (12). Indeed, Eq. (12) can be rewritten as

$$\lambda = \widehat{Z}\chi\frac{V_t}{V_{aw}}. \tag{27}$$

By virtue of the fact that the quantities V_t and V_{aw} depend on grain size, δ, [54, 64], differentiation of Eq. (12) is performed with respect to δ as

$$\frac{d\lambda}{d\delta} = \widehat{Z}\chi\frac{d}{d\delta}\left(\frac{V_t}{V_{aw}}\right) = \widehat{Z}\chi\left(\frac{V_{aw}dV_t/d\delta - V_t dV_{aw}/d\delta}{V_{aw}^2}\right). \tag{28}$$

With the proviso that $V_{aw} = \alpha\chi V_t/\lambda$, Eq. (28) can be rewritten as

$$d\lambda = \widehat{Z}\chi\left(\frac{dV_t}{d\delta V_{aw}} - V_t\frac{dV_{aw}}{d\delta V_{aw}^2}\right)d\delta = \left(\frac{dV_t}{d\delta V_t}\lambda - \frac{1}{\alpha\chi V_t}\frac{dV_{aw}}{d\delta}\lambda^2\right)d\delta, \tag{29}$$

or

$$d\lambda = \left(a_1\lambda - a_2\lambda^2\right)d\delta. \tag{30}$$

A solution of Eq. (30) yields Eq. (5). Taking into account Eq. (29), the coefficients of Eq. (5) take on the meaning: $a_1 = \frac{dV_t}{V_t d\delta} = \frac{d \ln V_t}{d\delta}$ and $a_2 = \frac{2 d V_{aw}}{\chi V_t d\delta}$.

Thus, Eqs. (3)–(5) follow from elastic-plastic invariant (12) and depend on the lattice properties of the deforming medium. It can thus be concluded that the elastic and plastic processes occurring in the deforming solid are closely related.

Plastic flow occurring in a medium is described by Eq. (7), which can be derived from invariant (12) as well. To do this, Eq. (12) is rewritten as

$$\frac{\lambda}{\chi} = \widehat{Z} \frac{V_t}{V_{aw}}, \tag{31}$$

where the term $\lambda / \chi \equiv \varepsilon$ is assumed to be deformation. By applying the differential operator $\partial / \partial t = D_\varepsilon \partial^2 / \partial x^2$ to the right and left sides of Eq. (31), we obtain

$$\frac{\partial \varepsilon}{\partial t} = \widehat{Z} D_\varepsilon \frac{\partial^2}{\partial x^2} (V_t / V_{aw}). \tag{32}$$

Differentiation of Eq. (32) yields

$$\frac{\partial \varepsilon}{\partial t} = \widehat{Z} D_\varepsilon \left(-V_t \frac{\partial^2 V_{aw}^{-1}}{\partial x^2} + V_{aw}^{-1} \frac{\partial^2 V_t}{\partial x^2} \right) \tag{33}$$

According to [56], the ultrasound rate V_t varies in an intricate fashion in the deforming medium, while the autowave propagates at a constant rate at the linear stage of work hardening. Thus, reduction of Eq. (33) yields

$$\frac{\partial \varepsilon}{\partial t} = -\widehat{Z} D_\varepsilon V_t \frac{\partial^2 V_{aw}^{-1}}{\partial x^2} + D_\varepsilon \frac{\partial^2 \varepsilon}{\partial x^2} = f(\varepsilon) + D_\varepsilon \frac{\partial^2 \varepsilon}{\partial x^2}, \tag{34}$$

which is equivalent to Eq. (7).

Thus, the main features of localized plastic flow are described by Eqs. (7) through (9), which are derived from invariant (12) obtained on the base of experimental evidence. Equation (12) states that the characteristics of localized plastic flow are determined by the lattice characteristics of the medium. This thesis in the form of logical implication would assert that the processes involved in elastic and plastic deformation in solids are closely related. It must be admitted that among the factors governing the processes of elastic deformation are not only the crystal lattice properties, but also its interparticle potential, whereas the processes of plastic deformation are governed by the behavior of lattice defects alone.

4   The Model of Localized Plastic Flow

By addressing the localization behavior of plastic deformation in solids, the acoustic characteristics manifested by the deforming medium also call for further investigation. Up to now, the acoustic characteristics of the deforming medium were addressed in terms of energy dissipation, in particular, in internal friction studies and related problems. In what follows, this subject is discussed at greater length.

4.1   Plastic and Acoustic Characteristics of the Deforming Medium

We are dealing here with the rate of transverse elastic waves, V_t, which appears in Eq. (12) for strain invariant and in the expression of acoustic resistance of the medium, which enters into Eq. (13). It is also shown above that the coefficient Ξ from Eq. (3), which describes the autowave propagation rate, depends on the rate of transverse elastic waves [60].

One must take into account another characteristic of the deforming medium, i.e. phonon gas viscosity [65] which appears in Eq. (16). At first glance, this quantity seems to be out of place in the analysis of slow processes of localized plasticity autowave propagation. This value is generally determined for a sample under impact loading by addressing high rate motion of dislocations. Nonetheless, phonon gas viscosity made its appearance in our discussion from the following considerations. The development of plastic deformation is due to dislocation motion. Dislocations move over the local obstacles [32], the rate of dislocation motion is given as $V_{disl} \approx (b/B)\sigma$ and is controlled by the phonon gas (see above). The appearance of the quantity B in the latter relation is accounted for by the occurrence of moving dislocations within the localized plasticity nuclei.

The mechanical and acoustic characteristics of the deforming medium are found to be closely related [28]. This finding is supported by the experimental evidence, which strongly suggests that acoustic processes play an important role in the development of localized plastic flow. Available acoustic emission data suggests that structural inhomogeneities would emerge in the deforming medium due to a traveling deformation front. Thus, the acoustic emission sources occurring in material bulk have to be linked to the localized plastic flow nuclei emergent in the deforming solid [56]. To address the nature of localized plasticity, a two-component model has been formulated, in which a key role is assigned to the acoustic properties of the deforming solid. Acoustic emission pulses play the role of information system and control the dynamics of form changing.

4.2 Two-Component Model of Localized Plasticity

In the conceptual framework used to address the basic problem of autowave formation is the nature of self-organization, which manifests itself in the deforming medium as a spontaneous emergence of autowave structure. Physical interpretation of Eqs. (7) and (9) might prove productive for elucidation of the problem. Kadomtsev [49] advanced the idea that a self-organizing system will separate spontaneously into dynamic and information subsystems, interacting with one another.

The idea of the proposed model is as follows. In the course of plastic deformation local stress concentrators would form and disintegrate; these are considered as slowed-down shears. Elementary stress relaxation act is due to breaking from a local obstacle, which involves acoustic emission [66]. These acoustic signals will activate other stress concentrators, to so that the same process is repeated over and over again. Thus acoustic emission signals propagating in the deforming medium play the role of information subsystem; dislocation shears are involved in the plastic deformation proper and operate as a dynamic subsystem. The model developed is made up of two components: acoustic emission and dislocation mechanisms of plasticity, which have been studied sufficiently, although in different contexts. The generation of acoustic signals was considered in connection with the initiation of dislocation shears, while the reverse process, i.e. initiation of shears due to acoustic pulses, has not been touched on thus far.

In what follows, the performance of the proposed model is assessed. Acoustic signal can propagate in non-uniform dislocation substructure, which forms by deformation and is observable by transmission electron microscopy, e.g. dislocation cell having size $R \approx 0.01$ mm. It is proposed by [56] that such cell be regarded as acoustic lens, which has focal length, f_l, given as

$$f_l \approx \frac{R}{V_t^{(def)}/V_t - 1},$$ (35)

where the ratio of ultrasound rates, $V_t^{(def)}/V_t$, observed for non-deformed and deformed volumes plays the role of acoustic refractive index. The initiation of plastic deformation is due to the ultrasound waves focusing at distance $\lambda \approx f_l \approx 10^{-2}$ m from the active localized plasticity nucleus.

Now lets estimate acoustic emission pulses in terms of energy expenditure required for activating dislocation shears. According to [67], the time needed for dislocations to move over barriers during thermally activated motion, is $\tau \approx \exp\left(\frac{U_0 - \gamma\sigma}{k_B T}\right)$. Here $U_0 - \gamma\sigma = H$ is the process enthalpy and k_B is the Boltzmann constant. Generally, $H \approx 1$ eV and $\tau \approx 10^{-6}$ s. For the case of $H = U_0 - \gamma\sigma - \varepsilon_{ph}$, where the phonon energy is given as $\varepsilon_{ph} = \hbar\omega_D \approx 0.3$ eV, $\tau \approx 5 \times 10^{-7}$ s. The generation of a new front is presented schematically in Fig. 6.

It is an established fact that the perfect crystal lattice is a source of crystal defects responsible for plastic form changing; therefore, its properties must be taken into account by addressing self-organization processes as well. Hence, the basic premise

Fig. 6 Scheme of autowave formation

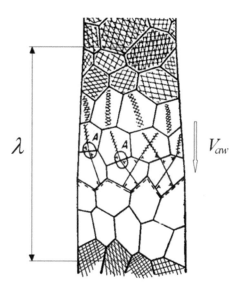

λ

V_{aw}

of the given paper is that the regular features of plastic flow macrolocalization are directly related to the lattice characteristics.

5 Plastic Flow Viewed as a Macroscopic Quantum Phenomenon

An innovative approach to the plasticity problem can be developed using elastic-plastic strain invariant (12), which has a deep physical meaning.

5.1 Localized Plastic Flow Autowaves and the Planck Constant

In the course of plastic flow the autowave processes are generated in the deforming medium. This mechanism has been established experimentally for all the plastic flow stages. The findings are giving us a clue to the most distinctive features of the plastic flow and thus provide additional insights into basic plasticity problems. Clearly, the next step in the development of this model is a quantum representation of the plastic flow. This necessitates introduction of a quasi-particle corresponding to the localized plastic flow autowave. The strong evidence that lends support to this idea is considered below.

Numerical analysis was performed for the experimental data on λ and V_{aw}, which produced an unexpected result. Thus it is found that the product $\lambda V_{aw} \rho r_{ion}^3$ (here ρ

Table 2 Estimation of the Planck constant with the help of Eq. (36)

	Metals										
	Cu	Zn	Al	Zr	Ti	V	Nb	γ-Fe	α-Fe	Ni	Co
$h \cdot 10^{34}$	11.9	9.3	2.8	6.1	4.9	3.5	4.9	4.6	4.6	6.1	7.1

	Metals						
	Sn	Mg	Cd	In	Pb	Ta	Mo	Hf
$h \cdot 10^{34}$	8.9	4.9	7.4	9.9	18.4	5.5	3.0	7.3

is metal density and $r_{ion} \approx \chi$ is ion radius of metal) is close to the Planck constant $h = 6.63 \times 10^{-34}$ J s [68]. Hence, we write

$$\lambda V_{aw} \rho r_{ion}^3 \approx h. \tag{36}$$

The validity of Eq. (36) is justified by the data listed in Table 2 (note: handbook values of ion radii are used herein). On the base of these data the average value of the Planck constant was calculated for thirteen metals; the resultant value $\langle h \rangle = (6.9 \pm 0.45) \times 10^{-34}$ J s, with the ratio $\langle h \rangle / h = 1.04 \pm 0.06 \approx 1$, i.e. $\langle h \rangle = h$.

Using a standard statistical procedure [69], the quantities $\langle h \rangle$ and h were matched. Let the value $\langle h \rangle$ be defined as the average of thirteen measurements ($n_1 = 19$). On the other hand, we operated on the premise that the value h was determined in a single measurement ($n_2 = 1$) in the absence of dispersion. The statistical significance of coincidence of the quantities $\langle h \rangle$ and h was determined with 95% confidence level by Student's t-test as

$$t = \frac{\langle h \rangle - h}{\hat{\sigma}} \cdot \sqrt{\frac{n_1 n_2}{n_1 + n_2}}, \tag{37}$$

where the value $\hat{\sigma}$ is the square root of the overall estimate of dispersion. This procedure shows that the values $\langle h \rangle$ and h are statistically identical, i.e. $\langle h \rangle = h$. It is pertinent to note that h is the fundamental constant; hence, its appearance in Eq. (36) is in no way accidental—this suggests that plastic deformation physics is related to quantum mechanics.

5.2 Introduction of a New Quasi-particle and Its Applications

Further on, the plasticity problem is approached using quantum ideas. By further elaborating the autowave concept of localized plasticity, we were guided by the fundamentals of modern condensed-state physics where quasi-particle concept is generally introduced and freely employed for simplifying description of solids [70]. Our consideration of localized plastic flow autowaves is based on the concept of wave-particle dualism.

In the first place, the mass of the hypothetical quasi-particle is of principal importance. Clearly, the characteristics of the quasi-particle have to be related to those of the autowave. Thus the quasi-particle mass is defined as follows. On the base of data obtained for dispersion autowaves in Al and γ-Fe, we write

$$m_{ef} = \left(\frac{d^2 U}{dp^2}\right)^{-1} = \hbar \frac{d^2}{dk^2}[\omega(k)], \tag{38}$$

where the values U and p denote, respectively, the energy and quasi-momentum of the quasi-particle. Another way of looking at it is proposed by [71, 72]. Thus the de Broglie formula can be employed to address localized plastic flow autowaves as

$$m_{ef} = \frac{h}{\lambda V_{aw}}. \tag{39}$$

An alternative method is the use of the term from Eq. (36)

$$\rho r_{ion}^3 = m_{ef}. \tag{40}$$

Apparently, the effective mass of a quasi-particle having size $\sim r_{ion}$ can also be defined from Eqs. (39) and (40).

Using Eqs. (38) through (40), the mass of the quasi-particle was found; the values obtained are in the range $0.5 \leq m_{ef} \leq 1.5$ a.m.u. (atomic mass unit). The averaged value $\langle m_{ef} \rangle \approx 1$ a.m.u. is a rough estimate of the quantity m_{ef}.

The hypothetic quasi-particle is named *auto-localizon*. The next task is equating the propagation rates of the autolocalizon and the autowave. The mobility of autolocalizon is affected by the phonon and electron gases in solids. Thus the effective mass of the auto-localizon is regarded as its virtual mass, which is defined by the resistance of both gases to the motion of auto-localizon. Strong evidence was recently obtained in support of this conjecture. The effective mass was calculated from Eq. (38) for a range of metals.

We will now look at some possibilities offered by this approach. Using Eq. (38), the formula for elastic-plastic strain invariant (12) can be rewritten as

$$\frac{h}{\lambda V_{aw}} \approx \hat{Z}^{-1} \frac{h}{\chi V_t}, \tag{41}$$

where $h/\chi V_t = m_{ph}$ and $h/\lambda V_{aw} = m_{a-l}$ are, respectively, the phonon and the auto-localizon masses. Evidently, Eq. (41) is equipotent to the equality $m_{a-l} \approx \hat{Z}^{-1} m_{ph}$. Hence, Eq. (41) accounts for the mechanism, which is responsible for the generation of dislocations due to phonon condensation [73].

This concept is elaborated in the frame of conventional approach adopted in the solids physics, which involves introduction of a quasi-particle for description of wave processes. By way of an example, it well suffices to mention elementary excitations in media [70]. One of the first attempts of this kind was made by [74] who introduced a quasi-particle named *crackon* in order to address mechanisms of brittle crack propagation.

It should be reminded that such ideas have long been in the air; therefore, attempts at introducing quantum concepts into the physics of plasticity are by no means scarce. Thus the quantum tunneling effect was used by [2, 75, 76] in descriptions of low-temperature processes involving dislocations breakaway from pinning points. Later on, [77] supplied a detailed explanation of this phenomenon for the case of dislocation motion in the Peierls-Nabarro potential relief. Steverding [78] made use of quantization of elastic waves propagating by material fracture. Zhurkov [79] introduced the notion of elemental excitation in crystals, which was termed as *dilaton*. Later on, the possible existence of a specific precursor of deformation or fracture was hypothesized by [80] who coined the name *frustron* for this phenomenon. It has been shown that an elementary act of interatomic bond rupture has activation volume close to that of an atom. Evidently, ideas of this kind are transparent enough; the underlying theoretical premises are based on the discreteness of crystal lattice in which generation and evolution of elementary acts of plasticity takes place. The autowave and quasi-particle concepts are distinct, though complementary and interrelated approaches.

In the frame of quasi-particle concept, the length of localized plasticity autowave can be estimated. With this aim in view, the motion of autolocalizon in the phonon gas is considered. In view of the fluctuations of phonon gas density, it is proposed that the autolocalizon be involved in the Brownian motion. In accordance with Einstein's theory, the free path of the Brownian particle is

$$s \approx \sqrt{\frac{k_B T}{\pi \hat{\eta} r_{a-l}} t}, \tag{42}$$

where $\hat{\eta}$ is the dynamic viscosity of the phonon gas and t is the time given as $t \approx 2\pi/\omega$ (here ω is the frequency of localized plastic flow autowave).

Hence, the free path of the quasi-particle is presented as autowave length λ and is given as follows. Assume that $T = 300$ K; the autolocalizon has size $r_{a-l} \approx 10^{-10}$ m; the autowave period is $t \approx 10^3$ s and $\hat{\eta} \approx 10^{-4}$ Pa s (the latter value was obtained by [62] in high-velocity dislocation motion tests). The resultant value $s \approx 10^{-2}$ m, which is evidently close to the autowave length, λ. Thus the application of Eqs. (35) and (42) yields equivalent numerical estimates.

5.3 Plasticity Viewed as a Macro-scale Quantum Phenomenon

A close relation has been established between the deformation and acoustic characteristics of the deforming medium, which suggests that the deformation processes can be described by a hybridized excitation spectrum (Fig. 7). Such a spectrum is

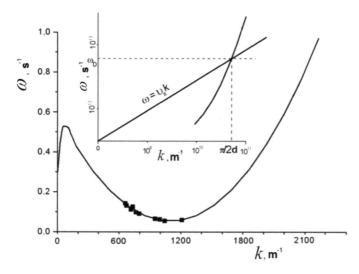

Fig. 7 Generalized dispersion curve obtained for elastically and plastically deforming solids (*insert* similar dependencies obtained for high-frequency oscillation spectrum)

obtained by the imposition of the linear dispersion relation $\omega \approx V_t k$ for elastic waves and quadratic the same $\omega = \omega_0 + \alpha(k - k_0)^2$ for localized plasticity autowaves.

For validation the coordinates $\hat{\omega}$ and $\hat{\lambda}$ were estimated for the point of intersection of the plots in a high-frequency spectral region. The frequency $\hat{\omega} \approx \omega_D$ and the wave number $\hat{k}$ correspond to the minimal length of elastic wave, which is of the order of distance between close packed planes, i.e. $\hat{k} \approx 2\pi/\chi$. The above evidence indicates that the generalized dispersion relation holds good for both the phonons and the auto-localizons.

The dispersion curve illustrated in Fig. 7 is suggestive of a remarkable analogy with the dispersion relation obtained for superfluid ^{4}He [70, 81]. The latter dispersion relation shows a minimum corresponding to the origination of '*rotons*', i.e. quasi-particles having effective mass $m_{rot} \approx 0.64$ a.m.u.; moreover, the quadratic dispersion law obtained for rotons is similar to Eq. (4). The above analogy is indicative of a similarity of localized plasticity and superfluidity. Whether this is a formal similarity or whether it has a physical meaning, remains to be seen.

Additional argument in favor of this attractive conjecture is as follows. The superfluidity of ^{4}He is attributed to the occurrence of normal and superfluid components in liquid ^{4}He at $T \leq 2.17$ K; the respective dynamic viscosities obtained for these components are $\hat{\eta}_n$ and $\hat{\eta}_{sf} \ll \hat{\eta}_n$. The plastic flow occurring in the deforming medium would involve both slow motion of individual material volumes, which undergo form changing, and fast dislocation motion. Slow process corresponds to high viscosity of material, $\hat{\eta}_{mat} = G\tau \approx 10^{10}$ Pa s (here $\tau \approx 10^3$ s is a characteristic time). The velocity of dislocation motion, $V_{disl} \approx (b/B) \cdot \sigma$, is controlled by the phonon gas viscosity, $B \approx 10^{-4}$ Pa s [62], i.e. $\hat{\eta}_{mat}/B \approx 10^{14}$.

Table 3 The macroscopic quantum phenomena

Phenomenon	Observable quantum macrocharacteristic	
Superconductivity	Magnetic flux	$\Phi = \frac{\pi \hbar c}{e} \cdot n$
Superfluidity	Rotation rate of ^{4}He	$v_s = \frac{\hbar}{A} \cdot \frac{1}{r} \cdot n$
The Hall quantum effect	Hall's resistance	$R_H = \frac{h}{e^2} \cdot \frac{1}{n}$
Localized plastic flow	Magnitude of deformation jump	$\delta L \approx \frac{h}{\rho \chi^3 V_{aw}} \cdot m \approx \frac{h}{m_{a-l} V_{aw}} \cdot n$

$\hbar = h/2\pi$ is the Planck constant; c—velocity of light; e—electron charge; A—atomic mass of ^{4}He; r—radius; $m = 1, 2, 3 \ldots$

Three macroscopic quantum effects are well-known in physics, i.e. supercon-ductivity, superfluidity and the quantum Hall effect [82]. The characteristics of these effects are listed in Table 3. On the base of data obtained in this study, the localized plasticity phenomenon might be included in the same 'short list'. In what follows, we shall attempt to apply a quasi-particle approach to the analysis of serrated plastic deformation similar to the Portevin-Le Chatelier effect [83–85]. Assume that autowaves having length λ are arranged along sample length, L. Then the number of autowaves is given as $\lambda = L/n$ where $n = 1, 2, 3 \ldots$. The deformed sample has length $L \approx L_0 + \delta L$ (here L_0 is initial length); hence, $\delta L \approx \lambda$. Thus from Eq. (35) follows

$$\delta L \approx \frac{h}{\rho \chi^3 V_{aw}} n \approx \frac{h}{m_{a-l} V_{aw}} n. \tag{43}$$

The autolocalizon mass $m_{a-l} \approx \rho \chi^3 \approx \rho r_{ion}^3$ appears in Eq. (43) which states that a jump-wise elongation of the tensile sample is necessary, i.e. $\delta L \sim n$. For the linear work hardening stage, $V_{aw} = const$; hence, $\frac{h}{\rho \chi^3 V_{aw}} = const$. Given sufficient instrumentation sensitivity, the recorded curves $\sigma(\varepsilon)$ will invariably exhibit a serrated behavior; moreover, accommodation of the sample length will occur to fit the general autowave pattern. Numerical estimates were made which suggest that for $n = 1$, $\rho \approx 5 \times 10^3$ kg/m^3 and $\chi \approx 3 \times 10^{-10}$ m hence, the elongation jump $\delta L_{m=1} \approx 10^{-4}$ m. For the sample length $L \approx 10^{-1}$ m the elongation jump corresponds to the deformation jump $\delta \varepsilon_{m=1} \approx 10^{-3}$, which is a close match of the experimentally obtained value.

It also follows from Eq. (43) that an increase in the loading rate would cause a decrease in the deformation jump value, i.e. $\delta L \sim V_{aw}^{-1}$. This inference is supported by the experimental results obtained for Al samples tested at 1.4 K at different loading rates [86]. Thus, the autowave rate was found to be proportional to the motion velocity of the testing machine crossheads, i.e. $V_{aw} \sim V_{mach}$. According to Eq. (43), the velocity V_{aw} will increase with rate V_{mach}, while the deformation jumps will grow smaller.

6   Conclusions

1. The discussion of factual evidence cited herein enables formulation of a new idea of the nature of plasticity. In the frame of proposed concept, the plastic flow localization is due not only to the formation and to redistribution of defects (dislocations) in the deforming medium, but also to the lattice and material characteristics related to quantum mechanics. It is found that the parameters of plastic flow localization are related to the quantities h, χ and V_t. This relation is a qualitative one, while material structure plays a subordinate, quantitative role.
2. An analysis of the plastic flow suggests that regular features, which are manifested in all deforming solids, distinguish the deformation process. The kinetics of plastic flow is determined by a regular changeover in the localization patterns (autowave modes).
3. Elastic-plastic strain invariant is introduced to relate the processes involved in plastic and elastic deformation. It is shown that the main laws of autowave plastic deformation are corollaries of this invariant.
4. A well-founded conjecture is proposed that the localized plasticity phenomenon belongs to the category of quantum effects manifested on the macro-scale level. Validation of this hypothesis is also provided. To advance this idea, a quasi-particle of localized plastic deformation (auto-localizon) is introduced.

Acknowledgements I recollect with the enormous gratitude the inestimable support of these studies, which were rendered on all their stages by Professor S. Psakhie, and very fascinating discussions with him of principal aspects of general plasticity problem.

As well as I thank Profs. S. A. Barannikova, V. I. Danilov, Yu. A. Khon and Yu. A. Alyushin as well as Dr. V. V. Gorbatenko for the fruitful discussion of the results and their interpretations.

The work was performed according to the Government research assignment for ISPMS SB RAS, project No. III.23.1.2.

References

1. Kuhlmann-Wilsdorf D (2002) Dislocations in solids. Elsevier, Amsterdam, pp 213–238 (The low energetic structures theory of solid plasticity)
2. Zbib YM, de la Rubia TD (2002) A multiscale model of plasticity. Int J Plast 18(9):1133–1163
3. Kubin L, Devincre B, Hoc T (2008) Toward a physical model for strain hardening in fcc crystals. Mater Sci Eng A 483–484:19–24
4. Lazar M (2013) On the non-uniform motion of dislocations: the retarded elastic fields, the retarded dislocation tensor potentials and the Lienard-Wiechert tensor potentials. Phil Mag 93(7):749–776
5. Lazar M (2013) The fundamentals of non-singular dislocations in the theory of gradient elasticity: dislocation loops and straight dislocations. Int J Solids Struct 50(2):52–362
6. Aifantis EC (1996) Nonlinearity, periodicity and patterning in plasticity and fracture. Int J Non-Linear Mech 31:797–809
7. Aifantis EC (2002) Handbook of materials behavior models. Academic Press, New York, pp 291–307 (Gradient plasticity)

8. Unger DJ, Aifantis EC (2000) Strain gradient elasticity theory for antiplane shear cracks. Part I Oscillatory Displacement Theor Appl Fract Mech 34(3):243–252
9. Pontes J, Walgraef D, Aifantis EC (2006) On dislocation patterning: multiple slip effects in the rate equation approach. Int J Plast 22(8):1486–1505
10. Ohashi T, Kawamukai M, Zbib H (2007) A multiscale approach for modeling scale-dependent yield stress in polycrystalline metals. Int J Plast 23(5):897–914
11. Langer JS, Bouchbinder E, Lookman T (2010) Thermodynamic theory of dislocation-mediated plasticity. Acta Mater 58(10):3718–3732
12. Lim H, Carroll JD, Battaile CC, Buchheit TE, Boyce BL, Weinberger CR (2014) Grain scale experimental validation of crystal plasticity finite element simulations of tantalum oligocrystals. Int J Plast 60:1–18
13. Aoyagi Y, Kobayashi R, Kaji Y, Shizawa K (2013) Modeling and simulation on ultrafine-graining based on multiscale crystal plasticity considering dislocation patterning. Int J Plast 47:13–28
14. Aoyagi Y, Tsuru T, Shimokawa T (2014) Crystal plasticity modeling and simulation considering the behavior of the dislocation source of ultrafine-grained metal. Int J Plast 55:18–32
15. Nicolis G, Prigogine I (1977) Self-organization in nonequilibrium systems. Wiley, New York
16. Nicolis G, Prigogine I (1989) Exploring complexity. An introduction. Freeman & Company, New York
17. Haken H (2006) Information and self-organization. A macroscopic approach to complex systems. Springer, Berlin
18. Zaiser M, Hähner P (1997) Oscillatory modes of plastic deformation: theoretical concepts. Phys Status Solidi B 199(2):267–330
19. Hähner P, Rizzi E (2003) On the kinematics of Portevin-Le Chatelier bands: theoretical and numerical modelling. Acta Mater 51:3385–4018
20. Rizzi E, Hähner P (2004) On the Portevin-Le Chatelier effect: theoretical modeling and numerical results. Int J Plast 20(1):121–165
21. Zaiser M, Aifantis EC (2006) Randomness and slip avalanches in gradient plasticity. Int J Plast 22:1432–1455
22. Borg U (2007) Strain gradient crystal plasticity effects on flow localization. Int J Plast 23:1400–1416
23. Voyiadjis GZ, Faghihi D (2012) Thermo-mechanical strain gradient plasticity with energetic and dissipative length scales. Int J Plast 30:218–247
24. Voyiadjis GZ, Faghihi D (2013) Gradient plasticity for thermo-mechanical processes in metals with length and time scales. Phil Mag 93(9):1013–1053
25. Zuev LB (2001) Wave phenomena in low-rate plastic flow in solids. Ann Phys 10(11–12):965–984
26. Zuev LB (2007) On the waves of plastic flow localization in pure metals and alloys. Ann Phys 16(4):286–310
27. Zuev LB (2012) Autowave mechanics of plastic flow in solids. Phys Wave Phenom 20:166–173
28. Zuev LB (2018) Autowave plasticity. Localization and collective modes. Fizmatlit, Moscow (in Russian)
29. Zuev LB, Danilov VI, Barannikova SA (2001) Pattern formation in the work hardening process of single alloyed γ-Fe crystals. Int J Plast 17(1):47–63
30. Rastogi PK (2001) Digital speckle interferometry and related techniques. Wiley, New York, pp 141–224
31. Kadić A, Edelen DGB (1983) A gauge theory of dislocations and disclinations. Springer, Berlin
32. Friedel J (1964) Dislocations. Pergamon Press, Oxford
33. Aydiner CC, Telemez MA (2014) Multiscale deformation heterogeneity in twinning magnesium investigated with in situ image correlation. Int J Plast 56:203–218
34. Taleb L, Cavallo N, Wäckel F (2001) Experimental analysis of transformation plasticity. Int J Plast 17:1–20
35. Fressengeas C, Beaudoin AJ, Entemeyer D, Lebedkina T, Lebyodkin M, Taupin V (2009) Dislocation transport and intermittency in the plasticity of crystalline solids. Phys Rev B 79(1):014108–014110

36. Mudrock RN, Lebyodkin MA, Kurath P, Beaudoin A, Lebedkina T (2011) Strain-rate fluctuations during macroscopically uniform deformation of a solid strengthened alloy. Scripta Mater 65(12):1093–1095
37. Lebyodkin MA, Kobelev NP, Bougherira Y, Entemeyer D, Fressengeas C, Gornakov VS, Lebedkina TA, Shashkov IV (2012) On the similarity of plastic flow processes during smooth and jerky flow: statistical analysis. Acta Mater 60(9):729–3740
38. Acharia A, Beaudoin A, Miller R (2008) New perspectives in plasticity theory: dislocation nucleation, waves, and partial continuity of plastic strain rate. Math Mech Solids 13:292–315
39. Roth A, Lebedkina TA, Lebyodkin MA (2012) On the critical strain for the onset of plastic instability in an austenitic FeMnC steel. Mater Sci Eng A 539:280–284
40. Zaiser M, Seeger A (2002) Dislocations in solids. Elsevier, Amsterdam, pp 1–100 (Long-range internal stress, dislocation patterning and work hardening in crystal plasticity)
41. Argon A (2008) Strengthening mechanisms in crystal plasticity. Oxford University Press, Oxford
42. Messerschmidt U (2010) Dislocation dynamics during plastic deformation. Springer, Berlin
43. Hill R (2002) The mathematical theory of plasticity. Oxford University Press, Oxford
44. Pelleg J (2012) Mechanical properties of metals. Springer, Dordrecht
45. Davydov VA, Davydov NV, Morozov VG, Stolyarov MN, Yamaguchi T (2004) Autowaves in the moving excitable media. Condens Matter Phys 7:565–578
46. Scott A (2003) Nonlinear sciences. Emergence and dynamics of coherent structures. Oxford University Press, Oxford
47. Krinsky VI (1984) Self-organization: autowaves and structures far from equilibrium. Springer, Berlin
48. Hallai JF, Kyriakides S (2013) Underlying material response for Lüders-like instabilities. Int J Plast 47:1–12
49. Kadomtsev BB (1994) Dynamics and information. UFN Phys Usp 164(5):449–530
50. Zacharov VE, Kuznetsov EA (2012) Solitons and collapses: two evolution scenarios of nonlinear wave systems. UFN Phys Usp 55(6):535–556
51. Zuev LB (2014) Using a crystal as a universal generator of localized plastic flow autowaves. Bull Russ Acad Sci Phys 78:957–964
52. Barannikova SA (2004) Dispersion of the plastic strain localization waves. Tech Phys Lett 30:338–340
53. Zuev LB, Barannikova SA (2010) Evidence for the existence of localized plastic flow autowaves generated in deforming metals. Nat Sci 2(05):476–483
54. Zuev LB, Barannikova SA (2010) Plastic flow macrolocalization: autowave and quasi-particle. J Mod Phys 1(01):1–8
55. Zuev LB, Khon YA, Barannikova SA (2010) Dispersion of autowaves in localized plastic flow. Tech Phys 55:965–971
56. Zuev LB, Semukhin BS, Zarikovskaya NV (2003) Deformation localization and ultrasound wave propagation rate in tensile Al as a function of grain size. Int J Solids Struct 40(4):941–950
57. Zuev LB (2005) Entropy of localized plastic strain waves. Tech Phys Lett 31:89–90
58. Nekorkin VI, Kazantsev VB (2002) Autowaves and solitons in a three-component reaction-diffusion system. Int J Bifurcat Chaos 12(11):2421–2434
59. Landau LD, Lifshitz EM (1987) Fluid mechanics. Butterworth-Heinemann, London
60. Zuev LB, Danilov VI, Barannikova SA, Gorbatenko V (2010) Autowave model of plastic flow of solids. Phys Wave Phenom 17(1):66–75
61. Newnham RE (2005) Properties of materials. Oxford University Press, Oxford
62. Al'shits VI, Indenbom VL (1986) Dislocations in solids, vol 7. Elsevier, Amsterdam, pp 43–111 (Mechanism of dislocation drag)
63. Counts WA, Braginsky MV, Battaile CC, Holm EA (2008) Predicting the Hall-Petch effect in fcc metals using non-local crystal plasticity. Int J Plast 24:1243–1263
64. Zuev LB, Barannikova SA (2011) Plastic deformation viewed as an autowave process generated in deforming metals. Solid State Phenom 172–174:1279–1283
65. Ziman JM (2001) Electrons and phonons. Oxford University Press, Oxford

66. Williams RV (1980) Acoustic emission. Adam Hilger, Bristol
67. Caillard D, Martin JL (2003) Thermally activated mechanisms in crystal plasticity. Elsevier, Oxford
68. Atkins PW (1974) Quanta. A handbook of conceptions. Clarendon Press, Oxford
69. Brownlee KA (1965) Statistical theory and methodology in science and engineering. Wiley, New York
70. Brandt NB, Kulbachinskii VA (2007) Quasi-particles in condensed state physics. Fizmatlit, Moscow ((in Russian))
71. Billingsley JP (2001) The possible influence of the de Broglie momentum-wavelength relation on plastic strain 'autowave' phenomena in 'active materials.' Int J Solids Struct 38:4221–4234
72. Zuev LB (2005) The linear work hardening stage and de Broglie equation for autowaves of localized plasticity. Int J Solids Struct 42(3–4):943–949
73. Umezava H, Matsumoto H (1982) Thermo field dynamics and condensed states. North-Holland Publ. Comp. (Elsevier), Amsterdam
74. Morozov EM, Polack LS, Fridman YB (1964) On variation principles of crack development in solids. Sov Phys Dokl 156:537–540
75. Gilman JJ (1968) Escape of dislocations from bound states by tunneling. J Appl Phys 39:6068–6090
76. Oku T, Galligan JM (1969) Quantum mechanical tunneling of dislocation. Phys Rev Lett 22(12):596–577
77. Petukhov BV, Pokrovskii VL (1972) Quantum and classic motion of dislocations in the potential Peierls relief. J Exp Theor Phys JETP (Zh Eksp Teor Fiz) 63:634–647
78. Steverding B (1972) Quantization of stress waves and fracture. Mater Sci Eng 9:185–189
79. Zhurkov SN (1983) Dilaton mechanism of the strength of solids. Phys Solid State 25:1797–1800 ((in Russian))
80. Olemskoi AI (1999) Theory of structure transformation in non-equilibrium condensed matter. Nova Science Pub Inc., New York
81. Landau LD, Lifshitz EM (1980) Statistical physics. Butterworth-Heinemann, London
82. Imry Y (1983) Introduction to mesoscopic physics. Oxford University Press, Oxford (UK)
83. Zhang Q, Jiang Z, Jiang H, Chen Z, Wu X (2005) On the propagation and pulsation of Portevin-Le Chatelier deformation bands: an experimental study with digital speckle pattern metrology. Int J Plast 21(11):2150–2173
84. Coër J, Manach PY, Laurent H, Oliveira MC, Menezes LF (2013) Piobert–Lüders plateau and Portevin–Le Chatelier effect in an Al–Mg alloy in simple shear. Mech Res Commun 48:1–7
85. Manach PY, Thuillier S, Yoon JW, Coër J, Laurent H (2014) Kinematics of Portevin–Le Chatelier bands in simple shear. Int J Plast 58:66–83
86. Pustovalov VV (2008) Serrated deformation of metals and alloys at low temperatures. Low Temp Phys 34(9):683–723

Three-Component Wear-Resistant PEEK-Based Composites Filled with PTFE and MoS$_2$: Composition Optimization, Structure Homogenization, and Self-lubricating Effect

Sergey V. Panin, Lyudmila A. Kornienko, Nguyen Duc Anh,
Vladislav O. Alexenko, Dmitry G. Buslovich, and Svetlana A. Bochkareva

Abstract The aim of this work was to design and optimize compositions of three-component composites based on polyetheretherketone (PEEK) with enhanced tribological and mechanical properties. Initially, two-component PEEK-based composites loaded with molybdenum disulfide (MoS$_2$) and polytetrafluoroethylene (PTFE) were investigated. It was shown that an increase in dry friction mode tribological characteristics in metal-polymer and ceramic-polymer tribological contacts was attained by loading with lubricant fluoroplastic particles. In addition, molybdenum disulfide homogenized permolecular structure and improved matrix strength properties. After that, a methodology for identifying composition of multicomponent PEEK-based composites having prescribed properties which based on a limited amount of experimental data was proposed and implemented. It was shown that wear rate of the "PEEK + 10% PTFE + 0.5% MoS$_2$" composite decreased by 39 times when tested on the metal counterpart, and 15 times on the ceramic one compared with neat PEEK. However, in absolute terms, wear rate of the three-component composite on the metal counterpart was 1.5 times higher than on the ceramic one. A three-fold increase in wear resistance during friction on both the metal and ceramic counterparts was achieved for the "PEEK + 10% PTFE + 0.5% MoS$_2$" three-component composite compared with the "PEEK + 10% PTFE". Simultaneous loading with two types of fillers slightly deteriorated the polymer composite structure compared with neat PEEK. However, wear rate was many times reduced due to facilitation of transfer film formation. For this reason, there was no microabrasive wear on both metal and ceramic counterpart surfaces.

S. V. Panin (✉) · L. A. Kornienko · V. O. Alexenko · D. G. Buslovich · S. A. Bochkareva
Institute of Strength Physics, Materials Science SB RAS, Tomsk 634055, Russia
e-mail: svp@ispms.ru

S. V. Panin · N. D. Anh · V. O. Alexenko · D. G. Buslovich
National Research Tomsk Polytechnic University, Tomsk 634030, Russia

© The Author(s) 2021
G.-P. Ostermeyer et al. (eds.), *Multiscale Biomechanics and Tribology of Inorganic and Organic Systems*, Springer Tracts in Mechanical Engineering,
https://doi.org/10.1007/978-3-030-60124-9_13

Keywords Wear · Permolecular structure · Mechanical properties ·
Polyetheretherketone · Molybdenum disulfide · Polytetrafluoroethylene · Friction
coefficient · Transfer film · Wear resistance · Control parameters

1 Introduction

Polyetheretherketone (PEEK) is one of the prospective structural polymeric materials due to the unique combination of operational characteristics: high strength and
toughness, thermal and chemical resistance, as well as biocompatibility. PEEK is
also stable during long-term operation at low and elevated temperatures (from −40
to 260 °C) while maintaining high mechanical properties. In addition, PEEK has
enough melt flow rate, which facilitates its processing and application, including in
additive manufacturing of complex-shaped parts [1, 2].

Varying composition of fillers changes PEEK characteristics and expands application areas. In particular, reinforcing fibers (carbon, glass, aramid, etc.) are loaded to
increase its mechanical properties [3–5]. PEEK-based composites containing about
30 wt% carbon or glass fibers are most widely used as a polymer structural material
[6, 7]. However, as has been shown in [8], metal counterparts wear out rapidly even
during friction on neat PEEK. If PEEK has been loaded with reinforcing fibers, wear
rate increases many times [9].

Traditionally, the problem of low PEEK antifriction properties has been solved
by loading with solid lubricant fillers. One of the most common among them is
polytetrafluoroethylene (PTFE), which in some cases reduce PEEK wear rate by
several orders of magnitude [10–14]. Recently, PEEK-based nanocomposites have
been actively designed as well [15, 16]. Meanwhile, some published data on effect of
the fillers on PEEK-based composite wear resistance during dry sliding friction have
been controversial [17, 18]. Nevertheless, loading with (nano)particles of various
compositions as solid lubricant inclusions have not caused a multiple increase in
their wear resistance. Moreover, improving some properties due to a change in the
compositions by loading with the fillers is usually accompanied by a deterioration
of their other characteristics. In this regard, various optimization methods have been
implemented to achieve the required properties of the polymer composites. They are
often difficult to use or imply obligatory presence of a pronounced extremum of an
objective function [19, 20], [etc.].

PEEK loading with fluoroplastic particles usually causes a decrease in its
deformation-strength properties [13]. Absence of interfacial adhesion due to the
non-polar nature of PTFE prevents formation of high-strength uniform structure.
Partial loss of its strength can be compensated by loading with reinforcing fibers
or improving of the polymer binder (matrix) structure (for example, by loading
with (nano)fillers). As mentioned above, application of the high-modulus reinforcing
fibers exerts very limited effect on metal-polymer tribological contacts. Therefore,
in the present work, an attempt was made to improve the polymer matrix structure by
loading with MoS_2 microparticles. This would provide solutions to several problems.

The first one was to ensure uniform structure formation during compression sintering of the polymer composite due to high thermal conductivity. The second problem was to implement dispersion hardening of the polymer, including through activation of processes at the "matrix–filler" interface. The last but not least was to provide inherent function of a solid lubricant, as complementary to action of PTFE particles. Obviously, it was difficult to increase in PEEK strength much like by loading with chopped carbon fibers, but there was a chance to reduce intense microabrasive wear of the metal counterparts. Based on the foregoing, the aim of this work was to design and optimize compositions of three-component PEEK-based composites with enhanced tribological and mechanical properties in an experimental-theoretical way that enabled to determine a range of possible filler contents.

2 Materials and Methods

The "Victrex" PEEK powder with an average particle size of 50 μm, as well as fillers: PTFE polytetrafluoroethylene (particle size of 6...20 μm, F4-PN20 grade, "Ruflon" LLC, Russia) and MoS_2 molybdenum disulfide (Climax Molybdenum, USA, particle size of 1...7 μm) were used in these studies.

The PEEK-based composites were fabricated by hot pressing at a specific pressure of 15 MPa and a temperature of 400 °C. Subsequent cooling rate was 2 °C/min. The polymer binder powders and the fillers were mixed through dispersing the suspension components in ethanol using a "PSB-Gals 1335-05" ultrasonic cleaner ("PSB-Gals" Ultrasonic equipment center). Processing time was 3 min; generator frequency was 22 kHz.

Shore D hardness was determined using an "Instron 902" facility in accordance with ASTM D 2240.

Tensile properties of the PEEK-based composite samples were measured using an "Instron 5582" electromechanical testing machine. The "dog-bone" shaped samples met the requirements of Russian state standard GOST 11262-80 and ISO 178:2010.

"Pin-on-disk" dry friction wear tests of the PEEK-based composites were performed using a "CSEM CH-2000" tribometer in accordance with ASTM G99 (load was 10 N; sliding speed was 0.3 m/s). Two ball-shaped counterparts 6 mm in diameter were made of GCr15 bearing steel and Al_2O_3 ceramics (distance was 3 km; radius of the rotation trajectory was 10 mm; rotation speed was 286 rpm). Wear rate was determined by measuring the volume of the friction track using an "Alpha-Step IQ" stylus surface profiler (KLA-Tencor, USA).

A "Neophot 2" optical microscope (Carl Zeiss, Germany) equipped with a digital camera (Canon EOS 550D, Canon Inc., Japan) was used to examine wear track surfaces after testing. Permolecular structure was studied on cleavage surfaces of notched specimens mechanically fractured after exposure in liquid nitrogen. A "LEO EVO 50" scanning electron microscope (Carl Zeiss, Germany) was used (accelerating voltage was 20 kV).

3 Results and Discussion

3.1 Two-Component "PEEK + MoS$_2$" Composites

Initially, two-component composites independently loaded with MoS$_2$ and PTFE were separately studied to evaluate effectiveness of each filler in changing PEEK mechanical and tribological properties (Sects. 3.1 and 3.2, respectively). Table 1 shows mechanical characteristics of the "PEEK + 1% MoS$_2$" and "PEEK + 10% MoS$_2$" composites (hereinafter all percentages are indicated by weight). The amount of the loaded filler was based on both published data and the results of preliminary experimental studies [10, 11, 13]. Elastic modulus increased after loading up to 10% of molybdenum disulfide particles into the polymer matrix, while tensile strength and elongation at break decreased by 11 and 54%, respectively (Fig. 1a). According to these data and taking into account an increase in Shore D hardness, it can be stated that loading with MoS$_2$ microparticles provided formation of a harder (and stiffer) composite.

SEM micrographs of the permolecular structure of neat PEEK and the PEEK-based composites are shown in Fig. 1b–d. They indicate that molybdenum disulfide was distributed quasi-uniformly mainly along the boundaries of the permolecular structure elements (Fig. 1c). The permolecular structure of neat PEEK possessed a fragmented pattern with the sizes of structural elements from units to tens of microns (Fig. 1a) which decreased after loading with 1% MoS$_2$ (Fig. 1b). Highly likely, finely dispersed MoS$_2$ particles had been crystallization centers. This effect was even more pronounced after loading with 10% MoS$_2$. The composite had a finely dispersed structure (Fig. 1c), most likely due to high thermal conductivity of the filler. In this case, disperse hardening (structure modification) caused an increase in hardness and elastic modulus, but, as expected, decreased elongation at break.

An increase in hardness of PEEK, modified by loading with a significant amount of filler (10%), dramatically reduced composite flexibility. Therefore, it did not contribute to improve wear resistance. The reasons were a higher material hardness and more hard wear particles, which had been formed during friction of the tribological contact parts, causing additional wear of both the polymer and the counterpart. These facts were confirmed by the results of the tribological tests presented below.

Table 1 Mechanical properties of the PEEK-based composites loaded with MoS$_2$

Filler composition, % (wt.)	Density ρ (g/cm^3)	Shore D hardness	Young module E, (MPa)	Tensile stress σ_U, (MPa)	Elongation at break ε, (%)
0	1.308	80.1 ± 1.17	2840 ± 273	106.9 ± 4.7	25.6 ± 7.2
+1% MoS$_2$	1.310	81.1 ± 0.7	3157 ± 56	108.9 ± 2.7	12.7 ± 1.6
+10% MoS$_2$	1.423	81.8 ± 0.3	3412 ± 25	96.8 ± 4.7	4.7 ± 1.4

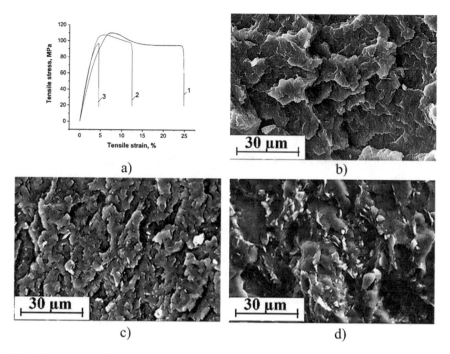

Fig. 1 Stress–strain diagram (**a**) and SEM-micrographs of the permolecular structure: neat PEEK (**b**); "PEEK + 1% MoS$_2$" composite (**c**); "PEEK + 10% MoS$_2$" composite (**d**)

Hardness of the used metal counterpart made of ball-bearing GCr15 steel was less than that of the Al$_2$O$_3$ ceramic one. In addition, the metal counterpart was able to chemically react with the polymer composite. Ceramics, in turn, was inert with respect to polymeric materials even under the conditions of tribological oxidation. As a result, chemical interaction was not supposed to occur between them. Tribological characteristics of the "PEEK + 1% MoS$_2$" and "PEEK + 10% MoS$_2$" composites are shown in Fig. 2 and Table 2. Their friction coefficient values were at the level of neat PEEK for the metal-polymer tribological contact (Fig. 2a). It is seen that MoS$_2$ particles in the polymer matrix did not exhibit a solid lubricant effect when slid on the softer (with respect to ceramic) metallic counterpart.

On the other hand, friction coefficient decreased by 13% in the ceramic-polymer tribological contact at a high filling degree (10% MoS$_2$), while at a low particle content (1%) it remained at the neat PEEK level (Fig. 2b). Thus, it was possible to realize separation of MoS$_2$ flakes under conditions of tougher interaction in the ceramic-polymer tribological contact, but only when filler content in PEEK was high. However, good adhesion between the polymer and the filler did not contribute to the more effective solid-lubricant action of MoS$_2$ particles in the two-component composite (regardless initially expected).

Wear rate of the composites increased in both metal- and ceramic-polymer tribological contacts (Table 2) despite revealed constancy or even a slight decrease in friction coefficient values. The reasons are discussed below when analyzing wear tracks/scars on the sample and counterpart surfaces. However, the pronounced trend was a multiple increase in wear rate of the metal-polymer tribological contact compared with the ceramic-polymer one (Fig. 3). The wear rate levels were approximately the same for neat PEEK and both PEEK-based composites. These data correlated well with optical images of wear track/scar surface topography on the samples and both counterparts (Figs. 4 and 5).

Initially, the metal-polymer tribological contact was considered. According to profilometry data (Fig. 4c), wear of neat PEEK caused formation of shallow microgrooves on the polymer friction surface (Fig. 4a). Their orientation was as usual in the sliding direction. The reason was, highly likely, micro-scratches and adherent separate debris fragments less than 200 μm in size on the metal counterpart friction surface (Fig. 4b).

PEEK loading with 1% MoS$_2$ solid lubricant particles caused formation of quite deep micro-grooves and scratches on the polymer friction surface (Fig. 4d and f). Surface roughness on the composite wear track was significantly greater than on neat PEEK (Ra was 0.707 μm versus 0.156 μm). Deep micro-grooves oriented along the sliding direction were also on the metal counterpart surface. Amount of

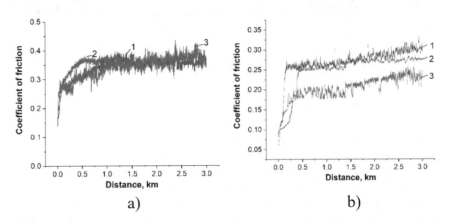

a) b)

Fig. 2 Friction coefficient versus test distance: neat PEEK (1); "PEEK + 1% MoS$_2$" composite (2); "PEEK + 10% MoS2" composite (3): **a**—on the metal counterpart; **b**—on the ceramic counterpart

Table 2 Tribological properties of the PEEK-based composites loaded with MoS$_2$

Filler composition, % (wt.)	Friction coefficient, f		Wear rate, 10^{-6} mm^3/N m	
	Metal counterpart	Ceramic counterpart	Metal counterpart	Ceramic counterpart
0	0.34 ± 0.03	0.27 ± 0.02	11.67 ± 1.00	3.00 ± 0.27
+1% MoS$_2$	0.35 ± 0.02	0.25 ± 0.02	13.67 ± 1.00	5.67 ± 0.17
+10% MoS$_2$	0.34 ± 0.03	0.20 ± 0.02	12.33 ± 0.33	4.00 ± 0.30

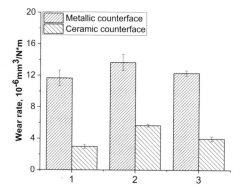

Fig. 3 Wear rate during dry sliding friction on the steel and ceramic counterparts: neat PEEK (1); "PEEK + 1% MoS$_2$" composite (2); "PEEK + 10% MoS$_2$" composite (3)

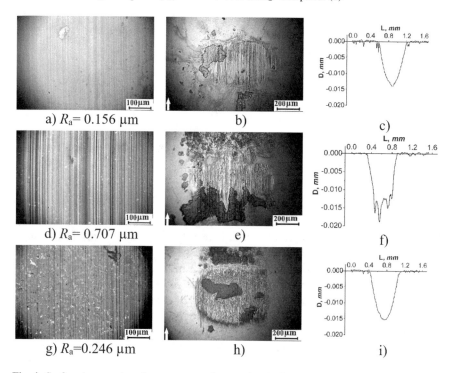

Fig. 4 Surface topography of wear scars on the samples (**a, d, g**), on the met-al counterpart (**b, e, h**), and wear track profiles (**c, f, i**) after 3 km test distance: neat PEEK (**a–c**); "PEEK + 1% MoS$_2$" composite (**d–f**); "PEEK + 10% MoS$_2$" composite (**g–i**)

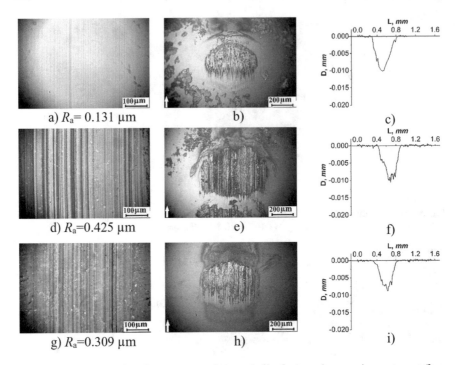

a) R_a= 0.131 µm   b)   c)

d) R_a=0.425 µm   e)   f)

g) R_a=0.309 µm   h)   i)

Fig. 5 Surface topography of wear scars on the samples (**a, d, g**), on the ceramic counterpart (**b, e, h**), and wear track profiles (**c, f, i**) after 3 km test distance: neat PEEK (**a–c**); "PEEK + 1% MoS$_2$" composite (**d–f**); "PEEK + 10% MoS$_2$" composite (**g–i**)

debris adhered to the metal counterpart friction surface and its wear were higher compared with neat PEEK (Fig. 4e). According to the authors, debris hardened by MoS$_2$ particles and oxidized during tribological loading had had a microabrasive effect on the polymer friction surface and increased its wear rate (Figs. 3 and 4d).

An increase in MoS$_2$ content caused composite hardness raising (Table 1). Nevertheless, the polymer composite wear track surface was smoother (Fig. 4g, i), and its roughness Ra decreased down to 0.246 µm, which was three times less than with a content of 1% MoS$_2$. Also, wear of the metal counterpart surface was not so heavy (Fig. 4h). Judged by the presence of rainbow colors on the surface, it can be concluded that a transfer film had been formed on it. The film had protected the metal surface from microabrasive wear by both debris and the polymer composite. Thus, after loading with 10% MoS$_2$, wear rate was at the level of neat PEEK despite friction coefficient was constant regardless of filling degree.

The ceramic-polymer tribological contact wear results were different. Microgrooves on the neat PEEK friction surface were also formed but their depth was much less comparing with the metal-polymer tribological contact (Figs. 4a, and 5a). However, rainbow colors were observed on the ceramic counterpart wear track (Fig. 5b) indicated that a polymer transfer film had been formed. Most likely, the film on the surface had been the reason for a fourfold decrease in wear rate in the ceramic-polymer tribological contact (Fig. 2).

At loading 1% MoS_2, deep micro-grooves were formed on the polymer composite wear track surface (Fig. 5d). They were the same as after the tests on the metal counterpart (Fig. 4d). This fact was confirmed by contact profilometry data (Fig. 5f). At the same time, there was more intensive wear of the counterpart. This result was unexpected for hard ceramics (Fig. 5e). A transfer film was revealed on the ceramic counterpart friction surface as well.

An increase in filler content up to 10% caused a decrease in microabrasive wear both of the polymer composite (Fig. 5g) and the ceramic counterpart (Fig. 5h). At the same time, a polymer transfer film was found on the ceramic counterpart wear track. However, there were no adherent debris particles as in Fig. 5b, e. This was probably due to a decrease in composite friction coefficient.

Thus, MoS_2 molybdenum disulfide, especially when it had been slightly loaded, was not a solid lubricant for the PEEK-based composites [5, 21]. However, MoS_2 particles, due to their high thermal conductivity, had contributed to a more uniform structure formation during compression sintering. This had increased strength properties of the composite with low filler content (up to 1%). Also, MoS_2 could act as a stabilizer of fragmentary structures of multicomponent composites due to distribution of its small amount mainly on the fragment boundaries, and, thereby increasing strength characteristics. The results of such studies are presented below in the section on three-component composites.

3.2 Two-Component "PEEK + PTFE" Composites

Changes in tribological and mechanical properties of the polymer composites were different after PEEK loading with PTFE (organic) filler particles. As is known, PTFE, being solid lubricant filler, formed a transfer film on counterpart surfaces and, due to this fact, transformed tribological contacts into a polymer-polymer type [22–25]. Below are the results of studies of PEEK-based composites loaded with various amounts of PTFE chosen on the basis of both published data and previous studies of the authors [10, 12, 26].

Table 3 shows mechanical properties of the PEEK-based composites loaded with 10, 20 and 30% PTFE. Compared with neat PEEK, all mechanical characteristics of the composites decreased with increasing filler content (hardness down to 1.1 times, elastic modulus down to 1.4 times, tensile strength down to 2 times, elongation at break down to 5 times).

Despite the fact that density of the composites increased, their permolecular structures were heterogeneous: PEEK matrix elements were separated by PTFE inclusions (Fig. 6). It was expected that the higher filling degree, the less uniform structure was formed.

The results of studies of tribological properties of the PEEK-based composites loaded with various amounts of PTFE are presented in Figs. 7 and 8, as well as Table 4.

Table 3 Mechanical properties of the PEEK-based composites loaded with PTFE

Filler composition, % (wt.)	Density ρ (g/cm³)	Shore D hardness	Young module E (MPa)	Tensile stress σ_U, (MPa)	Elongation at break ε (%)
0	1.308	80.1 ± 1.17	2840 ± 273	106.9 ± 4.7	25.6 ± 7.2
+10% PTFE	1.320	77.3 ± 0.2	2620 ± 158	83.9 ± 2.4	5.0 ± 0.8
+20% PTFE	1.408	75.9 ± 0.2	2159 ± 215	67.7 ± 1.8	5.0 ± 1.2
+30% PTFE	1.463	73.0 ± 0.5	2011 ± 108	55.1 ± 2.1	4.7 ± 1.4

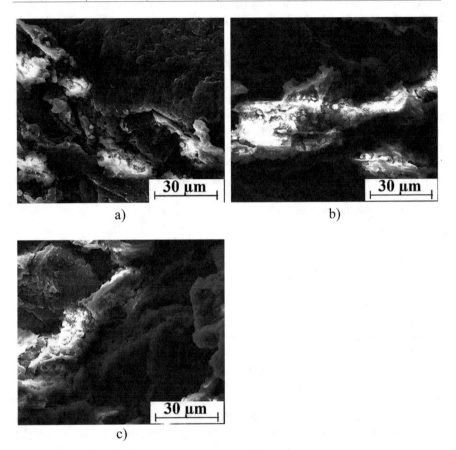

a)   b)

c)

Fig. 6 SEM-micrographs of the permolecular structure of the PEEK-based composites: "PEEK + 10% PTFE" (**a**), "PEEK + 20% PTFE" (**b**), and "PEEK + 30% PTFE" (**c**)

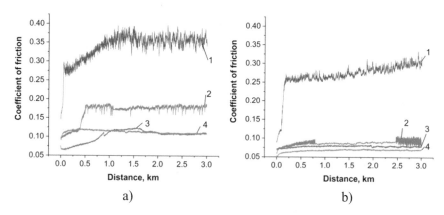

Fig. 7 Friction coefficient versus test distance: neat PEEK (1); "PEEK + 10% PTFE" composite (2); "PEEK + 20% PTFE" composite (3); "PEEK + 30% PTFE" composite (4): **a** on the metal counterpart; **b** on the ceramic counterpart

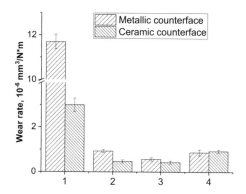

Fig. 8 Diagram of wear rate during dry sliding friction on the steel and ceramic counterparts: neat PEEK (1); "PEEK + 10% PTFE" composite (2); "PEEK + 20% PTFE" composite (3); "PEEK + 30% PTFE" composite (4)

Table 4 Tribological properties of the PEEK-based composites loaded with PTFE

Filler composition, % (wt.)	Friction coefficient, f		Wear rate, 10^{-6} mm³/N m	
	Metal counterpart	Ceramic counterpart	Metal counterpart	Ceramic counterpart
0	0.34 ± 0.03	0.27 ± 0.02	11.67 ± 1.00	3.00 ± 0.27
+10% PTFE	0.17 ± 0.02	0.09 ± 0.01	0.93 ± 0.07	0.47 ± 0.07
+20% PTFE	0.10 ± 0.01	0.08 ± 0.01	0.57 ± 0.07	0.43 ± 0.07
+30% PTFE	0.11 ± 0.01	0.07 ± 0.01	0.87 ± 0.13	$0.93 \pm 0.0.07$

Friction coefficient of the metal-polymer tribological contact gradually decreased by more than three times as PTFE content increased (Fig. 7a). In the ceramic-polymer tribological contact, it sharply decreased already at the minimum PTFE contents (of the studied); then it decreased slightly (Fig. 7b). This fact indicated heavier conditions of tribological loading during friction on the ceramic counterpart. As a result, PTFE inevitably acted as a solid lubricant.

Dynamics of wear resistance changes in various types of tribological contacts were significantly different. First of all, wear resistance of the "PEEK + 10% PTFE" composite increased 13.5 times in the metal-polymer and 6.5 times in the ceramic-polymer tribological contacts. However, wear rate during friction on the ceramic counterpart was two times lower in absolute terms. An increase in fluoroplastic content caused slight wear resistance rising for the "PEEK + 20% PTFE" composite. However, this improvement was not an attractive result taking into account significant deterioration of deformation-strength properties. The data from Table 4, also graphically presented in Fig. 8, enabled to conclude that the PEEK loading with 10% PTFE was sufficient to provide high wear resistance of the composites in both metal-polymer and ceramic-polymer tribological contacts.

Wear surface topographies and wear track profiles on the samples as well as counterparts' wear scars are presented and discussed below.

In the metal-polymer tribological contact, the metal counterpart was slightly worn after the "PEEK + 10% PTFE" composite test (Fig. 9b). PTFE particles were quasi-uniformly distributed in the form of rather large inclusions on the polymer composite surface (Fig. 9b) and micro-grooves almost had not been formed (Fig. 9c). On the other hand, a wear scar had been formed on the counterpart surface, whose area was smaller than that after the neat PEEK test (Figs. 4b and 9b). Also, a thin transfer film was found on the metal counterpart surface, as concluded based on the rainbow colors on the wear scar. The film, according to the authors, had protected both surfaces from (microabrasive) wear. In this case, roughness of the composite wear track surface decreased almost twofold compared with neat PEEK (Ra = 0.081 μm versus 0.156).

The polymer composite friction surface became smoother (Fig. 9d and g) and the friction track were less pronounced (Fig. 9f, i) as filling degree increased up to 20 and 30%. However, the amount of debris rose on the metal counterpart surface, and wear track area expanded compared with the composite loaded with 10% PTFE (Fig. 9e, h). However, separate micro-scratches on the metal counterpart surface were also found for the PEEK-based composites loaded with 10 and 20% PTFE (Fig. 9b, e).

In the ceramic-polymer tribological contact, pattern of wear was generally similar (Fig. 10). The higher PEEK loading with fluoroplastic, the wider was the wear scar area on the ceramic counterpart surface (or, more precisely, not "the wear scar" but scuffs, since its wear was minimal, Fig. 10b, e, and h). Micro-grooves had not been formed on the friction surface of the polymer-polymer composites, as suggested beforehand (Fig. 10a, d and g). Furthermore, the regularity was revealed for the PEEK-based composites loaded with 20 and 30% PTFE that the higher filling degree, the more debris had been transferred onto the ceramic counterpart surface (Fig. 10b, e, and h).

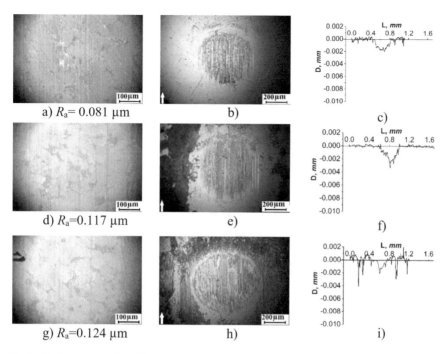

Fig. 9 Surface topography of the wear scars on the PEEK-based composites (**a, d, g**), on the metal counterpart (**b, e, h**), and wear track profiles (**c, f, i**) after 3 km test distance: "PEEK + 10% PTFE" (**a–c**); "PEEK + 20% PTFE" (**d–f**); and "PEEK + 30% PTFE" (**g–i**)

Accordingly, PTFE had formed the transfer film on the metal and ceramic counterparts, providing high wear resistance and low friction coefficient for PEEK, which in the initial state had had insufficient wear resistance for effective use in tribological contacts and a high friction coefficient of 0.34. However, PEEK loading with PTFE deteriorated structure and decreased mechanical properties. Therefore, it was suggested to additionally load with MoS_2 particles to increase mechanical and tribological properties of the "PEEK + PTFE" composites.

Presence of MoS_2 below 1% enabled to improve the process of composite formation during the sintering due to homogenization of the matrix permolecular structure. The following methodology was used to design the optimal three-component composite.

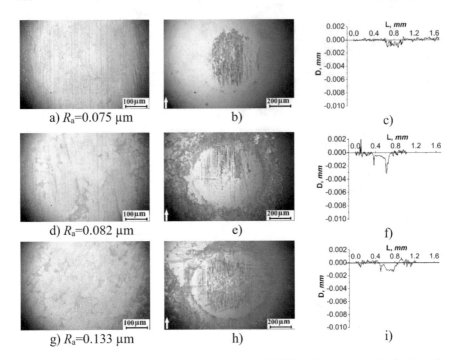

Fig. 10 Surface topography of the wear tracks on the PEEK-based compo-sites (**a, d, g**), on the ceramic counterpart (**b, e, h**), and wear track profiles (**c, f, i**) after 3 km test distance: "PEEK + 10% PTFE" (**a–c**); "PEEK + 20% PTFE" (**d–f**); and "PEEK + 30% PTFE" (**g–i**)

3.3 Three-Component PEEK-Based Composites Filled with PTFE and MoS₂

The previously developed experimental-theoretical approach [19, 20] was used to determine the optimal composition. Twelve three-component composites were made for this purpose; their compositions are presented in Table 5. To ensure the optimal content (range of contents) of both fillers was found, the amount of PTFE was chosen to be obviously lower (5%) and higher (20%) than previously studied in its two-component composites, while the maximum MoS_2 content was 1%.

The data from the physical experiments (Tables 6, 7, 8, 9, 10, 11, 12 and 13) were used as reference points. Control parameters were PTFE and MoS_2 filling degree. When drawing surfaces for each of the control parameters, normalization was used. The lower boundary was zero; the upper boundary was unit. Additional reference points for surface drawing had been obtained using linear interpolation of the experimental data by the Lagrange polynomial [19].

Properties of the three-component PEEK-based composites were specified (Table 14) on the basis of published data and neat PEEK characteristics [27].

Table 5 Composition of the designed PEEK-based three-component composites

No.	Filler composition, % (wt.)
1	PEEK + 5% PTFE + 0.25% MoS_2
2	PEEK + 5% PTFE + 0.50% MoS_2
3	PEEK + 5% PTFE + 1.00% MoS_2
4	PEEK + 10% PTFE + 0.25% MoS_2
5	PEEK + 10% PTFE + 0.50% MoS_2
6	PEEK + 10% PTFE + 1.00% MoS_2
7	PEEK + 15% PTFE + 0.25% MoS_2
8	PEEK + 15% PTFE + 0.50% MoS_2
9	PEEK + 15% PTFE + 1.00% MoS_2
10	PEEK + 20% PTFE + 0.25% MoS_2
11	PEEK + 20% PTFE + 0.50% MoS_2
12	PEEK + 20% PTFE + 1.00% MoS_2

Table 6 Friction coefficient f of the PEEK-based composites having different MoS_2 and PTFE filling degrees during the tribological test on the metal counterpart

φMoS_2	φ PTFE			
	5% PTFE	10% PTFE	15% PTFE	20% PTFE
0.25% MoS_2	0.17	0.15	0.11	0.08
0.50% MoS_2	0.14	0.05	0.045	0.08
1.00% MoS_2	0.15	0.12	0.16	0.09

Table 7 Friction coefficient f of the PEEK-based composites having different degrees MoS_2 and PTFE filling degrees during the tribological test on the ceramic counterpart

φMoS_2	φ PTFE			
	5% PTFE	10% PTFE	15% PTFE	20% PTFE
0.25% MoS_2	0.14	0.08	0.06	0.047
0.50% MoS_2	0.10	0.03	0.07	0.06
1.00% MoS_2	0.11	0.08	0.06	0.06

Table 8 Wear rate (I, 10^{-6} mm^3/N·m) of the PEEK-based composites having different MoS_2 and PTFE filling degrees during the tribological test on the metal counterpart

φMoS_2	φ PTFE			
	5% PTFE	10% PTFE	15% PTFE	20% PTFE
0.25% MoS_2	2.53 ± 0.10	1.53 ± 0.07	0.83 ± 0.10	0.60 ± 0.10
0.50% MoS_2	2.23 ± 0.33	0.30 ± 0.03	0.33 ± 0.10	0.73 ± 0.03
1.00% MoS_2	2.97 ± 0.33	2.67 ± 0.17	2.00 ± 0.20	0.80 ± 0.03

Table 9 Wear rate (I, 10^{-6} mm^3/N·m) of the PEEK-based composites having different MoS$_2$ and PTFE filling degrees during the tribological test on the ceramic counterpart

φMoS$_2$	φ PTFE			
	5% PTFE	10% PTFE	15% PTFE	20% PTFE
0.25% MoS$_2$	1.00 ± 0.07	0.32 ± 0.10	0.90 ± 0.10	0.73 ± 0.10
0.50% MoS$_2$	0.53 ± 0.07	0.20 ± 0.023	0.43 ± 0.07	0.43 ± 0.13
1.00% MoS$_2$	1.13 ± 0.07	0.27 ± 0.07	0.37 ± 0.07	0.92 ± 0.20

Table 10 Elastic modulus (E, MPa) of the PEEK-based composites having different MoS$_2$ and PTFE filling degrees

φMoS$_2$	φ PTFE			
	5% PTFE	10% PTFE	15% PTFE	20% PTFE
0.25% MoS$_2$	3.08 ± 0.08	2.77 ± 0.10	2.40 ± 0.05	2.16 ± 0.07
0.50% MoS$_2$	3.05 ± 0.06	2.76 ± 0.08	2.52 ± 0.07	2.07 ± 0.09
1.00% MoS$_2$	2.97 ± 0.08	2.74 ± 0.04	2.73 ± 0.04	2.08 ± 0.02

Table 11 Tensile strength (σ_U, MPa) of the PEEK-based composites having different MoS$_2$ and PTFE filling degrees

φMoS$_2$	φ PTFE			
	5% PTFE	10% PTFE	15% PTFE	20% PTFE
0.25% MoS$_2$	94.2 ± 3.4	88.5 ± 4.7	71.2 ± 1.5	49.2 ± 2.2
0.50% MoS$_2$	90.3 ± 0.7	84.9 ± 1.8	68.1 ± 1.8	45.1 ± 3.0
1.00% MoS$_2$	86.5 ± 3.0	79.0 ± 0.5	66.8 ± 4.0	42.0 ± 1.4

Table 12 Elongation at break (ε, %) of the PEEK-based composites having different MoS$_2$ and PTFE filling degrees

φMoS$_2$	φ PTFE			
	5% PTFE	10% PTFE	15% PTFE	20% PTFE
0.25% MoS$_2$	9.9 ± 1.4	7.6 ± 0.9	7.5 ± 0.3	6.0 ± 0.6
0.50% MoS$_2$	8.4 ± 0.8	9.8 ± 0.7	6.1 ± 0.3	5.2 ± 1.1
1.00% MoS$_2$	7.3 ± 2.8	6.3 ± 0.4	4.3 ± 1.0	4.8 ± 0.2

Table 13 Shore D hardness of the PEEK-based composites having different MoS$_2$ and PTFE filling degrees

φMoS$_2$	φ PTFE			
	5% PTFE	10% PTFE	15% PTFE	20% PTFE
0.25% MoS$_2$	77.9 ± 0.4	76.7 ± 0.3	76.3 ± 0.4	74.7 ± 0.5
0.50% MoS$_2$	77.6 ± 0.1	77.0 ± 0.7	76.6 ± 0.5	75.5 ± 0.2
1.00% MoS$_2$	78.5 ± 0.5	77.6 ± 0.3	76.9 ± 0.4	76.2 ± 0.5

Table 13.14 Specified properties for the designed three-component PEEK-based composites

Properties	Values
Shore D hardness	>75
Elastic modulus, MPa	>2500
Tensile strength, MPa	>70
Elongation at break (%)	>5
Friction coefficient on metal counterpart	<0.1
Friction coefficient on ceramic counterpart	<0.1
Wear rate on metal counterpart, 10^{-6} mm^3/N m	<1.0
Wear rate on ceramic counterpart, 10^{-6} mm^3/N m	<0.5

As a result, dependences of operational properties (friction coefficient, wear rate, Shore D hardness, elastic modulus, tensile strength, and elongation at break) on composition were obtained in the form of continuous functions. Regular data arrays reflecting the listed dependences on the control parameter discrete values were formed. Then, 3D surfaces and their corresponding contours were drawn (Figs. 11, 12, 13, 14, 15, 16, 17, 18 and 19).

The contours were overlapped to determine the values of the control parameters corresponding to the specified operational properties. The obtained range of the control parameter values, presented in Fig. 19 as a filled region, ensured that all the operational properties of the composites corresponded to the specified limits. Based on the presented data, it can be concluded that the optimal amount of MoS$_2$ loading was in the range from 0.4 to 0.6%, while PTFE was from 8 to 14%.

Based on the obtained data, the "PEEK + 10% PTFE + 0.5% MoS$_2$" composite was chosen and studied in more detail. Table 15 presents its mechanical properties and (for comparison) that of the "PEEK + 10% PTFE" one. Figure 20a shows a stress–strain diagram for these materials. The results of the analysis of these data enabled

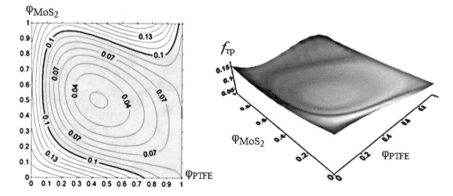

Fig. 11 Friction coefficient on the metal counterpart versus PEEK-based composite filling degree with MoS$_2$ and PTFE

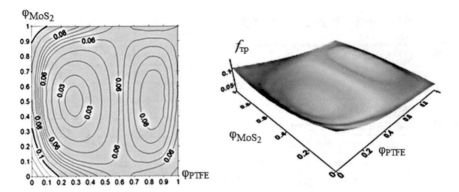

Fig. 12 Friction coefficient on the ceramic counterpart versus PEEK-based composite filling degree with MoS$_2$ and PTFE

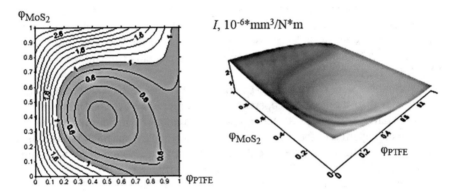

Fig. 13 Wear rate on the metal counterpart versus PEEK-based composite filling degree with MoS$_2$ and PTFE

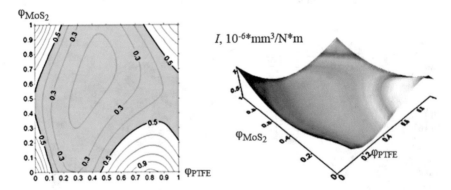

Fig. 14 Wear rate on the ceramic counterpart versus PEEK-based composite filling degree with MoS$_2$ and PTFE

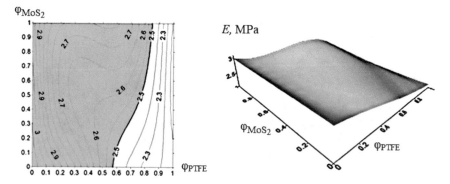

Fig. 15 Elastic modulus versus PEEK-based composite filling degree with MoS_2 and PTFE

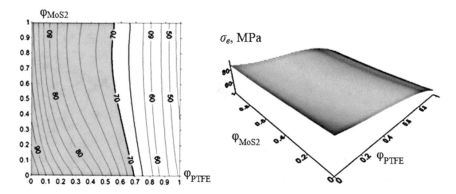

Fig. 16 Tensile strength versus PEEK-based composite filling degree with MoS_2 and PTFE

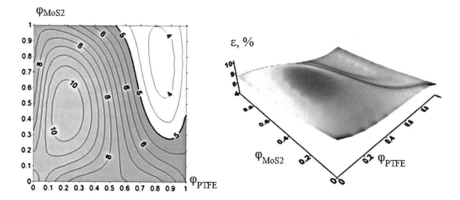

Fig. 17 Elongation at break versus PEEK-based composite filling degree with MoS_2 and PTFE

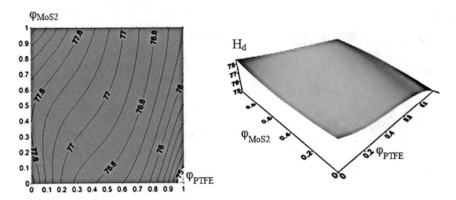

Fig. 18 Shore D hardness versus PEEK-based composite filling degree with MoS$_2$ and PTFE

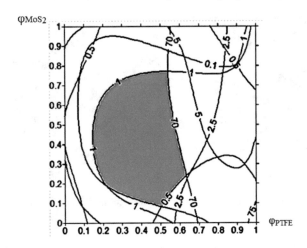

Fig. 19 Diagram of the control parameters to ensure that the mechanical properties meet the specified limits for the materials

Table 15 Mechanical properties of the optimal composition PEEK-based composite

Filler composition, % (wt.)	Density ρ (g/cm^3)	Shore D hardness	Young module E, (MPa)	Tensile stress σ_U, (MPa)	Elongation at break ε (%)
+10% PTFE (comparison)	1.324	77.3 ± 0.2	2620 ± 158	83.9 ± 2.4	4.4 ± 0.7
+ 10% PTFE + 0.50% MoS$_2$	1.371	76.7 ± 0.3	2760 ± 85	84.9 ± 1.8	9.8 ± 0.2

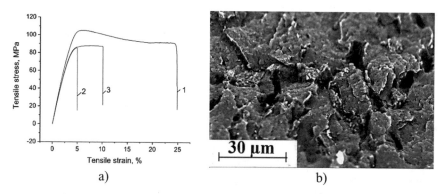

Fig. 20 Stress–strain diagram (**a**): 1—neat PEEK; 2—PEEK + 10% PTFE; 3—PEEK + 10% PTFE + 0.5% MoS2; SEM-micrographs of the permolecular structure of the "PEEK + 10% PTFE + 0.5% MoS$_2$" composite (**b**)

to conclude that strength properties of the three-component composite increased slightly compared with the two-component ones. On the other hand, elongation at break doubled (Fig. 20a). Highly likely, this was due to favorable homogenization effect of 0.5% MoS$_2$ loaded particles on permolecular structure formation.

SEM-micrographs of the permolecular structure of the "PEEK + 10% PTFE + 0.5% MoS$_2$"composite are shown in Fig. 20b. It is seen that the structure was slightly loose; although there were no pronounced signs of cracking or agglomeration of each filler particles as in the case of the "PEEK + 10% PTFE" composite. According to the authors, loading with MoS$_2$ particles homogenized the permolecular structure due to their location along the boundaries of polymer composite structural elements. In addition to improve deformation-strength characteristics (in comparison with the "PEEK + 10% PTFE" composite), it also contributed to an increase in wear resistance. More details are discussed below.

Table 16 shows tribological characteristics of the three-component composite for dry friction on the metal and ceramic counterparts. Friction coefficient decreased by more than three times in both metal-polymer and ceramic-polymer tribological contacts. Wear resistance increased by 3.1 and 2.3 times, respectively, compared with the "PEEK + 10% PTFE" composite. Wear rate of the "PEEK + 10% PTFE + 0.5% MoS$_2$" composite decreased by 39 times when testing on the metal counterpart, and

Table 16 Tribological properties of the PEEK-based composites

Filler composition, % (wt.)	Friction coefficient, f		Wear rate, 10^{-6} mm^3/N m	
	Metal counterpart	Ceramic counterpart	Metal counterpart	Ceramic counterpart
+10% PTFE	0.17 ± 0.02	0.10 ± 0.02	0.93 ± 0.07	0.47 ± 0.07
+10% PTFE + 0.50% MoS$_2$	0.05 ± 0.01	0.03 ± 0.01	0.30 ± 0.03	0.20 ± 0.02

15 times on the ceramic one compared with neat PEEK. However, in absolute terms, wear rate of the three-component composite on the metal counterpart was 1.5 times higher than on the ceramic one.

Figure 21 shows friction surfaces of the samples and counterparts, as well as wear track profiles on the three-component composite. These results explain the data presented in Table 15. The counterparts did not wear out in both cases (Fig. 21b, e). Based on all the previously obtained data, this was most relevant for the metal one. Wear scars had been formed on both counterpart surfaces, but their area were less than that in the case of neat PEEK (Figs. 9b and 21). Micro-grooves and other damages were expectedly absent on the polymer composite friction surface, although inclusions of both fluoroplast and MoS$_2$ were visible.

Accumulation of a significant amount of debris in the form of a continuous film was on the metal counterpart surface, in contrast to the similar test results of the "PEEK + 10% PTFE" composite (Figs. 9b and 21b). This means that simultaneous presence of a significant content of PTFE particles and a small amount of MoS$_2$ in the polymer matrix had facilitated formation of a transfer film that protected the metal counterpart from microabrasive wear.

This was even more clearly shown on the ceramic counterpart surface, where the wear scar was covered with a clearly distinguishable transfer film (in the "classical" sense) which was evidenced by its rainbow reflection. The effect was most pronounced precisely in the analyzed tribological contact. It should be noted that the polymer debris clusters in the form of a uniform layer was on the ceramic counterpart surface (Fig. 21e), in contrast to the "PEEK + 10% PTFE" composite test results (as well as on the metal counterpart, Fig. 21b).

Summarizing the above, we note that a three-fold increase in wear resistance during testing both on metal and ceramic counterparts was achieved for the "PEEK

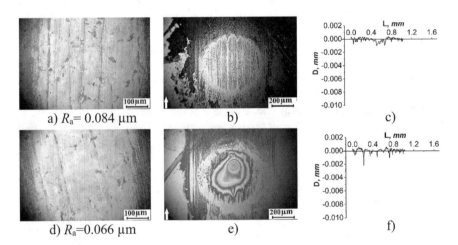

Fig. 21 Surface topography of the wear tracks on the "PEEK + 10% PTFE + 0.5% MoS$_2$"composite (**a, d**), on the metal (**b**) and ceramic (**c**) counterparts, and wear track profiles (**c, f**) after 3 km test distance

+ 10% PTFE + 0.5% MoS$_2$" three-component composite compared with the "PEEK + 10% PTFE". Simultaneous loading with two types of the fillers slightly deteriorated the polymer composite structure compared with neat PEEK. However, wear rate was many times reduced due to facilitation of transfer film formation. For this reason, no microabrasive wear on both metal and ceramic counterpart surfaces developed. Besides self-lubricating effect of the three-component composite, an additional (probable) cause for metal counterpart wear eliminating was protective action of the transfer film which suppressed oxidation processes in the tribological contact of PEEK and ball-bearing steel [28].

4 Conclusions

To improve tribological and mechanical properties of polymer materials, two- and three-component PEEK-based composites loaded with molybdenum disulfide (MoS$_2$) and polytetrafluoroethylene (PTFE) were investigated. It was shown that an increase in dry friction mode tribological characteristics in the metal-polymer and ceramic-polymer tribological contacts was attained by loading with lubricant fluoroplastic particles. In addition, molybdenum disulfide homogenized permolecular structure and improved matrix strength properties.

A methodology for identifying composition of multicomponent PEEK-based composites having prescribed properties which based on a limited amount of experimental data was proposed and implemented. It could be used to design similar dispersion hardened composites based on prospective thermoplastic matrixes. Advantages of the methodology were shown by analysis of the experimental results on mechanical and tribological tests of the PEEK-based composites.

It was shown that wear rate of the "PEEK + 10% PTFE + 0.5% MoS$_2$" composite decreased by 39 times when testing on the metal counterpart, and 15 times on the ceramic one compared with neat PEEK. However, in absolute terms, wear rate of the three-component composite on the metal counterpart was 1.5 times higher than on the ceramic one. A three-fold increase in wear resistance during testing both on metal and ceramic counterparts was achieved for the "PEEK + 10% PTFE + 0.5% MoS$_2$" three-component composite compared with the "PEEK + 10% PTFE". Simultaneous loading with two types of fillers slightly deteriorated the polymer composite structure compared with neat PEEK. However, wear rate was many times reduced due to facilitation of transfer film formation. For this reason, there was no microabrasive wear on both metal and ceramic counterpart surfaces. Besides self-lubricating effect of the three-component composite, an additional (possible) cause for metal counterpart wear eliminating was protective action of the transfer film, which suppressed oxidation processes in the tribological contact of PEEK and ball-bearing steel.

Acknowledgements This research was performed according to the Government research assignment for ISPMS SB RAS, project No. III.23.1.3, and RFBR grant number 20-58-00032 Bel_a and

19-38-90106. The work was also supported by the RF President Council Grant for the support of leading research schools NSh-2718.2020.8.

References

1. Haleem A, Javaid M (2019) Polyether ether ketone (PEEK) and its manufacturing of customised 3D printed dentistry parts using additive manufacturing. Clin Epidemiol Glob Health 7(4):654–660
2. Stepashkin AA, Chukova DI, Senatova FS, Salimonac AI, Korsunskybc AM, Kaloshkina SD (2018) 3D-printed PEEK-carbon fiber (CF) composites: structure and thermal properties. Compos Sci Technol 164:319–326
3. Rasheva Z, Burkhart Th, Zang G (2010) A correlation between the tribological and mechanical properties of SCF reinforced PEEK materials with different fiber orientation. Tribol Int 43(8):1430–1437
4. Sumer M, Mimaroglu A, Unal H (2008) Evaluation of tribological behavior of PEEK and glass fiber reinforced PEEK composite under dry sliding and water lubricated conditions. Wear 265(7–8):1061–1065
5. Lu ZP, Friedrich K (1995) On sliding friction and wear of PEEK and its composites. Wear 181–183:624–631
6. Ünal H, Mimaroglu A (2006) Friction and wear characteristics of PEEK and its composite under water lubrication. J Reinf Plast Compos 25(16):1659–1667
7. Kumar D, Rajmohan T, Venkatachalapathi T (2018) Wear behavior of PEEK matrix composites: a review. Materialstoday: Proc 5(6):14583–14589
8. Kurtz SM (2012) PEEK biomaterials handbook, 1st edn. Plast Des Lib, Waltham
9. Kandemir G, Joyce TJ, Smith S (2019) Wear behavior of CFR PEEK articulated against CoCr under varying contact stresses: low wear of CFR PEEK negated by wear of the CoCr counterface. J Mech Behav Biomed Mater 97:117–125
10. Bijwe J, Ghosh A, Sen S (2005) Influence of PTFE content in PEEK–PTFE blends on mechanical properties and tribo-performance in various wear modes. Wear 258(10):1536–1542
11. Zalaznik M, Kalin M, Novak S, Jakša G (2016) Effect of the type, size and concentration of solid lubricants on the tribological properties of the polymer PEEK. Wear 364–365:31–39
12. Burris DL, Sawyer WG (2006) A low friction and ultra-low wear rate PEEK/PTFE composite. Wear 261(3–4):410–418
13. Panin SV, Anh ND, Kornienko LA, Ivanova LR, Ovechkin BB (2018) Comparison on efficiency of solid-lubricant fillers for polyetheretherketone-based composites. AIP Conf Proc 2051:020232. https://doi.org/10.1063/1.5083475
14. Burris DL, Sawyer WG (2007) Tribological behavior of PEEK components with composition graded PEEK/PTFE surfaces. Wear 262(1–2):220–224
15. Werner P, Altstädt V, Jaskulka R, Jacobs O, Sandler JKW, Shaffer MSP, Windle AH (2004) Tribological behavior of carbon nanofibre reinforced PEEK. Wear 257(9–10):1006–1014
16. Kuo MC, Tsai CM, Huang JC, Chen M (2005) PEEK composites reinforced by nano-sized SiO2 and Al2O3 particulates. Mater Chem Phys 90:185–195
17. Kalin M, Novak S, Zalaznik M (2015) Wear friction behavior of Poly-ether-ether-ketone (PEEK) filled with graphene, WS2 and CNT nanoparticles. Wear 332–333:855–862
18. Wang N, Yang Z, Wang Y, Thummavicahi K, Xia Y, Ghita O, Zhu Y (2017) Interface and properties of inorganic fullerene tungsten sulphide nanoparticle reinforced poly (ether ether ketone) nanocomposites. Results Phys 7:2417–2424
19. Panin SV, Kornienko LA, Qitao H, Buslovich DG, Bochkareva SA, Alexenko VO, Panov IL, Berto F (2020) Effect of adhesion on mechanical and tribological properties of glass fiber composites based on ultra-high molecular weight polyethylene powders having various initial particle sizes. Materials 13:1602. https://doi.org/10.3390/ma13071602

20. Bochkareva SA, Grishaeva NY, Lyukshin BA, Lyukshin PA, Matolygina NY, Panov IL (2017) Obtaining of specified effective mechanical, thermal, and electrical characteristics of composite filled with dispersive materials. Inorg Mater: Appl Res 8(5):651–661

21. Vail JR, Krick BA, Marchman KR, Sawyer WG (2011) Polytetrafluoroethylene (PTFE) fiber reinforced polyetheretherketone (PEEK) composites. Wear 270(11–12):737–741

22. Salamov AK, Mikitaev AK, Beev AA, Beeva DA, Kumysheva YuA (2016) Polyetherether ketone (PEEK) as a representative of aromatic polyarylene. Fundam Res 1(1):63–66

23. Hoskins TJ, Dearn KD, Chen YK, Kukureka SN (2014) The wear of PEEK in rolling-sliding contact-simulation of polymer gear applications. Wear 309:35–42

24. Lu Z, Liu H, Zhu C, Song H, Yu G (2019) Identification of failure modes of a PEEK-steel gear pair under lubrication. Int J Fatigue 25:342–348

25. Berer M, Major Z, Pinter G (2013) Elevated pitting wear of injection molded polyetheretherketone (PEEK) rolls. Wear 297:1052–1063

26. Panin SV, Anh ND, Kornienko LA, Ivanova LR (2019) Antifriction multi-component polyetheretherketone (PEEK) based composites. AIP Conf Proc 2141(1). https://doi.org/10.1063/1.5122124

27. https://www.victrex.com/~/media/literature/en/material-properties-guide_us-4-20.pdf

28. Puhan D, Wong JSS (2019) Properties of polyetheretherketone (PEEK) transferred materials in a PEEK- steel contact. Tribol Int 135:189–199

Regularities of Structural Rearrangements in Single- and Bicrystals Near the Contact Zone

Konstantin P. Zolnikov, Dmitrij S. Kryzhevich, and Aleksandr V. Korchuganov

Abstract The chapter is devoted to the analysis of the features of local structural rearrangements in nanostructured materials under shear loading and nanoindentation. The study was carried out using molecular dynamics-based computer simulation. In particular, we investigated the features of symmetric tilt grain boundary migration in bcc and fcc metals under shear loading. The main emphasis was on identifying atomic mechanisms responsible for the migration of symmetric tilt grain boundaries. We revealed that grain boundaries of this type can move with abnormally high velocities up to several hundred meters per second. The grain boundary velocity depends on the shear rate and grain boundary structure. It is important to note that the migration of grain boundary does not lead to the formation of structural defects. We showed that grain boundary moves in a pronounced jump-like manner as a result of a certain sequence of self-consistent displacements of grain boundary atomic planes and adjacent planes. The number of atomic planes involved in the migration process depends on the structure of the grain boundary. In the case of bcc vanadium, five planes participate in the migration of the $\Sigma 5(210)[001]$ grain boundary, and three planes determine the $\Sigma 5(310)[001]$ grain boundary motion. The $\Sigma 5(310)[001]$ grain boundary in fcc nickel moves as a result of rearrangements of six atomic planes. The stacking order of atomic planes participating in the grain boundary migration can change. A jump-like manner of grain boundary motion may be divided into two stages. The first stage is a long time interval of stress increase during shear loading. The grain boundary is motionless during this period and accumulates elastic strain energy. This is followed by the stage of jump-like grain boundary motion, which results in rapid stress drop. The related study was focused on understanding the atomic rearrangements responsible for the nucleation of plasticity near different crystallographic surfaces of fcc and bcc metals under nanoindentation. We showed that a wedge-shaped region, which consists of atoms with a changed symmetry of the nearest environment, is formed under the indentation of the (001) surface of the copper crystallite. Stacking faults arise in the (111) atomic planes of the contact zone under the indentation of the (011)

K. P. Zolnikov (✉) · D. S. Kryzhevich · A. V. Korchuganov
Institute of Strength Physics and Materials Science of Siberian Branch of the Russian
Academy of Sciences, 2/4 Akademicheskii Ave, Tomsk 634055, Russia
e-mail: kost@ispms.ru

© The Author(s) 2021
G.-P. Ostermeyer et al. (eds.), *Multiscale Biomechanics and Tribology of Inorganic and Organic Systems*, Springer Tracts in Mechanical Engineering,
https://doi.org/10.1007/978-3-030-60124-9_14

surface. Their escape on the side free surface leads to a step formation. Indentation of the (111) surface is accompanied by nucleation of partial dislocations in the contact zone subsequent formation of nanotwins. The results of the nanoindentation of bcc iron bicrystal show that the grain boundary prevents the propagation of structural defects nucleated in the contact zone into the neighboring grain.

Keywords Nanocrystalline materials · Plastic deformation · Grain boundary migration · Atomic displacements · Structural defects · Shear loading · Nanoindentation · Molecular dynamics

1 Introduction

The behavior of the material in contact zones is a complex multiscale process [1], which depends on a number of parameters: the roughness of the contacting surfaces, the chemical composition of materials, loading parameters, etc. [2]. The features of fracture and wear processes in the surface layer of the materials during friction are largely determined by the shear stresses. Note that the nucleation of structural changes in materials always begins at the atomic scale. Moreover, the features of the internal structure of the material, in particular, the grain boundaries (GB), can have a significant effect on structural changes in the contact zone [3, 4]. The role of GBs in the processes of friction and wear is most significant for nanostructured metallic materials having a high GB density [5]. Note that large interest in such materials is due to their high operational properties and therefore broad prospects for their use in mechanical engineering, technology, medicine, as well as in the creation of structures for various purposes. Nanostructured metallic materials have high strength at low temperatures due to GB hardening (Hall-Petch effect). At the same time, they become superplastic at high temperatures due to GB softening, which facilitates and improves their treatment in different technological processes.

Dislocation glide is substantially suppressed in nanostructured materials [6–8]. Furthermore, the role of different modes of GB deformation or twinning is enhanced [9–12]. The main mechanism of GB deformation becomes intergranular sliding, which largely determines superplastic deformation. The experimental data and the results of computer simulations confirm the significant contribution of intergranular sliding to the plastic behavior of nanostructured metallic materials under high strain-rate loading, which lead to the formation of high local stresses [13].

GB sliding leads to the nucleation of various defects in triple junctions. These defects become sources of internal stresses and can lead to crack nucleation and brittle fracture of the material [14, 15]. The physical nature and dynamics of accommodation processes in nanostructured materials under mechanical loading are extensively studied in materials science. Typical examples of accommodation mechanisms include the emission of lattice dislocations from the zone of triple junctions, diffusion processes, rotational deformation, splitting, and migration of GBs [16]. At that, GB migration and splitting caused by GB sliding are often realized in the form of collective self-consistent displacement of atoms in the interface region, which significantly enhances the efficiency of material accommodation on the applied loading.

A significant part of the studies on the atomic mechanisms of friction, wear, and plasticity in nanocrystalline materials is carried out using various computer simulation approaches. Still many issues related to the nucleation and development of plastic deformation, structural transformations, and wear in the contact zones in nanostructured metallic materials remain debatable. This is due to both the great variety of chemical composition and internal structure of nanostructured materials, and the difficulties in their experimental studying at the microscopic level, associated with small spatial and temporal intervals of the processes.

The considerable interest in the study of the tribological properties of nanostructured metallic materials is the elucidation of structural transformations in bicrystals with different types of GBs under mechanical loading. In [17] it was shown that some types of GBs in fcc materials can migrate with abnormally high speeds under shear loading. This can lead to a change in the structure and tribological parameters of the material in the friction zone. Therefore, the identification of mechanisms responsible for GB migration is important for the development of new approaches to stabilize the internal structure of materials in the friction zone.

The results of molecular dynamics simulation showed that the GB migration along the normal often occurs together with the tangential displacement of the grains, which leads to the shear deformation of the lattice intersected by the GB [18]. In turn, the shear stresses applied to the GB can cause its normal displacement, i.e. during GB migration, one grain will grow at the expense of another. Depending on the direction of the applied shear stresses, the GB can shift either in one direction or in the opposite direction along the normal vector.

High-rate shear loading of the crystallite can lead to the formation of vortex motion of atoms in the region of symmetric tilt GBs [19, 20]. The dimensions of the vortices in diameter are several lattice parameters and are characterized by significant atomic displacements not only in the direction of loading, but also in the GB plane. This process is dynamic, and the accommodation of the material is carried out on the basis of an abnormally high GB velocity. GB migration is based on self-consistent collective atomic displacements. Despite the fact that the displacement of each individual atom in the GB region is small, self-consistent vortex atomic displacements result in reconstruction of a significant region of one grain into the structure of the neighboring grain.

In this chapter, the regularities of structural rearrangements in the region of symmetric tilt GBs in bcc vanadium and fcc nickel initiated by high-speed shear loading are considered. Features of the behavior of materials during contact interaction are presented by studying atomic mechanisms of nucleation and development of plasticity in fcc copper and bcc vanadium metals with an ideal structure and GBs under nanoindentation.

2 Materials and Methods

Studies were carried out on the base of the molecular dynamics method using the
LAMMPS software package [21]. The interatomic interaction in bcc vanadium and
iron was described using many-body potentials [22, 23] calculated with the embedded
atom method in the Finis-Sinclair approximation. In the case of fcc nickel and
copper, the interatomic interaction was described using many-body potentials [24,
25]. Parallelepiped-shaped samples were modeled. The gamma-surface minimization algorithm [26] was used to construct GBs. Visualization of the simulation results
was carried out using the OVITO package [27].

We used two parameters to identify the atoms involved in the generation of structural changes, the reduced slip vector [28] and the topological parameter that takes
into account the nature of the relative positions of the nearest neighbors for each
atom (common neighbor analysis) [29]. The reduced slip vector P_i is a dimensionless quantity, which is determined by the formula: where j is the nearest neighbor of
atom i, N_s is the number of neighboring atoms, $\bar{r}^{ij}$ and $\bar{r}_0^{ij}$ are the vectors between
the positions of atoms i and j in the current and initial positions, respectively.

$$P_i = \frac{1}{N_S} \sum_{i \neq j} \frac{\left(\bar{r}^{ij} - \bar{r}_0^{ij} \right)}{\left| \bar{r}_0^{ij} \right|},$$

3 Features of Symmetric Tilt Grain Boundary Migration in Metals

The objects of the study were vanadium bicrystals containing about 40, 000 atoms.
Periodic boundary conditions were simulated in the X and Z directions, rigid
boundary conditions were set in the Y direction (Fig. 1). The initial temperature

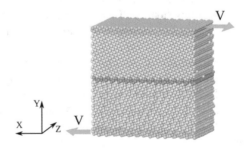

Fig. 1 The initial structure and the loading scheme of the sample containing the $\Sigma 5(210)[001]$
GB, and the loading scheme. The grips are marked in yellow, the GB region is highlighted in gray,
directions of the grip shift are indicated by arrows

of the samples was 300 K. The calculations were performed for samples containing two types of symmetric tilt GBs: Σ5 (310) [001] and Σ5 (210) [001]. The shear loading rate in the X direction in different calculations varied from 1 to 100 m/s.

The simulation results showed that shear loading of bicrystals leads to a high-speed GB migration. We found that the velocity of GB motion is determined by the shear rate and the GB structure. The GB velocity increases with an increase in the shear rate. The average velocity of the Σ5 (210) [001] GB for the considered shear rates is within the range from 3 to 280 m/s. The average velocity in the case of the Σ5 (310) [001] GB is significantly lower and is in the range from 2 to 180 m/s. It is important to note that plasticity does not nucleate in the samples despite the high GB migration velocity.

During the loading, a periodic increase and drop of stresses occur. The GB migrates in a pronounced jump-like manner due to the crystallinity of the sample. Upon stress drop, the instant GB migration velocity rapidly increases, reaches a maximum value, and drops to almost zero. This is clearly seen from a comparison of the corresponding curves in Fig. 2. Note that the dependence of stresses on time in the interval of growth and drop is linear. This indicates that no structural defects are generated in the loaded sample.

Simulation results showed that atomic rearrangements in the Σ5 (310) [001] and Σ5 (210) [001] GB regions responsible for GB migration are significantly different. The displacement of the Σ5 (210) [001] GB is a result of atomic rearrangements in three atomic planes: two planes of the GB and the adjacent upper grain plane. Green color shows atomic planes belonging to the GB, and blue shows the plane of the upper grain in Fig. 3 This plane adjusts to the structure of the lower grain

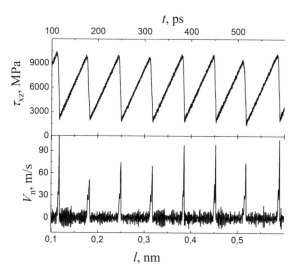

Fig. 2 The velocity of the Σ5 (210) [001] GB migration in the normal direction (V_n), and stress τ_{xz} depending on the grip displacement. Shear rate is 1 m/s

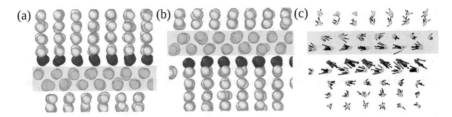

Fig. 3 Fragment of the sample with the Σ5 (310) [001] GB before loading (**a**) and after the GB displacement on three interplanar distances (**b**). The projection of the displacements on the YZ plane (**c**). The shear rate is 1 m/s. The GB region is marked with gray

during shear loading (Fig. 3b). The resulting atomic displacements in the YZ plane after such a rearrangement are shown in Fig. 3c. The value of these displacements is about 0.08 nm. Analysis of simulation results showed that the incorporation of the atomic plane of the upper grain into the structure of the lower grain is realized as a sequence of three successive displacements in different directions. The duration of each displacement for the grip velocity of 1 m/s is about 3.6 ps. The values of these three displacements are approximately equal to 0.07, 0.03, and 0.06 nm. As a result of such displacements, the atoms of the blue plane successively occupy the positions of the atoms of the upper and then the lower GB planes and, finally, adjust to the lattice of the lower grain.

Three atomic planes are simultaneously involved in the displacement of the Σ5 (210)[001] GB. The four atomic planes make up the GB (highlighted in gray), and the three atomic planes (highlighted in blue, green, and red) belong to the upper grain in Fig. 4. As a result of a certain sequence of displacements the upper grain planes first rearranged into the structure of GB planes and then adjust to the structure of the lower grain. The value of the resulting displacements of the atoms of the blue plane during its transformation to the lower grain is 0.05 nm. During this interval, two jump-like displacements occur, each with a duration of 4.0 ps. The value of the first displacement is 0.04 nm, the second is 0.05 nm. After the first displacement, the

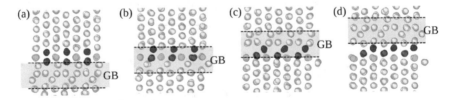

Fig. 4 Fragment of the structure with the Σ5 (210)[001] GB before loading (**a**), structural configuration of colored planes after first (**b**) and second (**c**) GB displacement, and after the GB was displaced on four interplanar distances (**d**). The GB region is located between the dashed lines and is highlighted in gray. The shear rate is 1 m/s

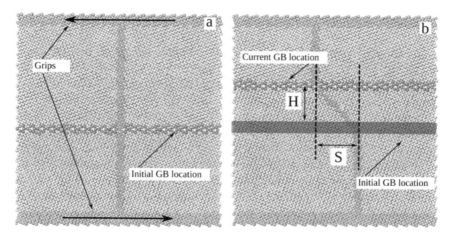

Fig. 5 Crystallite structure and the GB position at different points in time after the start of loading: 0 ps (**a**); 300 ps (**b**). Arrows in **a** show the displacement direction of the grips. The green line shows the position of the atoms forming a vertical line along the middle of the sample before loading

selected planes are rearranged into the GB structure (Fig. 4b). The second displacement changes the stacking order of green and red planes and then adjusts the green plane to the lower grain (Fig. 4c, d).

Note that the atomic planes far from the GB also have a pronounced periodicity of motion in the normal direction. It is due to the crystallinity of the sample structure and the constant velocity of the grips. However, the jump-like motion of these planes is less pronounced in comparison with atomic planes belonging to GB.

The structural transformations leading to high-speed migration of GBs in fcc metals under shear loading were studied for nickel bicrystals containing the Σ5 (310)[001] symmetric tilt GB. The simulated sample was composed of about 70,000 atoms. Periodic boundary conditions were simulated in directions parallel to GB, and rigid boundary conditions were set in the third one. The loading scheme and the position of the GB are shown in Fig. 5a. The initial temperature of the sample was 300 K. The shear loading rate in different simulations varied from 1 to 100 m/s.

The simulation results showed that shear loading of nickel sample causes a high-speed GB motion along the normal to its plane. To analyze the peculiarities of the GB motion in the bicrystal, a vertical layer was selected with a thickness of several lattice parameters normal to the GB plane. It was found that the atoms of this layer in the interval between the initial and final positions of the GBs have a pronounced displacement gradient (Fig. 5b). Atoms outside this interval are displaced by equal distances with grips. This is a shear-induced displacement of the GB. Such character of the GB displacement was revealed experimentally in [30]. It is one of the main mechanisms of grain growth and is quite common in the processes of recrystallization of the structure. For a quantitative description of this GB displacement, a coupling factor is introduced. It is defined as the ratio of the lateral (S) and normal (H) displacement values:

$$b = S/H.$$

The coupling factor depends on the structure of the GB region and the sample temperature [18]. For simulated GB, the coefficient b is approximately equal to 1.

The change in the shear stress τ_{xy} during the GB motion is shown in Fig. 6. One can see that the GB moves in a jump-like manner, which is due to the sawtooth nature of the change in shear stress. Note that the segments of the curve on which the shear stress increases correspond to the flat segments on the GB displacement curve. The segments of the τ_{xy} curve at which the drop occurs correspond to the GB displacement. The pronounced periodicity of the curves in Fig. 6 is associated with the discreteness of the crystallite structure and the symmetry of the GB. At the same time, thermal fluctuations of the atomic system cause deviations from this periodicity.

Analysis of the simulation results showed that the GB motion is realized through a certain sequence of transformations of typical structural elements in the intergranular region. Figure 7 shows a fragment of the structure in the GB region at different points in time. This fragment contains two atomic planes along the normal to the plane of the figure. The Roman numerals denote the atoms in the GB region, which belonged to the upper grain at the initial moment. Note that the atoms with numbers I and III belong to the same plane in the normal direction to the figure, and II and IV belong to the other plane. We revealed that the transformation of the GB atomic layer of the upper grain into the structure of the lower grain occurs through three characteristic jump-like displacements of numbered atoms in the XY plane. The directions and values of these three displacements for numbered atoms are shown in Table 1. The GB motion in the lateral and normal direction to its plane during loading is always provided by the indicated displacements of the boundary atoms.

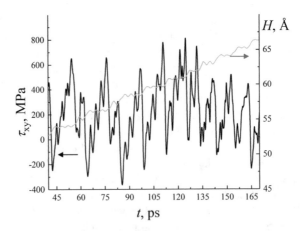

Fig. 6 Dependence of the shear stress and coordinates of the $\Sigma 5$ (310)[001] GB position in nickel bicrystal on time. The black curve is shear stress; the gray curve is the GB position

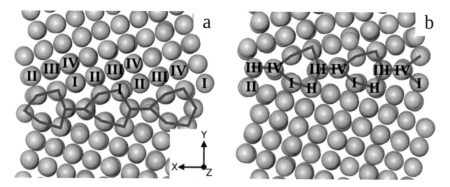

Fig. 7 Fragment of the structure containing the Σ5 (310)[001] GB in nickel at the following points in time after the start of the shear: **a** 22.5 ps; **b** 43 ps. Lines indicate the GB structural elements

Table 1 Atomic displacements providing GB motion under shear loading

Atom numbers	First displacement, Å (X; Y)	Second displacement, Å (X; Y)	Third displacement, Å (X; Y)
I	(0.7; 0.1)	(0.4; 0.0)	(−0.5; 0.0)
II	(0.6; 0.0)	(0.3; −0.5)	(−0.4; −0.4)
III	(0.5; 0.0)	(0.5; 0.1)	(0.6; 0.0)
IV	(0.6; 0.0)	(0.4; 0.0)	(0.4; −0.4)

Note that the motion of the symmetric tilt GB is realized without nucleation of structural defects. The use of periodic boundary conditions does not allow grain rotation during the displacement process. Since a tilt symmetric boundary is simulated, both grains have the same shear moduli in the direction of applied loading. Therefore, the GB motion is completely due to the coupling effect.

4 Peculiarities of Plasticity Nucleation in Metals Under Nanoindentation

One of the most informative and effective methods for studying the physical and mechanical properties of materials during contact interaction is nanoindentation. A change in the indentation conditions allows a systematic study of the influence of various factors on the processes occurring in the contact zones of the materials. As a rule, the aim of works related to computer simulation of the material behavior under indentation is to study the mechanisms of plastic deformation in the zone of a spherical or pyramidal indenter, visualize defect structures, and interpret load-indentation curves [31–34]. Despite the high information content of such studies, it is difficult to analyze the results due to the complex deformation pattern. For the clearer and simpler interpretation of the indentation results, it is convenient to use an extended indenter of a cylindrical shape [35–37]. For this choice of the indenter, the contact region is linearly extended from one face of the sample to the other. The loading scheme of a copper crystallite with such an indenter is shown in Fig. 8. The axis of the indenter was parallel to the loaded crystallite surface. Free boundary conditions were set along this axis. The loaded face was a free surface, while atomic positions of several layers of the opposite face were fixed in the indentation direction. The lateral faces of crystallites were simulated as free surfaces. The indentation rate was 25 m/s. The simulated crystallites were loaded at 300 K.

To study the behavior of simulated crystallite under indentation the loading force was calculated as a function of the indentation depth. The loading force (F) was defined as the total force acting on the indenter from the loaded crystallite. The indentation depth (d) was calculated as the distance from the lower boundary of the indenter surface to the level of the crystallite surface in the initial state. The results of the calculation of the loading force - indentation depth dependence are shown in Fig. 9.

The indenter and crystallite begin to interact as soon as the distance between them becomes smaller than the cutoff radius of the interatomic interaction. Initially, an attractive force arises between the indenter and the loaded surface (this corresponds to a negative value of the loading force in Fig. 9). This effect is called "jump-to-contact" [38]. During the indenter displacement, the attractive force changes into repulsive. In accordance with this, the loading force in Fig. 9 has a pronounced minimum. The dependence of the indentation force on the indentation depth can be divided into four stages. The first stage is characterized by a linear dependence of the loading force on the indentation depth and corresponds to the elastic response of the material. At the beginning of the second stage, local structural transformations are generated in the contact zone (Fig. 10a). The generation of such local structural transformations

Fig. 8 Indentation scheme and crystallographic orientation of the simulated system

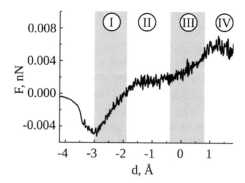

Fig. 9 Dependence of the loading force on the indentation depth

in an ideal crystallite leads to a partial relaxation of excess stresses and a decrease in the slope of the curve with its subsequent transition to plateau in Fig. 9. The change in the behavior of the loading force during the transition from the first stage to the second is also associated with the discreteness of the indenter structure. In particular, during the indenter penetration, new atomic layers of the indenter start to interact with the free surface of the crystallite. The number of defects in the contact zone at the second stage quickly reaches saturation, and their number at the third stage changes insignificantly. This is due to the fact that the mechanism of excess stress relaxation by the generation of structural defects exhausts itself, resulting in an increase in the slope of the loading curve. An analysis of the simulation results shows that further indenter penetration (the beginning of the fourth stage) leads to an intensive increase in the number of local structural transformations. This leads not only to a slowdown in the growth of the loading force but to its superseeding by a decrease. Moreover, local structural transformations lead to the nucleation and development of structural defects of a higher rank, in particular, intrinsic and extrinsic stacking faults (SFs) (Fig. 10b). Structural defects in the contact zone are generated in the {111} atomic planes. Formed defects can spread along the indicated planes to the free surfaces. Their escape to the free surface leads to the formation of steps that change the crystallite shape. The defect structure in the loaded crystallite at the moment of the SF escape to the free surface is shown in Fig. 10c. The step on the free surface can be seen in Fig. 10d.

Note that the crystallite structure, its crystallographic orientation, indenter shape, loading scheme, and boundary conditions have a significant effect on the simulation results. To study the effect of the orientation of the loaded surface on the response of the material during indentation, the calculations were carried out for three surfaces with small Miller indices: $(01\bar{1})$, (001) and (111). The simulated samples had the shape of a cube with a side of 165 Å and consisted of 350,000 atoms. The loading scheme was similar to that described earlier, except that periodic boundary conditions were set along the axis of the cylindrical indenter. The indentation speed was 50 m/s. The influence of the orientation of the loaded surface on the formation of structural defects was studied based on the calculation of the dependences of the indentation force $F(d)$ and the fraction of atoms n involved in local structural changes on the indentation depth d (Fig. 9).

The calculation results of the $F(d)$ and $n(d)$ curves for the $(01\bar{1})$ orientation of the loaded surface are shown in Fig. 11. Marks on the $F(d)$ curve indicate the indentation depths at which the loading curve abruptly changes the angle of inclination or has pronounced kinks. Projections of the crystallite structure for indentation depths corresponding to these points are shown in Fig. 12. We found that the features of the $F(d)$ and $n(d)$ curves correlate well with each other. In particular, the calculations show that the local structural changes lead to a decrease in the slope of the indentation force curve or to the appearance of an extremum in the loading curve due to relaxation of internal stresses.

Initially, an attractive force begins to act on indenter as it approaches the sample (the region of negative values in Fig. 11), and reaches its absolute maximum at a distance of 3.0 Å from the loaded surface. This "dip" in the curve of the loading force is accompanied by a bending of the loaded surface towards the indenter.

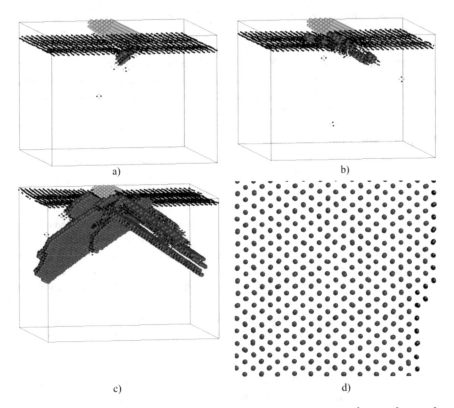

a)

b)

c)

d)

Fig. 10 Fragment of the simulated crystallite for indentation depths: **a** −1.0 Å, **b** 1.5 Å, **c** 8.7 Å. Atoms with the fcc symmetry of the nearest environment are not shown. Green spheres indicate indenter atoms, red large spheres, and dark points indicate atoms with hcp and undefined symmetry of the nearest environment, correspondingly. **d** a step on the free surface

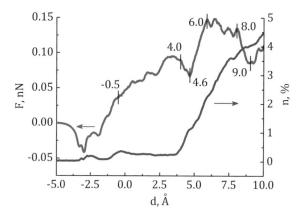

Fig. 11 Dependences of the loading force (red curve) and the fraction of atoms involved in local structural changes (blue curve), on the indentation depth in the sample with the (01 $\bar{1}$) loaded surface

A change in the shape of the loaded surface leads to the generation of local structural changes in the region under the indenter. An analysis of the structure showed that the atoms in this region have 12 nearest neighbors, but their environment does not correspond to any of the known lattices. With further indentation, the surface returns to its previous position, the number of defects decreases, and the loading force remains practically unchanged when the indenter moves in the range from −2.5 to − 1.8 Å. Then the loading force begins to grow with a constant number of defects. After passing the depth of −0.5 Å, its slope decreases. This is due to an increase in the number of local structural changes that form a/6 < 112 > {111} partial dislocations, and then SFs. They are located in adjacent (11$\bar{1}$) and (1$\bar{1}$1) planes (Fig. 12a). Further, the number of local structural changes slightly decreases. In this case, dislocations remain motionless, which leads to a further increase in the loading force.

Starting from an indentation depth of 3.6 Å, new SFs are generated and grow in the crystallite. They are located in the area under the indenter in planes of the same type, but deeper than previously formed SFs. The latter also begin to increase in size (Fig. 12b). In this case, a significant decrease in the loading force is observed, which continues until the indenter depth reaches 4.6 Å. The number of defects continues to increase at further loading, but the loading force begins to grow. This is due to the fact that the trailing partial dislocations move from the contact zone towards the free surface (Fig. 12c, d). At an indentation depth of 6.0 Å, the maximum value of the loading force is reached. The escape of dislocations to free surfaces leads to change of the shape of the crystallite, in particular, to the formation of steps on the free surface (Fig. 12e, f).

The simulation results of the copper crystallite behavior upon indentation of the (001) free surface are shown in Fig. 13. The peculiarities of the loading force curve in this figure correlate well with the peculiarities of the curve describing the number of atoms involved in local structural changes.

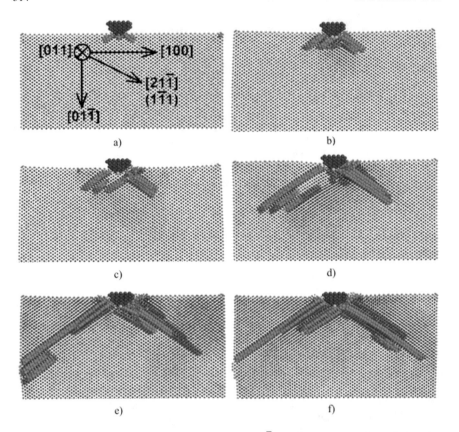

Fig. 12 Fragment of the crystallite structure with the (01 $\bar{1}$) loaded surface at different indentation depths: **a** −0.5 Å, **b** 4.0 Å, **c** 4.6 Å, **d** 6.0 Å, **e** 8.0 Å, **f** 9.0 Å. Large blue and green spheres show the atoms with hcp and undefined symmetry of the nearest environment, respectively. Indenter atoms are marked in red

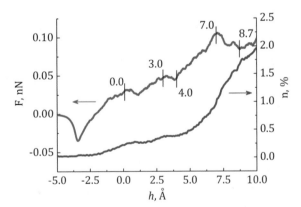

Fig. 13 Dependences of the loading force (red curve) and the fraction of atoms involved in local structural changes (blue curve) on the indentation depth in the sample with the (001) loaded surface

The attractive force of the indenter to the surface reaches its maximum value at a distance of -3.4 Å. The number of defects increases due to the bending of the free surface. At further indentation, the loading force and the number of atoms involved in local structural changes begin to grow. Note that the curve of the loading force has features at the indentation depths of 0 and 3 Å. They are related to the discrete structure of the indenter. As the indenter penetrates the material, the atomic rows of the indenter alternately interact with the crystallite. Since distant atomic rows are initially attracted to the loaded surface, this leads to a slowdown in the growth of the loading force.

An analysis of the structure of the simulated crystallite showed that at the indentation depth of 0.8 Å, a region containing atoms with either hcp or undefined symmetry of the nearest environment is formed in the contact zone (Fig. 14a). The sizes of this region are comparable with the sizes of the indenter. With further indentation, the width of the region along the [010] direction doubles and then does not change. At the same time, the size of the defective region along the [001] direction increases with the penetration depth of the indenter. Note that a more intensive increase in the density of defects leads to a decrease in the loading force of the sample at an indentation depth of 7.0 Å.

Due to the inertia of accommodation processes, structural changes in the crystallite continue for some time after the indenter stops. During relaxation after the indenter stops at a depth of 10.0 Å, dislocation loops continue to move from the top of the region containing defects to the side free surfaces of the sample (Fig. 14c–e). The defect escape on free surfaces leads to the formation of steps (Fig. 14f), and the fraction of local structural changes decreases from 2.0 to 0.3%.

The simulation results of the copper crystallite behavior under indentation of the (111) free surface are shown in Fig. 15. The curves presented in this figure correlate quite well with each other.

Calculations showed that local structural changes begin to form in the sample at the indentation depth of -1.7 Å (Fig. 16a). Their formation slows down the growth of the loading force. The SF starts to grow in the $(11\bar{1})$ free surface at the indentation depth of 0.8 Å (Fig. 16b), and the loading force changes slightly.

The leading and trailing a/6 < 112 > {111} partial dislocations are generated and move in adjacent $(11\bar{1})$ planes during loading of the crystallite (Fig. 16c–e). As a result of this process, a twin is formed in the crystallite. Atoms with hcp symmetry of the nearest environment are located on its boundaries (Fig. 16f). Note the formation of a fragmented region in the contact zone. This region consists of hcp atoms and grows towards the right side free surface (Fig. 16e, f).

In the case of bcc iron, the indenter was modeled by a repulsive force field in the form of a cylinder. The use of such an indenter reduces the effect of the structure discreteness on the structural response of the material. The sample dimensions with an ideal structure were 170 × 170 × 170 Å (Fig. 17a), and the dimensions of sample the with a GB was 230 × 170 × 170 Å (Fig. 17b). The projection of the structure of the Σ13 (320)[001] symmetric tilt GB is shown in Fig. 18. The axis of the cylindrical indenter was oriented parallel to the loaded surface of the crystallites. The indenter force field was described by the formula: where R is the indenter radius, r is the

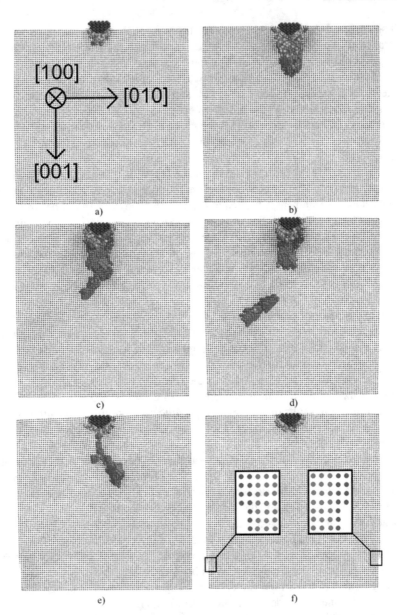

Fig. 14 Crystallite structure under indentation of the (001) free surface at different indentation depths: **a** 4.0 Å, **b** 8.7 Å. The crystallite structure at different points in time after the indenter was stopped at the depth of 10 Å: **c** 40 ps, **d** 60 ps, **e** 80 ps, **f** 120 ps. Large blue and green spheres show the atoms with hcp and undefined symmetry of the nearest environment, respectively. Indenter atoms are marked in red

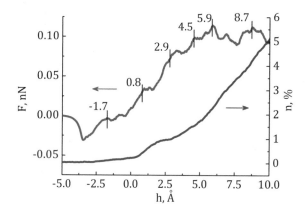

Fig. 15 Dependences of the loading force (red curve) and the fraction of atoms involved in local structural changes (blue curve), on the indentation depth in the sample with the (111) loaded surface

distance from the indenter axis to the atom. The loading scheme was similar to that described previously for fcc samples of different orientations. The indenter was pressed in at a constant speed of 1 m/s. The kinetic temperature of the samples was 300 K.

$$
U = \begin{cases} -\frac{(R-r)^4}{4}, & r < R \\ 0, & r > R \end{cases},
$$

The dependences of the loading force on the indentation depth are shown in Fig. 19. The loading force curves for both crystallites are similar. The regions of linear growth of the loading force correspond to the elastic response of crystallites. The onset of plastic deformation (the formation of local structural changes) can be determined by an abrupt decrease in the loading force. An analysis of the indentation results showed that atoms involved in local structural changes have the value of the reduced slip vector exceeding 0.2. Note that, for a crystallite with the GB, plastic deformation nucleates at smaller indentation depths. Such a response is associated not only with the GB presence in the crystallite but also with the fact that the indenter contact line is oriented differently with respect to the loaded surface than in the case of a crystallite with an ideal lattice. The growth and drop of the loading force in Fig. 19 correlate well with curves showing a change in the number of atoms forming local structural changes.

An analysis of the simulation results shows that local structural changes initially nucleate in the region of contact of the indenter with the surface, and then propagate along the slip planes towards the lateral faces of the crystallite (Fig. 20). Their escape to the side face leads to the formation of a step on the free surface. Note that the GB prevents the propagation of local structural changes in the neighboring grain, which is clearly seen in Fig. 20b.

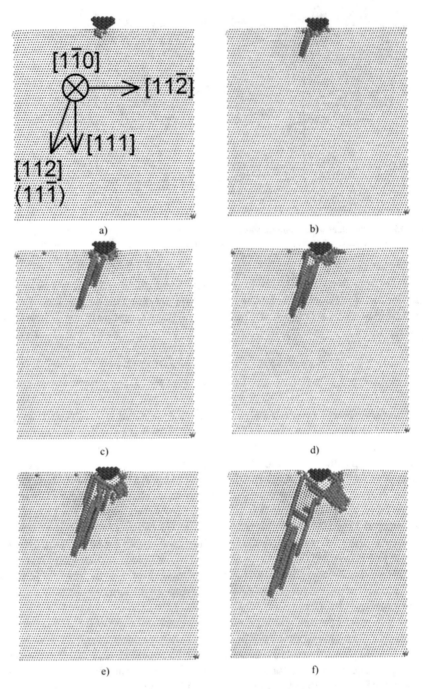

Fig. 16 Crystallite structure under indentation of the (111) free surface at different indentation depths: **a** −1.7 Å, **b** 0.8 Å, **c** 2.9 Å, **d** 4.5 Å, **e** 5.9 Å, **f** 8.7 Å. Large blue and green spheres show atoms with hcp and undefined symmetry of the nearest environment, respectively. Indenter atoms are marked in red

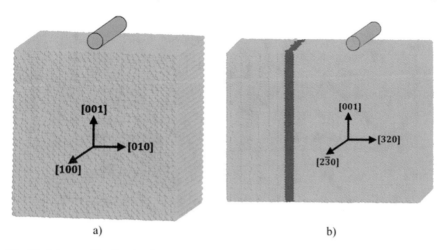

Fig. 17 Scheme of loading for the sample with an ideal lattice (**a**) and with the GB (**b**). The GB region is marked in blue

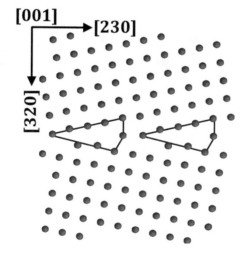

Fig. 18 Projection of the $\Sigma 13$ (320)[001] symmetric tilt GB structure onto the (001) plane after relaxation. Solid lines indicate GB structural elements

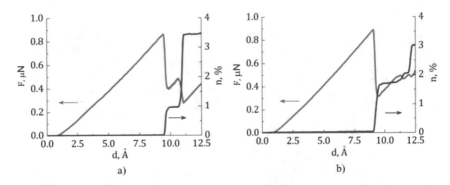

Fig. 19 Loading force (F) and the fraction of structural defects (n) depending on the indentation depth (d) for the single crystal (**a**) and the sample with the GB (**b**)

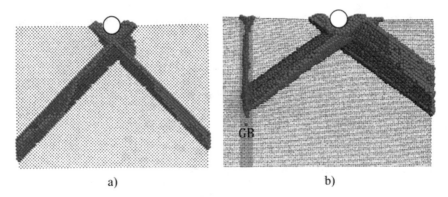

a)   b)

Fig. 20 Projections of the sample structure with an ideal structure (**a**) and with the GB (**b**) for an indentation depth of 12.5 Å. Dark gray spheres show atoms which reduced slip vector is greater than 0.2

Acknowledgements The work was performed according to the Government research assignment for ISPMS SB RAS, project No. III.23.1.4.

References

1. Panin VE, Pinchuk VG, Korotkevich SV, Panin SV (2017) Multiscaling of lattice curvature on friction surfaces of metallic materials as a basis of their wear mechanism. Phys Mesomech 20:69–77. https://doi.org/10.1134/S1029959917010064
2. Popov VL (2016) What does friction really depend on? Robust governing parameters in contact mechanics and friction. Phys Mesomech 19:115–122. https://doi.org/10.1134/S102995916020016
3. Zolnikov KP, Korchuganov AV, Kryzhevich DS, Psakhie SG (2018) Dynamics of the formation and propagation of nanobands with elastic lattice distortion in nickel crystallites. Phys

Mesomech 21:492–497. https://doi.org/10.1134/S1029959918060036
4. Dmitriev AI, Nikonov AY, Shugurov AR, Panin AV (2019) The role of grain boundaries in rotational deformation in polycrystalline titanium under scratch testing. Phys Mesomech 22:365–374. https://doi.org/10.1134/S1029959919050035
5. Yan J, Lindo A, Schwaiger R, Hodge AM (2019) Sliding wear behavior of fully nanotwinned Cu alloys. Friction 7:260–267. https://doi.org/10.1007/s40544-018-0220-z
6. Ovid'ko IA (2007) Review on the fracture processes in nanocrystalline materials. J Mater Sci 42:1694–1708. https://doi.org/10.1007/s10853-006-0968-9
7. Korchuganov AV, Tyumentsev AN, Zolnikov KP, Litovchenko IY, Kryzhevich DS, Gutmanas E, Li S, Wang Z, Psakhie SG (2019) Nucleation of dislocations and twins in fcc nanocrystals: Dynamics of structural transformations. J Mater Sci Technol 35:201–206. https://doi.org/10.1016/j.jmst.2018.09.025
8. Wolf D, Yamakov V, Phillpot SR, Mukherjee A, Gleiter H (2005) Deformation of nanocrystalline materials by molecular-dynamics simulation: Relationship to experiments? Acta Mater 53:1–40. https://doi.org/10.1016/j.actamat.2004.08.045
9. Naik SN, Walley SM (2020) The Hall-Petch and inverse Hall-Petch relations and the hardness of nanocrystalline metals. J Mater Sci 55:2661–2681. https://doi.org/10.1007/s10853-019-04160-w
10. Koch CC (2007) Structural nanocrystalline materials: an overview. J Mater Sci 42:1403–1414. https://doi.org/10.1007/s10853-006-0609-3
11. Wu XL, Zhu YT (2008) Inverse grain-size effect on twinning in nanocrystalline Ni. Phys Rev Lett 101:025503. https://doi.org/10.1103/PhysRevLett.101.025503
12. Zhu YT, Wu XL, Liao XZ, Narayan J, Mathaudhu SN, Kecskés LJ (2009) Twinning partial multiplication at grain boundary in nanocrystalline fcc metals. Appl Phys Lett 95:031909. https://doi.org/10.1063/1.3187539
13. Monk J, Hyde B, Farkas D (2006) The role of partial grain boundary dislocations in grain boundary sliding and coupled grain boundary motion. J Mater Sci 41:7741–7746. https://doi.org/10.1007/s10853-006-0552-3
14. Kou Z, Yang Y, Yang L, Huang B, Luo X (2018) Twinning-assisted void initiation and crack evolution in Cu thin film: an in situ TEM and molecular dynamics study. Mater Sci Eng, a 737:336–340. https://doi.org/10.1016/j.msea.2018.09.069
15. Ovid'ko IA, Sheinerman AG (2006) Nanovoid generation due to intergrain sliding in nanocrystalline materials. Philos Mag a 86:3487–3502. https://doi.org/10.1080/14786430600643290
16. Ovid'ko IA, Valiev RZ, Zhu YT (2018) Review on superior strength and enhanced ductility of metallic nanomaterials. Prog Mater Sci 94:462–540. https://doi.org/10.1016/J.PMATSCI.2018.02.002
17. Psakhie SG, Zolnikov KP, Dmitriev AI, Smolin AY, Shilko EV (2014) Dynamic vortex defects in deformed material. Phys Mesomech 17:15–22. https://doi.org/10.1134/S1029959914010020
18. Cahn JW, Mishin Y, Suzuki A (2006) Coupling grain boundary motion to shear deformation. Acta Mater 54:4953–4975. https://doi.org/10.1016/j.actamat.2006.08.004
19. Psakh'e SG, Zol'nikov KP (1997) Anomalously high rate of grain boundary displacement under fast shear loading. Tech Phys Lett 23:555–556. https://doi.org/10.1134/1.1261742
20. Psakh'e SG, Zol'nikov KP (1998) Possibility of a vortex mechanism of displacement of the grain boundaries under high-rate shear loading. Combust Explosion Shock Waves 34:366–368. https://doi.org/10.1007/BF02672735
21. Plimpton S (1995) Fast parallel algorithms for short-range molecular dynamics. J Comput Phys 117:1–19. https://doi.org/10.1006/jcph.1995.1039
22. Mendelev MI, Han S, Son WJ, Ackland GJ, Srolovitz DJ (2007) Simulation of the interaction between Fe impurities and point defects in V. Phys Rev B 76:214105. https://doi.org/10.1103/PhysRevB.76.214105
23. Mendelev MI, Han S, Srolovitz DJ, Ackland GJ, Sun DY, Asta M (2003) Development of new interatomic potentials appropriate for crystalline and liquid iron. Phil Mag 83:3977–3994. https://doi.org/10.1080/14786430310001613264

24. Mishin Y, Farkas D, Mehl MJ, Papaconstantopoulos DA (1999) Interatomic potentials for monoatomic metals from experimental data and ab initio calculations. Phys Rev 59:3393–3407. https://doi.org/10.1103/PhysRevB.59.3393
25. Mendelev MI, King AH (2013) The interactions of self-interstitials with twin boundaries. Phil Mag 93(10–12):1268–1278. https://doi.org/10.1080/14786435.2012.747012
26. Mishin Y, Farkas D (1998) Atomistic simulation of [001] symmetrical tilt grain boundaries in NiAl. Philos Mag a 78(1):29–56. https://doi.org/10.1080/014186198253679
27. Stukowski A (2010) Visualization and analysis of atomistic simulation data with OVITO-the Open Visualization Tool. Modell Simul Mater Sci Eng 18(1):015012. https://doi.org/10.1088/0965-0393/18/1/015012
28. Zimmerman JA, Kelchner CL, Klein PA, Hamilton JC, Foiles SM (2001) Surface step effects on nanoindentation. Phys Rev Lett 87(16):165507. https://doi.org/10.1103/PhysRevLett.87.165507
29. Honeycutt JD, Andersen HC (1987) Molecular dynamics study of melting and freezing of small Lennard-Jones clusters. J Phys Chem 91:4950–4963. https://doi.org/10.1021/j100303a014
30. Rupert TJ, Gianola DS, Gan Y, Hemker KJ (2009) Experimental observations of stress-driven grain boundary migration. Science 326:1686–1690. https://doi.org/10.1126/science.1178226
31. Wen W, Becker AA, Sun W (2017) Determination of material properties of thin films and coatings using indentation tests: a review. J Mater Sci 52:12553–12573. https://doi.org/10.1007/s10853-017-1348-3
32. Ruestes CJ, Alhafez IA, Urbassek HM (2017) Atomistic studies of nanoindentation—a review of recent advances. Crystals 7(10):293. https://doi.org/10.3390/cryst7100293
33. Voyiadjis GZ, Yaghoobi M (2017) Review of nanoindentation size effect: Experiments and atomistic simulation. Crystals 7(10):321. https://doi.org/10.3390/cryst7100321
34. Gouldstone A, Chollacoop N, Dao M, Li J, Minor AM, Shen YL (2007) Indentation across size scales and disciplines: recent developments in experimentation and modeling. Acta Mater 55(12):4015–4039. https://doi.org/10.1016/j.actamat.2006.08.044
35. Psakhie SG, Zolnikov KP, Kryzhevich DS, Korchuganov AV (2019) Key role of excess atomic volume in structural rearrangements at the front of moving partial dislocations in copper nanocrystals. Sci Rep 9(1):3867. https://doi.org/10.1038/s41598-019-40409-9
36. Zeng FL, Sun Y, Liu YZ, Zhou Y (2012) Multiscale simulations of wedged nanoindentation on nickel. Comput Mater Sci 62:47–54. https://doi.org/10.1016/j.commatsci.2012.05.011
37. Hu X, Ni Y (2017) The effect of the vertex angles of wedged indenters on deformation during nanoindentation. Crystals 7(12):380. https://doi.org/10.3390/cryst7120380
38. Khajehvand M, Sepehrband P (2018) The effect of crystallographic misorientation and interfacial separation on jump-to-contact behavior and defect generation in aluminum. Mod Simul Mater Sci Eng 26(5):055007. https://doi.org/10.1088/1361-651X/aac427

Fault Sliding Modes—Governing, Evolution and Transformation

Gevorg G. Kocharyan, Alexey A. Ostapchuk, and Dmitry V. Pavlov

Abstract A brief summary of fundamental results obtained in the IDG RAS on the mechanics of sliding along faults and fractures is presented. Conditions of emergence of different sliding regimes, and regularities of their evolution were investigated in the laboratory, as well as in numerical and field experiments. All possible sliding regimes were realized in the laboratory, from creep to dynamic failure. Experiments on triggering the contact zone have demonstrated that even a weak external disturbance can cause failure of a "prepared" contact. It was experimentally proven that even small variations of the percentage of materials exhibiting velocity strengthening and velocity weakening in the fault principal slip zone may result in a significant variation of the share of seismic energy radiated during a fault slip event. The obtained results lead to the conclusion that the radiation efficiency of an earthquake and the fault slip mode are governed by the ratio of two parameters—the rate of decrease of resistance to shear along the fault and the shear stiffness of the enclosing massif. The ideas developed were used to determine the principal possibility to artificially transform the slidding regime of a section of a fault into a slow deformation mode with a low share of seismic wave radiation.

Keywords Fault · Earthquake · Slow slip event · Stick-slip · Seismic waves · Rock friction

1 Introduction

One of the important areas of Professor S. G. Psakhie's activities was studying regularities of deformation of hierarchical blocky media. His interest in these problems arose, among other things, under the influence of Academician S. V. Goldin. Sergey Grigorievich took active part in his seminar. Special attention in his works was paid to the possibility of altering the regime of blocky medium deformation to prevent powerful seismic impacts [1–3]. Starting with the application of the method

G. G. Kocharyan (✉) · A. A. Ostapchuk · D. V. Pavlov
Sadovsky Institute for Dynamics of Geospheres, Russian Academy of Sciences, Moscow, Russia
e-mail: gevorgkidg@mail.ru

© The Author(s) 2021
G.-P. Ostermeyer et al. (eds.), *Multiscale Biomechanics and Tribology of Inorganic and Organic Systems*, Springer Tracts in Mechanical Engineering, https://doi.org/10.1007/978-3-030-60124-9_15

of movable cellular automata [4] to the problems of deformation of contact areas between rock blocks, S. G. Psakhie and his colleagues conducted a series of experimental works, including investigations of the reaction of a natural fault to dynamic disturbances and watering [5–7]. Especially a big cycle of works should be noted that had been performed on the ice cover of the Lake of Baikal, which was used as a model to study the tectonic processes in the Earth's crust [8, 9].

Authors of this paper were lucky to take part in experiments that were held under the guidance of S. G. Psakhie at a segment of the Angarskiy seismically active fault (Baikal Rift zone) [10]. The device constructed in IDG RAS—the Borehole Generator of Seismic Waves (BGSW, Fig. 1)—was mounted there to disturb the fault. The device was used to produce periodical explosions of the air-fuel mixture in a specially drilled borehole [11]. The results of observations showed that after the active disturbance the nature of movements along the fault changed. Precise measurements of the parameters of displacement along the fault showed that the disturbance triggered movements along strike, which manifested as a left-side shear. The quasi-dynamic micro-displacements which were detected many times in the

Fig. 1 The Borehole Generator of Seismic Waves (BGSW) mounted at one of the sides of the Angarsky fault

records after actions with a drop-hammer, explosions and BGSW coincided with the macro-displacement in direction. Approximately in a day after the action of BGSW had been terminated the background direction of the creep and the average velocity of sliding (~4÷5 μm/day) restored [12].

Approaches that were developed in those works had a clear physical sense. Injecting water and the effect of vibrations can both change the parameters of friction during the shear along the fault and increase the pore pressure. These changes may result in emergence of conditions that correspond to the Mohr–Coulomb failure criterion. It means that a movement can be provoked at a fault that hasn't reached the ultimate state yet. Though the mechanics of the process seemed evident, it remained unclear whether the triggered movement would be a "quasi-viscous" one (in terms of V. V. Ruzhich with co-authors [7]) and no dynamic failure would occur. Let us cite Sh. Mukhamediev: "Here the clarity in fault's behavior ends. Further development of the artificially triggered movement (velocity of its propagation and its size) is not so evident. Here the non-uniformities of properties and stresses along the existing fault or its future trajectory play the most important role" [13]. The macroscopic conditions of different sliding modes on faults remained uncertain at that time. For the last 10–15 years the geophysical community has essentially advanced in many components of fault mechanics—cumulating information about the structure of segments where slip localizes [14–16], conditions of dynamic slip to be initiated [17, 18], in situ observations [19, 20] and laboratory modeling of slip episodes on faults [21–23]. We are going to present some of these data in this chapter.

2 Fault Slip Modes

In the very beginning of instrumental observations over deformations of the Earth's surface it became clear that stresses cumulated in tectonically active regions relax not only through dynamic failure of some sections of the Earth's crust, but through continuous aseismic sliding (creep) along existing faults, too. Earthquakes were interpreted as a quasi-brittle failure of rock, while creep—as a plastic deformation. It was believed that in the areas where the rate of deformation is high enough, accumulation of elastic stresses occurs with further dynamic failure of rock accompanied by intensive emission of seismic waves. In case the rate of deformation of a limited volume of the medium is so low that stresses have time to relax on all the structural inhomogeneities, regimes of deformation at a constant velocity without destruction (creep) occur [24]. Thus, it was believed that the earthquake and the aseismic sliding are two opposite phenomena that take place under different loading conditions in the medium.

As the observation data were cumulated and measuring facilities were upgraded, qualitative and quantitative differences between seismic events of one and the same rank were detected. For example, it turned out that seismic energies emitted in earthquakes with approximately equal seismic moments can differ by several orders of magnitude [25, 26].

Sensitive deformographs and tiltmeters periodically registered displacements and deformations at velocities several orders of magnitude higher than the background ones, but essentially slower than the velocity of rupture propagation in an "ordinary" earthquake. However, the low installation density of such devices didn't allow to summarize the data being obtained, especially as the attention of investigators was concentrated primarily on post-seismic and pre-seismic deformations.

The situation changed qualitatively when dense networks of GPS sensors and broadband sensitive seismic stations were launched to operation in a continuous regime [27, 28]. As a result, in the last 25–30 years the fault slip modes were detected and classified as what can be treated as transitional from the stable sliding (creep) to the dynamic failure (earthquake). Discovering these phenomena changed to a great extent the understanding of how the energy cumulated during the Earth's crust deformation releases—slow slippage along faults are apprehended not as a special sort of deformation, but they span a continuum of slip modes from creep to earthquake [29].

Studying the conditions of occurrence and evolution of transitional slip modes can give new important information about structure and laws of fault behavior. That is why investigations of these "unusual" movements on faults have become one of the leading trends. Detecting the phenomenon of episodic tremor and slip (ETS) in many subduction zones is believed to be one of the most important advances of geophysics in recent history [30].

Developing observation systems has allowed to reveal a number of new deformation phenomena associated with discontinuities of the Earth's crust—subduction zones [31], continental fault zones [32], tectonic fractures [10], fractures in large ice masses [33] and even with micro-cracks in hydrocarbon reservoirs [34].

The classification of deformation events along tectonic discontinuities adopted currently is based primarily on duration of the process in the source [29, 35]. Only several percent of the released deformation energy are emitted in the form of seismic waves in a 'normal' earthquake (duration of the process in the source 0.1–100 s). This turns out to be enough for the strongest macroscopic manifestations of powerful earthquakes to occur. The ratio of the emitted seismic energy E_s to the seismic moment M_0 varies in the range of $E_s/M_0 \sim 10^{-6}$–10^{-3}, the average value being ~2×10^{-5} [25, 26].

Under some conditions, the slip velocity may not reach seismic slip rates, but nevertheless, low amplitude low-frequency seismic waves are emitted. These are the so called Low Frequency Earthquakes (LFE) and Very Low Frequency Earthquakes (VLFE) [29]. The spectrum of these vibrations is depleted with high frequencies which testifies a longer (than it follows from standard relations) duration of the process in the source—up to hundreds of seconds. The ratio of the emitted seismic energy to seismic moment, specific for LFEs, is about $E_s/M_0 \sim 5 \times 10^{-8} - 5 \times 10^{-7}$, and the velocity of rupture propagation is $V_r \sim 100$–500 m/s [36–38]. VLFEs have durations in the source of about tens to hundreds of seconds, a velocity of rupture propagation of about $V^r \sim 10$–100 m/s and an energy-to-moment ratio of $E_s/M_0 \sim 10^{-9}$–10^{-7} [39].

In some cases, the peak slip velocity is so low that seismic waves which could be recorded instrumentally are not emitted at all. Nevertheless, the slip velocity in these deformation phenomena noticeably exceeds typical velocities of aseismic creep along faults which is, on average, several centimeters a year. Such deformation events that can last from several hours to several years are called Slow Slip Events (SSE).

First reports in which the phenomena of aseismic sliding were for the first time interpreted as "slow earthquakes" came after observations at the Izu Peninsula in Japan [40]. But perhaps for the first time, an episode of slow slip as a self-contained event that had a start and a termination was described by A. T. Linde with co-authors [32]. The authors presented a recorded deformation event about a week in duration. They called it a "slow earthquake" and proposed to characterize such events quantitatively, just like ordinary earthquakes, with the value of seismic moment M_0 or moment magnitude M_w, which is linked to the seismic moment by the well known relation [41]:

$$M_w = 2/3(\lg M_0 - 9.1) \tag{1}$$

The velocity of rupture propagation along the fault strike for SSEs lies in the range from several hundreds of meters a day to 20–30 km a day [39]. Despite small displacements, an appreciable seismic moment accumulates at the expense of a large fault area, at which the displacements take place. An essential part of energy cumulated in the course of deformation releases through slow movements. For example, in New Zeeland about 40% of seismic moment release through SSEs [42]. Slow displacements are registered there with durations from several months to a year and with moment magnitudes up to $M_w \sim 7$. Such SSEs repeat with a period of 5 years. Less scale events with durations of several weeks have a recurrent time of 1–2 years [43, 44].

Large scale SSEs last for months and even years. The seismic moment released in them is comparable to the one of the most powerful earthquakes. For example, from 1995 to 2007 more than 15 events were registered in different regions around the world, each of them with the released seismic moment of more than $M_0 \sim 5 \times 10^{19}$ Nm, which corresponds to the moment magnitude of $M_w \sim 7$ [39]. Their durations were from a month to one year and a half, and the amplitude of displacement along the fault reached 300 cm. It should be noted that some of these SSEs were not independent events, but episodes of post-seismic sliding.

The results of studying slow movements along faults show that these specific deformations are widespread all around the world and to a great extent at their expense stress conditions of many segments of the Earth's crust are regulated. Though initially it was thought that periodical slow slip is specific mainly for depths of several tens of kilometers in the subduction zones, installation of dense networks of seismic and geodetic observations allowed to detect similar phenomena at shallow sections of submerging plates and continental faults [45]. It is not impossible, that as the density and sensitivity of installed devices increase, sections of periodic slow slips with small moment magnitudes will be detected at numerous tectonic structures, including areas of high anthropogenic activities. Slope phenomena have also much in common with slow tectonic slip along faults [46].

The importance of studying the slow slip modes bases on several reasons. Investigating mechanisms and driving forces of these processes will allow to advance essentially in understanding regularities of interactions of blocks in the Earth's crust, and, consequently, in assessing risks of natural and man-caused catastrophes linked to movements along interfaces—earthquakes, fault-slip rock bursts, landslides, etc.

Slow movements along faults can, beyond all doubt, be triggers of dynamic events. Detecting transitional slip modes all around the world has led many investigators into the attempt of linking the slow slip phenomena to powerful earthquake triggering [47–50], but the mechanics of this process is not developed so far. There are only few works, in which sequences of deformation events of different modes were registered instrumentally, and their interrelations were soundly demonstrated [51, 52]. Manifestations of seismicity were registered most reliably after slow slip [53], or manifestations of seismicity in the form of non-volcanic tremor against the background slow slip [54]. Geodetic and seismic observations can give only a confined insight into the physical mechanism of slow sliding. For example, though it is admitted that an asesimic slip preceded the Tohoku 2011 earthquake [55], it remains uncertain, how and why the sliding regime altered just before the main shock. Taking measurements near the surface, it is actually impossible to detect small spots of "accelerated slip" at a seismogenic depth [56].

Last but not least, a problem which regularly attracts the attention of the scientific community shall be mentioned—the possibility to alter the seismic regime of some area or the deformation regime of a specific fault segment through some external actions [57]. Therefore, it is important to investigate the conditions under which different slip modes emerge and evolve in fault zones.

3   Localization of Deformations and Hierarchy of Faults

Applying the ideas of self-similar blocky structure of Earth's crust [58] inevitably leads to the necessity of introducing a hierarchy of interblock gaps—tectonic fractures and faults. Meanwhile the situation seems unobvious for these objects. At first glance, it is hard to reveal a similarity between a closed crack in a rock mass and a large fault zone, as opposed to the rock blocks they bound.

Unlike blocks whose linear sizes can be reliably measured, it is often impossible to estimate unambiguously the geometric characteristics of discontinuities. Analysis of any tectonic map shows that prolonged linear structures can be considered to be single objects only partially. Each of these structures is a concatenation of separate sections adjoining distinctly detectable blocks of certain ranks. Widening of the zones of active faults, as well as delta-wise and trapezium-wise sections of fault splitting can be observed in tectonic junctions.

The size of an active zone can be limited to a local section, even though a rather long linear structure may be available. As a rule, sizes of active zones match well with the sizes of structural blocks, which corresponds to the classical concept that the energy of an earthquake is controlled by a certain size of a block being unloaded. Actually, it means that the length of a linear structure should be characterized by the length of its active section, which in its turn manifests quasi-independently in the geodynamic sense. The hierarchical rank of a fault zone is determined by the rank of the blocks it separates.

Despite many publications, the relationship between parameters of fault zones such as length, width (the size across strike), amplitude of displacement are being discussed actively. Empirical scale relations linking fault length L, fault width W and amplitude of displacement along rupture are widely used in describing structural characteristics of fault zones [59, 60]. Power relations of the following types are often used to establish links between these parameters:

$$W = \alpha \cdot D^a, \quad D = \beta \cdot L^b, \quad W = \chi \cdot L^c \tag{2}$$

In many publications the indexes in Eqs. (2) are more often close to one, while the factors α, β and χ vary in a wide range. Some authors expressed essential doubts in applicability of Eqs. (2). The doubts based mainly on a noticeable dispersion of experimental data [61, 62]. Closeness of indexes in Eqs. (2) to one means fulfillment of similarity relations for the process of faulting—all the linear sizes are linked to each other through direct proportionalities.

More detailed investigations of the last years [63, 64] have shown that there are several hierarchical levels, in which alterations of parameters of the events with scale occur according to different laws, which differ very often strongly from the similarity relations. Figure 2a presents maximum displacement along a fault versus fault length. The plot uses the data of several investigations.

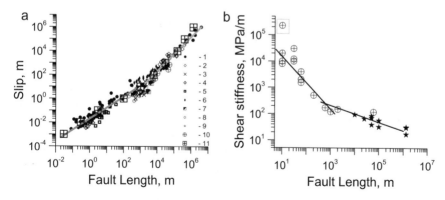

Fig. 2 Structural and mechanical characteristics of faults versus fault length. **a** Cumulative slip along a fault. References [1–11] are presented in [65]. Lines—best fit of the data in the range $L < 500$ m and $L > 500$ m. **b** Shear stiffness of a fault. Crosses in circles—the data were obtained with the method of seismic illumination. The point in the frame was not used in constructing the regression dependence. Black stars—the data were obtained with the method of trapped waves [65]

By all appearances the linear size L ~ 500–1000 m is a transitional zone, a specific boundary between two bands, in which the scale relations differ. Faults that have reached this stage of evolution may be called the "mature" ones. In [63] for example, a relation linking the width of the influence zone to the fault length was suggested:

$$W = \begin{cases} 0.15 \cdot L^{0.63}, & L \leq 500 \div 1000\,\text{m} \\ 0.85 \cdot L^{0.42}, & L > 500 \div 1000\,\text{m} \end{cases} \tag{3}$$

Here, the width of the influence zone is the width of the cross-section with higher fracturing.

It is important to emphasize that the change of mechanical characteristics of faults (fault stiffness) with scale demonstrates the presence of approximately the same transitional zone (Fig. 2b).

Further alteration of scaling relations is observed for the most powerful deformation events with characteristic sizes exceeding the thickness of the crust.

Investigations performed for the last 20–30 years have allowed to essentially widen and clarify the knowledge about the inner structure of fault zones. New data have been acquired in the frames of the international program of "fast drilling" of a fault zone after an earthquake [66]. Similar projects of drilling through fault zones have already been fulfilled in several regions [67, 68]. Together with the results of traditional geologic explorations at the surface and in mines [14–16, 69, 70] these data allow to acquire a rather orderly notion about the structure of faults' central parts.

The *damage zone* is located at the periphery of the fault. Its width may vary from meters to hundreds of meters. It is usually associated with a higher fracture density, if compared to the intact massif. The damage zone contains distributed fractures of a wide range of sizes. It is structured to a great extent and usually contains a standard set of discontinuity types [71].

Cataclastic metamorphism becomes more intensive in the direction to the *fault core*. One or several sub-zones of intensive deformations can usually be detected there. Their widths may be from centimeters to meters. They are usually composed of gouge, cataclasite, ultracataclasite or of their combination. Deformations can either be distributed uniformly over the fault core, or be localized inside a narrow shear zone. This zone of intensive grain grinding is the *principal slip zone* (PSZ). Its width is usually from one millimeter to decimeters [14, 16].

The structure of a fault zone depends on depth, properties of enclosing rock, tectonic conditions (shear, compression, tension), cumulated deformations, hydrogeological conditions and type of the deformation process. In slow aseismic creep the principal slip zone is often represented by a set of individual slip zones and zones of distributed shear deformations. Some secondary shears (they are often of oscillatory origin) can be localized along discrete fracture plains. The width of shear zones in the sections of aseismic creep of such faults as Hayward fault and San Andreas lie in the range of meters—tens of meters, the average value at the surface being 15 m

[14]. There are suggestions that this zone becomes narrower with depth and its width reduces to about 1 m deep in the rock massif.

An essentially higher degree of localization is observed in seismically active fault zones, where most deformations are, presumably, of coseismic origin. For example, investigations of shears in Punchbowl and San Gabriel faults in California have detected the thickness of the principal slip zone to be not wider than 1–10 cm. Note that cumulative displacements along these faults reach tens of kilometers [16, 71–73].

Chester and Chester [16] showed that it is in the principal slip zone that displacements of sides of large faults localize. According to their data, concerning one of the segments of the Punchbowl fault, only 100 m of displacement (of the total displacement length of 10 km) localized in the zone of fracturing about 100 m thick, while the rest occurred inside a narrow ultracataclasite layer from 4 cm to 1 m thick. A continuous, rather plane interface about 1 mm thick was detected inside this core. This interface was the principal slip zone of the last several kilometers of displacement [74].

In fault zones, whose cores consist of cataclastic rocks, coseismic ruptures often occur along one and the same interface, formed of ultracataclasites that have emerged at previous deformation stages [14]. Displacements along secondary, novel discontinuities are small and have a negligible contribution to the cumulative amplitude of fault side displacement.

Individual zones of the PSZ can rarely be traced longer than for several hundred meters, though it is suggested that their lengths can reach several kilometers [14]. In all likelihood PSZs may interact at some deformation stages through the zones of distributed cataclastic deformations without clear signs of a single rupture in latter. An analogy with laboratory sample destruction comes to mind here [75]. Such linear conglomerates of separate PSZs and sections of heterogeneous fracturing can make up an integrated fault core.

Geophysical investigations in wells that penetrate through fault zones at appreciable depths have also demonstrated an extreme degree of localization not only of deformation structure, but such parameters as porosity, permeability, velocity of propagation of elastic vibrations [19, 68].

Thus, the results of geological description of exhumed fault segments, the data on deep drilling of fault zones, as well as detailed investigations of seismic sources located with high accuracy [76] allow to speak about an extreme degree of localization of coseismic displacements. Macroscopic interblock displacements are not distributed over the thickness of material crushed in the course of shear, but are localized along a narrow interface of sliding. It means that with some conditionality, a dynamic movement along a fault can be considered as a relative displacement of two blocks, and their interaction is determined mainly by forces of friction.

4   Frictional Properties of Geomaterial and the Slip Mode

Investigating exhumed segments of fault zones has shown that the structural hetero-geneity of large fault zones results in appreciable spatial variations in rheology and deformation rate at one and the same segment [77]. The evolution of frictional prop-erties of rocks composing the massif and their spatial distribution play an important role in the processes of nucleation, localization and propagation of rupture in seismic events [78]. It has been shown in several works that the mineral composition affects the frictional strength and the sliding regime of the fault [79], and frictional stability depends on the evolution of structural properties of the fault during deformation [80].

Models that interpret emergence of different slip modes base on dependences of frictional properties of the sliding interface on velocity, displacement and P-T conditions revealed in laboratory experiments [81, 82].

These dependences are different for different geomaterials. Over the past years, a great number of experiments have been performed with materials collected while drilling fault zones. These experiments have shown that there exist materials with a pronounced property of velocity strengthening (VS). For example, saponite (a mate-rial with a low friction factor, which increases as sliding velocity grows), determines the deformation behavior of the creeping segment of the San Andreas fault [83]. Judging by the results of laboratory experiments weak materials rich with phyllosil-icates manifest only stable sliding (corresponds to velocity strengthening), at least until the mineral composition of geomaterial in the principal slip zone alters with time. Stronger materials rich with quartz and feldspar, after creeping for a while, become 'velocity weakening' (VW) and provide unstable sliding [84].

VS- and VW-wise behavior can take place at different segments of one and the same fault zone. A complex topography of the fault interface leads to emergence of areas of stress concentration and rather unloaded areas. The probability of stick-slip to realize increases essentially in the areas of stress concentration. Deposition of minerals drawn by fluids takes place in unloaded areas, which in many cases promotes formation of layers composed of weak materials rich with phyllosilicates that manifest velocity strengthening. Thus, in many cases it is the contacts of rough surfaces (areas of stress concentration) that turn out to be dynamically unstable in sliding along a fault, while fault segments located between rough surfaces in contact manifest frictional properties of stable sliding. It is likely that these segments preserve their properties during, at least, several seismic cycles. This is supported by, the so called, repeated earthquakes (events identical in locations, energy and waveforms) [85]. These events repeatedly break again and again one and the same spot or asperity on the fault. These repeaters, whose manifestations are now found all around the world, have in most cases small magnitudes, but there are powerful ones too, with magnitudes higher than M6. These observations lead to the conclusion that different sliding regimes are determined by the non-uniform distribution of frictional properties and stress conditions over the fault interface.

It is convenient to demonstrate the effect of spatial non-uniformity of frictional properties on integral characteristics of the sliding process by the example of a numerical calculation of the relative shear of two elastic blocks separated by a sliding interface. The friction between blocks was described by a rate-and-state friction law [81, 82]. According to relations of this empirical model, the coefficient of friction μ depends on the running velocity of sliding V and on the variable of state θ:

$$\mu = \mu_0 - a \ln\left(\frac{V_0}{V} + 1\right) + b \ln\left(\frac{V_0\theta}{D_c} + 1\right) \tag{4}$$

Here, μ_0 is the constant corresponding to stable sliding at a low velocity V_0; a, b, D_c are empirical constants, V is the running velocity of displacement, θ is the variable of state, which is determined by the kinetic equation:

$$\dot{\theta} = 1 - \left(\frac{|V|\theta}{D_c}\right) \tag{5}$$

When $(b - a) > 0$ the regime of velocity strengthening realizes. The case of $(b - a) < 0$ leads to velocity weakening and provides conditions for stick-slip to occur.

One or several spots with "true" condition of velocity weakening but different values of the friction parameter $\Delta = (b - a) < 0$ were "installed" into the sliding interface when the boundary conditions were set. A typical computation scheme is given in Fig. 3.

The rate-and-state law used here differed from the traditional one by the fact that after the instability has arisen the factors a and b in Eq. (4) were considered to be

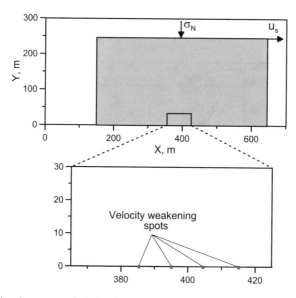

Fig. 3 Simulating the process of relative shear of two elastic blocks separated by a sliding interface: computation scheme. The block is pressed to the half-space by the normal stress. The upper right corner of the block is being pulled at a constant velocity u_s

zero. It was done to avoid repeated dynamic failures triggered by the waves reflected from the boarders of the mesh.

At the rest sliding interface the force of friction was described either by the Coulomb's law with the same friction factor μ_0 (i.e., there was no dependence on velocity and displacement) or by law (4) with constants that provided velocity strengthening $\Delta = b - a > 0$. Kinematic parameters of motion, components of stress tensor, spatial distribution of alteration of power density of shear deformation of blocks, kinetic energy at different moments of time from the rupture start were controlled.

The size of the upper block (width, length) varied from 50 × 100 m to 200 × 600 m. The lower block was a half-space with the same elastic characteristics (the density 2.5 g/cm³, velocity of P-wave propagation $C_p = 3000$ m/s, shear modulus $G = 52$ MPa). The coefficients used in the main series of computations provided the regime of velocity weakening: $\mu_0 = 0.3$, $a = 0.0002$, $b = 0.0882$, $D_c = 1$ μm, $V_0 = 0.002$ mm/s.

Figure 4 shows the hodograph of a rupture propagating along a fault segment including 1 weakening spot. One can see that the rupture starts at a non-uniformity and propagates away from both sides of the spot. The velocity of rupture propagation along the interface of Coulomb friction is close to the one of P-wave, while at the interface with strengthening the velocity of rupture propagation is noticeably lower than that of the S-wave.

In the case of a contact with strengthening the amplitude of the velocity of displacement decays essentially faster in both directions. For example, at a distance of 100 diameters of the spot in the direction normal to the fault the maximum velocity of displacement in the case of interface with strengthening approximately 6 times lower than in the case of Coulomb friction.

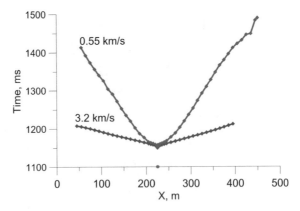

Fig. 4 Hodograph of the first onset of relative block motion. The size of the upper moveable block is 50 × 100 m. The lower block with the size of 400 × 800 m is immovable. Blue line—$\Delta = 0$ at the surface outside the spot; red line—velocity strengthening ($\Delta = 0.024$) is "switched on" at the surface outside the spot

Figure 5 shows the seismic moment $M_0 = G \cdot L \cdot U$ and kinetic energy of the block versus time for several variants of simulations. In the relation of seismic moment: L is the block length and U is the relative displacement. When friction between blocks increases as velocity and displacement grow (variant 2, 3), then the value of kinetic energy (the analogue of emitted energy) becomes lower than in the case of Coulomb friction. The final values of seismic moment are close for all the three variants, though the rate of growth of the value of M_0 is noticeably lower in the case of strengthening. It means that the energy of elastic deformation cumulated in the course of the interseismic period releases to a great extent during a rather slow sliding. It is this slow sliding that the dynamic rupture degenerates to, when it reaches the contact segment exhibiting velocity strengthening. This is well seen in the computation variant with several spots of weakening (Fig. 6). Again, the rupture starts at one of the spots of velocity weakening and propagates away from both sides of the spot. Outside the spot of weakening the velocity decays rapidly even in absence of segments with strengthening, speeding up again at neighboring VW spots. Though the maximum velocity of sliding decreases rapidly outside the spot of weakening, the integral value of relative 'fault side' displacement (the sum of dynamic slip and slow pre- and post-seismic slips) in this computation statement remains almost the same. The higher the total share of the VW spots is, the higher is the share of deformation energy spent to emission of the elastic wave in the high-frequency band of the spectrum.

For a great number of "spots-asperities" the rupture may start almost simultaneously at several spots. As far as the particle velocity is concerned, two types of segments can be clearly detected, just like in the case of several spots. The first one includes the spots themselves and their closest vicinity. Maximum velocity there reaches the values of 0.1–0.2 m/s. All over the rest contact the maximum particle velocity is 10–20 times lower.

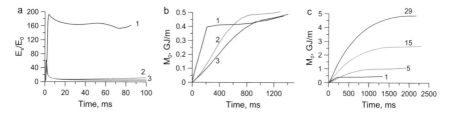

Fig. 5 Kinetic energy of the block and seismic moment versus time. The upper block is of the size of 150 × 450 m. **a, b** The centre of the single spot of velocity weakening of the size of $l = 1$ m is located in the point x = 225 m. 1—friction outside the spot doesn't depend on velocity ($\Delta = 0$). 2, 3—friction outside the spot increases as the velocity and displacement grow ($\Delta = -0.008$ and $\Delta = -0.024$, relatively), **c** the number of velocity weakening spots is shown by corresponding numbers near the curves; parameters a and b in the R&S law vary in a random way in the limits: $a = 0.0002$–0.0006, $b = 0.0071$–0.0084

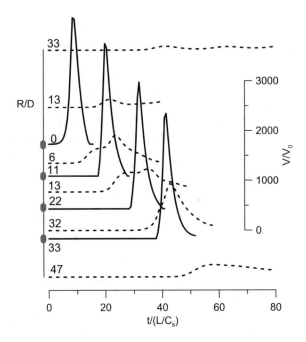

Fig. 6 Waveforms of particle velocity in the direction parallel to the sliding interface for the computation variant with four identical spots of velocity weakening. Locations of the VW spots are shown with bold segments at the left axis. Coulomb friction acts outside the spots. Numbers near the curves are the distances from the point of rupture start scaled by the spot diameter D. Solid lines are the epures corresponding to points inside the spots; dashed lines are the epures corresponding to points outside the spots. The amplitude of particle velocity is scaled by the velocity, at which the upper side of the block is pulled. Only first phases of motion are shown for better readability

Increasing quantity and density of asperities leads to an abrupt growth of the kinetic energy of block gained for the first 100 ms, i.e., the energy emitted in the high-frequency band of the spectrum. Meanwhile the integral value of seismic moment increases essentially slower.

Thus, the presence of VW spots determines the possibility of dynamic rupture to emerge. Their density and mutual locations govern the amount of energy emitted in the high-frequency band. Location and size of zones exhibiting velocity strengthening affect the velocity of rupture propagation, the scaled seismic energy (the ratio of E_s/M_0) and the size of the "earthquake". Under the adopted statement of computations, termination of the dynamic rupture corresponds to a radical decrease of the velocity of sliding. The rupture propagates along a stressed tectonic fault to the zone, in which the contact exhibits velocity strengthening—velocity of displacement, emitted energy and rate of seismic moment growth harshly decrease. If the size of the strengthening zone is big enough, the "earthquake" degenerates into a "slow slip event" or the rupture terminates completely. The rupture crosses small zones, then again speeding up at VW spots. If there are no big strengthening segments at fault interface, in this statement all the length of the block becomes involved in sliding, though in nature the rupture is usually terminated at some structural barrier [86].

5   Generating Different Slip Modes in Laboratory Experiments

Experiments on investigating regularities of emergence of different slip modes on a fracture with filler were performed on a well known slider-model set-up, in which a block under normal and shear loads slides along an interface (Fig. 7). A granite block (*B*) $8 \times 8 \times 3$ cm in size was put on an immoveable granite base. The contact between rough surfaces (the average depth of roughness was 0.5–0.8 mm) was filled with a layer of discrete material (*S*), imitating the PSZ of a natural fault. The layer thickness was about 2.5 mm. The normal load σ_n was applied to the block through a thrust bearing. σ_n varied in the range of 1.2×10^4 to 1.5×10^5 Pa. The shear load τ_s was applied to the block through a spring block. Its stiffness could vary. The set-up was equipped with an electromotor with a reducer that allowed to maintain the velocity of loading u_s with high accuracy in the range of 0.08–25 μm/s. The shear force was controlled with a force sensor. Displacements of the block relative the base were measured with LVDT sensors with the accuracy of 1 μm and with laser sensors (*D*) in the frequency range of 0–5 kHz and with the accuracy of 0.1 μm.

Filling the contact with mixtures of different materials, we managed to realize a wide spectrum of slip modes in the experiments, which correspond qualitatively to all types of interblock movements observed in nature—from aseismic creep to earthquakes. Examples of slip episodes realized in experiments are shown in Fig. 8. The figure also shows the recorded pulses of acoustic emission radiated during sliding.

Let's consider the conditions for slip to occur. A necessary condition for a slip to occur is closeness of effective stresses, tangential to fault plane, to the local or running ultimate strength:

$$\tau \geq \tau_0 \tag{6}$$

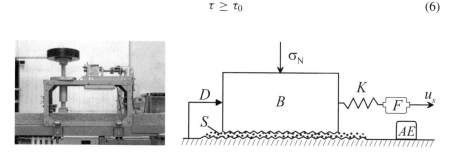

Fig. 7 Photo of the slider-model setup and the scheme of the experiments. *B*—movable block, *S*—layer of filler, *D*—laser sensor, *K*—spring element, *F*—force sensor, *AE*—acoustic emission sensor

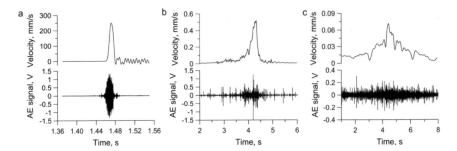

Fig. 8 Examples of diagrams of block velocity corresponding to different deformation events and their acoustic signal portraits. **a** fast mode (corresponds to normal earthquake); **b** medium mode (corresponds to slow earthquake); **c** slow mode (corresponds to slow slip event)

We use the term "local ultimate strength" because slip can occur under tangential stresses knowingly lower than the Coulomb strength τ_p.

Another necessary condition is the weakening of the sliding area as the velocity v and/or amplitude D of fault side displacement grows:

$$\frac{\partial \tau}{\partial v} < 0 ; \quad \frac{\partial \tau}{\partial D} < 0 \tag{7}$$

It is clear that if the contact strength will not decrease during shear, a dynamic slip will be actually impossible.

And, at last, the third condition: the rate of decrease of stresses in the enclosing massif, tangential to sliding interface (this is the shear stiffness of the massif K) should be lower than the rate of decrease of resistance to shear (this is the modulus of fault stiffness k_s at the post-critical section of the rheologic curve):

$$|k_s| = \left| \frac{\partial \tau}{\partial D} \right| \geq K = \eta \frac{G}{\hat{L}} \tag{8}$$

G is the shear modulus of the enclosing massif, $\eta \sim 1$ is the shape factor [75], and $\hat{L}$ is a specific size linked to the magnitude of the earthquake.

It is convenient to demonstrate the meaning of relations (6)–(8) at the scheme shown in Fig. 9. After the stress has reached the running strength of the contact (τ_p corresponds to the maximum of rheological curve $\tau(D)$ in Fig. 9, though it is not obligatory) the contact strength starts to decrease as the relative displacement and velocity grow. If condition (8) is true (black solid line in Fig. 9), a dynamic instability occurs and the energy is emitted outside the system. The amount of emitted energy in this simple example corresponds to the area bounded by the rheologic curve and the solid line of unloading of the enclosing massif. In the case when condition (8) is not true (dashed line in Fig. 9) the dynamic slip and, consequently, emission of energy are impossible.

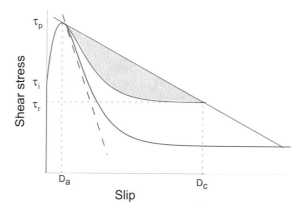

Fig. 9 Scheme of the slip emerging on a fault. Red line is the rheologic dependence stress-displacement; black straight lines are the lines of enclosing massif unloading in the course of relative fault side displacement. Black solid line—the rate of massif unloading is lower than the rate of decrease of resistance to shear along the fault. This case corresponds to emergence of dynamic slip. The energy emitted by a unit area of the sliding interface corresponds to the hatched area. Black dashed line—the rate of massif unloading is higher than the rate of decrease of resistance to shear along the fault. The fault remains stable in this case. Blue line is the decrease of effective friction during sliding at the expense of frictional melting, thermal effects, etc. It is well seen that these processes can change the parameters of an earthquake, but not the moment of dynamic slip start

In the course of sliding at velocities of ~1–10 m/s and in P-T conditions specific for seismogenic depths, a number of processes severely affecting the parameters of resistance to shear can occur at the interacting surfaces. These are the effects of friction lowering either because of thermal effects or due to effects produced by the high velocity of sliding: frictional melting [87], dynamic lubrication with solid materials [88], localization of heating in the area of "real" contact [89], macroscopic rise of temperature and the effect of velocity weakening [90], thermal decomposition of minerals leading to growth of pore pressure and generation of weak material [91], generation of silica gel during quartz amorphization under high pressure and large deformations [92], neo-mineralization of the sliding interface at the nano-crystal level [93].

All these effects are very important because they lead to a decrease of residual friction τ_r (blue line in Fig. 9) and, consequently, a decrease of the amplitude of stress drop $\Delta\tau$ and of the amount of emitted energy (magnitude of the earthquake). However, as these phenomena have no effect on fulfillment of conditions (6)–(8), they are of no use in searching for the signs of dynamic failure preparation.

Thus, it seems reasonable to select the modulus of the rate of decrease of resistance to shear (fault stiffness $|k_s|$ at the beginning of the post-critical section of the rheologic curve) as the characteristics which controls the initial stage of earthquake rupture nucleation. The ratio $\psi = |k_s| / K$ determines not only the possibility of sliding, but its character as well.

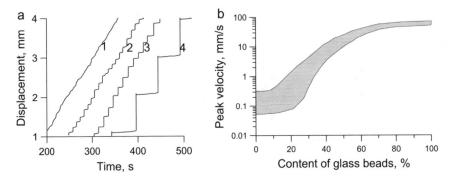

Fig. 10 Transformation of sliding regime. **a** Displacement versus time for experiments with filler consisting of quartz sand with different admixtures of glass beads. (1) pure sand, (2) 20% of glass beads, (3) 40% of glass beads, (4) pure glass beads. **b** The variations in peak velocity versus changes in the gouge texture

A certain slip mode realized in fault deformation is determined by structural, physical and mechanical properties of the filler. In a regular stick-slip repeated slip events take place with close parameters, while in irregular regimes stochastic events are observed, and their statistics obeys a power law.

For example, increasing the share of smooth grains (glass beads) leads to jamming of the granular layer and transition from stable sliding to stick-slip (Fig. 10). Earlier a similar result was obtained in [94]. In absence of smooth grains 90% of all the events have peak velocity of only 0.1–0.2 mm/s, while events with peak velocities exceeding 0.5 mm/s are absolutely absent. Thus, we can say that the steady sliding mode is realized.

When the share of glass beads reaches 40–50% of filler mass the motion becomes the stick slip with a relatively small value of stress drop. A further increase of the amount of smooth grains changes only the amplitude of displacement during a failure. When the filler consists only of glass beads the value of stress drop $\Delta\tau$ reaches approximately 15% of the maximum value of shear strength τ_0, and the peak velocity of blocks reaches the value of 60–80 mm/s.

The emergence of a certain slip mode is determined not only by the grain geometry, but also by their chemical and physical properties. For example, for mono-component low dispersion fillers consisting of angulated grains with ionic bonds between molecules (sodium chloride, corundum, magnesium oxide) realization of dynamic failures is much more probable than for fillers with covalent or metallic bonds between molecules (dry quartz sand, graphite and others) under similar loading parameters [95]. Probably the ionic bonds provide a stronger adhesive interaction of the filler grains in contact.

Moistening of the filler has an essential effect on the slip mode. It is well known that adding even a small amount of liquid to a granular media changes its collective stability [96, 97]. It is interesting that the value of stress drop changes non-monotonically as the viscosity of the moistening fluid grows (Fig. 11).

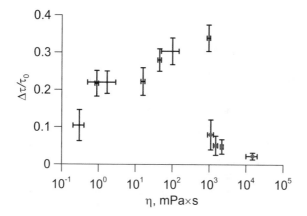

Fig. 11 Variations of shear stress drop versus fluid viscosity. Red symbols correspond to experiments with post-critical values of viscosity

Varying interstitial fluid viscosity, we traced the transformation of the sliding regime. The mass humidity of gouge was 1%. Variation of liquid viscosity η from 3×10^{-4} to 2×10^{1} Pa s caused a 15 times change of static stress drop. At the same time relative change of τ_s was only 25%. The value of $\Delta\tau/\tau_s$ gradually increases with increasing liquid viscosity up to the critical value $\eta_{cr} \approx 1$ Pa·s and then sharply drops by almost one order of magnitude, after which the value of $\Delta\tau/\tau_s$ gradually decreases with increasing viscosity up to 20 Pa s. At the range of after-critical values of viscosity the peak velocity decrease is approximately inversely proportional to the value of viscosity, and growth of slip duration is observed.

The presence of a small amount of interstitial liquid promotes jamming of the model fault. Probably, fault jamming is caused by the emergence of an additional intergrain force. The higher the force is, the higher is the value of elastic energy accumulated and, consequently, the higher are the stress drop and peak velocity during the slip episode. The cohesive force between gouge grains initially increases sharply for very small volumes of liquid because of the roughness of grains, further cohesiveness is effectively constant at least up to the stage when the liquid occupies about 35 percent of the available pore space [98, 99].

During the slip episode a reorganization of the meso-scale fault structure takes place, manifested in intergrain slipping. [95, 100]. At the stage of rest, when velocity is low, the intergrain contact dewets, which is accompanied by the accumulation of excess liquid in the pore space. But during the slip episode the liquid can penetrate into the grain contact at "critical" slip velocity conditions [101]. Probably, in the presented experiments, in the case of pre-critical viscosity ($\eta \leq 1$ Pa s) the slip velocity is not high enough for the "lubrication effect" to occur. But when $\eta > 1$ Pa s liquid penetration takes place, provoking damping of stick-slip events and the higher value of viscosity corresponds to the lower value of peak velocity. According to [102], when the fluid viscosity is about 10^5 Pa s, the occurrence of steady sliding must be observed.

The effect of filler properties on the sliding regime can be traced if one uses low dispersion mixtures of different grain materials as fracture filler. Variations of clay content in the moistening sand/clay mixture provided considerable changes of the fault slip modes. Figure 12 shows the dependences of slip episode parameters for different content of clay in the gouge moistened with water and glycerol. As clay content was increasing, transformation from stick-slip to steady sliding was observed. For the clay content of less than 15% a regular regime of fast slip episodes took place. Increasing clay content up to 20% led to the formation of irregular regime consisting of fast and slow slip episodes. The average values of slip episode parameters varied relatively slow (about an order of magnitude) in the range of clay content of $\nu = 0$–25%. When clay content was increased from 25 to 28% of mass, the decrease of average value of peak velocity by more than two orders of magnitude (from ~7 mm/s to ~0.04 mm/s) occurred and shear stress drop decreased 5 times. For mixtures with clay content of 30% rare slow slip episodes were observed, and for mixtures with clay content exceeding 35% the sliding became steady.

A similar pattern of changes in the fault behavior, but more drastic, was observed in experiments with glycerol. In the range of clay content of $\nu = 0$–1% the peak velocity decreased 3 times, further increase of clay content from 1 to 3% resulted in the drop of peak velocity by approximately 3 orders of magnitude. For mixtures with clay content exceeding 5% sliding became steady.

A wide spectrum of shear deformation regimes was realized in the presented experiments by changing the filler properties or the stiffness of loading—from dynamic failures to stable sliding. It should be emphasized that slow slip modes have all the phases intrinsic for stick-slip—acceleration, prolonged sliding, braking, stopping and the phase of rest. This suggests the idea that all the slip modes on faults span a continuum—they may be considered as a single set of phenomena.

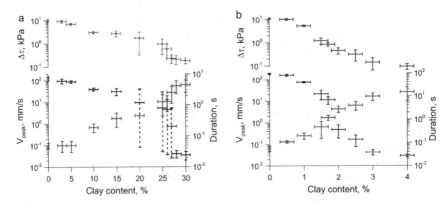

Fig. 12 Variation of shear stress drop (blue), peak slip velocity (black) and the slip episode duration (red) versus clay content in the gouge moistened with water (**a**) and glycerol (**b**). Vertical solid lines are standard deviations, vertical dashed lines show the range of variation of slip event parameters for irregular slip regimes

Interesting consequences follow from comparison of the laboratory experimental data to results of numerical simulation basing on the rate-and-state frictional law— Eqs. (4) and (5). By overrunning the constants in the rate-and-state relations a, b and D_c the best fit of simulated and experimental data was provided.

Figure 13a compares the simulated and experimental dependences of block slip velocity on time for the contact filled with quartz sand. For convenience, here and below the time is counted from the moment of velocity maximum. The following parameters of the rate-and-state model were used in simulations: $a = 0.0002$, $b = 0.00109$, $D_c = 10$ μm. The characteristics of slider model corresponded to experiment. The figure confirms that in the case of a pronounced stick-slip the simulated epure reproduces the experimental one rather well. Comparing the results of simulations to the experiment at the diagram force-displacement also demonstrates good correspondence. Calculations made for numerous experiments have shown that failure episodes during stick-slip ("laboratory earthquakes") can be satisfactorily simulated with the canonical rate-and-state law (4).

Attempts to simulate the slow sliding regimes (laboratory "slow earthquakes" and "episodes of slow slip events") have met certain difficulties. This is well seen in the example presented in Fig. 13b, which shows the results of calculation of the block slipping slowly along the contact filled with dry clay, the stiffness of loading element being equal to 17 kN/m. In this experiment the duration of sliding has increased, if compared to the contact filled with dry quartz sand, approximately by an order of the magnitude, along with a corresponding decrease of both the maximal and the average slip velocities (curve 1 in Fig. 13b).

We failed to reproduce such a signal in simulations involving only the canonical rate-and-state model (curve 2 in Fig. 13b). Varying the parameters, one can only fit the amplitude of the signal, but the "fullness" of the epure, i.e., the value of displacement, is controlled mainly by block mass and spring stiffness (parameters

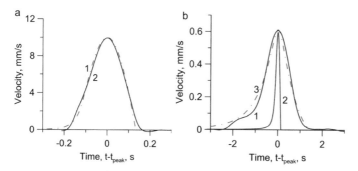

Fig. 13 **a** Block velocity versus time for the contact filled with quartz sand. R&S parameters of simulation: $K = 17$ kN/m; $\mu = 0.61$; $D_c = 10$ μm; $a = 2 \times 10^{-4}$; $b = 1.09 \times 10^{-3}$. Solid line is the simulation, dashed line is the experiment. **b** Block velocity versus time. The contact is filled with dry fire clay. The stiffness of the spring is $K = 17$ kN/m; (1)—experiment; (2)—simulation with $\eta_d = 0$ in Eq. (9); (3)—simulation with $\eta_d = 1000$ Pa s. Parameters of the R&S model: $\Delta_c = 6.5 \times 10^{-4}$; $\mu = 0.7$; $D_c = 50$ μm; $a = 1.2 \times 10^{-4}$; $b = 1.35 \times 10^{-3}$

rigorously defined for each experiment). We could only "dilute" the onset of the peak by changing D_c. The simulated peak of the dynamic failure for a given amplitude always remained essentially narrower than the one obtained in experiment.

We managed to better fit the results of simulating "slow" movements to experiment by introducing a term to the canonical R&S Eq. (4) that takes into account the emergence of additional resistance to shear produced by the "dynamic viscosity" of the contact:

$$F_s = \sigma_N \cdot S \cdot \left[\mu_0 + a \ln\left(\frac{|\dot{x}|}{u^*}\right) + b \ln\left(\frac{u^*\theta}{D_c}\right) \right] + \frac{\eta_d \cdot S \cdot \dot{x}}{d} \qquad (9)$$

where η_d is the factor of dynamic viscosity of interblock contact; S and d are the area and thickness of the contact zone. Results of simulations with account for the dynamic viscosity are presented in Fig. 13b by the line 3. One can see that introducing the "dynamic viscosity" of the contact into the equation of motion allows to reproduce the slow slip mode in simulations with satisfactory accuracy, too.

Equation (9) allowed to satisfactorily reproduce the character of motion in all the experiments by fitting the effective viscosity. Figure 14 gives examples of epures of slip velocities of the block in two experiments—with the contact filled with watered fire clay and with the contact filled with quartz sand with an admixture of 25% of talc. In both cases the slippages have long phases (~30–40 s) of gradual velocity increase, and then phases of deceleration of about the same duration. The maximum velocity of displacement decreases to 10–100 μm/s. For best fitting of simulated and experimental results the factor of effective viscosity η was increased by more than an order of a magnitude if compared to the simulation of experiment with dry fire clay (Fig. 13b).

It is well known that the generalized viscosity of a rigid body is not the property of the material itself, as for example, the viscosity of Newtonian fluid. This parameter

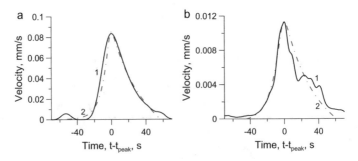

Fig. 14 Block velocity versus time for the slow slip mode. **a** contact filled with watered fire clay; (1) experiment; (2) simulation; Parameters of modified R&S model: $K = 10$ kN/m; $\Delta_c = 6.5 \times 10^{-4}$; $\mu = 0.56$; $D_c = 90$ мкм; $a = 1.0 \times 10^{-4}$; $b = 3.5 \times 10^{-3}$; $\eta_d = 2.76 \times 10^4$ Pa s. **b** contact filled with mixture of quartz sand (75%) and talc (25%). (1) experiment; (2) simulation; Parameters of modified R&S model (9): $K = 16.56$ kN/m; $\Delta_c = 2.3 \cdot 10^{-4}$; $\mu = 0.61$; $D_c = 18$ μm; $a = 1.0 \times 10^{-5}$; $b = 8.5 \times 10^{-4}$; $\eta_d = 3.94 \times 10^4$ Pa s

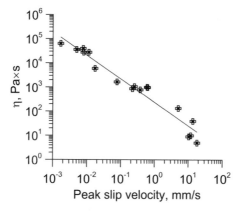

Fig. 15 Effective viscosity versus peak velocity of block displacement. The line shows the best fit by the Eq. (10) with $R^2 = 0.94$

is a characteristics of the rigid body rheology, depending on the specific time of deformation process, or, to be more precise, on the deformation rate.

Figure 15 shows the dependence of the factor of effective viscosity η_d on the peak velocity of block displacement. The effective viscosity was obtained by fitting the results of simulations to the ones of experiments, the simulations being performed according to the modified Rate-and-State law (9). Symbols show the results for different fillers and stiffnesses of the loading device. The line shows the best fit by the following power function:

$$\eta_d = 235 \cdot u_{max}^{-0.97} \tag{10}$$

Thus, the conducted laboratory and numerical experiments have demonstrated that the effective viscosity is a conventional parameter with the dimensionality of Pa s. This parameter is convenient to describe the alterations of fault slip modes. The factor of dynamic viscosity depends both on the properties of the filler and on the stiffness of the loading system. A verification of the results of simulations involving the rate-and-state model allows to conclude that to have the opportunity of modeling all the spectrum of fault slip modes the empirical Dieterich's law [81, 82] should be supplemented by a term that takes into account the emergence of additional resistance to shear produced by the dynamic viscosity of the fault interface, or, to be more precise, by the fact that the force of resistance to shear depends on the velocity of interblock motion. After this the episodes of slow slip observed in experiment can be reproduced with an appropriate accuracy. The obtained results agree with the data presented in the works [103, 104], where the dependence of viscosity factor on specific time of deformation is close to a linear one.

6 Radiation Efficiency of Slip Episodes

It is evident that in different slip modes different shares of deformation energy are emitted as seismic waves. It is a common thing that seismic events produced by the processes of deformation and destruction of a rock massif are described by two parameters: seismic moment (M_0) and seismic energy (E_s). The scalar seismic moment:

$$M_0 = \mu \cdot D_S \cdot S \tag{11}$$

is a generally recognized measure of the event size. This value does not depend on details of the process in the source, because it is determined by the asymptotics of the spectrum of displacements in the low-frequency band. It is proportional to the amplitude of the low-frequency band of the spectrum, and, provided that up-to-date equipment and processing methods are used, it can be estimated rather reliably. In Eq. (11), μ is the shear modulus of rock in the source, D_s is the displacement along the rupture, S is the area of the source. The divergence of estimations made by different authors for powerful earthquakes rarely exceeds 2–3 times.

The seismic energy E_s, i.e., the part of the deformation energy emitted in the form of seismic waves, on the contrary, is determined by the dynamics of rupture development and depends on velocity of rupture propagation, balance of energy in the source, etc. The value of seismic energy is usually determined by integrating the recorded vibrations.

In laboratory experiments the following product can be considered as the analogue of seismic moment: $M_{lab} = K \cdot D \cdot L$. Indeed, in nature the seismic moment is the product of shear force drop ΔF_s acting at the fault plane by the fault length Ls:

$$M_0 = \mu \cdot D_S \cdot L_s^2 = \mu \frac{D}{L_S} L_S^3 = \Delta\tau \cdot S \cdot L_S = \Delta F_s \cdot L_S \tag{12}$$

In this relation D_s is the displacement produced by slip events, $S \approx L_s^2$ is the area of rupture, $\Delta\tau$ is the tangential stress drop at the fault due to slip. For laboratory experiments with the spring-block model the single shear force drop can be written as $\Delta F = K \cdot D$, where K is the spring stiffness, D is the block displacement.

Then, comparing the energy budgets for earthquake and for laboratory slip event [23], we may consider the relation $e_{lab} = E_k / (K \cdot D \cdot l)$ as the analogue of the energy/moment ratio $e = E_s/M_0$, which is used in seismology to characterize the seismic efficiency of an earthquake. Here l is the block size.

It is convenient to estimate the share of deformation energy that transited to the kinetic energy of block motion (emitted in a laboratory "earthquake") using the experimental dependence of the measured shear force on block displacement. Examples of such dependences are given in Fig. 16.

When stresses, tangential to fault interface, reach the ultimate strength of the contact τ_0 and the condition (6) is true, the resistance to shear $\tau_{fr}(x) = R(x)/l^2$

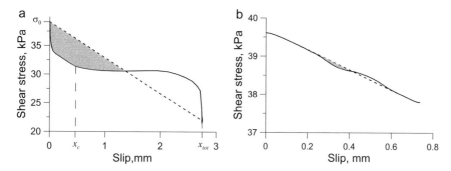

Fig. 16 Examples of experimental diagrams 'shear stress—displacement'. Solid line is the frictional resistance of the contact. Dashed line is the force applied by the spring. Shaded areas correspond to the values of kinetic energy of the block. **a** Fast mode (laboratory earthquake). The filler is quartz sand moistened with glycerol (1% of the mass). **b** Slow mode (laboratory slow slip event). The filler is watered mixture of quartz sand (70%) and clay (30%)

starts to decrease with displacement faster than the applied load $\tau_s(x) = F_s(x)/l^2$. As a result a slip starts, which is described by the equation:

$$m \cdot \frac{\partial^2 x}{\partial t^2} = \left[\tau_s(x) - \tau_{fr}(x)\right] \cdot l^2 \tag{13}$$

where m is the mass of moveable block, x is the relative displacement of blocks. After a certain displacement D_c has been reached (its value depends on roughness of fault walls, filler properties, etc.) the value of τ_{fr} comes to the residual (dynamic) value and stops changing.

After the following condition becomes true:

$$\int_0^x \left[\sigma_s(\chi) - \tau_{fr}(\chi)\right]d\chi = 0 \tag{14}$$

block slippage along the interface ceases ($x_{tot} = D_{tot}$ in Fig. 16a) and a new cycle of cumulating the deformation energy starts.

If the character of the dependence $\tau_{fr}(x)$ is such that the difference $\tau_s(x) - \tau_{fr}(x)$ becomes negative earlier than the minimum possible value of the friction force is reached (Fig. 16b), the contact actually doesn't come to the slip regime fully, which results in a small stress drop. Such an effect is most pronounced for fractures with high content of ductile grains (clay, talc).

The energy E_s, emitted during a slip event, was determined by integrating the difference of experimental dependences of the applied load $\tau_s(x)$ and resistance to shear $\tau_{fr}(x)$:

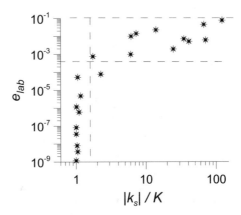

Fig. 17 Scaled emitted energy versus ratio of the stiffnesses of fault to enclosing massif. Symbols are the results of laboratory experiments. The area bounded by two horizontal dashed lines corresponds to dynamic failures. Slow events are to the left from the vertical dashed line

$$E_s = \int_0^{D_0} \left[\tau_s(\varsigma) - \tau_{fr}(\varsigma) \right] \cdot l^2 \cdot d\varsigma \tag{15}$$

where D_0 is the displacement, for which $\tau(D_0) - \tau_{fr}(D_0) = 0$. Shaded areas in the examples presented in Fig. 16 correspond to the values of E_s.

Thus, the ratio of the stiffness of fault to the one of enclosing massif (fracture and spring in laboratory experiment) $\psi = |k_s|/K$ determines not only the possibility of slippage, but its character as well. The dependence of scaled emitted energy e_{lab} on this parameter, plotted according to the results of laboratory experiments is shown in Fig. 17.

It is well seen that stick-slip takes place in a rather wide range of the values of ψ, while slow slip modes realize in a narrow area of the values of $|k_s|/K \sim 1 \div 2$. It means that at "brittle" faults, whose stiffnesses (the rate of decrease of resistance to shear) are rather high, the deformation energy releases solely in dynamic failures—normal earthquakes. Slow slip events can take place at faults with low stiffnesses. As massif stiffness in the crust alters slightly for different regions and different depths, it is the fault stiffness k_s that should be taken as governing parameter.

7 On Artificial Transformation of the Slip Mode

The main anthropogenic factors that can trigger movements on a prepared fault are variations of fluid-dynamic regime in the fault zone, effect of seismic vibrations, excavation and displacement of large amounts of rock in mining. Irrespective of what factor we are speaking about, the geomechanical criteria (6)–(8) formulated above should be true at a certain fault segment and in the enclosing rock massif. It makes sense to assume that the specific size of the segment should exceed the size of the, so called, "zone of earthquake nucleation"—the section where the rupture rate reaches the dynamic value [75]. Currently this value can be estimated only roughly. According to seismological data [103–107], the size of nucleation zone L_n can reach about 10% of the length of future rupture, i.e., for $M = 6$ $L_n \leq 1000$ m. Let us consider possibilities of realization of the formulated criteria under anthropogenic factors.

7.1 Changing the Fluid Dynamics

The effect of injection/withdrawal of fluid in/out of rock masses on seismicity has been studied in numerous works. One can find citations of corresponding publications, for example, in monographs and reviews [108, 109]. Rising pore or formation pressure and corresponding decrease of the effective Coulomb strength of faults and fractures is considered to be the main physical mechanism. However, there are many evidences [109, 110] that very weak variations of hydrostatic pressure (about millibar) affect seismicity. It is unlikely that the Coulomb model can be applicable here. It has been established in in situ experiments that the size of the area where parameters of sliding regime along a fault change can exceed several times the radius of the zone of pore pressure alterations [57]. It means that injection or withdrawal of fluid can change the characteristics of geomaterial.

The frictional parameter—the difference (a − b) from Eq. (4)—decreases abruptly, i.e., velocity weakening becomes more pronounced even when a small portion of fluid is injected. In the laboratory experiments described above adding fluid weighing 0.1% of the mass of laboratory fault filler is enough for a radical change of the character of sliding from creep to pronounced stick-slip [65].

Injecting fluid is, probably, one of the few possibilities to change the frictional parameter in situ. Such an effect was observed in the above laboratory experiments, in which the stick-slip of a granite block on a thin layer of granular material was investigated. Increasing fluid content, when its volume share had already reached $\zeta \approx 0.1\%$, resulted in a rather abrupt transition from stable sliding to stick-slip. In the presence of glycerol the maximum velocity of sliding increased more than 300 times. In case the humidity was further increased the regime stabilized and up to $\zeta \approx 10\%$ the deformation regime exhibited almost no dependence on content of fluid in the filler. Most likely, this phenomenon results from the character of interaction of particles in the fracture filler. After adding a small amount of fluid a thin film of sub-micron thickness emerged at the surfaces of filler grains. This film smoothed roughness and promoted good contacts between separate grains. The deformation regime on the fracture depends essentially on the effective viscosity of the fluid (Fig. 11). The performed estimations show that in nature colloidal films covering filler grains may emerge during the processes of aggregation—formation of enlarged structural elements as a result of adhesion of separate grains. Judging by the results of the performed experiments, viscosity of these films, i.e., the chemical content of clays, may affect the regime of fault deformation.

In laboratory tests described in [111], fast injection of fluid into the contact zone resulted in alteration of such sliding parameters as velocity of movement, stress drop, emitted energy. Alteration of pore pressure was negligible in comparison to the normal stress on the fault, i.e., it is the changed frictional properties of the contact that led to alteration of the slip mode. It should be emphasized that in this case the effect was observed after an appreciable amount of fluid had been injected (~20% of fracture volume, the porosity of the filler being about 35%), so, the fluid spread to about 80% of the contact area.

Injecting noticeably less amount of water in experiments on loading a mono-lithic heterogeneous sample [112] resulted in appreciable variations of the regime of acoustic emission and kinetics of the process of macro-destruction. These processes, probably, caused by physical and chemical interactions in fracture snouts of the type of Rehbinder effect, had no relation to the effect of alteration of parameters of velocity weakening during sliding, being discussed here.

Thus, though an anthropogenic change of fluid dynamics hypothetically can lead to triggering a dynamic movement, one should keep in mind that this change should involve a rather big fault area.

7.2 Effect of Seismic Vibrations

Triggering seismic events by vibrations of earthquakes that occurred at distances of hundreds and thousands of kilometers is an admitted example of the trigger effect [18]. As far as the data of numerous investigations of the so called "dynamic trigger-ing" is concerned, we should note that in most cases the minimum strain level needed for initiation is estimated to be about $\sim 5 \times 10^{-7}$–10^{-6}, though some authors give less estimations [113]. In most cases occurrence of dynamically triggered seismicity is linked to the effect of low-frequency surface waves with periods of 20–40 s. It is generally admitted that triggering with high-frequency body waves seems hardly probable. A detailed review of this topic is given, for example, in [18] and in the monograph [65].

Explosions turn out to be less effective from the point of view of triggering dynamic movements, than powerful distant earthquakes. The results of measuring parameters of seismic vibrations produced by ripple-fired explosions (for example, in [112]) allow to estimate the maximum particle velocity of the wave at different distances. For delay-fired explosions with typical parameters used in mining, maximum particle velocity at the distance of $R \sim 3 \div 5$ km does not exceed the value of $V_m \sim 0.3$–0.6 mm/s. The characteristic frequency of vibrations is about $f \sim 0.1$–0.5 Hz, while the duration of the wave-train may reach 100 s. It should be emphasized that increasing the integral energy of a delay-fired explosion leads (at distances of several kilometers) only to an increase of duration of the signal, but actually has no effect on the values of peak ground velocity (PGV). Therefore, dynamic stresses in

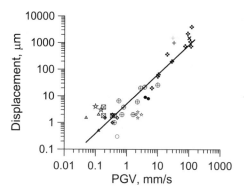

Fig. 18 Residual displacements registered on discontinuities versus the value of peak ground velocity (PGV) in seismic wave

seismic waves at the depth of 3–5 km can reach the values of only several kPa, and strains—$\sim 10^{-7}$.

Estimations [65] and the data of precise measurements of residual displacements on faults, produced by seismic vibrations from explosions, show (Fig. 18) that the expected value of a residual displacement on a fault under the effect of such a disturbance can be from sub-microns to tens of microns, and only in extreme cases it can reach 1 mm. Under such displacements a direct triggering of earthquake of an appreciable magnitude by seismic waves from explosions is hardly to be expected, because according to seismological data the value of critical displacement for an earthquake of average size (M ~ 6) is about 10 cm [75, 105]. It can be seen in Fig. 18 that for a noticeable effect to occur the PGV at an appreciable part of the fault area should be about 10 cm/s and even more. It means that the charge should be installed rather close to a fault that have reached the ultimately stressed state. An example of such an event is the April 16, 1989 earthquake in Khibiny Mountains [113].

7.3 Excavation and Displacement of Rock in Mining

Perhaps the most powerful anthropogenic triggering factor is the displacement of rock in mining. Considering this question is out of the frames of this article. We should only note that excavation of material from a large operating quarry with the sizes of kilometers in plane and hundreds of meters in depth leads to a reduction of Coulomb stresses up to 1 MPa at the planes of faults located at depths of several kilometers [114]. This value is negligible compared to the level of lithostatic stresses. But it may turn to be enough to trigger a seismogenic movement along a stressed fault. This is supported by the well known calculations of the field of static stresses in the vicinity of hypocenters of aftershocks of powerful earthquakes [115]. It is important to emphasize that for large quarries the size of the zone, in which the change of Coulomb stresses at fault plane exceeds 10^{-1} MPa, is essentially bigger than the size of the nucleation zone of an earthquake with magnitude $M \leq 6$.

Open (surface) mining operations in most cases only bring the moment of the earthquake closer, while underground developing of deposits changes effective elastic moduli of the rock massif in the vicinity of an active fault [116]. Therefore, it seems probable that without the anthropogenic effect the cumulated deformation energy would have released not through a dynamic movement (earthquake), but in another way, e.g., through slow creep or a series of slow slip events.

8 Conclusion

Discovery and classification of sliding regimes on faults and fractures, that are transitional from stable sliding (creep) to dynamic failure (earthquake), alter to a great extent the understanding of how the energy cumulated in the process of the Earth's crust deformation is released. Slow movements along faults are now perceived not as a special sort of deformations, but together with earthquakes span a continuum of slip modes.

Judging by the results of laboratory tests, small variations of material content of the fault principal slip zone can lead to an appreciable change of the part of seismic energy emitted in dynamic unloading of the adjacent section of the rock massif. Regimes of interblock sliding with values of scaled kinetic energy differing by several orders of magnitude, while differences of contact strengths are small, have been reproduced in experiments. The obtained results allow to conclude that the sliding regime and, in particular, the part of deformation energy that goes to seismic emission, is determined by the ratio of two parameters – stiffness of the fault and stiffness of the enclosing massif. A particular consequence of this statement is the well known condition of stick-slip occurrence.

It means that for episodes of slow movements to occur it is not obligatory that the fault is in a transitional state from brittle to plastic, as it happens either at large depths (25–45 km) between the seismogenic zone (beneath the zone) and the zone of stable sliding in subduction zones, where slow slip events are observed most often, or at shallow depths (~5 km) between the seismogenic zone (above the zone) and the surface zone of continuous creep [43]. Presence of watered clays in the principal slip zone, or of some amount of talc, which often substitutes the minerals of serpentite group along fracture walls in chemical reaction of serpentite with the silicon dioxide contained in thermal fluids, decreases harshly the shear stiffness of the fault, so that its value can be essentially lower than 10% of the normal one. In this situation the effective gradient of shear strength of the fault may turn to be close to the stiffness of the massif, which, can lead to occurrence of slow movements on faults. That is why similar effects can be observed at all depths in the crust.

As far as the possibility of artificial change of the slip mode on a fault is considered, the aim of external action should be not the removal of excessive stresses, but the decrease of fault zone stiffness. The change of sliding conditions should involve a rather big area, appreciably bigger that the size of the zone of earthquake nucleation. For example, pumping a clay-containing suspension into a fault zone may provide such a result, but this difficult scientific and engineering problem demands the development of a detailed technique of fulfilling the operation and methods to estimate its consequences.

Acknowledgements The work was carried out within the framework of the Russian State task No. AAAA-A17-117 112350020-9 (A.O.) and was supported by Russian Foundation for Basic Research (project no. 19-05-00378) (G.K. and D.P.).

References

1. Psakh'e SG, Popov VL, Shil'ko EV, Astafurov SV, Ruzhich VV, Smekalin OP, Bornjakov SA (2004) Method for controlling shifts mode in fragments of seismic-active tectonic fractures, Patent of invention RUS № 2273035. Application № 2004108514/28, 22.03.2004
2. Filippov AE, Popov VL, Psakh'e SG et al (2006) On the possibility of the transfer of the displacement dynamics in the block-type media under the creep conditions. Pis'ma v ZhTF 32(12):77–86 (in Russian)
3. Ruzhich VV, Smekalin OP, Shilko EV, Psakhie SG (2002) About nature of slow waves and initiation of displacements at fault regions. In: Proceedings of the international conference on new challenges in mesomechanics, Aalborg University, Aalborg, Denmark, pp 311–318
4. Psakhie SG, Horie Y, Ostermeyer GP, Korostelev SYu, Smolin AYu, Shilko EV, Dmitriev AI, Blatnik S, Špegel M, Zavšek S (2001) Movable cellular automata method for simulating materials with mesostructured. Theor Appl Fract Mech 37(1–3):311–334. https://doi.org/10.1016/S0167-8442(01)00079-9
5. Astafurov SV, Shilko EV, Psakhie SG (2005) Study into the effect of the stress state of block media on the response of active interfaces under vibration. Phyzicheskaya Mezomechanika 8(4):69–75 (in Russian)
6. Ruzhich VV, Psakhie SG, Chernykh EN, Federyaev OV, Dimaki AV, Tirskikh DS (2007) Effect of vibropulse action on the intensity of displacements in rock cracks. Phyzicheskaya Mezomechanika 10(1):19–24
7. Psakh'e SG, Ruzhich VV, Shil'ko et al (2005) On the influence of the interface state on the character of local displacements in the fracture-block and interface media. Pis'ma v ZhTF 31(16):80–87 (in Russian)
8. Dobretsov NL, Psakhie SG, Shilko EV, Astafurov SV, Dimaki AV, Ruzhich VV, Popov VL, Starchevich Y, Granin N G, Timofeev VYu (2007) Ice cover of lake Baikal as a model for studying tectonic processes in the Earth's crust. Doklady Earth Sci 413(1):155–159. https://doi.org/10.1134/S1028334X07020018
9. Ruzhich VV, Chernykh EN, Bornyakov SA, Psakhie SG, Granin NG (2009) Deformation and seismic effects in the ice cover of lake Baikal. Russian Geol Geophys 50(3):214–221. https://doi.org/10.1016/j.rgg.2008.08.005
10. Psakhie SG, Ruzhich VV, Shilko EV, Popov VL, Astafurov SV (2007) A new way to manage displacements in zones of active faults. Tribol Int 40(6):995–1003. https://doi.org/10.1016/j.triboint.2006.02.021
11. Kocharyan GG, Benedik AL, Kostyuchenko VN, Pavlov DV, Pernik LM, Svintsov IS (2004) The experience in affecting the fractured collector by the low-amplitude seismic vibration. Geoekologiya 4:367–377
12. Kocharyan GG, Vinogradov EA, Kishkina SB, Markov VK, Pavlov DV, Svintsov IS (2006) Deformation measurements on the Angara fault fragment (preliminary results). Dynamicheskie process vo vzaimodeystvuyushchich geospherach, 104–114 (in Russian)

13. Mukhamediev SA (2010) Prevention of strong earthquakes: Goal or utopia? Izvestiya. Phys Solid Earth 46:955–965. https://doi.org/10.1134/s1069351310110054
14. Sibson RH (2003) Thickness of the Seismic Slip Zone. Bull Seismol Soc Am 93(3):1169–1178. https://doi.org/10.1785/0120020061
15. Shipton ZK, Cowie PA (2001) Damage zone and slip-surface evolution over μm to km scales in high-porosity Navajo sandstone, Utah. J Struct Geol 23(12):1825–1844. https://doi.org/10.1016/S0191-8141(01)00035-9
16. Chester FM, Chester JS (1998) Ultracataclasite structure and friction processes of the Punchbowl fault, San Andreas system, California. Tectonophysics 295(1–2):199–221. https://doi.org/10.1016/S0040-1951(98)00121-8
17. Gomberg J, Blanpied ML, Beeler NM (1997) Transient triggering of near and distant earthquakes. Bull Seismol Soc Am 87(2):294–309
18. Hill DP, Prejean SG (2007) Dynamic triggering. In: Treatise on geophysics. Elsevier, Amsterdam, pp 257–291
19. Jeppson TN, Bradbury KK, Evans JP (2010) Geophysical properties within the San Andreas Fault Zone at the San Andreas Fault Observatory at Depth and their relationships to rock properties and fault zone structure. J Geophys Res 115:B12423. https://doi.org/10.1029/2010jb007563
20. Li YG, Chen P, Cochran ES, Vidale JE, Burdette T (2006) Seismic evidence for rock damage and healing on the San Andreas fault associated with the 2004 M6 Parkfield Earthquake. Bull Seismol Soc Am 96(4B):349–363. https://doi.org/10.1785/0120050803
21. Ikari MJ, Carpenter BM, Marone C (2016) A microphysical interpretation of rate- and statedependent friction for fault gouge. Geochem Geophy Geosyst 17(5):1660–1677. https://doi.org/10.1002/2016gc006286
22. Wu BS, McLaskey GC (2019) Contained laboratory earthquakes ranging from slow to fast. J Geophy Res—Solid Earth 124(10):270–210, 291. https://doi.org/10.1029/2019jb017865
23. Kocharyan GG, Novikov VA, Ostapchuk AA, Pavlov DV (2017) A study of different fault slip modes governed by the gouge material composition in laboratory experiments. Geophys J Int 208:521–528. https://doi.org/10.1093/gji/ggw409
24. Rodionov BN, Sizov IA, Tzvetkov VM (1986) Basics of geomechanics. Nedra, Moscow, p 301 (in Russian)
25. Choy GL, Boatwright JL (1995) Global patterns of radiated seismic energy and apparent stress. J Geophys Res-Solid Earth 100:18205–18228. https://doi.org/10.1029/95JB01969
26. Kocharyan GG, Ivanchenko GN, Koshkina SB (2016) Energy radiated by seismic events of different scales and geneses, Izvestiya. Phys Solid Earth 52:606–620. https://doi.org/10.1134/s1069351316040030
27. Savage JC, Svarc JL, Yu SB (2007) Postseismic relaxation and aftershocks. J Geophys Res-Solid Earth 112:B06406. https://doi.org/10.1029/2006JB004584
28. Nettles M, Ekstrom G (2004) Long-period source characteristics of the 1975 Kalapana, Hawaii, earthquake. Bull Seismol Soc Am 94:422–429. https://doi.org/10.1785/0120030090
29. Peng Z, Gomberg J (2010) An integrated perspective of the continuum between earthquakes and slow-slip phenomena. Nat Geosci 3:599–607. https://doi.org/10.1038/ngeo940
30. Lay T (ed) (2009) Seismological grand challenges in understanding earth's dynamic systems. Report to the National Science Foundation, IRIS Consortium, p 76
31. Dragert H, Wang K, James TS (2001) A silent slip event on the deeper Cascadia subduction interface. Science 292:1525–1528. https://doi.org/10.1126/science.1060152
32. Linde AT, Gladwin MT, Johnston MJS, Gwyther RL, Bilham RG (1996) A slow earthquake sequence on the San Andreas fault. Nature 383:65–68. https://doi.org/10.1038/383065a0
33. Ekström G, Nettles M, Abers GA (2003) Glacial earthquakes. Science 302:622–624
34. Das I, Zoback MD (2013) Long-period long-duration seismic events during hydraulic stimulation of shale and tight gas reservoirs—Part 1: Waveform characteristics. Geophysics 78(6). https://doi.org/10.1190/geo2013-0164.1
35. Ide S, Beroza GC, Shelly DR, Uchide T (2007) A scaling law for slow earthquakes. Nature 447:76–79. https://doi.org/10.1038/nature05780

36. Kanamori H, Hauksson E (1992) A slow earthquake in the Santa Maria Basin, California. Bull Seismol Soc Am 82:2087–2096
37. Thomas AM, Beroza GC, Shelly DR (2016) Constraints on the source parameters of low-frequency earthquakes on the San Andreas Fault. Geophys Res Lett 43:1464–1471. https://doi.org/10.1002/2015gl067173
38. Walter JI, Svetlizky I, Fineberg J, Brodsky EE, Tulaczyk S, Barcheck CG, Carter SP (2015) Rupture speed dependence on initial stress profiles: insights from glacier and laboratory stick-slip. Earth Planet Sci Lett 411:112–120
39. Gao H, Schmidt DA, Weldon RJ (2012) Scaling relationships of source parameters for slow slip events. Bull Seismol Soc Am 102(1):352–360. https://doi.org/10.1785/10120110096
40. Sacks IS, Suyehiro S, Linde AT, Snoke JA (1978) Slow earthquakes and stress redistribution. Nature 275:599–602
41. Hanks T, Kanamori H (1979) A moment magnitude scale. J Geophys Res 84:2348–2350
42. Little C (2013) M7 slow release earthquake under Wellington, GeoNet (Monday, 27 May 2013, 4:04 pm), Available from: https://Info.geonet.org.nz/display/quake/2013/05/27/
43. Wallace LM, Beavan J (2010) Diverse slow slip behavior at the Hikurangi subduction margin, New Zealand. J Geophys Res-Solid Earth 115(B12402)
44. Douglas A, Beavan J, Wallace L, Townend J (2005) Slow slip on the northern Hikurangi subduction interface, New Zealand. Geophys Res Lett 32(16):L16305. https://doi.org/10.1029/2005gl023607
45. Gomberg J (2018) Unsettled earthquake nucleation. Nat Geosci 11:463–464. https://doi.org/10.1038/s41561-018-0149-x
46. Handwerger AL, Rempel AW, Skarbek RM, Roering JJ, Hilley GE (2016) Rate-weakening friction characterizes both slow sliding and catastrophic failure of landslides. PNAS 113(37):10281–10286. https://doi.org/10.1073/pnas.1607009113
47. Kocharyan G, Ostapchuk A, Pavlov D, Markov V (2018) The effects of weak dynamic pulses on the slip dynamics of a laboratory fault. Bull Seismol Soc Am 108(5B):2983–2992. https://doi.org/10.1785/0120170363
48. Schurr B, Asch G, Hainzl S, Bedford J, Hoechner A, Palo M, Wang R, Moreno M, Bartsch M, Zhang Y, Oncken O, Tilmann F, Dahm T, Victor P, Barrientos S, Vilotte J (2014) Gradual unlocking of plate boundary controlled initiation of the 2014 Iquique earthquake. Nature 512:299–302. https://doi.org/10.1038/nature13681
49. Uchida N, Iinuma T, Nadeau R, Bürgmann R, Hino R (2016) Periodic slow slip triggers megathrust zone earthquakes in northeastern Japan. Science 351(6272):488–492. https://doi.org/10.1126/science.aad3108]
50. Obara K, Kato A (2016) Connecting slow earthquakes to huge earthquakes. Science 353(6296):253–257. https://doi.org/10.1126/science.aaf1512
51. Guglielmi Y, Cappa F, Avouac J-P, Henry P, Elsworth D (2015) Seismicity triggered by fluid injection-induced aseismic slip. Science 348(6240):1224–1226. https://doi.org/10.1126/science.aab0476
52. Wei S, Avouac J-P, Hudnut KW, Donnellan A, Parker JW, Graves RW, Helmberger D, Fielding E, Liu Z, Cappa F, Eneva M (2015) The 2012 Brawley swarm triggered by injection-induced aseismic slip. Earth Planet Sci Lett 422:115–122
53. Radiguet M, Perfettini H, Cotte N, Gualandi A, Valette B, Kostoglodov V, Lhomme T, Walpersdorf A, Cano EC, Campillo M (2016) Triggering of the 2014Mw7.3 Papanoa earthquake by a slow slip event in Guerrero, Mexico. Nat Geosci 9:829–833
54. Frank WB (2016) Slow slip hidden in the noise: the intermittence of tectonic release. Geophys Res Lett 43(19):10,125–10,133. https://doi.org/10.1002/2016GL069537
55. Kato A, Obara K, Igarashi T, Tsuruoka H, Nakagawa S, Hirata N (2012) Propagation of slow slip leading up to the 2011 Mw 9.0 Tohoku-Oki earthquake. Science 335:705–708. https://doi.org/10.1126/science.1215141
56. Johnston MJS, Borcherdt RD, Linde AT, Gladwin MT (2006) Continuous borehole strain and pore pressure in the near field of the 28 September 2004 M 6.0 Parkfield, California, earthquake: implications for nucleation, fault response, earthquake prediction, and tremor. Bull Seismol Soc Am 96(4B):S56–S72. https://doi.org/10.1785/0120050822

57. Ruzhich VV, Psakhie SG (2006) Challenge earthquake. Science first-hand 6(12):54–63 (in Russian)
58. Sadovsky MA, Bolkhovitinov LG, Pisarenko VF (1987) Deformation of the geophysical environment and the seismic process. Nauka, Moscow, 100 p (in Russian)
59. Hull J (1988) Thickness displacement relationships for deformation zones. J Struct Geol 10(4):431–435. https://doi.org/10.1016/0191-8141(88)90020-X
60. Sherman SI, Seminsky KZh, Bornyakov SA, Buddo VYu, Lobatskaya RM, Adamovich AN, Truskov VA, Babichev AA (1991) Faulting in the lithosphere. Strike slip zones. "Nauka", Siberian Branch, Novosibirsk 1:261 (in Russian)
61. Blenkinsop TG (1989) Thickness—displacement relationships for deformation zones: discussion. J Struct Geol 11(8):1051–1054. https://doi.org/10.1016/0191-8141(89)90056-4
62. Evans JP (1990) Thickness displacement relationships for fault zones. J Struct Geol 12(8):1061–1065. https://doi.org/10.1016/0191-8141(90)90101-4
63. Kocharyan GG (2014) Scale effect in seismotectonics. Geodyn Tectonophys 5(2):353–385. https://doi.org/10.5800/gt-2014-5-2-0133 (in Russian)
64. Kolyukhin D, Torabi A (2012) Statistical analysis of the relationships between faults attributes. J Geophys Res-Solid Earth 117:B05406. https://doi.org/10.1029/2011JB008880
65. Kocharyan GG (2016) Geomechanics of faults. GEOS, Moscow, p 424 (in Russian)
66. Brodsky EE, Ma KF, Mori J, Saffer DM et al (2009) Rapid response drilling: past, present, and future. In: ICDP/SCEC International Workshop of Rapid Response Fault Drilling in Tokyo, November 17–19, 2008, 30 p
67. Tanaka H, Fujimoto K, Ohtani T, Ito H (2001) Structural and chemical characterization of shear zones in the freshly activated Nojima fault, Awaji Island, southwest Japan. J Geophys Res Atmos 106(B5):8789–8810. https://doi.org/10.1029/2000jb900444
68. Zoback M, Hickman S, Ellsworth W (2010) Scientific drilling into the San Andreas fault zone. Eos Trans Am Geophys Union 91(22):197–199. https://doi.org/10.1029/2010eo220001
69. Chester FM, Chester JS, Kirschner DL, Schulz SE, Evans JP (2004) Rheology and deformation of the lithosphere at continental margins. Columbia University Press, New York, pp 223–260
70. Faulkner DR, Jackson CAL, Lunn RJ, Schlische RW, Shipton ZK, Wibberley CAJ, Withjack MO (2010) A review of recent developments concerning the structure, mechanics and fluid flow properties of fault zones. J Struct Geol 32(11):1557–1575. https://doi.org/10.1016/j.jsg.2010.06.009
71. Tchalenko JS (1970) Similarities between shear zones of different magnitudes. Geol Soc Am Bull 81:1625–1640. https://doi.org/10.1130/0016-7606(1970)81%5b1625:sbszod%5d2.0.co;2
72. Schulz SE, Evans JP (2000) Mesoscopic structure of the Punchbowl Fault, Southern California and the geologic and geophysical structure of active strike-slip faults. J Struct Geol 22:913–930
73. Evans JP, Chester FM (1995) Fluid-rock interaction in faults of the San Andreas system: inferences from San Gabriel fault rock geochemistry and microstructures. J Geophys Res Atmos 1001:13007–13020
74. Chester JS, Chester FM, Kronenberg A. K. (2005). Fracture surface energy of the Punchbowl fault, San Andreas system. Nature 437:133–136
75. Scholz CH (2002) The mechanics of earthquakes and faulting. Cambridge University Press, Cambridge, 496 p
76. Kocharyan GG, Kishkina SB, Ostapchuk AA (2010) Seismic picture of a fault zone. What can be gained from the analysis of the fine patterns of spatial distribution of weak earthquake centers? Geodyn Tectonophys 1(4):419–440. http://dx.doi.org/10.5800/GT2010140027 (in Russian)
77. Fagereng A, Sibson RH (2010) Mélange rheology and seismic style. Geology 38:751–754. https://doi.org/10.1130/G30868.1
78. Saffer DM, Wallace LM (2015) The frictional, hydrologic, metamorphic and thermal habitat of shallow slow earthquakes. Nat Geosci 8:594–600. https://doi.org/10.1038/ngeo2490
79. Kato N, Hirono T (2016) Heterogeneity in friction strength of an active fault by incorporation of fragments of the surrounding host rock. Earth, Planets and Space 68:134. https://doi.org/10.1186/s40623-016-0512-3

80. Scuderi MM, Collettini C, Marone C (2017) Frictional stability and earthquake triggering during fluid pressure stimulation of an experimental fault. Earth Planet Sci Lett 477:84–96. https://doi.org/10.1016/j.epsl.2017.08.009
81. Dieterich JH (1979) Modeling of Rock Friction: 1. Experimental results and constitutive equations. J Geophys Res-Solid Earth 84(B5):2161–2168
82. Ruina AL (1983) Slip instability and state variable friction laws. J Geophys Res-Solid Earth 88(B12):10359–10370
83. Scholz CH, Campos J (2012) The seismic coupling of subduction zones revisited. J Geophys Res-Solid Earth 117(B5):B05310. https://doi.org/10.1029/2011JB009003
84. Ikari MJ, Marone C, Saffer DM, Kopf AJ (2013) Slip weakening as a mechanism for slow earthquakes. Nat Geosci 6(6):468–472. https://doi.org/10.1038/ngeo1818
85. Uchida N, Bürgmann R (2019) Repeating earthquakes. Ann Rev Earth Planet Sci 47(1):305–332. https://doi.org/10.1146/annurev-earth-053018-060119
86. Wesnousky SG (2006) Predicting the endpoints of earthquake ruptures. Nature 444(7117):358–360. https://doi.org/10.1038/nature05275
87. Di Toro G, Hirose T, Nielsen S, Pennacchioni G, Shimamoto T (2006) Natural and experimental evidence of melt lubrication of faults during earthquakes. Science 311:647–649
88. Brodsky EE, Kanamori H (2000) Elastohydrodynamic lubrication of faults. J Geophys Res-Solid Earth 106:16357–16374
89. Rice JR (2006) Heating and weakening of faults during earthquake slip. J Geophys Res-Solid Earth 111(B5):B05311. https://doi.org/10.1029/2005JB004006
90. Noda H (2008) Frictional constitutive law at intermediate slip rates accounting for flash heating and thermally activated slip process. J Geophys Res-Solid Earth 113:B09302. https://doi.org/10.1029/2007jb005406
91. Brantut N, Schubnel A, Rouzaud J-N, Brunet F, Shimamoto T (2008) High-velocity frictional properties of a clay bearing, fault gouge and implications for earthquake mechanics. J Geophys Res-Solid Earth 113:B10401. https://doi.org/10.1029/2007jb005551
92. Kirkpatrick JD, Rowe CD, White JC, Brodsky EE (2013) Silica gel formation during fault slip: evidence from the rock record. Geology 41(9):1015–1018. https://doi.org/10.1130/G34483.1
93. Sobolev GA, Vettegren VI, Mamalimov RI, Shcherbakov IP, Ruzhich VV, Ivanova LA (2015) A study of nanocrystals and the glide-plane mechanism. J Volcanol Seismol 9(3):151–161. https://doi.org/10.1134/S0742046315030057
94. Mair K, Frye KM, Marone C (2002) Influence of grain characteristics on the friction of granular shear zones. J Geophys Res-Solid Earth 107(B10):2219. https://doi.org/10.1029/2001jb000516, 2002
95. Ostapchuk AA, Pavlov DV, Markov VK, Krasheninnikov AV (2016) Study of acoustic emission signals during fracture shear deformation. Acoust Phys 62:505–513. https://doi.org/10.1134/S1063771016040138
96. Hornbaker D, Albert R, Albert I, Barabási A-L, Schiffer P (1997) What keeps sandcastles standing? Nature 387:765. https://doi.org/10.1038/42831
97. Bocquet L, Charlaix É, Restagno F (2002) Physics of humid granular media. Comptes Rendus Physique 3(2):207–215. https://doi.org/10.1016/s1631-0705(02)01312-9
98. Kurdolli A (2008) Sticky sand. Nat Mater 7:174–175. https://doi.org/10.1038/nmat2131
99. Scheel M, Seemann R, Brinkmann M, Di Michiel M, Sheppard A, Breidenbach B, Herminghaus S (2008) Morphological clues to wet granular pile stability. Nat Mater 7:189–193. https://doi.org/10.1038/nmat2117
100. Morgan J, Boettcher M (1999) Numerical simulations of granular shear zones using the distinct element method. J Geophys Res-Solid Earth 104(B2):2703–2719. https://doi.org/10.1029/1998JB900056
101. Martin A, Clain J, Buguin A, Broachard-Wyart F (2002) Wetting transitions at soft, sliding interfaces. Phys Rev E 65:031605. https://doi.org/10.1103/PhysRevE.65.031605
102. Reber J, Hayman NW, Lavier LL (2014) Stick-slip and creep behavior in lubricated granular material: insights into the brittle-ductile transition. Geophys Res Lett 41:3471–3477

103. Qi Ch, Wang M, Qian Q, Chen J (2007) Structural hierarchy and mechanical properties of rocks. Part 1. Structural hierarchy and viscosity. Phys Mesomech 10(1–2):47–56
104. Khristoforov BD (2010) Rheological properties of solids in a wide range of deformation times. Fiz Mesomech 13(3):111–115 (in Russian)
105. Ellsworth WL, Beroza GC (1995) Seismic evidence for an earthquake nucleation phase. Science 268:851–855
106. Ide S, Takeo M (1997) Determination of constitutive relations of fault slip based on seismic wave analysis. J Geophys Res 102(B12):27.379–27.391
107. Papageorgiou AS, Aki KA (1983) Specific barrier model for the quantitative description of inhomogeneous faulting and the prediction of strong ground motion, part II, Applications of the model. Bull Seismol Soc Am 73(3):953–978
108. Adushkin VV, Turuntaev SB (2015) Technogenic seismicity—induced and trigger. IDG RAS, Moscow, 364 p (in Russian)
109. Foulger GR, Wilson MP, Gluyas JG, Julian BR, Davies RJ (2017) Global review of human-induced earthquakes. Earth-Sci Rev 178:438–515. https://doi.org/10.1016/j.earscirev.2017.07.008
110. Djadkov PG (1997) Induced seismicity at the Lake Baikal: principal role of load rate, The 29th General Assembly of the International Association of Seismology and Physics of the Earth's Interior, August 18–28, 1997, Thessaloniki, Greece, Abstracts, p 359
111. Kocharyan GG, Ostapchuk AA, Martynov VS (2017) Alteration of fault deformation mode under fluid injection. J Min Sci 53:216–223. https://doi.org/10.1134/S1062739117022043
112. Sobolev GA, Ponomarev AV (2011) Dynamics of fluid-triggered fracturing in the models of a geological medium, Izvestiya. Phys Solid Earth 47(10C):902–918. https://doi.org/10.1134/s1069351311100119
113. Sobolev GA, Zakrzhevskaya NA, Sobolev DG (2016) Triggering of repeated earthquakes, Izvestiya. Phys Solid Earth 52(2C):155–172
114. Kocharyan GG, Kulikov BI, Pavlov DV (2019) On the influence of mass explosions on the stability of tectonic faults. J Min Sci (in press)
115. Kremenetskaya EO, Trjapitsin VM (1995) Indused seismicity in the Khibiny Massif (Kola Peninsula). PAGEOPH (Pure Appl Geophy) 145(1):29–37
116. Kocharyan GG, Kishkina SB (2018) Initiation of tectonic earthquakes caused by surface mining. J Min Sci 54:744–750. https://doi.org/10.1134/s1062739118054844

Multilayer Modelling of Lubricated Contacts: A New Approach Based on a Potential Field Description

Markus Scholle, Marcel Mellmann, Philip H. Gaskell, Lena Westerkamp, and Florian Marner

Abstract A first integral approach, derived in an analogous fashion to Maxwell's use of potential fields, is employed to investigate the flow characteristics, with a view to minimising friction, of shear-driven fluid motion between rigid surfaces in parallel alignment as a model for a lubricated joint, whether naturally occurring or engineered replacement. For a viscous bilayer arrangement comprised of immiscible liquids, it is shown how the flow and the shear stress along the separating interface is influenced by the mean thickness of the layers and the ratio of their respective viscosities. Considered in addition, is how the method can be extended for application to the more challenging problem of when one, or both, of the layers is a viscoelastic material.

Keywords Lubrication theory · Finite elements · Complex variable analysis · Shear flow · Immiscible liquids · Viscoelasticity · Surface contouring · Joints · Bearings

1 Introduction and Model Assumptions

Hydrodynamic lubrication [1, 2] is one of the classic topics contributing to the field of fluid mechanics that is of considerable relevance; for example, in technological terms in connection with the design of lubricated contacts such as plain bearings or ball joints [3] and more specifically, in bio-engineering terms, in the context of joint replacement [4]. In the present work, a model, based on a potential field description is

M. Scholle (✉) · M. Mellmann · L. Westerkamp
Institute for Flow in Additively Manufactured Porous Media, Heilbronn University, Heilbronn, Germany
e-mail: markus.scholle@hs-heilbronn.de

P. H. Gaskell
Department of Engineering, Durham University, Durham, UK

F. Marner
Inigence GmbH, Bretzfeld, Germany

G.-P. Ostermeyer et al. (eds.), *Multiscale Biomechanics and Tribology of Inorganic and Organic Systems*, Springer Tracts in Mechanical Engineering, https://doi.org/10.1007/978-3-030-60124-9_16

presented for the case of a lubricated contact consisting of two contiguous immiscible layers, located one on top of the other. In a general sense, both layers can be considered as purely viscous liquids and/or viscoelastic layers, depending on the application of interest.

Different aspects of film flows involving two or more immiscible liquid layers have been investigated in [5–7] with a focus on both confined shear-driven flows, the topic of interest here, and free-surface flows due to their relevance in the production and deposition of functional coatings [8]. Current studies addressing the material modelling of articular cartilage, see for example [9], reveal an appreciable complexity of material behaviour, which among other things includes chemo-elastic effects and anisotropy; here standard simplified viscoelastic models are considered as a first step.

The model problem considered is that of an idealised system of two-dimensional steady Couette flow, as illustrated in Fig. 1: the lower, flat surface translates with speed v_0 while the upper, corrugated, one remains stationary. The region separating the surfaces, which are in parallel alignment, is taken to be filled with contiguously contacting, immiscible liquids or viscoelastic layers, having different dynamic viscosity η and Young's modulus E; the case shown is for a bilayer system, mimicking the more general case of a joint in which the synovia meets a protective layer exhibiting viscoelastic properties. While at outset the general case is formulated, the focus of the results presented and discussed subsequently is restricted to the simpler case of two Newtonian liquid layers.

The two-phase system is defined in terms of a number of non-dimensional parameters, the most relevant of these being the ratios of the two layer thicknesses, H_1/H_2, and of their fluid properties, namely the viscosities η_1/η_2 and the densities ρ_1/ρ_2. The shape of the periodic profile defining the upper corrugated surface is given by the function:

$$b(x) = -2a\frac{\ln(2 - 2s\cos x) - \ln\left(1 + \sqrt{1 - s^2}\right)}{\ln(1 + s) - \ln(1 - s)}. \tag{1}$$

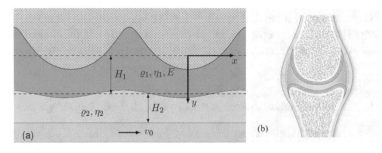

(a)   (b)

Fig. 1 Model of a periodic two-phase system (**a**), consisting of a layer of viscous liquid lying on top of a viscoelastic one, both confined between non-compliant rigid surfaces; the lower one is flat and translating while the upper one is profiled/contoured and stationary. The model is based on the natural form of biological joints (**b**), [10] and is a key feature of the planned investigation

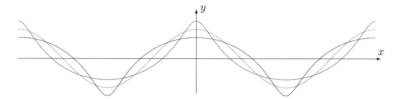

Fig. 2 Surface profile shapes obtained for different values of the shape parameters. The red curve, with $s = 0.9$, results in peak asperities while the blue one, with $s = -0.9$, (phase shift π) leads to smoother ones. The cosine function (green curve) is shown as a reference for the limit $s \to 0$

It depends on two parameters: the dimensionless amplitude $a = 2\pi A/\lambda$ and $s \in (-1, 1)$ determining the shape. Figure 2 illustrates three potential shapes, demonstrating the role of the shape parameter s. In the limiting case $s \to 0$ the surface shape becomes a cosine function, $b(x) = -a \cos x$, for positive values of s the corrugations form pronounced peak asperities while for negative values they result in a smoother levelling.

If the upper layer is assumed to be viscoelastic, the Deborah number $De = \eta_1 v_0/E\lambda$ enters the problem as an additional parameter, while the Reynolds number is so small that it can be taken to be zero.

The focus here, from a bio-engineering viewpoint, is the determination of the normal and shear stresses along the interface separating the two layers: normal stresses, especially when periodically varying with time, have a positive influence on the nutrient supply to the articular cartilage by promoting the exchange of substances between the nutrient-containing synovial fluid and the partially porous articular cartilage; shear stresses, on the other hand, cause wear and thus have a detrimental effect [4].

2 Mathematical Formulation

The field equations for the different layer types, together with the boundary and interface conditions, are formulated below, making use of the first integral approach [11–13]. The benefits of the latter are: (i) an elegant implementation for arbitrary rheological models; (ii) a beneficial form for the dynamic condition at the interface separating the two layers.

2.1 Field Equations for Newtonian Layer Types

When one, or both layers, is assumed to be an incompressible viscous Newtonian liquid, resolving the associated flow requires a solution of the governing Navier-Stokes equations and accompanying continuity equation:

$$\rho(\vec{v} \cdot \nabla)\vec{v} = -\nabla p + \eta \nabla^2 \vec{v}, \tag{2}$$

$$\nabla \cdot \vec{v} = 0, \tag{3}$$

respectively, to obtain the velocity $\vec{v} = v_x(x, y)\vec{e}_x + v_y(x, y)\vec{e}_y$ and scalar pressure $p = p(x, y)$ fields. Defining a complex coordinate and complex velocity as:

$$\xi = x + iy, \tag{4}$$

$$v = v_x + iv_y, \tag{5}$$

and introducing the scalar potential Φ as an auxiliary unknown, facilitates integration of Eq. (2), leading finally to the following two complex field equations [14]:

$$\frac{\rho}{2}v^2 = 2\eta \frac{\partial v}{\partial \bar{\xi}} - 4\frac{\partial^2 \Phi}{\partial \bar{\xi}^2}, \tag{6}$$

$$\frac{\rho}{2}\bar{v}v = -p + 4\frac{\partial^2 \Phi}{\partial \bar{\xi} \partial \xi}. \tag{7}$$

The continuity equation is fulfilled identically on introduction of the streamfunction Ψ, according to $v = -2i\partial\Psi/\partial\bar{\xi}$.

Note that the above approach originates from two-dimensional elasticity theory [15, 16], where the scalar potential Φ plays the role of Airy's stress function. This prominent complex variable approach was adopted subsequently by several authors, e.g. [17], for the solution of Stokes flow problems, before being generalised for the case of inertial flows in [11].

2.2 Field Equations for Viscoelastic Layer Types

A complex variable formulation of the governing evolution equations, as in the case of Newtonian liquids and Hookean materials, is also available for any generalised material [18] with an elegant embodiment of the respective rheological model equations in the first integral approach. By applying the complex transformations (4) and (5) to the general momentum balance in place of the Navier-Stokes equations and following the procedure described in [14], one obtains the following complex equations:

$$\frac{\rho}{2}v^2 = \underline{\sigma} - 4\frac{\partial^2 \Phi}{\partial \bar{\xi}^2}, \tag{8}$$

Table 1 Substitution rules for the implementation of generalised rheological models

Rheological models	σ	$\dot\sigma$	E	$\dot\varepsilon$
w.r.t. Eq. (8)	$\underline{\sigma}$	$v\frac{\partial\underline\sigma}{\partial\xi}+\bar v\frac{\partial\underline\sigma}{\partial\bar\xi}$	$2\frac{\partial u}{\partial\xi}$	$2\frac{\partial}{\partial\xi}$
w.r.t. Eq. (9)	σ_0	$v\frac{\partial\sigma_0}{\partial\xi}+\bar v\frac{\partial\sigma_0}{\partial\xi}$	$\frac{\partial u}{\partial\xi}+\frac{\partial\bar u}{\partial\bar\xi}$	$\frac{\partial v}{\partial\xi}+\frac{\partial\bar v}{\partial\bar\xi}$

$$\frac{\rho}{2}\bar v v = \sigma_0 + 4\frac{\partial^2\Phi}{\partial\bar\xi\partial\xi}, \tag{9}$$

where $\sigma_0 = \frac{\sigma_x+\sigma_y}{2}$ is the isotropic part of the stress tensor of the respective material, while the complex quantity $\underline\sigma = \frac{\sigma_x-\sigma_y}{2}+i\tau_{xy}$ is its traceless part. The adoption of a corresponding rheological model, given as a relationship between the stress σ, the deformation ε, together with their time derivatives, can be implemented by formal substitutions according to the rules listed in Table 1.

Here $u = u_x + iu_y$ denotes the complex displacement field. Note also, the kinematic constraint $v = v\frac{\partial u}{\partial\xi} + \bar v\frac{\partial u}{\partial\bar\xi}$ between the velocity and displacement fields.

Implementation of the above methodology is demonstrated for a Kelvin-Voight model $\sigma = E\varepsilon + \eta\dot\varepsilon$ with Young's modulus E and viscosity η [19]. Following substitution, according to Table 1, Eqs. (8) and (9) become:

$$\frac{\rho}{2}v^2 = 2E\frac{\partial u}{\partial\bar\xi} + 2\eta\frac{\partial v}{\partial\bar\xi} - 4\frac{\partial^2\Phi}{\partial\bar\xi^2}, \tag{10}$$

$$\frac{\rho}{2}\bar v v = E\left(\frac{\partial u}{\partial\xi} + \frac{\partial\bar u}{\partial\bar\xi}\right) + \eta\left(\frac{\partial v}{\partial\xi} + \frac{\partial\bar v}{\partial\bar\xi}\right) + 4\frac{\partial^2\Phi}{\partial\bar\xi\partial\xi}, \tag{11}$$

which are generalised forms of Eqs. (6) and (7), respectively; the latter equations result in the limit case of a viscous liquid when $E = 0$.

Note that the above procedure can be applied to any arbitrary rheological model, using the formal substitution rules in Table 1 with respect to the general complex momentum Eqs. (8) and (9).

2.3 Boundary and Interface Conditions

Along both the stationary profiled surface, given by the function $b(x)$, and the moving flat surface no-slip/no-penetration conditions have to be fulfilled:

$$u(x, b(x)) = 0, \tag{12}$$

$$v(x, H_1 + H_2) = v_0. \tag{13}$$

Periodic boundary conditions at inflow and outflow, to the left and right, are enforced. At the interface separating the layers, $y = f(x)$, the shape of which is unknown a priori, the velocity field has to be continuous:

$$[[v]] = 0, \tag{14}$$

with the double square brackets denoting the discontinuity of the associated term. Moreover, the kinematic boundary condition there:

$$v_y - f'(x)v_x = 0, \tag{15}$$

can be used to determine the shape of the interface. Finally, the dynamic interface condition:

$$\left[\underline{\underline{T}}_2 - \underline{\underline{T}}_1\right] \cdot \vec{n} = \sigma_S \kappa \vec{n}, \tag{16}$$

accounts for the equality in stress at the interface; σ_S is the interfacial tension, κ the curvature, $\vec{n}$ the vector normal to the interface and $\underline{\underline{T}}_{1,2}$ the stress tensor associated with the materials forming the respective layers. Using a conventional description, the treatment of the dynamic interface condition is a challenging task, since the viscous or viscoelastic stresses present lead to combinations of different derivatives of different components of the displacement and velocity field and therefore to a mathematically unfavourable form. Using the first integral approach, the unfavourable terms involved in the interface condition (16) can be replaced by second order derivatives of the scalar potential Φ and the interface condition integrated [14], leading finally to the simple jump condition:

$$\left[\left[\frac{\partial \Phi}{\partial \bar{\xi}}\right]\right] = \frac{\sigma_S n}{4}, \tag{17}$$

where $n = n_x + i n_y$ is the complex equivalent of the normal vector. It is shown in [14] that, after re-transformation to a real-valued representation, the two conditions resulting from (17) can be formulated in standard Dirichlet/Neumann form. Among various other benefits, the reduction of the complicated dynamic interface condition (16) to the significantly simpler jump condition (17) for the potential Φ justifies its introduction as an additional auxiliary field and demonstrates its use.

3 Methods of Solution

The problem is formulated in three different ways: via a Lubrication approximation allowing for an analytical solution; a numerical finite-element FE approach; a semi-analytical one benefitting from the use of complex variables. The FE approach enables Reynolds number effects to be investigated and, more generally, the validity of the two simpler models to be assessed. In terms of the adopted first integral approach, the standard mathematical form of the jump conditions (17) is advantageous, since the resulting friction coefficient can be calculated conveniently from the auxiliary potential field Φ based on the first integral formulation without the need to approximate velocity derivatives in a post-processing step as would be the case if a primitive variable formulation had been adopted.

3.1 Lubrication Approximation

The first integral Eq. (6) can be simplified based on a Lubrication approximation, noting that the same applies to the Navier-Stokes equations, [20–23], leading to a single equation for the local film thickness; the velocity field to leading order being locally parabolic. The requirement underpinning its applicability in the case of surface contours exhibiting rapid changes is that the commensurate interface disturbance is slowly varying.

Applying the Lubrication approximation to the real-valued decomposed Eq. (6) leads to:

$$\frac{\rho}{2}\left(v_x^2 - v_y^2\right) + \frac{\partial}{\partial x}\left[\frac{\partial \Phi}{\partial x} - \eta v_x\right] - \frac{\partial}{\partial y}\left[\frac{\partial \Phi}{\partial y} - \eta v_y\right] = 0,$$

$$\rho v_x v_y + \frac{\partial}{\partial y}\left[\frac{\partial \Phi}{\partial x} - \eta v_x\right] + \frac{\partial}{\partial x}\left[\frac{\partial \Phi}{\partial y} - \eta v_y\right] = 0. \tag{18}$$

Next, by neglecting $\partial v_y/\partial x$ compared to $\partial v_x/\partial y$ and omitting inertial terms, the set of PDEs is reduced to one that can be solved by successive integration, as shown in detail in [24], leading to the following general solution:

$$v_x = \frac{1}{2}F_1''(x)y^2 + F_2'(x)y + F_3(x), , \tag{19}$$

$$\psi = F_1''(x)\frac{y^3}{6} + F_2'(x)\frac{y^2}{2} + F_3(x)y + F_4(x), \tag{20}$$

for the velocity v_x and the streamfunction ψ, involving four integration functions $F_1(x)$, $F_2(x)$, $F_3(x)$ and $F_4(x)$; while the gradient of the potential results in:

$$\frac{2}{\eta}\frac{\partial \Phi}{\partial x} = \frac{1}{2}F_1''(x)y^2 + F_2'(x)y + F_3(x) + F_1(x),$$

$$\frac{2}{\eta}\frac{\partial \Phi}{\partial y} \approx F_1'(x)y + F_2(x). \tag{21}$$

Note that four integration functions occur with respect to each layer m, thus for a bilayer system eight functions $F_{mi}(x)$ with $m = 1, 2$ and $i = 1, \ldots, 4$ have to be considered. Together with the shape of the interface, $f(x)$, nine unknown functions have to be determined from the boundary and interface conditions (12)–(15) and (17), considering that each of the complex conditions delivers two real-valued conditions after decomposition into real and imaginary parts.

By successive elimination of unknown functions, the resulting set of nine ODEs can be reduced to three nonlinear ODEs for the functions $F_{11}(x)$, $F_{21}(x)$ and $f(x)$. In non-dimensional form, taking $L = \lambda/2\pi$ as the characteristic length with λ the wavelength and v_0 as a characteristic velocity, the resulting equations read:

$$\frac{2Q_2}{h - f(x)} - \frac{2Q_1}{f(x) - b(x)} - \frac{[f(x) - b(x)]^2}{6} F_{11}''(x) + \frac{[h - f(x)]^2}{6} F_{21}''(x) = 1 \tag{22}$$

$$(1 - n)\left[\frac{2Q_1}{f(x) - b(x)} + \frac{[f(x) - b(x)]^2}{6} F_{11}''(x)\right] + F_{21}(x) - nF_{11}(x) = 0 \tag{23}$$

$$\frac{Q_2}{[h - f(x)]^2} + \frac{nQ_1}{[f(x) - b(x)]^2} + n\frac{f(x) - b(x)}{3} F_{11}''(x)$$
$$+ \frac{h - f(x)}{3} F_{21}''(x) = \frac{1}{h - f(x)} \tag{24}$$

where $Q_1 = \psi(x, f(x))$ and $Q_2 = \psi(x, h) - Q_1$ are the constant partial flow rates of the two different layers and $n = \eta_1/\eta_2$ is the viscosity ratio. In contrast to the conventional Lubrication model derived starting from the original Navier-Stokes equations leading to the well-known Reynold's equation [3], the first integral approach yields the equation set (22)–(24) comprising three equations for the interface shape $f(x)$ and the two functions $F_{11}(x)$, $F_{21}(x)$ which are connected with the curvatures of the respective velocity profiles within the two layers. Having once determined these three functions by solving (22), (23) and (24), the streamfunction results for the two layers, as:

$$\psi^{(1)} = \frac{[y - b(x)]^2}{2}\left[\frac{y - f(x)}{3} F_{11}''(x) + \frac{2Q_2}{[f(x) - b(x)]^2}\right],$$
$$\psi^{(2)} = Q_1 + \left[1 - \frac{(h - y)^2}{[h - f(x)]^2}\right][Q_2 - h + f(x)]$$
$$+ \frac{(h - y)^2}{6}[y - f(x)]F_{21}''(x), \tag{25}$$

where the numbers 1, 2 in the brackets denote the respective layer.

The above set of nonlinear equations, (22), (23) and (24), can be solved numerically in their given form or asymptotically after linearization, as shown below. The latter approach is briefly demonstrated in the following. By introducing:

$$\varphi(x) := f(x) - h_1, \tag{26}$$

as the deviation of the interface from its mean value and assuming that the functions $b(x)$, $\varphi(x)$, $F_{11}''(x)$ and $F_{21}''(x)$ depend linearly on the amplitude a of the profiled upper surface, an asymptotic expansion of Eqs. (22) and (24) with respect to powers of a leads, to zeroth order, to:

$$Q_1 = \frac{h_1}{h_1 + nh_2} \frac{h_1}{2},$$
$$Q_2 = \frac{2h_1 + nh_2}{h_1 + nh_2} \frac{h_2}{2}, \tag{27}$$

as solutions for the flow rates in the case of bilayer flow between two parallel planar walls. The first order contribution of the same equations resulting from linearization with respect to a:

$$\frac{h_1}{3}(h_1 + nh_2)F_{11}''(x) = \frac{2h_1 + nh_2}{h_1(h_1 + nh_2)}[\varphi(x) - b(x)] + \frac{3h_1 + 2nh_2}{h_2(h_1 + nh_2)}\varphi(x),$$
$$\frac{h_2}{3}(h_1 + nh_2)F_{21}''(x) = -\frac{n[\varphi(x) - b(x)]}{h_1 + nh_2} - \frac{h_1^2 + 4nh_1h_2 + 2n^2h_2^2}{h_2^2(h_1 + nh_2)}\varphi(x), \tag{28}$$

allow the two functions $F_{11}''(x)$ and $F_{21}''(x)$ to be expressed in terms of $\varphi(x)$ and $b(x)$. By computing the linear part of (23), taking the second order derivative with respect to x and eliminating $F_{11}''(x)$ and $F_{21}''(x)$ by means of (28), one ends up with a second order ODE of the form:

$$A_1\varphi''(x) + A_2\varphi(x) = A_3b''(x) + A_4b(x), \tag{29}$$

with the four coefficients given by:

$$A_1 = -(1 - n)\left(3h_1^2 + 2nh_1h_2 - nh_2^2\right)$$
$$A_2 = 6\left(\frac{h_1}{h_2} + \frac{nh_2}{h_1}\right)^2 + 12n\left[2\frac{h_1}{h_2} + (1 + n)\left(1 + \frac{h_2}{h_1}\right)\right]$$
$$A_3 = n(1 - n)h_2^2$$
$$A_4 = 6n\left[1 + 2\frac{h_2}{h_1} + n\frac{h_2^2}{h_1^2}\right] \tag{30}$$

Since the above ODE (29) is linear and of second order, it allows for a closed form analytic solution for any prescribed profile shape $b(x)$. Note that the latter need not to be periodic at this stage; apart from the examples considered in the results section, any profile shape can be considered, including step and trench geometries as in [24].

3.2 Finite Elements Approach

For validation purposes numerical calculations are obtained, in the case of two viscous layers, by using existing and established practices based on classical FE formulations for the original Eqs. (2) and (3) in terms of "primitive variables", namely velocity and pressure, see e.g. [25, 26] or [27]; an alternative approach would have been to use a streamfunction and vorticity formulation as adopted by [28, 29] or [30].

A challenging task is the a priori unknown shape $f(x)$ of the interface separating the two layers, requiring an iterative approach in which the calculation for the two system components, no matter whether they are fluid or viscoelastic, is carried out separately, assuming a starting value for $f(x)$ and calculating the velocity/displacement field for both layers without considering the kinematic boundary condition (15). Following this, a new interface shape $f(x)$ is calculated separately as the limiting streamline. The iteration process is repeated until either the change in the interface shape from one iteration step to the next falls below a prescribed tolerance or if the ratio of the normal velocity to the tangential one along the current interface shape is smaller than a tolerable value, typically 0.25% [31]. Implementation of the methodology was performed using standard libraries for efficient FE Galerkin solvers, making use of the packages 'numpy', 'scipy' and 'matplotlib', accessed via Python and the 'Triangle' mesh generator [32].

3.3 Complex-Variable Approach with Spectral Solution Method

For completeness and different to the above aforementioned approaches, direct use can be made of the complex formulation (10) and (11) of the field equations. On neglecting the nonlinear inertial terms, Eq. (10) becomes integrable with respect to $\bar{\xi}$, implying:

$$Eu + \eta v = 2\frac{\partial \Phi}{\partial \bar{\xi}} - 4g_0(\xi), \tag{31}$$

containing the integration function $g_0(\xi)$. After inserting this into Eq. (11), the identity:

$$g_0'(\xi) + \overline{g_0'(\xi)} = \frac{\partial^2 \Phi}{\partial \bar{\xi} \partial \xi}, \tag{32}$$

is obtained, which is again integrable. Further integration and noting that the potential Φ is real-valued, finally leads to:

$$\Phi = 2\Re\left[\bar{\xi}g_0(\xi) + g_1(\xi)\right], \tag{33}$$

yielding a second integration function $g_1(\xi)$. Thus, the entire problem has been reduced to one of determining the two holomorphic functions $g_0(\xi)$ and $g_1(\xi)$, frequently referred to as Goursat functions.

For the incompressible flow of a Newtonian liquid $v = -2i\partial\psi/\partial\bar{\xi}$ and $E = 0$, in which case Eq. (31) can be integrated a second time leading, together with expression (33), to the advantageous form:

$$\Phi + i\eta\psi = \bar{\xi}g_0(\xi) + g_1(\xi),\tag{34}$$

of the solution for both the streamfunction and the scalar potential [12].

Since the Goursat functions are functions of only one complex variable, the mathematical problem is elegantly reduced from a two- to a one-dimensional problem. While in the classical literature a purely analytical approach via conformal mappings is preferred, which is limited to finding approximate solutions for simple geometries, a spectral method based on a Fourier expansion of either the Goursat functions directly or their boundary values enables fulfillment of the boundary conditions for arbitrary profile shapes with arbitrary accuracy, depending of the truncation order of the Fourier series. Although not considered in the context of the result presented and discussed below, further details of the use of this elegant semi-analytical method and its application to free surface flows and Couette flows for Newtonian liquids can be found in [31, 33, 34].

4 Results

In the present work bilayer flow is explored, for a stationary upper surface having a particular contoured shape, utilising the Lubrication approach with the results validated via corresponding FE-calculations.

4.1 Sinusoidal Upper Surface Shapes

Assuming a sinusoidally shaped profile for the upper stationary surface, $b(x) = -a\cos x$, resulting as the limit case $s \to 0$ of the more general shape (1), the solution of the ODE (29) is obtained straightforwardly by assuming a corresponding harmonic form for the interface shape, i.e.: $\varphi(x) = -\hat{\varphi}a\cos x$, with the amplitude factor $\hat{\varphi}$ resulting in:

$$\hat{\varphi} = \frac{A_4 - A_3}{A_2 - A_1},\tag{35}$$

and with the coefficients $A_1, \ldots, A_4$ given according to (30). From the above solution the interface function $f(x)$ and the two functions $F_{11}(x)$, $F_{21}(x)$ are obtained from (26) and (28) and finally the streamfunction from (25).

This closed form analytic solution for a viscous bilayer flow is visualised via streamlines in the left column of Fig. 3 for a fixed thickness ratio of $h_1/h_2 = 2/3$ and a fixed amplitude $a = 1/2$ for varying viscosity ratio n. Corresponding FE solutions are provided in the right column.

The calculations reveal a clear dependence of the interface shape on the viscosity ratio n: if the viscosity η_1 of the layer adjacent to the profiled surface is much larger than the viscosity η_2 of the layer adjacent to the planar translating one, the interface shape mimics the upper surface profile. Comparing the resulting flow for $n = 2$ with the corresponding one for $n = 500$, a minimal change only is apparent, indicating that for a very large ratio a limit case exists. If, vice versa, the viscosity of the lower layer is larger than the viscosity of the upper one, the interface becomes smoother. For n with a very small value, the second layer acts effectively as a continuation of the translating planar surface and the interface approaches a straight line. For the

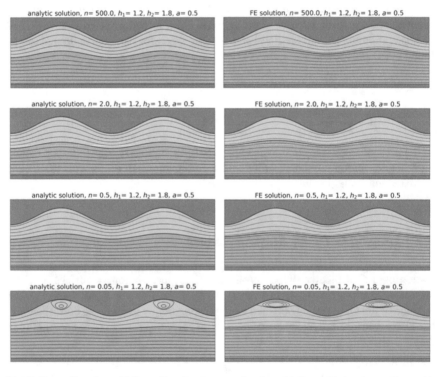

Fig. 3 Streamline plots of bilayer flow for the case of a sinusoidally profiled upper surface, with fixed geometry and varying viscosity ratio. The analytical results stem from the linearized Lubrication approximation (left column) and are compared to their corresponding FE solutions (right column)

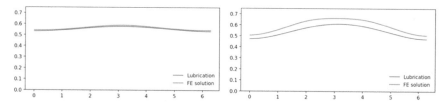

Fig. 4 Resulting shear stress along the interface of a bilayer flow over one wavelength; the layer thicknesses are $h_1 = 1.2$, $h_2 = 1.8$ and $n = 1/3$, for two different amplitudes, $a = 0.1$ (left) and $a = 0.3$ (right), calculated via the Lubrication and FE approaches

present geometry this induces the onset of a small eddy in the troughs of the profiled surface, which is a well-known observation in monolayer Couette flow, see e.g. [35], or [36].

The above results show that the solution obtained via an asymptotic analysis, although slightly overestimating the slope of the interface shape, leads to a good approximation for bilayer flow in the presence of a sinusoidally profiled surface; its limitations are due primarily to the prerequisite of a small corrugation amplitude a. Additionally, one has to keep in mind that the lubrication analysis is effectively a long-wave approximation [21] requiring the film thickness not to exceed the wavelength of the upper surface profile.

As mentioned in the introduction, the shear stress τ_f along the interface $y = f(x)$ is of particular interest and can, in general, be calculated as:

$$\tau_f = 2\eta \left[\frac{\partial v}{\partial y} - \frac{\partial u}{\partial x} \right] \frac{f'(x)}{1 + f'(x)^2} + \eta \left[\frac{\partial u}{\partial y} + \frac{\partial v}{\partial x} \right] \frac{1 - f'(x)^2}{1 + f'(x)^2}. \tag{36}$$

Two examples of the resulting distribution of the shear stress along the interface over one wavelength of the bilayer flow are presented in Fig. 4, for the same layer thicknesses $h_1 = 1.2$ and $h_2 = 1.8$ considered when generating the streamline patterns for the flows in Fig. 3 but with a viscosity ratio $n = 1/3$ and for two different profile amplitudes.

As expected, the maximum shear stress coincides with the peak value of the upper surface profile where the local film thickness is a minimum. Furthermore, for the smaller of the two amplitudes, $a = 0.1$, the agreement between the analytically calculated shear stress and the comparative FE solution is excellent; while for the larger amplitude, $a = 0.3$, the shear stress is underestimated by the Lubrication approach.

4.2 Inharmonic Periodic Upper Surface Profiles

Due to the linearity of the ODE (29), the result obtained for sinusoidally profiled upper surfaces can be adapted to generally periodic profiles by utilizing a spectral decomposition:

$$b(x) = a \sum_{k=1}^{\infty} \beta_k \cos(kx), \tag{37}$$

of the respective profile shape function. For the shape given by Eq. (1), the Fourier coefficients read [37]:

$$\beta_k = \frac{4\left(1 - \sqrt{1 - s^2}\right)^k}{k[\ln(1 + s) - \ln(1 - s)]s^k}. \tag{38}$$

This allows the ODE (29) to be solved separately for each spectral component and the writing of the solution for the interface shape as the superposition:

$$\varphi(x) = a \sum_{k=1}^{\infty} \varphi_k \beta_k \cos(kx), \tag{39}$$

where:

$$\varphi_k = \frac{A_4 - k^2 A_3}{A_2 - k^2 A_1}. \tag{40}$$

FE solutions were generated in the same manner as above. Figure 5 shows the streamline patterns obtained for a bilayer flow in the presence of an upper surface profile given by the analytic form (1), with $a = 0.5$ and a shape parameter $s = 0.9$ and as before layer thickness of $h_1 = 1.2$ and $h_2 = 1.8$, for three different values of n. The corresponding interface shape given by the Lubrication approximation is shown as a dashed line in each case.

As can be seen the results obtained are qualitatively similar to those of Fig. 3 for the case of a sinusoidally shaped upper surface profile: for the larger of the three n values the interface disturbance is greater, while for the smaller of the two n values the interface tends to a straight line when $n = 0.05$, and for which case distinct eddies are observed to exist in the troughs of the profiled surface. As for the case of a sinusoidally profiled upper surface, it can be seen that the curvature of the interface is overestimated by the Lubrication approach.

5   Conclusions and Perspectives

Three different approaches are presented as solutions to the problem of bilayer flow for the case of two immiscible Newtonian liquids confined between an upper profiled

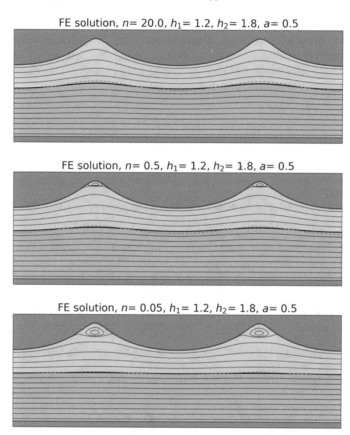

FE solution, $n= 20.0$, $h_1= 1.2$, $h_2= 1.8$, $a= 0.5$

FE solution, $n= 0.5$, $h_1= 1.2$, $h_2= 1.8$, $a= 0.5$

FE solution, $n= 0.05$, $h_1= 1.2$, $h_2= 1.8$, $a= 0.5$

Fig. 5 Streamline plots for bilayer flow, obtained using the FE approach, in the presence of an inharmonic upper surface profile, with fixed geometry and three different viscosity ratios. The interface shape in each case, obtained analytically via the Lubrication approximation with linearization, is shown for comparison purposes as a dashed line

surface at rest and a lower translating planar surface. The FE approach enables sufficiently accurate solutions of the Navier-Stokes equations to be obtained, that provide a reliable means of validating the predictions of the other two methods. The latter originate from a potential-based first integral formulation of Navier-Stokes equation. In the present work only the Lubrication approach in combination with linearization of the resulting ODEs is considered. This allows for closed form analytic solutions, which are compared in detail with corresponding FE solutions. Although overestimating the curvature of the interface between the two fluid layers and underestimating the shear stress along the latter, the Lubrication approach provides results of quantitatively acceptable accuracy if the amplitude of the profiled upper surface is sufficiently small. A potential improvement of the method could be realised by solving the nonlinear Eqs. (22), (23) and (24) without linearization.

For both biomedical and advanced technological applications, the consideration of non-Newtonian materials is an essential next step. In this context the complex variable approach outlined above provides an interesting perspective towards the implementation of viscoelastic models within a first integral framework; further details of which will appear in forthcoming articles.

References

1. Hamrock BJ, Schmid SR, Jacobson BO (2004) Fundamentals of fluid film lubrication, 2nd edn. Marcel Dekker Inc., New York
2. Szeri AZ (2011) Fluid film lubrication. Cambridge University Press, Cambridge (UK)
3. Dowson D, Higginson GR (1966) Elastohydrodynamic lubrication. The fundamentals of roller and gear lubrication. Pergamon Press, Oxford (UK)
4. Popov VL (2019) Active bio contact mechanics: concepts of active control of wear and growth of the cartilage in natural joints. AIP Conf Proc 2167(1):020285
5. Abdalla AA, Veremieiev S, Gaskell PH (2018) Steady bilayer channel and free-surface isothermal film flow over topography. Chem Eng Sci 181:215–236
6. Papageorgiou DT, Tanveer S (2019) Analysis and computations of a non-local thin-film model for two-fluid shear driven flows. Proc R Soc A: Math, Phys Eng Sci 475(2230):20190367
7. Lenz RD, Kumar S (2007) Steady two-layer flow in a topographically patterned channel. Phys Fluids 19(10):102103
8. Kistler SF, Schweizer PM (1997) Liquid film coating: scientific principles and their technological implications. Chapman and Hall, New York
9. Linka K, Schäfer A, Hillgärtner M, Itskov M, Knobe M, Kuhl C, Hitpass L, Truhn D, Thuering J, Nebelung S (2019) Towards patient-specific computational modelling of articular cartilage on the basis of advanced multiparametric MRI techniques. Sci Rep 9(1):7172
10. Müller J (2016) Retrieved 01 2020, from https://encrypted-tbn0.gstatic.com/images?q=tbn: ANd9GcQ_1ms-f6vkQxA2ii5my-BNLht5L16E3D7g4jpcxVoPaTo72SNO
11. Ranger KB (1994) Parametrization of general solutions for the Navier-Stokes equations. Q J Appl Math 52:335–341
12. Marner F, Gaskell PH, Scholle M (2017) A complex-valued first integral of Navier-Stokes equations: unsteady Couette flow in a corrugated channel system. J Math Phys 58(4):043102
13. Scholle M, Gaskell PH, Marner F (2018) Exact integration of the unsteady incompressible Navier-Stokes equations, gauge criteria, and applications. J Math Phys 59(4):043101
14. Marner F, Gaskell PH, Scholle M (2014) On a potential-velocity formulation of Navier-Stokes equations. Phys Mesomech 17(4):341–348
15. Muskhelishvili NI (1953) Some basic problems of the mathematical theory of elasticity. Noordhoff Ltd., Groningen (NL)
16. Mikhlin SG (1957) Integral equations and their applications to certain problems in mechanics, mathematical physics and technology. Pergamon Press, New York
17. Coleman CJ (1984) On the use of complex variables in the analysis of flows of an elastic fluid. J Nonnewton Fluid Mech 15(2):227–238
18. Irgens F (2014) Rheology and non-Newtonian fluids. Springer International Publishing, Cham (Schweiz)
19. Malkin AY, Isayev AI (2012) Rheology: concepts, methods, and applications, 3rd edn. ChemTec Publishing, Toronto
20. Stillwagon LE, Larson RG (1990) Leveling of thin films over uneven substrates during spin coating. Phys Fluids A 2(11):1937–1944
21. Oron A, Davis SH, Bankoff SG (1997) Long-scale evolution of thin liquid films. Rev Mod Phys 69(3):931–980

22. Matar OK, Sisoev GM, Lawrence CJ (2006) The flow of thin liquid films over spinning discs. Can J Chem Eng 84(6):625–642
23. Craster RV, Matar OK (2009) Dynamics and stability of thin liquid films. Rev Mod Phys 81(3):1131–1198
24. Scholle M, Gaskell PH, Marner F (2019) A potential field description for gravity-driven film flow over piece-wise planar topography. Fluids 4(2):82
25. Boffi D, Brezzi F, Demkowicz L, Durán R, Falk R, Fortin M (2008) Mixed Finite Elements, Compatibility Conditions, and Applications. Springer, Berlin-Heidelberg
26. Elman HC, Silvester DJ, Wathen AJ (2014) Finite elements and fast iterative solvers: with applications in incompressible fluid dynamics. Oxford University Press, Oxford (UK)
27. John V (2016) Finite element methods for incompressible flow problems. Springer International Publishing, Cham (Schweiz)
28. Gaskell PH, Summers JL, Thompson HM, Savage MD (1996) Creeping flow analyses of free surface cavity flows. Theoret Comput Fluid Dyn 8(6):415–433
29. Gaskell PH, Thompson HM, Savage MD (1999) A finite element analysis of steady viscous flow in triangular cavities. Proc Inst Mech Eng, Part C: J Mech Eng Sci 213(3):263–276
30. Girault V, Raviart P-A (2012) Finite element methods for Navier-Stokes equations: theory and algorithms, vol 5. Springer Science & Business Media, Berlin
31. Marner F (2019) Potential-based formulations of the Navier-Stokes equations and their application. Doctoral thesis, Durham (UK): Durham University
32. Shewchuk JR (1996) Triangle: engineering a 2D quality mesh generator and Delaunay triangulator. In: Applied computational geometry towards geometric engineering, FCRC'96 workshop, WACG'96 Philadelphia, PA, 27–28 May 1996, pp 203–222
33. Scholle M, Wierschem A, Aksel N (2004) Creeping films with vortices over strongly undulated bottoms. Acta Mech 168(3):167–193
34. Scholle M (2007) Hydrodynamical modelling of lubricant friction between rough surfaces. Tribol Int 40(6):1004–1011
35. Scholle M, Haas A, Aksel N, Wilson MC, Thompson HM, Gaskell PH (2009) Eddy genesis and manipulation in plane laminar shear flow. Phys Fluids 21(7):073602
36. Esquivelzeta-Rabell FME, Figueroa-Espinoza B, Legendre D, Salles P (2015) A note on the onset of recirculation in a 2D Couette flow over a wavy bottom. Phys Fluids 27(1):014108
37. Rund A (2006) Optimierung des M aterialtransports bei schleichenden Filmströmungen, Diploma Thesis, University of Bayreuth

Microstructure-Based Computational Analysis of Deformation and Fracture in Composite and Coated Materials Across Multiple Spatial Scales

Ruslan R. Balokhonov and Varvara A. Romanova

Abstract A multiscale analysis is performed to investigate deformation and fracture in the aluminum-alumina composite and steel with a boride coating as an example. Model microstructure of the composite materials with irregular geometry of the matrix-particle and substrate-coating interfaces correspondent to the experimentally observed microstructure is taken into account explicitly as initial conditions of the boundary value problem that allows introducing multiple spatial scales. The problem in a plane strain formulation is solved numerically by the finite-difference method. Physically-based constitutive models are developed to describe isotropic strain hardening, strain rate and temperature effects, Luders band propagation and jerky flow, and fracture. Local regions experiencing bulk tension are found to occur during compression that control cracking of composites. Interrelated plastic strain localization in the steel substrate and aluminum matrix and crack origination and growth in the ceramic coating and particles are shown to depend on the strain rate, particle size and arrangement, as well as on the loading direction: tension or compression.

Keywords Composites · Coated materials · Constitutive modeling · Plastic strain localization · Fracture · Multiscale numerical simulation

1 Introduction

Actual materials have essentially inhomogeneous microstructure (Fig. 1). According to the concepts of physical mesomechanics, stress concentrators of different physical origin are a major factor influencing the deformation pattern in nonhomogeneous materials. The effects are most conspicuous in composite materials (metal-matrix composites, coated and surface-hardened materials, doped alloys, etc.) because of differences in the mechanical properties (density, elastic moduli, and strength and plasticity characteristics) of their constituent elements. Thus, basic research along

R. R. Balokhonov (✉) · V. A. Romanova
Institute of Strength Physics and Materials Science, Siberian Branch of the Russian Academy of Sciences, pr. Akademicheskii 2/4, 634055 Tomsk, Russia
e-mail: rusy@ispms.tsc.ru

© The Author(s) 2021
G.-P. Ostermeyer et al. (eds.), *Multiscale Biomechanics and Tribology of Inorganic and Organic Systems*, Springer Tracts in Mechanical Engineering, https://doi.org/10.1007/978-3-030-60124-9_17

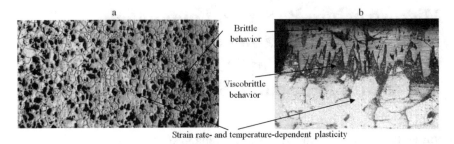

Fig. 1 Microstructures of an Al/Al$_2$O$_3$ composite (**a**) and steel coated by diffusion borating (**b**) [16, 34]

these lines is of great practical importance for development of advanced structural and functional materials.

The foundations of the physical mesomechanics of materials as a consistent methodology were laid more than 35 years ago [1]. At this time, the basic principles underlying the scientific approach have been formulated and developed [2, 3]. Early theoretical studies helped outline the range of immediate tasks and determined modeling and simulation techniques capable of solving these problems [2, 4–10]. New deformation and fracture mechanisms operative at macro-, meso- and microscales in solids were identified and accounted for [4]. Development of techniques for computer-aided design of materials and software enabled deformation and fracture processes to be simulated [2, 4–10].

At present, the problem of an adequate consideration of the multiscale nature of solids is recognized internationally as the first-priority line of investigations aimed at developing new-generation materials, and there is an increasing interest in theoretical studies in this research area—see e.g. [11–22]. This has been due to a general awareness that a correct prediction of the macroscopic properties of solids is hardly possible without hierarchy of structural levels and scales in the materials under study.

Nowadays, there are a large number of studies addressing multiscale numerical simulation and modeling, with an explicit consideration of the microstructure being taken into account. Material components are associated with proper constitutive models. For instance, some papers involve artificial models of real microstructures in studying micro-, meso- and homogenized macromechanical response of materials [2], [4, 9–11, 15, 17–19, 23–32]. Other authors reported results on the stress-strain analysis of experiment-based microstructure models [2, 6, 8, 12–14, 16, 20–22, 33–39]. All these and related works extend our understanding of the relationship between the microstructure and mechanical properties of materials. A special attention is given to interfaces. A majority of contributions devoted to the interfacial problem considers the interfacial fracture, decohesion and debonding [40–48]. Nevertheless, a comprehensive study of the phenomena related to the irregular interface geometry effects is often neglected.

The main purpose of this contribution is to show that, from the standpoint of mechanics stress concentration in local regions of a composite at different scale levels could be of the same origin. It is controlled by the irregular geometry of microstructural elements making up the composition (ductile matrix/substrate, brittle ceramic particles/coatings/hardened interlayers, etc.) and the difference in their mechanical properties.

It is found out that the value of local stresses in composites might, by a large factor, exceed the average level of the load applied. The evolution of this effect is attributed to the presence of inhomogeneities with characteristic sizes corresponding to different scales—macro, meso and micro. It is demonstrated that the regions of stress concentration might undergo both compressive and tensile stresses irrespective of the type of external loading. The larger the difference in the mechanical properties of the constituents, the higher is the level of stress concentration developed in the vicinity of inhomogeneities of certain geometry.

The main aim of the paper is to investigate mechanisms of deformation and fracture which are related to complex geometry of interfaces in a composite material. Multiscale analysis of deformation and fracture in composites is performed. A dynamic boundary-value problem is solved numerically by the finite-difference method. Constitutive models for the elastic-plastic deformation and elastic-brittle fracture are developed to describe the mechanical response of steel substrate/aluminum matrix and boride coating/alumina particles. Interface geometries correspond to the configurations found experimentally and are accounted for explicitly in the calculations.

2 Numerical Modelling Across Multiple Spatial Scales

In the frame of the proposed formulation, a multiscale numerical analysis implies at least two factors (Fig. 2): (1) the use of different models to describe the mechanical response of different constituent elements of a composite material under load in order to characterize the physical processes developing in the components and their

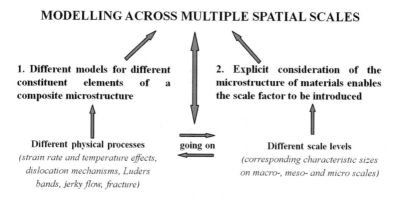

Fig. 2 Schematics of multiscale numerical simulation

interplay (Fig. 3), and (2) an explicit consideration of the microstructure of the material that provides information on the characteristic scales for which the models used are valid (Fig. 4).

It is suggested that deformation of composite materials can be described by a system of equations using the laws of conservation of mass and momentum, strain equations, and constitutive relations for the material constituents complemented with initial and boundary conditions (Fig. 3). The models presented in Fig. 3 enable us to handle only particular problems. Notably, the models under discussion have

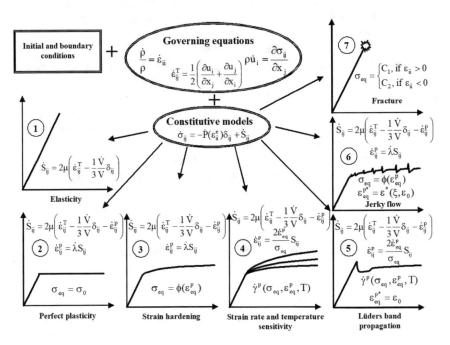

Fig. 3 A set of constitutive models for composite material constituents

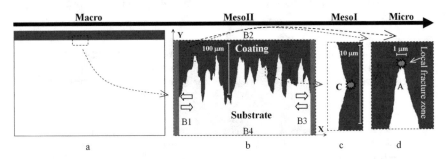

Fig. 4 Microstructure of steel with a hardened boride surface layer to be simulated and characteristic sizes of structure inhomogeneities at different scale levels

already been tested. The list of those accounting for the great diversity of the physical phenomena and processes observed in loaded solids is by no means complete.

The choice of particular elastoplastic response models numbered 1 through 6 in Fig. 3 depends on the material used as a substrate or a matrix (aluminum-based alloys, different types of steel, etc.) and on the applied loading conditions (high strain rate deformation, quasi-static loading, etc.). Fracture models, such as model 7, are required to describe the mechanical behavior of brittle and viscobrittle inclusions, coatings, and intermediate subsurface layers. Purely elastic or perfectly plastic descriptions (model 1 and 2) can be used as a first approximation.

An explicit consideration of the material microstructure makes it possible to introduce a scale factor and specify the length scales where one or another model can successfully be employed (Fig. 4). Figure 1b shows the mesostructure of a coated material, which was used in the calculations and corresponds to that observed experimentally (see Fig. 1). For the case in question, where the interface has a serrated profile and irregular geometry, we might single out certain types of inhomogeneities at different scale levels and their respective characteristic sizes. In particular, boride teeth proper, whose size is ~50–100 μm (Fig. 4b), are independent stress concentrators. A quasiperiodic alternation of boride and steel teeth results in the formation of a peculiar stress-strained state at mesoscale II.

The shape of an individual tooth, is not perfect and exhibits fine structure at a lower scale level. Throughout the interface profile, there are convexities and concavities. The characteristic size of these inhomogeneities is within 5–10 μm (Fig. 4c). The regions of "intrusion" of ductile steel into a more brittle and strong boride material are sources of geometrical stress concentration at mesoscale I. Let us single out two types of such regions with respect to the direction of applied loading: types C and A, along (Fig. 4c) and perpendicular (Fig. 4d) to the x-direction.

The local fracture zone has a characteristic size of ~1 μm (Fig. 4d) and gives rise to stress concentration at the microscale.

A homogenized stress-strain curve reflects the mechanical behavior of the mesovolume at the macroscale. Thus, an introduction of the mesovolume with a boride hardened layer of a complicated geometry in an explicit form allows us to prescribe the length scales—a scale hierarchy of inhomogeneities, whose characteristic sizes might differ by two orders of magnitude.

3 Governing Equations and Boundary Conditions

Let us formulate the governing equations (Fig. 3) in terms of plane strain. In this case there are the following non-zero components of the strain rate tensor:

$$\dot{\varepsilon}_{xx} = \dot{u}_{x,x}, \ \dot{\varepsilon}_{yy} = \dot{u}_{y,y} \ \dot{\varepsilon}_{xy} = \frac{1}{2}\left(\dot{u}_{x,y} + \dot{u}_{y,x}\right), \tag{1}$$

where u_x and u_y are the components of the displacement vector, ε_{xx}, ε_{yy} and ε_{xy} are the strain tensor components, the upper dot and comma in the notations stand for the time and space derivatives, respectively.

The mass conservation law and the equations of motion take the forms

$$\dot{V}/V = \dot{\varepsilon}_{xx} + \dot{\varepsilon}_{yy}, \tag{2}$$

$$\sigma_{xx,x} + \sigma_{yx,y} = \rho\ddot{u}_x, \quad \sigma_{xy,x} + \sigma_{yy,y} = \rho\ddot{u}_y, \tag{3}$$

where σ_{xx}, σ_{yy} and σ_{xy} are the stress tensor components, V is the specific volume and ρ is the mass density.

Taking into account the resolution of the stress tensor in the spherical and deviatoric parts

$$\sigma_{ij} = -P\delta_{ij} + S_{ij} \tag{4}$$

the pressure and the stress deviator components are written as follow

$$\dot{S}_{xx} = 2\mu\left(\dot{\varepsilon}_{xx} - \frac{1}{3}\dot{\varepsilon}_{kk} - \dot{\varepsilon}_{xx}^P\right), \quad \dot{S}_{yy} = 2\mu\left(\dot{\varepsilon}_{yy} - \frac{1}{3}\dot{\varepsilon}_{kk} - \dot{\varepsilon}_{yy}^P\right),$$

$$\dot{S}_{zz} = 2\mu\left(-\frac{1}{3}\dot{\varepsilon}_{kk}\right) = -(\dot{S}_{xx} + \dot{S}_{yy}), \quad \dot{S}_{xy} = 2\mu(\dot{\varepsilon}_{xy} - \dot{\varepsilon}_{xy}^P), \quad \dot{P} = -K\dot{\varepsilon}_{kk}, \tag{5}$$

where K and μ are the bulk and shear moduli, $\dot{\varepsilon}_{ij}^P$ is the plastic strain rate tensor, and δ_{ij} is the Kronecker delta.

To eliminate an increase in stress due to rigid rotations of medium elements, we define deviatoric stresses through the Jaumann derivative

$$\dot{S}_{ij}^* = \dot{S}_{ij} - S_{ik}\omega_{jk} - S_{jk}\omega_{ik}, \tag{6}$$

where $\omega_{ij} = \frac{1}{2}(\dot{u}_{i,j} - \dot{u}_{j,i})$ is the material spin tensor.

The strain tensor is the sum of elastic and plastic strain $\varepsilon_{ij} = \varepsilon_{ij}^e + \varepsilon_{ij}^P$, and $\dot{\varepsilon}_{kk}^P = 0$ is the hypothesis of plastic incompressibility. Unloading is elastic.

The boundary conditions on the surface B_1 and B_3 simulate uniaxial tension parallel to the X-axis, whereas on the bottom and top surfaces, they correspond to symmetry and free surface conditions, respectively (Fig. 4b). We obtain

$$\dot{u}_x = const = -v \text{ for } (x, y) \in B_1 \; \dot{u}_x = const = v \text{ for } (x, y) \in B_3, \; \dot{u}_y$$

$$= 0 \text{ for } (x, y) \in B_4, \; \sigma_{ij} \cdot n_j = 0 \text{ for } (x, y) \in B_2 \; \sigma_{xy}$$

$$= 0 \text{ for } (x, y) \in B_1 \cup B_3 \cup B_4 \tag{7}$$

Here $B = B_1 \cup B_2 \cup B_3 \cup B_4$ is the boundary of the computational domain, t is the computation time, v is the load velocity and n_j is the normal to the surface B_2.

The system of Eqs. (1–7) have to be completed with a formulation for plastic strain rates $\dot{\varepsilon}_{ij}^p$.

4 Constitutive Modelling for Plasticity of the Substrate and Matrix Materials

4.1 Physically-Based Strain Hardening

Depending on the loading strain rate and material to be used as a substrate or matrix, different formulations of the constitutive models (models 2–6 in Fig. 3) have to be applied. To describe plasticity (model 3 in Fig. 3), use was made of the plastic flow law

$$\dot{\varepsilon}_{ij}^p = \dot{\lambda} \frac{\partial f}{\partial S_{ij}} \tag{8}$$

associated with the yield condition

$$f(S_{ij}) = \sigma_{eq} - \sigma_A(\varepsilon_{eq}^p) = 0. \tag{9}$$

Here λ is a scalar parameter, σ_{eq} and ε_{eq}^p are the equivalent stress and accumulated equivalent plastic strain

$$\sigma_{eq} = \frac{1}{\sqrt{2}} \sqrt{(S_{11} - S_{22})^2 + (S_{22} - S_{33})^2 + (S_{33} - S_{11})^2 + 6(S_{12}^2 + S_{23}^2 + S_{31}^2)} \tag{10}$$

$$\varepsilon_{eq}^p = \frac{\sqrt{2}}{3} \int_0^t \sqrt{\left(\varepsilon_{11}^p - \varepsilon_{22}^p\right)^2 + \left(\varepsilon_{22}^p - \varepsilon_{33}^p\right)^2 + \left(\varepsilon_{33}^p - \varepsilon_{11}^p\right)^2 + 6\left(\varepsilon_{12}^{p2} + \varepsilon_{23}^{p2} + \varepsilon_{31}^{p2}\right)} dt \tag{11}$$

The function $\sigma_A(\varepsilon_{eq}^p)$ prescribes isotropic strain hardening in the steel substrate or aluminium matrix. The following phenomenological function can be selected:

$$\sigma_A = \sigma_s - (\sigma_s - \sigma_0) \exp(-\varepsilon_{eq}^p / \varepsilon_r^p), \tag{12}$$

where σ_0 is the yield point and σ_s is the strength, ε_r^p is the reference value of plastic strain. Model 2 in Fig. 3 describes perfect plasticity at $\sigma_A = \sigma_0$.

For a more detailed description, σ_A can be obtained from physical consideration of dislocation dynamics. σ_A is the athermal part of the stress σ_{eq} and associated with long-range obstacles to the dislocation motion. It is independent of the strain rate and mainly depends on the microstructure of the material: the dislocation density and substructures, grain sizes, point defects and various solute atoms. The current yield stress is proposed to be defined in the following way:

$$\sigma_A = \sigma_0 + \alpha\mu b\sqrt{N(\varepsilon_{eq}^p)} + \sum_j \alpha_{1j} P_j(\varepsilon_{eq}^p), \tag{13}$$

with the Hall-Perth dependence being introduced

$$\sigma_0 = \sigma_0^0 + kd^{-1/2} \tag{14}$$

where σ_0^0 is the yield point of the single crystal and d is the grain size.

The addend in Eq. (13) is a familiar dependence from physics of plasticity that accounts for a microscopic contribution from a forest of dislocations, where N is the dislocation density. The third summand is associated with the formation of substructures: dislocation cells, band and fragmented substructures, where α_{1i} denotes coefficients accounting for dislocation substructure contributions to strengthening, and the probability functions $P_j(\varepsilon_{eq}^p)$ are connected with volume fractions of the substructures. Based on the experimental evidence the following form of the exponential law is proposed:

$$P_j(\varepsilon_{eq}^p) = \int_0^{\varepsilon_{eq}^p} \lambda_j \exp(\eta - \exp\eta) d\varepsilon_{eq}^p. \tag{15}$$

Here $\eta = -\lambda_j(\varepsilon_{eq}^p - \varepsilon_{eqj}^p)$, ε_{eqj}^p is a parameter associated with deformation giving rise to formation of the i-th substructure, and λ_j specifies the strain range wherein the substructure exists. The volume fraction of the substructure of interest is determined via the distribution function (15) to give

$$P_j^v(\varepsilon_{eq}^p) = \exp(1 + \eta - \exp\eta). \tag{16}$$

For instance, for the only dislocation cell substructure Eq. (13) takes the form

$$\sigma_A = \sigma_0 + \alpha\mu b\sqrt{N(\varepsilon_{eq}^p)} + \alpha_c P_c(\varepsilon_{eq}^p). \tag{17}$$

Comparing Eq. (17) with the well-known experimental evidence $\sigma_A \propto d_c^{-1}$ [49], which is similar to Eq. (14), and taking into account the fact that the dislocation cell diameter d_c decreases during plastic deformation reaching the saturation point where its value does not depend on the stacking fault energy ($d_c^{sat} \approx 0.2\,\mu\text{m}$ for many materials [50, 51]), the following expression can be obtained

$$d_c(\varepsilon_{eq}^p) = \frac{d_c^{sat}}{P_c(\varepsilon_{eq}^p)}. \tag{18}$$

The physically-based expression (13) includes microscopic parameters such as elastic moduli, dislocation density, grain size and dislocation cell diameter, module of Burgers vector b, empirical strength coefficient α, while purely phenomenological Eq. (12) operates with macroscopic yield point and strength.

4.2 Strain Rate and Temperature Effects

To describe strain rate and temperature sensitivity of steel or aluminum, it is necessary to develop a relaxation constitutive equation (model 4 in Fig. 3). Substituting (8) in the equation for the equivalent plastic strain (11), it can be found

$$\dot{\varepsilon}_{ij}^p = \frac{3}{2} \frac{\dot{\varepsilon}_{eq}^p}{\sigma_{eq}} S_{ij}. \tag{19}$$

In order to describe the equivalent plastic strain rate $\dot{\varepsilon}_{eq}^p$, let us continue proceeding from a dislocation concept of plastic flow. Kinetic equations for the plastic strain rate based on the motion of dislocations are the subject of a considerable amount of literature. Theoretical concepts relevant to the discussion are in part summarized in [52]. Following the model let us define

$$\dot{\varepsilon}_{eq}^p = \dot{\varepsilon}_r^p \exp\left\{ -\frac{G_0}{kT} \left[1 - \left(\frac{\sigma_{vis}}{\tilde{\sigma}} \right)^w \right]^z \right\}, \tag{20}$$

$$\sigma_{vis} = \sigma_{eq} - \sigma_A(\varepsilon_{eq}^p) \quad T = T_0 + \int_0^{\varepsilon_{eq}^p} \frac{\beta}{\rho C_v} \sigma_{eq} d\varepsilon_{eq}^p.$$

Here G_0 is the energy that a dislocation must have to overcome its short-range barrier solely by its thermal activation, $\tilde{\sigma}$ is the stress above which the barrier is crossed by a dislocation without any assistance from thermal activation, k is the Boltzmann constant, $\dot{\varepsilon}_r^p$ is the reference value of the plastic strain rate, T_0 is the test temperature, C_v is the heat capacity, β is the fraction of plastic work which is converted into heat. $\beta \cong 1$, $z = 2/3$ and $w = 2$ for many metals [52].

Parameters of the model were derived by solving an initial value problem and by fitting the calculation results to the experimental stress-strain curves under tension at different strain rates and temperatures. For uniaxial loading in the X-direction $\sigma_{xx} = \sigma_{eq}$ and $\varepsilon_{xx} = \varepsilon_{eq}$. In this case the constitutive Eq. (4) takes the form

$$\sigma_{eq} = E(\varepsilon_{eq} - \varepsilon_{eq}^p), \tag{21}$$

where E is the Young's module. Equations (20), (21) and (12) were solved numerically by a fourth-order Runge–Kutta method.

The results obtained are presented in Fig. 5. In order to validate the model, Eqs. (19) and (20) were introduced into the commercial software ABAQUS and the tension of steel H418 plates was simulated in a plane stress formulation. Here and in what follows, stress $\langle \sigma \rangle$ was computed as the equivalent stress σ_{eq} averaged over the mesovolume $\langle \sigma \rangle = \sum_{k=1,n} \sigma_{eq}^k s^k / \sum_{k=1,N} s^k$, where n is the number of computational mesh cells and s^k is the k-th cell area. Strain ε corresponds to the relative elongation of the computational domain in the X-direction $\varepsilon = (L - L_0)/L_0$, where L_0 and L are the initial and current lengths of the computational domain along X. The results show good agreement between the calculations and experiment (Fig. 5d). Model parameters are presented in Table 1. Strain rate effects were not taken into account for Al6061 alloy.

For the investigated steels, $G_0/k = 10.6 \times 10^{-5}\,\text{K}^{-1}$ and $\tilde{\sigma} = 1450$ MPa are the constants which reflect the temperature sensitivity of the material. They were extracted from experimental mechanical tests at different test temperatures. Since corresponding experiments for the austenitic steels were not available, the values of these parameters are suggested to be the same as for HSLA-65 steel defined in [52].

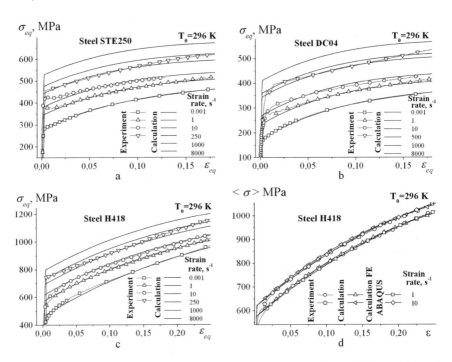

Fig. 5 Predicted mechanical properties of austenitic steels in comparison with the experimental data (**a–c**) and comparison of the calculation results for H418 steel with those obtained by ABAQUS (**d**)

Table 1 Material constants and model parameters for different steels and an Al6061 alloy

	σ_s, MPa	σ_0, MPa	ε_r^p	$\dot{\varepsilon}_r^p$, s^{-1}
HSLA	713	422	0.21842	4·108
H418	1300	466	0.27456	2·109
STE250	497	278	0.09978	3·1010
DC04	395	173.7	0.09312	5·1010
Al6061	184	62	0.054	–

Experimental and calculated stress-strain curves for HSLA-65 steel are presented in Fig. 6. It can be seen that for $\varepsilon < 10 \div 20\%$ Eq. (20) overestimates the current stress and fails to give a correct description of the shape of the stress-strain curve. The overestimation may be due to the fact that the parameter $\dot{\varepsilon}_r^p$ is assumed as the constant value which is proportional to the density of mobile dislocations $N_m \cong 10^{11}$ cm^{-2} [52], although it is well known that N_m changes with the plastic deformation development and reaches its saturation at total strains of $10 \div 40\%$ for different steels.

In order to take into account the evolution of the dislocation continuum, the plastic strain rate is proposed to be as follows

$$\dot{\varepsilon}_{eq}^p = \dot{\varepsilon}_r^p F(\varepsilon_{eq}^p) \exp\left\{-\frac{G_0}{kT}\left[1 - \left(\frac{\sigma_{vis}}{\tilde{\sigma}}\right)^w\right]^z\right\} \quad (22)$$

where

$$F(\varepsilon_{eq}^p) = F^* + (1 - F^*) \cdot \exp\left(-\frac{B}{|g|b}\varepsilon_{eq}^p\right) \quad (23)$$

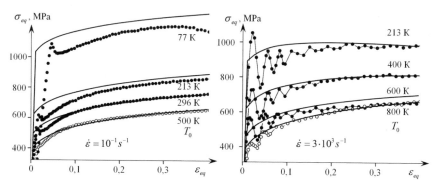

Fig. 6 Stress-strain curves for steel HSLA-65 [52]. Dots—experiment, lines—calculations, using Eq. (20)

is the fraction of mobile dislocations (Kelly and Gillis 1974) which decreases at the initial stages of plastic flow due to stalemating events. Here $|g| = 0.5$ is the orientation factor, $b \cong 3.3\text{Å}$ is the magnitude of the Burgers vector, F^* is the minimum value of F.

It is suggested for F^* and B to be connected with the reciprocal mean free path for stalemating events as follows. Parameter B is attributed to the initial state of the material, when the free path is a half of an average grain size d

$$B = \frac{2}{dN^0}, \tag{24}$$

where N^0 is the initial dislocation density.

In the equation for F^*, the free part is calculated from some intermediate state, using the dislocation cell diameter d_c^{sat} formed inside the grains

$$F^* = \frac{N^0 d}{N^* d_c^{sat}}, \tag{25}$$

where N^* is the maximum dislocation density. In this formulation the density of mobile dislocations

$$N_m^* = N^* F^* \tag{26}$$

is a measure for the saturation density. Defining $N^* = 5 \times 10^{12}\,\text{cm}^{-2}$, we obtain from Eq. (26) $F^* = 0.02$. The grain size $d = 15\,\mu\text{m}$ for HSLA-65 steel [52], hence $N^0 \cong 1.33 \times 10^9\,\text{cm}^{-2}$ from Eq. (25) and $B = 0.01\,\mu\text{m}$ from Eq. (24).

The system of Eqs. (1)–(7), (12) and (22) for a rectangular homogeneous region was solved numerically by the finite-difference method (Sect. 6). Figure 7 shows the results of plane strain calculations for varying strain rate and temperature. For reference, a dotted curve (296 K, 8000/s) is plotted in the figure to present calculations according to the model, where $\dot{\varepsilon}_{eq}^p$ was computed, using Eq. (20). Relation (22) is

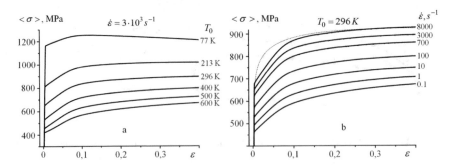

Fig. 7 Stress strain curves calculated using Eq. (22) for different temperatures (**a**) and strain rates (**b**)

seen to provide a more accurate description of stress-strain curves than Eq. (20) for small deformations ($\varepsilon \leq 20\%$).

4.3   Lüders Band Propagation and Jerky Flow

Experimental stress-strain curves for HSLA-65 steel are characterized by the upper and lower yield stresses (Fig. 6), which could be an evidence of Lüders band propagation.

The jerky-flow phenomenon in alloys has been well studied experimentally, for instance in [52–59], and attributed to the formation of localized deformation bands at the mesoscale. As a special case of such an anomalous behavior, Lüders band propagation is characterized by single displacement of macroscopic localization zone along the test piece. It is generally agreed that the microscopic essence of the discontinuous yielding is the dynamic aging of dislocations by diffusing solute atoms. There are some physically based attempts to simulate the propagation of bands of localized plastic deformation [60–62].

In this paper, to describe Lüders band propagation (model 5 in Fig. 3), use was made of a phenomenological approach [63]. It is suggested that the methods of continuum mechanics and discrete cellular automata can be used in combination. The approach relies on the experimentally established fact that the plastic deformation originates near surfaces and interfaces and subsequently propagates from the surface sources as localized deformation front.

Each cell of the computational grid is treated as a cellular automaton which can be either in elastic or plastic state. Initially all cells are elastic. Elastic-to-plastic transition of a certain computational cell is controlled by both the stress value in this local point and the deformation behavior of the neighboring computational cells. Noteworthy, different yield criteria are formulated for the surface cells and for local regions in the bulk of the material. In the former case, a local region near the surface becomes plastic if the equivalent stress acting there reaches its critical value; the stress-based criterion is used for the surface and interface cells. The response of an internal region is elastic until two conditions are satisfied: the equivalent stress in this cell achieves the yield limit and the plastic deformation accumulated in any of the neighboring cells amounts to its critical value. In such a way, the internal regions are successively involved in plastic deformation by flows propagating from the surface and interface sources.

Mathematically, the stress-based yield criterion given by a physically-based constitutive model for any local internal region D is complemented with the necessary condition whereby there must be plastic deformation at least in one of the regions D^* adjacent to D:

$$\varepsilon_{eq}^{p*} = \varepsilon_0. \tag{27}$$

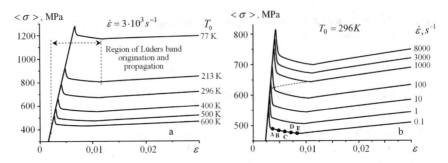

Fig. 8 Initial portions of stress-strain curves calculated using Eqs. (27) and (22) for different temperatures (**a**) and strain rates (**b**)

Here, ε_0 is the new parameter that reflects material properties associated with the strain aging effects. This is a critical value of the equivalent plastic strain accumulated in the D^* region, which is necessary for the onset of plastic flow in the region D.

Using the criterion (27) and the constitutive model (22) in combination, we have performed numerical simulations of Lüders band propagation in a wide range of strain rates and temperatures. Figure 8 demonstrates calculated stress-strain curves at early deformation stages ($\varepsilon \leq 3\%$). The computational domain is approximated by a regular mesh consisting of 600 cells, $n_x = 80, n_y = 20$. The parameter $\varepsilon_0 = 8 \times 10^{-4}$. For comparison, the dashed curve in Fig. 8b ($T_0 = 296$ K, the strain rate is $1000 \, s^{-1}$) shows the initial portion of the curve calculated in the assumption of homogeneous deformation (see Fig. 7b).

The combined approach is seen to provide a more accurate description of the experimental stress-strain relations for HSLA-65 steel (see Fig. 6). Plastic flow first originates near the left boundary of the computational domain where load is applied and propagates along the specimen in the form of a localized plastic deformation front (Fig. 9a). The material ahead of the front is elastic and the plastic strain is accumulated behind the moving Lüders band front. Similar behavior was observed in experiments. Presented in Fig. 9b are the experimental data on Lüders band propagation in a steel plate surface-hardened by the electron-beam-induced deposition [64]. This process leads to the occurrence of the upper and lower yield points in the macroscopic stress-strain curve and a zone of slow variation in the current resistance to load—the yield plateau (Fig. 8).

On further loading, the stress relaxation in the elastic region slows down (Fig. 10). Simultaneously, the strain hardening in the expanding plastic flow region makes an increasingly greater contribution to the macroscopic stress. Therefore, the yield plateau appears which is characterized by a slow variation in the current resistance to deformation (Fig. 8). In this stage, the relative elongation of the specimen occurs primarily by plastic deformation of the zone located behind the front.

The approach discussed was further developed to account for the Portevin-Le Chatellier effect (model 6 in Fig. 3) associated with sequential propagation of multiple

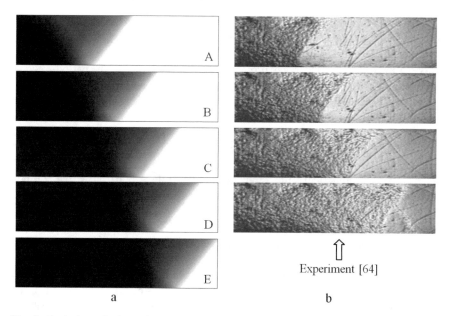

Fig. 9 Equivalent plastic strain distributions for Lüders band propagation: calculation results (**a**) where A–E are the material states corresponding to the points A–E in Fig. 8b and experimental data [64] (**b**)

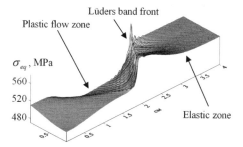

Fig. 10 Equivalent stress filed at a compressive strain corresponding to the point A in Fig. 8b

localized deformation bands from the specimen ends (model 6 in Fig. 3). Experimental observations of unstable deformation effects [55–59] show that propagation of a localized deformation band corresponds to each drop in stress seen in the stress-strain curve. As a rule, the stress amplitude at that point and the average quasi-homogeneous deformation during time intervals between the sequential band formation are found to increase with strain hardening. Therefore, condition (27) was modified to

$$\varepsilon_{eq}^{p*} = \varepsilon^*(\xi, \varepsilon_0), \ \xi = \sigma_A(\varepsilon_{eq}, \sigma_0)/\sigma_0. \tag{28}$$

The localized-deformation bands are formed near specimen loaded ends at regular intervals with the proviso that $\Delta\varepsilon_{eq}^{p\,\min} = \varepsilon^{\Delta}(\xi, \varepsilon_0)$. Here, $\Delta\varepsilon_{eq}^{p\,\min}$ is a minimum increase in the equivalent plastic deformation as the result of propagation of the previous band. Thus, in the model proposed, the drop in stress and the periodic generation pattern of the localized deformation bands are expressed by some dimensionless parameter ξ accounting for the strain hardening. The simple relations

$$\varepsilon^* = \varepsilon_0 \exp\left(\frac{\xi}{1-\xi}\right), \quad \varepsilon^{\Delta} = \varepsilon_0(\xi - 1), \tag{29}$$

were derived in numerical simulations of loading of an Al6061 alloy that exhibits unstable plastic flow [55]. The parameters for a hardening function of the type given in Eq. (12) were chosen according to the experiments performed in [55] (see Table 1). The calculated results demonstrated in Figs. 11 and 12 are an evidence for an essentially nonhomogeneous stressed-strained state at the yield point. The stress-strain curve is a serrated line (Fig. 11).

Each drop in the curve (Fig. 11d) corresponds to the formation and propagation of one or two localized deformation bands (Fig. 12). According to Eq. (29), the quantities $\varepsilon^*(\xi, \varepsilon_0)$ and $\varepsilon(\xi, \varepsilon_0)$ are rather small in early plastic flow stages, as a result of which the stress amplitude during the drastic decrease in the equivalent stress is

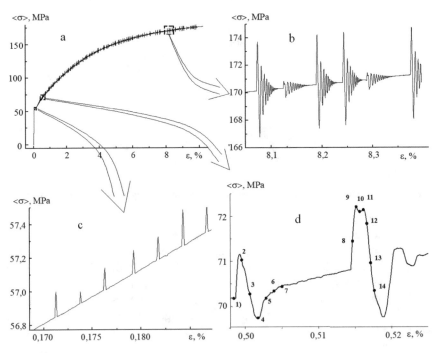

Fig. 11 Calculated stress-strain curve for an Al6061 alloy (**a**) and three selected sections shown at enlarged scales (**b–c**)

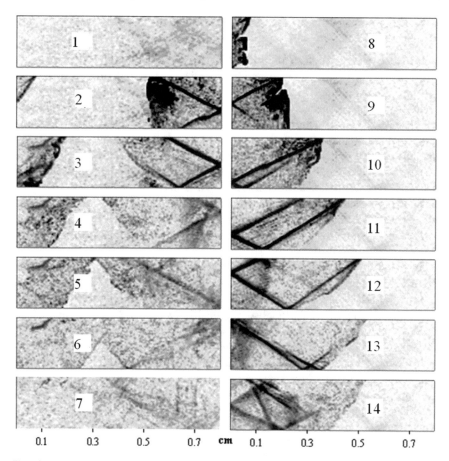

Fig. 12 Equivalent plastic strain rate distributions for strain states 1–14 in Fig. 11

low (Fig. 11c). On further loading, the deformation fronts are set in a regular motion as a consequence of hardening and nonhomogeneous deformation due to propagation of the previous localized-deformation band. The band can move faster or slower or can even cease to move (Figs. 12 and 14), which causes oscillations in the stress-strain curves (Fig. 11b and d). Propagation of one deformation band is responsible for oscillations of larger amplitude (Figs. 12 (8–14), 11d), whereas formation of two deformation fronts propagating from the boundaries of the computational domain in opposite directions is associated with oscillations of smaller amplitude (Figs. 12 (1–7) and 11d).

4.4   Brittle Fracture of Ceramic Particles and Coatings

A distinctive feature of deformation of brittle materials is the fact that under compression their fracture can occur along planes where the macroscopic stresses are thought to be zero [65], so that the crack can propagate along the direction of the applied loading (the X-direction in Fig. 13). For instance, for particle-reinforced metal matrix composites and coated materials it was experimentally shown that cracks in the particles and in the coating under compression are largely oriented along the direction of compression [66, 67]. For the case of modeling of a homogenous specimen, however, the stress tensor components in the transverse direction, Y, which are supposed to open this crack, are identically equal to zero. Thus, in experiments, cracks propagate under stresses that are zero from the standpoint of mechanics. To describe fracture in this case use is made of strain-based criteria. The simplest one is the criterion of positive elongation along the planes normal to the above-mentioned cross-section, [65], i.e., along Y.

In what follows, we are going to show that in simulation of composites the stress tensor components along Y are nonzero in contrast to the case of a homogeneous material. Moreover, at the interfaces there are localized regions oriented with respect to the direction of externally applied compression so that they experience tensile stresses. It is these stresses that can give rise to crack opening and propagation along the direction of external loading.

To describe fracture of the boride coating and corundum particles (model 7 in Fig. 3) use was made of the maximum distortion energy criterion. The criterion is thought to poorly describe fracture of brittle materials. In this work, we show that when it is applied to composite materials with realistically simulated interface geometry, where the calculations contain localized regions of tensile stresses under any type of external loading, the maximum distortion energy criterion works fairly well and provides a correct description of brittle materials and composites. We have modified the criterion to account for the difference in strength values of the tensile and compressive regions:

$$\sigma_{eq} = \begin{cases} C_{ten}, & if\ \varepsilon_{kk} > 0 \\ C_{com}, & if\ \varepsilon_{kk} < 0, \end{cases} \tag{30}$$

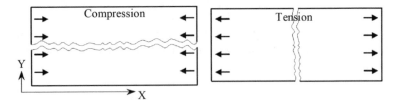

Fig. 13 Fracture under tension and compression

Table 2 Elastic modules, densities and strength constants [15, 68]

	ρ, kg/m^3	K, GPa	μ, GPa	C_{ten}, GPa	C_{com}, GPa
Steels	7900	133	80	–	–
Al6061	2700	76	26	–	–
FeB	7130	200	140	1	4
Al$_2$O$_3$	3990	438	141	0.26	4

where C_{com} and C_{ten} are the values of the tensile and compressive strengths. According to the criterion, Eq. (30), fracture occurs in the local regions undergoing bulk tension. The following fracture conditions are prescribed for any local region of the coating: if the cubic strain ε_{kk} has a negative value and σ_{eq} reaches its critical value C_{com}, then all components of the deviatoric stress tensor in this region are taken to be zero, and in the case of $\varepsilon_{kk} > 0$ and $\sigma_{eq} \geq C_{ten}$, pressure P is equal to zero as well. Elastic modules for the composite constituents and strength parameters for the boride coating and corundum particles are presented in Table 2.

5 Finite-Difference Numerical Procedure

The boundary-value problems in terms of the plane strain were solved numerically by the finite difference method (FDM) [69, 70]. In contrast to the finite-element method (FEM), where the solution of the system is approximated, the FDM approximates the derivatives entering this system.

Let us look at a microstructure region near the base of one of the teeth (Fig. 14a). The region is approximated by a mesh containing N uniform rectangular cells $N = N_x \times N_y$ (Fig. 14b). The mesh is "frozen" into the material and is deformed together with it. The system of equations for this mesh is replaced by a difference analog. Use is made of an explicit conditionally stable scheme of the second order of accuracy. For the time step, it is necessary that the Courant criterion is satisfied as follows:

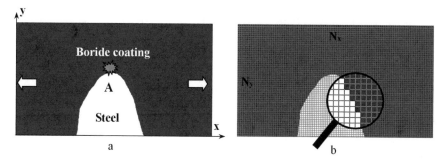

Fig. 14 A-type region of the composite structure (**a**) and discretization of the region (**b**)

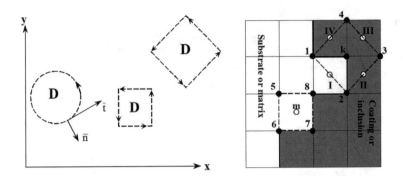

Fig. 15 Schematic representation of approximating the space derivatives

$$\Delta t = k_C \frac{h_{\min}}{C_l}, \tag{31}$$

where $h_{\min}$ is the minimum step of the mesh, C_l is the longitudinal velocity of sound, and $0 < k_C < 1$ is the Courant ratio. The stability condition, Eq. (31), implies that an elastic wave within one time step does not cover the distance longer than the minimum mesh step.

The values of stress σ_{ij}, strain ε_{ij}, and density ρ are computed in the cell centers (points I, II …. in Fig. 15), while those of the displacements u_i and velocities $\dot{u}_i$ correspond to the mesh nodes (points 1, 2 …. in Fig. 15). Let us use the following definition of partial derivatives:

$$\frac{\partial F}{\partial x} = \lim_{D \to 0} \frac{\int_B F(\bar{n} \cdot \bar{i}) ds}{D}, \frac{\partial F}{\partial y} = \lim_{D \to 0} \frac{\int_B F(\bar{n} \cdot \bar{j}) ds}{D}, \tag{32}$$

where B is the boundary of region D, s is the arc length, $\bar{n}$ is the normal vector, and $\bar{t}$ is the tangent vector (Fig. 15).

$$\bar{n} = \frac{\partial x}{\partial n} \bar{i} + \frac{\partial y}{\partial n} \bar{j} = \frac{\partial y}{\partial S} \bar{i} - \frac{\partial x}{\partial S} \bar{j}. \tag{33}$$

Applying Eqs. (32)–(33) to the rectangular regions bounded by dashed lines (Fig. 15), we obtain for the case of stress derivatives corresponding to, e.g., node k

$$\int_C \sigma_{ij}(\bar{n} \cdot \bar{i}) ds = \int_C \sigma_{ij} \frac{\partial y}{\partial S} ds,$$

from which

$$\left(\sigma_{ij,x}\right)^k = \frac{1}{D}\left(\sigma_{ij}^I(y_2 - y_1) + \sigma_{ij}^{II}(y_3 - y_2) + \sigma_{ij}^{III}(y_4 - y_3) + \sigma_{ij}^{IV}(y_1 - y_4)\right) \tag{34}$$

and

$$\int_C \sigma_{ij}(\bar{n} \cdot \bar{j}) ds = \int_C \sigma_{ij} \frac{\partial x}{\partial S} ds,$$

whence

$$\left(\sigma_{ij,y}\right)^k = \frac{1}{D}\left(\sigma_{ij}^I(x_2 - x_1) + \sigma_{ij}^{II}(x_3 - x_2) + \sigma_{ij}^{III}(x_4 - x_3) + \sigma_{ij}^{IV}(x_1 - x_4)\right). \tag{35}$$

Derivatives $u_{i,j}$, in their turn, correspond to the cell centers and are calculated from the values of u_i in the surrounding nodes. In particular, for cell m, we have

$$\left(u_{i,x}\right)^m = \frac{1}{2D}\left(\begin{array}{c}(u_i^5 + u_i^6)(y_6 - y_5)(u_1^6 + u_i^7)(y_7 - y_6) + \\ (u_i^7 + u_i^8)(y_8 - y_7)(u^8 + u_i^5)(y_5 - y_8)\end{array}\right) \tag{36}$$

$$\left(u_{i,y}\right)^m = \frac{1}{2D}\left(\begin{array}{c}(u_i^5 + u_i^6)(x_6 - x_5)(u_1^6 + u_i^7)(x_7 - x_6) + \\ (u_i^7 + u_i^8)(x_8 - x_7)(u^8 + u_i^5)(x_5 - x_8)\end{array}\right) \tag{37}$$

where D is the area of the respective quadrangle. The computation is performed in time steps, moving from one layer n to another $n + 1$.

$$\dot{u}_i = \frac{u_i^{n+1} - u_i^n}{\Delta t} \tag{38}$$

In modeling multi-phase materials, the interface between the microstructure constituents goes across the computational mesh nodes (Fig. 15, thick solid line), with the properties of the two materials prescribed on either side of this interface. The continuity of displacements and normal stresses at this interface is, therefore, preserved, i.e., the conditions of an ideal mechanical contact are satisfied. It is shown in [69] that in the case where the densities of the two materials differ only slightly, the approximating equations, Eqs. (34) and (35), at this interface could be used without any changes. Inaccuracies might appear if we consider, e.g., a solid–liquid interface. In the latter case, it is reasonable to use a formula taking into account both left and right limits in the approximation.

6 Coated Materials

6.1 Overall Plastic Strain and Fracture Behavior Under Tension of the Coated Material with Serrated Interface

In this Section a model microstructure of coated steel DC04 (Fig. 5) with a toothed interface is investigated under tension (Fig. 16a). Figure 16b illustrates a macroscopic stress-strain curve for the mesovolume. The stages in the stress-strain curve are due to the fact that the substrate and coating materials behave in different ways in the corresponding deformation stages: 1—the substrate and coating experience elastic strain, 2—plastic flow develops in the substrate, whereas the coating is still in the elastic state, 3—a cracking of the coating.

Because of the difference in the elastic moduli between the coating and the substrate, the stress and strain distributions are nonuniform even at stage 1. At stage 2, in turn, two stages of plastic deformation can be distinguished. Initially, plastic strains are initiated near the stress concentrators at the root of boride teeth (Fig. 17a). As the loading increases, plastic deformation propagates deep into the specimen, gradually covering the steel-base regions between the boride teeth (Fig. 17b). Thus,

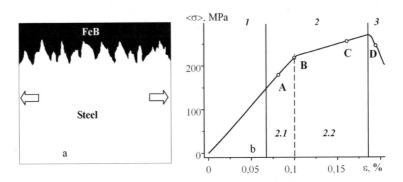

Fig. 16 Computational mesovolume (**a**) and its stress-strain curve under tension (**b**)

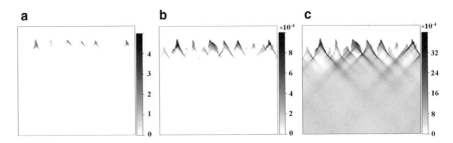

Fig. 17 Equivalent plastic strain distributions for tensile strains of 0.08 (**a**), 0.1, (**b**) and 0.16% (**c**) shown in Fig. 16

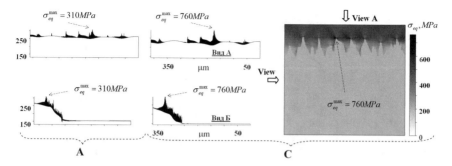

Fig. 18 Equivalent stress distributions at tensile strains lettered by A and A in Figs. 16 and 17

at stage 2.1 (Fig. 16b), plastic flow is localized near the teeth-concentrators and the main part of the base material is still in the elastic state. When an average stress level in the steel base exceeds the yield point, the main part of the base material becomes plastic—stage 2.2 is realized (Fig. 16b) at which the macroscopic stress-strain curve sharply changes the slope. As the plastic flow develops from stage 2.1 to stage 2.2, localized shear bands are formed from the stress concentrators near the boride teeth (Fig. 17c). The bands develop at an angle of about 45° to the axis of loading. Under further loading the average level of plastic strain in the substrate, as well as the degree of strain localization in the bands is increased.

A similar conclusion can be drawn relative to a value of the stress concentration at the "coating–base material" interface. Local concentrations of stresses arise at stage 1. Their distribution is due to the geometry of boride teeth. At the elastic stage, values of local stresses relative to an average level of the equivalent stress do not change in the coating. As plastic flow in the base material develops, the stress concentration increases. At stage 2.1, this effect is less pronounced—the patterns given in Fig. 18a are close to the corresponding distributions at stage 1. An intensive straining of the base material at stage 2.2 leads to a sharp nonlinear increase of local stresses (Fig. 18c).

In this case, the rate of increase of the equivalent stresses can differ for various concentrators. For example, we can see from Fig. 18 that at the stage close to elasticity three stress concentrators lying at the center of the investigated region have almost the same power (Fig. 18a). As plastic flow develops, the stresses in one of these concentrators increase relatively quicker and at the prefracture stage reach a maximum (Fig. 18c).

Stage 3 is the stage of composite failure. Plastic deformation in the base material and cracking in the coating develop simultaneously. These processes are interrelated and interconsistent. When σ_{eq}^{max} exceeds a value of C_{ten}, a first local fracture zone forms in the coating. The surrounding regions of the material begin to intensely deform and the crack propagates toward the free surface perpendicular to the direction of tension (Fig. 19). The crack formation is a dynamic process: there appears a new free surface from which release waves propagate, causing an unloading of the coating material (Fig. 19a). A descending portion is observed on the stress-strain

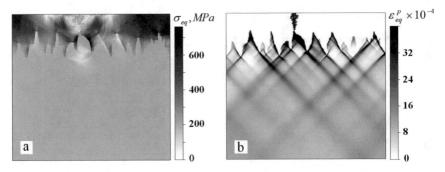

Fig. 19 Equivalent stress (**a**) and plastic strain distributions (**b**) (cracked regions in the coating are given in black) for the state D presented in Fig. 16. Total strain—0.19%

curve (Fig. 16, stage 3). A localized plastic flow enhances in the steel base near the place of crack insipience (Fig. 19b).

6.2 Interface Asperities at Microscale and Mesoscale I. Convergence of the Numerical Solution

Let us examine the loading of two different-scale regions of the coated H418 steel (Fig. 5). The first region is shown in Fig. 14. It is an A-type region at mesoscale I (see Fig. 4). The results of calculations under tensile conditions are given in Figs. 20 and 21. The crack nucleates at the hump of the convexity and propagates towards the specimen surface. Crack propagation is a purely dynamic process occurring at velocities approximating that of sound. The time of crack propagation from the interface to the specimen surface is small compared to the characteristic time of quasistatic loading. The cracking can be regarded as a formation of new free surfaces, from which elastic release waves begin to propagate. The wave dynamics of crack propagation is clearly shown in Fig. 20, where equivalent stress patterns are presented.

To verify the solution convergence, a set of calculations was performed with the step varying in space (Figs. 20 and 21). The total number of cells N in the computational mesh in each case was: 1222, 5035, 8820, 12,768, 19,950, 35,420 and 79,800. The computations showed that the fracture region has a physically-based size that is controlled by the interface geometry—its curvature, and only weakly depends on the size of a local fracture region.

Shown in Fig. 21a are the respective stress-strain curves. The drooping part of the curves corresponds to the initiation and propagation of a unit crack. Figure 21b depicts the dependence of the maximum value of $\langle \sigma \rangle$, corresponding to the onset of crack propagation, on the computational mesh size. It is evident that the solution is convergent and well approximated by the exponential law (dotted line)

$$\langle\sigma\rangle_{\max} = \sigma_{act} + \sigma_{mesh}\exp\!\left(N_x/N_{ref}\right), \tag{39}$$

where $\sigma_{act} = 556$ MPa is the actual stress, $\sigma_{mesh} = 140$ MPa is the mesh dependent overestimation and $N_{ref} = 79$ is the reference number of cells.

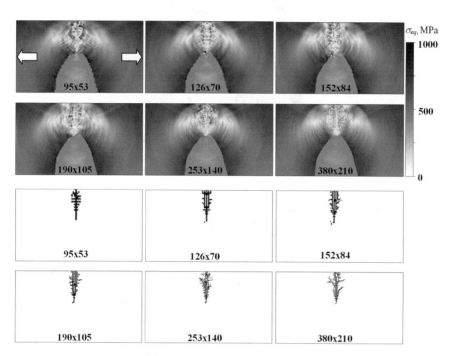

Fig. 20 Mesh-dependent equivalent stress and fracture patterns in the region presented in Fig. 14 in tension

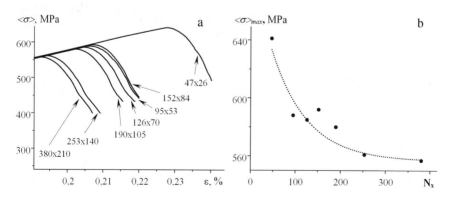

Fig. 21 Homogenized stress-strain curves (**a**) and the stress maximum values versus the number of cells in the X-direction (**b**)

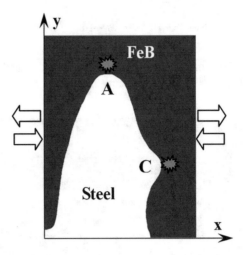

Fig. 22 Calculated region containing characteristic A- and C-type inhomogeneities

For the second computational run, we selected a larger region (Fig. 22). This region is part of the structure given in Fig. 4; it includes the region in Fig. 14, and contains both A- and C-type inhomogeneities. Figure 23 is an illustration of the computation results under different types of external loading: tension and compression. It is evident that initiation and propagation of cracks under tensile and compressive loading occurs in different places. Under tension, cracks primarily propagate from the teeth to the specimen surface, while under compression—from one lateral face of a tooth to the other. This phenomenon is discussed in Sect. 7.4 in details.

The investigations of mesh convergence showed that in the case where a step in space is quite small, in other words when an A-type characteristic convexity of the

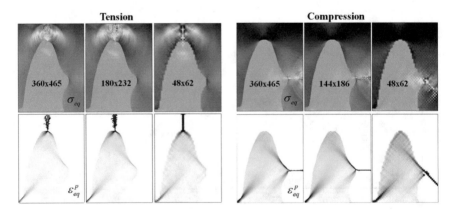

Fig. 23 Mesh-dependent equivalent stress and fracture patterns for a region containing A- and C-type inhomogeneities under tension and compression

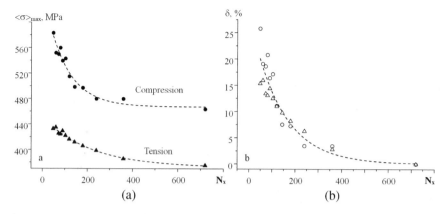

Fig. 24 Stress maximum values under tension and compression (**a**) and an average error (**b**) versus the number of cells in the X direction

Table 3 Convergence exponential parameters for the crack initiation in the coated steel

	σ_{act}, MPa	σ_{mesh}, MPa	N_{ref}
Tension	372	79	217
Compression	467	182	97

least curvature is approximated by as many as 10 computational cells, the character of fracture changes, but only slightly. In Fig. 24a the curves under both tension and compression are presented with the convergence parameters shown in Table 3. Figure 24b shows an averaged estimation of the convergence. An average error δ connected with the mesh size effects is calculated as

$$\delta = \frac{\langle\sigma\rangle_{max} - \sigma_{act}}{\sigma_{act}} \cdot 100\%. \tag{40}$$

The dependence is approximated by the following formula (dotted line in Fig. 24b)

$$\delta = 28.5\exp\left(-\frac{N_x}{136}\right). \tag{41}$$

The calculations discussed in this paragraph yield the following conclusions. Interface convexities are sources of unit cracks in the coating. The mesh-convergence analysis showed that the solution for this system of primary cracks converges when the step in space is decreased. The finer the mesh, the more detailed the fracture pattern, while the general behavior is similar for different meshes. There are limiting real stress and strain values for the onset of fracture, which are controlled by the physical geometry of concavities and their curvature. For the investigated microstructure and properties of contacting materials, the maximum mesh size error is about 28%.

6.3 Fracture of the Coating with Plane Interface. Macroscale Simplification

In this Section, the fracture criterion, Eq. (30), is shown to provide an adequate description of directions of crack propagation under different types of external loading: tension and compression. Let us simulate tension and compression of a specimen with a plane interface, i.e., excluding the interface curvature factor. Since the interface is plane, there is no concentration of tensile stresses. Fracture would not appear locally as we have excluded the cause for its local initiation. In this case, we have to artificially nucleate cracking by initially introducing a single fractured zone at the steel–boride coating interface. Even if an interface does not exhibit any serrated structure and appears to be regular at a certain scale, in the places of its adhesion to the substrate there could be various inhomogeneities, including cracks and discontinuities.

In Fig. 25, we show the results of computations under tensile and compressive loading. The differences in the direction of crack propagation are clearly seen: propagation is along the interface in the case of compression and perpendicular to it in the case of tension. For the latter, the coating separates along the interface due to the absence of boride teeth grown into the steel substrate. Thus, the fracture criterion, Eq. (30), correctly describes the direction of crack propagation in a brittle material, and its use, combined with an explicit interface of a complicated geometry, would allow us (as it will be shown later in Sect. 7.4) to interpret a possible mechanism of fracture initiation. This mechanism, according to our simulation results, is associated with the mechanical concentration of tensile stresses in the places where the interface curvature changes.

Fig. 25 Crack propagation in the coating with a plane interface under tension and compression

6.4   Plastic Strain Localization and Fracture at Mesoscale II. Effects of the Irregular Interfacial Geometry Under Tension and Compression of Composites

In the following sections we consider a real microstructure of a coated material with an interface of irregular geometry (Fig. 4b). Steel STE250 (Fig. 5) was selected as a substrate to be coated by diffusion borating. A macroscopic response of the composite microstructure under different external loading conditions is presented in Fig. 26.

According to the calculated dependences, the coated material shows higher tolerance to compressive stresses than to external tensile loading: the macroscopic yield strength, both in terms of stress and strain, is higher in the case of compression (see Fig. 26). This mechanical behavior is typical for composite materials and, as the calculations show, is associated with the principal difference of fracture processes in the coating under different external loading conditions.

Figure 27 shows the distribution of the stress tensor components under external compressive loading for the cases of needle-like and plane "coating-substrate" interfaces. It can be seen that under uniaxial loading of a coated material with the plane interface along the X-direction, only the values of σ_{xx} remain nonzero throughout the computational domain. In contrast, the serrated shape of the substrate–coating interface favors the development of a complex stressed state with nonzero values of the stress component σ_{yy}. Note that it is in the Y-direction that the material experiences both compressive and tensile stresses which are in their absolute values comparable to the values of external compressive loads. Thus, the regions of the steel substrate

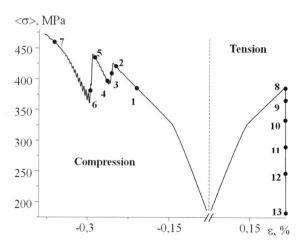

Fig. 26 Calculated stress-strain curves of the coated material under tension and compression. The respective stress and strain distributions in the mesovolume for states 1–13 are given in Figs. 27, 28 and 29

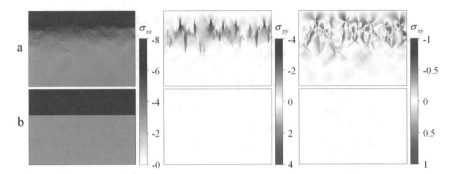

Fig. 27 Distributions of stress tensor components ($\times 100$ MPa) for needle-like (**a**) and plane "coating-substrate" interfaces (**b**) at a strain corresponding to point (1) in Fig. 26

located between the boride teeth are subjected to compressive stresses while the teeth themselves experience tensile stresses.

Should we change the direction of external loading and address tension rather than compression, the pattern presented in Fig. 27 would be the same both qualitatively and quantitatively, the difference being in the sign of the stress tensor components. Thus, the local tensile stresses develop in different places under tension and compression. This fact is responsible for the difference in fracture processes under tension and compression (Fig. 28).

Both tension and compression cracks originate in the local tensile regions. Under compression, the regions are situated at the lateral side of the boron teeth (red color regions in Fig. 27). Cracks successively nucleate on boride tooth sides and propagate along the axis of compression (Fig. 28, states 2–7). No formation of the main longitudinal crack is, however, observed. The upper coating layer maintains its stressed state and resists to the load, while multiple cracking of a boride tooth unloads the composite in the intermediate sublayer (Fig. 28, state 7). Thus, the presence of a serrated structure grown into the steel substrate prevents the coating from spalling. The stress-strain curve in this case exhibits local drops of the averaged stress, whose general level, however, continues to increase, and no catastrophic loss of strength is observed (see Fig. 26b).

A different fracture pattern is found under external tension (Fig. 28, states 8–13). The crack nucleates in the local region of highest concentration of tensile stress, which is situated at a boron tooth base, and propagates in the boride coating towards the free surface of the specimen. This unloads the material along the direction of applied tension. A descending portion appears in the stress-strain curve (points 8–13 in Fig. 26).

The formation of longitudinal and transverse cracks was found experimentally during nanoindentation (Fig. 29b), which, due to its specific geometry, gives rise both to tensile and compressive conditions in the coating within one and the same experiment. Cracks in this experiment propagate in different ways: perpendicular and parallel to the direction of applied tensile and compressive loading, respectively.

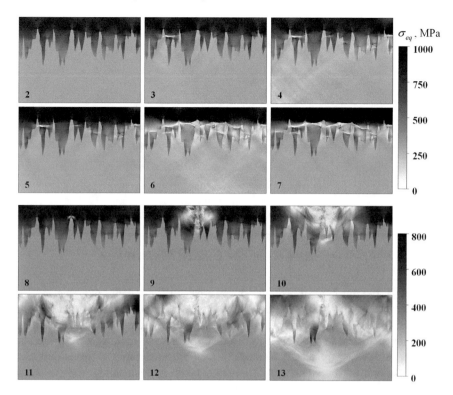

Fig. 28 Equivalent stress distributions for compressive (2–7) and tensile (8–13) strains (cf. Fig. 26)

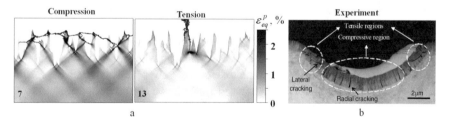

Fig. 29 Equivalent plastic strain distributions for states 2 and 13 (see Figs. 26 and 28) (**a**) and microscopic section of a TiN coating deposited on stainless steel after nanoindentation [67] (**b**)

The same result was obtained in the discussed simulations of two different experiments on uniaxial compression/tension (Fig. 29a). Note that the key role belongs to the complicated geometry of the interface and the presence of regions undergoing localized tensile stresses.

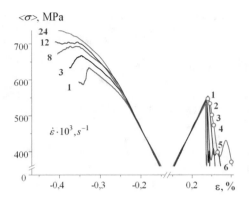

Fig. 30 Homogenized stress-strain curves under compression and tension at different strain rates

6.5 Dynamic Deformation of the Coated Material

To describe the mechanical response of the steel substrate under dynamic loading use was made of the constitutive Eq. (20) taking into account plastic strain rate and temperature in an explicit form. This equation allows a prediction of the mechanical properties of austenitic steels (Fig. 5). Boron coating is elastic-brittle. Experimental evidence shows that the elastic modules as well as the strength of brittle ceramics weakly depend on the strain rate. Therefore, strain rate sensitivity for the boron material was not introduced into the model formulation.

A series of numerical compression and tension tests were carried out by varying the strain rate externally applied (Fig. 30). The calculations show a difference between the composite responses under different types of loading: tension and compression. The following conclusions can be made.

First, under the high-strain-rate tension the fracture process intensifies in comparison with that observed under quasistatic loading. Figure 31 shows the simulation of crack initiation and growth under strain rate of $24 \times 10^3 \, s^{-1}$. Comparing the results with those presented in Fig. 28 (states 8–13), it can be seen that under quasistatic loading only one crack propagates, whereas under a high strain rate multiple cracks arise. The stress under a high strain rate increases rapidly, and release waves from the first crack formation are not in time for unloading the nearby A-type stress concentration regions, i.e. the equivalent stress in one of these regions reaches the strength value C_{ten} before the release wave arriving (see Fig. 31, states 2–3).

The second conclusion is that the macroscopic strength under tension changes, but only slightly with the strain rate increasing, while both the total strain and homogenized stress of the fracture onset strongly depend on the value of the compression strain rate (Fig. 30). Figure 32 shows the stress, plastic strain and fracture patterns under compression at different strain rates. The simulations show that the higher the strain rate, the less intensive the cracking of the coating and, as a result, the higher the dynamic strength of the coated material (see Fig. 30). The explanation is the

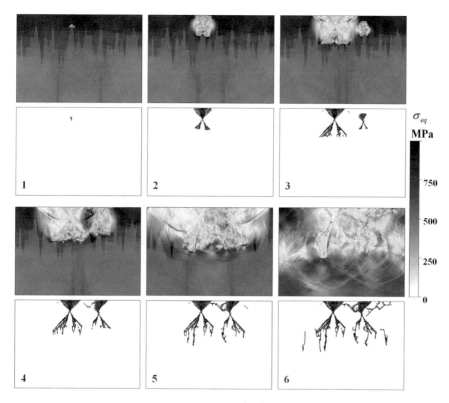

Fig. 31 Coating cracking at a strain rate of 24×10^3 s^{-1}. Equivalent stress and fracture patterns correspond to states 1–6 shown in Fig. 30

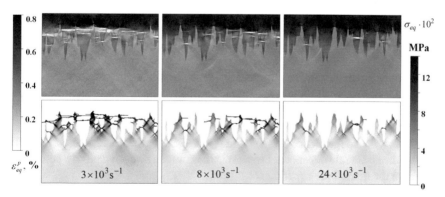

Fig. 32 Distributions of equivalent stresses and the equivalent plastic strains (fractured regions in the coating are marked by black color) for different strain rates of compression. Total strain—0.37% (see Fig. 30)

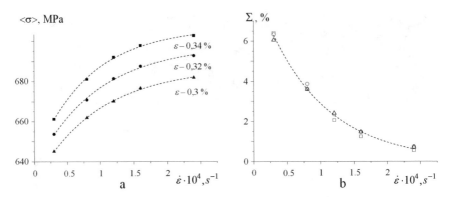

Fig. 33 Homogenized stress values for different compressive strains (**a**) and relative increase in the stress (**b**) versus the strain rate of compression

following. The value of stress concentration in C-type regions depends on the difference between mechanical properties of the steel and boron ceramics. According to the model formulation, the stress in the boron material does not change but the current dynamic yield stress of the steel increases with the strain rate increasing (Fig. 5). Therefore, the stress difference at a certain macroscopic strain decreases, and this arises only due to plasticity of the steel substrate. As the analysis of the calculation results showed, the fracture of the coating under tension develops at the elastic stage of composite deformation. The difference discussed does not change with the strain rate increasing at the elastic stage, and there is no change in the macroscopic strength.

The dependence of the macroscopic stress on the compression strain rate is shown in Fig. 33a for different strains (Fig. 30). The curves are well approximated by the exponential law (dotted lines)

$$\langle \sigma \rangle = \sigma_{dyn} - (\sigma_{dyn} - \sigma_{stat}) \exp\left(\dot{\varepsilon}/\dot{\varepsilon}_{ref}\right), \qquad (42)$$

where σ_{dyn} is the saturation dynamic stress, σ_{stat} is the stress under quasistatic loading, and $\dot{\varepsilon}_{ref}$ is the reference strain rate. Considering the relative value, the formula can be rewritten as

$\Sigma = \Delta\Sigma \exp\left(\dot{\varepsilon}/\dot{\varepsilon}_{ref}\right)$, where $\Sigma = \frac{\sigma_{dyn} - \langle \sigma \rangle}{\sigma_{dyn}}$, $\Delta\Sigma = \frac{(\sigma_{dyn} - \sigma_{stat})}{\sigma_{dyn}}$.

The obtained dependence presented in Fig. 33b shows that there is a possibility to predict the strain rate dependent increase in the compressive strength of a composite based on the quasistatic experiments.

7 Metal-Matrix Composites

A mesovolume of a metal matrix composite is represented schematically in Fig. 34a. The microstructure was chosen according to the experiments reported in [35].

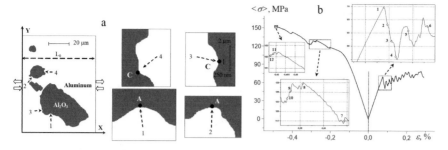

Fig. 34 Calculated composite microstructure (**a**) and its tensile and compressive stress-strain curves (**b**)

Figure 34b illustrates the macroscopic response of the mesovolume in tension and compression. Similar to the coated material, the composite under study is seen to withstand a higher load in compression than in tension. At first glance, this difference could be due to the substantial difference in the tensile and compressive strength values of the particles: since $C_{com} \gg C_{ten}$, it must take a much higher average stress level for local regions to fail in compression than in tension. However, a proportional increase in the macroscopic yield stress (relative to the ratio between aluminum and alumina volume fractions) does not occur for two reasons. First, plastic deformation in aluminum causes the total stress level to decrease (stress deviator is constrained), and free surfaces are a hindrance to a build-up of pressure. Second, the calculations for Al/Al$_2$O$_3$ under different types of applied loading have shown that C_{com} is not reached.

Again, for the composite the microstructural inhomogeneity and interfacial effect are responsible for the formation of tensile regions, while the mesovolume is subjected to compression, and it is essential that the required values of tensile stresses are obtained in these regions. This conclusion is of critical importance for an analysis of the resulting plastic deformation of the matrix and of crack growth in brittle particles. It is necessary to stress that in simulation of uniaxial loading of a homogeneous material the stress tensor components along the direction perpendicular to the loading axis are assumed to be zero over the entire region of interest.

In the case of the metal-matrix composite the following types of inhomogeneities referred to the meso II, meso I, and microscales by their characteristic sizes can be singled out: (1) $\approx 20\,\mu m$ for the particles, (2) $\approx 2\,\mu m$ for the matrix-particle interfacial asperities, and (3) $\approx 250\,nm$ for a local fracture zone. The particle is a mesoscale stress concentrator responsible for the formation of macroscopic localized shear bands in the matrix. The type 2 inhomogeneities will result in the stress concentration at the mesoscale I. As a consequence, plastic deformation will be localized in the matrix in the vicinity of interfacial asperities, and primary fracture zones will be formed in the particle.

Ideally the inhomogeneities can be assumed to take the form of a true circle. Hence, the stress concentration and corresponding types of the stressed state can be estimated using an analytical solution obtained in [71] for a round inclusion embedded in a matrix. The materials for the matrix and inclusion are chosen arbitrarily. For inhomogeneities of type 1, a solution for a stiff inclusion surrounded by a comparatively soft matrix material is valid. For inhomogeneities of type 2, on the contrary, a relatively soft inclusion will be found in a rigid matrix. Finally, in the limiting case 3, we will have to solve the classical elasticity theory problem on the influence of a round hole on the stress distribution in a plate (Fig. 35). Our analytical and numerical estimations have shown that the maximum values of σ_{eq} for $\varepsilon_{kk} > 0$ are obtained at points of the A-type in tension, and at points of the C-types in compression (Fig. 35).

The first fracture zone nucleates in the vicinity of a hump of the interface concavities (Fig. 35). The fracture zone is a new stress concentrator. A new interface between fractured material and ceramics is characterized by a higher curvature due to a smaller area and by a larger difference in mechanical characteristics than the aluminum–alumina interface, since the fractured material no longer resists shear. This is a more powerful stress concentrator at the microscale than a concavity at the mesoscale I but it is formed following the same principle. Under external tension, the maximum value of the equivalent stress is observed in the A type tensile regions near the new fractured material—alumina interface (Fig. 35a), while under external compressive loading it is the C regions that undergo tensile stresses, in whose vicinity the second fracture zone is formed (Fig. 35b). Further, the process is repeated, and the crack propagates perpendicular to the direction of applied loading in the case of tension, and parallel to it—under compression.

Deformation and fracture of the composite are illustrated in Fig. 36, where the equivalent stress and strain distributions and velocity fields superimposed on a

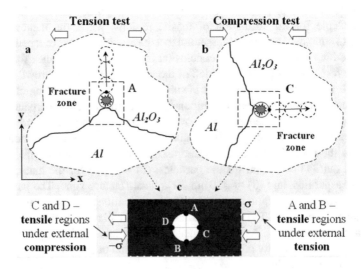

Fig. 35 Schematics of crack propagation in aluminum-alumina composite in different loading conditions

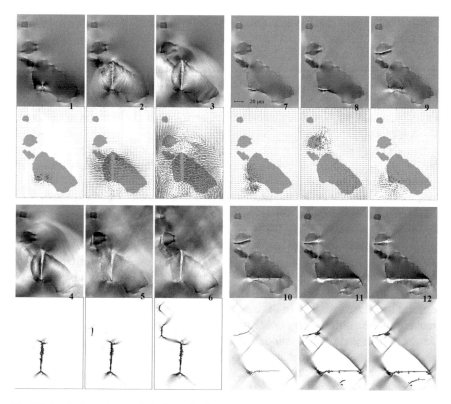

Fig. 36 Equivalent stress and plastic strain distributions and velocity fields in tension (1–6) and in compression (7–12) for the stress-strain curves in Fig. 34b

microstructure map are shown. The calculated results are presented for compression (1–6) and tension (7–12). The corresponding states (7–12) in the stress-strain curves are shown in Fig. 34b on an enlarged scale.

It is obvious from Fig. 36 (1) that in tension, the primary crack is initiated in the largest particle in the neighborhood of point A (Fig. 34a), where the stress concentration σ_{eq} is at its maximum. The crack propagates in the direction normal to that of tensile load and reaches the opposite side of the interface (Fig. 36 (1–4)). As this takes place, the adjacent regions are unloaded due to release waves propagating from newly formed free surfaces. This is supported by the descending portion of the stress-strain curve (Fig. 34(1–4)). The velocity fields in Fig. 36 (1–3) illustrate the crack opening process. Interactions of elastic waves with interfaces give rise to formation of a complex stressed state and may be responsible for generation of vortex structures. Rotations of local regions contribute to further increase in the stress concentration (Fig. 36 (3)). On further loading the average stress level in the region of interest is increased. Another stress concentrator is formed in a particle of smaller size, and new cracks are initiated (Fig. 36 (4–6)) with the resulting unloading of the material (Fig. 34 (2, 5, 6)). Figure 36 (4–6) shows the equivalent plastic strain distributions.

The fracture regions corresponding to a maximum value of plastic strain are shown in black. The tension crack nucleation and growth in the mesovolume are seen to occur essentially in the elastic deformation stage. This is attributed to the fact that the average stress level in the matrix is below the yield point $\sigma_0 = 62$ MPa (see Table 1), whereas the stress concentration in the local regions of the particles (points A in Fig. 34a) may be above a critical point $C_{ten} = 260$ MPa (see Table 2). Moderate plastic strain is seen in the neighborhood of points where local fracture zones are generated localized (Fig. 36 (1–4)).

In compression (Fig. 36 (7–12)), cracks are initiated at points of the maximum tensile stress concentration (points C in Fig. 34a). Notably, the equivalent stress at points of local compression (points A) is much higher than at points C. Thus, the compression fracture of the particle occurs at a much higher total stress level than that involved in tension (Fig. 34b) and is accompanied by high-intensity plastic deformation in the matrix (Fig. 36 (10–12)).

The compression cracks propagate in the loading direction. However, unlike the case of tensile load the crack propagation exhibits an oscillatory pattern. Severe plastic deformation of the matrix hinders fast increasing of the stress concentration. That is why the average velocity of compression-induced crack propagation through particles is much lower than in the case of tension-induced cracks. In consequence, cracking of the particles may occur in a switching mode. The regime in question can easily be traced by examining field velocities (Fig. 36 (7–9)). The first crack is initiated to produce an unloading effect of the mesovolume (Fig. 36 (7)). Then, the crack ceases to propagate, and the average stress level is seen to rise thereafter for a fairly long period of time due to strain hardening of the matrix (Fig. 34 (2, 7)). A new local fracture zone is formed in another particle. When a second crack ceases to grow the primary crack resumes its growth, propagating slowly (Fig. 34 (9–11)) to the opposite side of the interface (Fig. 36 (9–11)). Next, a third crack is formed (Fig. 36 (12)) and the process recurs.

In the calculations presented in Fig. 36, both for tension and for compression, the largest, medium-size, and the smallest particles are involved successively in the fracture process. Similar situations where the largest grains undergo the largest deformations and, in consequence, suffer the greatest damage have been observed experimentally. However, for an arbitrary microstructure of the type studied here (Fig. 34a), the sequence of events may be due to other reasons in addition to the size factor: different shape of the interface segments at points of crack nucleation and different types of the stressed-strained state of particles by virtue of their particular relative positions. We looked for ways to avoid the effects of the geometry and the loading conditions. To this end, a series of calculations were performed for composite microstructures subjected to tension. In the calculations, particles of the same shape and varying size (centers of masses of the particles) were aligned at the center of the computational domain. There are six independent combinations of relative positioning of particles according to this principle (Fig. 37). As in the former calculation presented in Fig. 36, the order of failure of the particles (the largest— medium-size—the smallest particles) is the same in all cases at hand. Such a fracture pattern is accounted for on the basis of the foregoing analysis of the calculated results

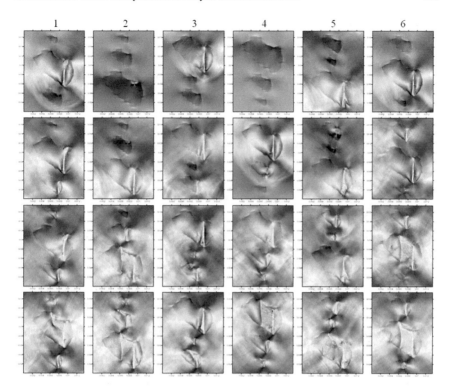

Fig. 37 Fracture patterns in composite mesovolumes with particles of the same shape and different size for 6 independent combinations of relative positions of the particles

and analytical estimation of the stress concentration: the larger the particle, the closer the stress concentration near the type II inhomogeneity approaches a maximum value obtained from an analytical solution for an infinite domain, and hence the sooner a local fracture zone is formed.

Acknowledgements This work was performed according to the Government research assignment for ISPMS SB RAS, and was supported by the Russian Science Foundation (Project No. 18-19-00273).

References

1. Panin VE, Elsukova TF, Ivanchin AG (1982) Structural levels of deformation of solids. Russ Phys J (Sov Phys J) 25(6):479–497
2. Panin VE (1998) Physical mesomechanics of heterogeneous media and computer-aided design of materials. Cambridge International Science Publishing Ltd., Cambridge

3. Panin VE, Egorushkin VE (2015) Basic physical mesomechanics of plastic deformation and fracture of solids as hierarchically organized nonlinear systems. Phys Mesomech 18(4):377–390
4. Needleman A, Asaro RJ, Lemonds J, Peirce D (1985) Finite element analysis of crystalline solids. Comput Methods Appl Mech Eng 52(1–3):689–708. https://doi.org/10.1016/0045-782 5(85)90014-3
5. Sih GC, Chao CK (1989) Scaling of size/time/temperature part 1 + 2. Theoret Appl Fract Mech 12(2):93–119
6. Psakhie SG, Korostelev SYu, Negreskul SI, Zolnikov KP, Wang Z, Li S (1993) Vortex mechanism of plastic deformation of grain boundaries—computer simulation. Physica Status Solidi B—Basic Solid State Phys 176(2):K41–K44. https://doi.org/10.1002/pssb.2221760227
7. Needleman A (2000) Computational mechanics at the mesoscale. Acta Mater 48(1):105–124. https://doi.org/10.1016/S1359-6454(99)00290-6
8. Psakhie SG, Zavshek S, Jezershek J, Shilko EV, Smolin AYu, Blatnik S (2000) Computer-aided examination and forecast of strength properties of heterogeneous coal-beds, Comput Mater Sci 19(1–4):69–76. https://doi.org/10.1016/S0927-0256(00)00140-3
9. Balokhonov RR, Makarov PV, Romanova VA, Smolin IYu, Savlevich IV (2000) Numerical modelling of multi-scale shear stability loss in polycrystals under shock wave loading. J de Physique IV France 10(9):515–520. https://doi.org/10.1051/jp4:2000986
10. Psakhie SG, Horie Y, Ostermeyer GP, Korostelev SYu, Smolin AYu, Shilko EV, Dmitriev AI, Blatnik S, Špegel M, Zavšek S (2001) Movable cellular automata method for simulating materials with mesostructured. Theoret Appl Fract Mech 37(1–3):311–334. https://doi.org/10. 1016/S0167-8442(01)00079-9
11. Nicot, F., Darve, F., RNVO Group (2005) A multi-scale approach to granular materials. Mech Mater 37(9):980–1006. https://doi.org/10.1016/j.mechmat.2004.11.002
12. Balokhonov RR (2005) Hierarchical numerical simulation of nonhomogeneous deformation and fracture of composite materials. Phys Mesomech 8(3–4):99–120
13. Romanova V, Balokhonov R, Panin A, Kazachenok M, Kozelskaya A (2017) Micro- and mesomechanical aspects of deformation-induced surface roughening in polycrystalline titanium. Mater Sci Eng, A 697:248–258
14. Ghosh S, Bai J, Raghavan P (2007) Concurrent multi-level model for damage evolution in microstructurally debonding composites. Mech Mater 39(3):241–266. https://doi.org/10.1016/ j.mechmat.2006.05.004
15. Psakhie SG, Shilko EV, Smolin AYu, Dimaki AV, Dmitriev AI, Konovalenko IS, Astafurov SV, Zavshek S (2011) Approach to simulation of deformation and fracture of hierarchically organized heterogeneous media, including contrast media. Phys Mesomech 14(5–6):224–248. https://doi.org/10.1016/j.physme.2011.12.003
16. Balokhonov RR, Romanova VA, Schmauder S, Schwab E (2012) Mesoscale analysis of deformation and fracture in coated materials. Comput Mater Sci 64:306–311. https://doi.org/10. 1016/j.commatsci.2012.04.013
17. Psakhie SG, Shilko EV, Grigoriev AS, Astafurov SV, Dimaki AV, Smolin AYu (2014) A mathematical model of particle-particle interaction for discrete element based modeling of deformation and fracture of heterogeneous elastic-plastic materials. Eng Fract Mech 130:96–115. https://doi.org/10.1016/j.engfracmech.2014.04.034
18. Popov VL, Dimaki A, Psakhie S, Popov M (2015) On the role of scales in contact mechanics and friction between elastomers and randomly rough self-affine surfaces. Sci Rep 5:11139. https://doi.org/10.1038/srep11139
19. Shilko EV, Psakhie SG, Schmauder S, Popov VL, Astafurov SV, Smolin A (2015) Overcoming the limitations of distinct element method for multiscale modeling of materials with multimodal internal structure. Comput Mater Sci 102:267–285. https://doi.org/10.1016/j.commatsci.2015. 02.026
20. Schmauder S, Schäfer I (2016) Multiscale materials modeling: approaches to full multiscaling. De Gruyter, Berlin, Boston. https://doi.org/10.1515/9783110412451

21. Patil RU, Mishra BK, Singh IV (2019) A multiscale framework based on phase field method and XFEM to simulate fracture in highly heterogeneous materials. Theoret Appl Fract Mech 100:390–415. https://doi.org/10.1016/j.tafmec.2019.02.002

22. Balokhonov RR, Romanova VA, Schmauder S, Emelianova ES (2019) A numerical study of plastic strain localization and fracture across multiple spatial scales in materials with metal-matrix composite coatings. Theoret Appl Fract Mech 101:342–355. https://doi.org/10.1016/j. tafmec.2019.03.013

23. Llorca J, Needleman A, Suresh S (1991) An analysis of the effects of matrix void growth on deformation and ductility in metal-ceramic composites. Acta Metall Mater 39(10):2317–2335. https://doi.org/10.1016/0956-7151(91)90014-R

24. Ghosh S, Nowak Z, Lee K (1997) Quantitative characterization and modeling of composite microstructures by Voronoi cells. Acta Mater 45(6):2215–2234. https://doi.org/10.1016/S1359-6454(96)00365-5

25. Romanova V, Balokhonov R, Makarov P, Schmauder S, Soppa E (2003) Simulation of elasto-plastic behaviour of an artificial 3D-structure under dynamic loading. Comput Mater Sci 28(3–4):518–528. https://doi.org/10.1016/j.commatsci.2003.08.009

26. Diard O, Leclercq S, Rousselier G, Cailletaud G (2005) Evaluation of finite element based analysis of 3D multicrystalline aggregates plasticity: application to crystal plasticity model identification and the study of stress and strain fields near grain boundaries. Int J Plast 21:691–722. https://doi.org/10.1016/j.ijplas.2004.05.017

27. Pierard O, LLorca J, Segurado J, Doghri I (2007) Micromechanics of particle-reinforced elasto-viscoplastic composites: Finite element simulations versus affine homogenization. Int J Plast 23(6):1041–1060. https://doi.org/10.1016/j.ijplas.2006.09.003

28. Romanova V, Balokhonov R (2019) A method of step-by-step packing and its application in generating 3D microstructures of polycrystalline and composite materials. Eng Comput. https://doi.org/10.1007/s00366-019-00820-2

29. Romanova VA, Balokhonov RR, Schmauder S (2013) Numerical study of mesoscale surface roughening in aluminum polycrystals under tension. Mater Sci Eng, A 564:255–263. https:// doi.org/10.1016/j.msea.2012.12.004

30. Donegan SP, Rollett AD (2015) Simulation of residual stress and elastic energy density in thermal barrier coatings using fast Fourier transforms. Acta Mater 96:212–228. https://doi.org/ 10.1016/j.actamat.2015.06.019

31. Josyula SK, Narala SKR (2018) Study of TiC particle distribution in Al-MMCs using finite element modeling. Int J Mech Sci 141:341–358. https://doi.org/10.1016/j.ijmecsci.2018.04.004

32. Pachaury Y, Shin YuC (2019). Assessment of sub-surface damage during machining of additively manufactured Fe-TiC metal matrix composites. J Mater Process Technol 266:173–183. https://doi.org/10.1016/j.jmatprotec.2018.11.001

33. Sørensen N, Needleman A, Tvergaard V (1992) Three-dimensional analysis of creep in a metal matrix composite. Mater Sci Eng, A 158(2):129–137. https://doi.org/10.1016/0921-509 3(92)90001-H

34. Soppa E, Schmauder S, Fischer G, Brollo J, Weber U (2003) Deformation and damage in Al/Al$_2$O$_3$. Comput Mater Sci 28(3–4):574–586. https://doi.org/10.1016/j.commatsci.2003. 08.034

35. Chawla N, Sidhu RS, Ganesh VV (2006) Three-dimensional visualization and microstructure-based modeling of deformation in particle-reinforced composites. Acta Mater 54(6):1541–1548. https://doi.org/10.1016/j.actamat.2005.11.027

36. Balokhonov RR, Romanova VA (2009) The effect of the irregular interface geometry in deformation and fracture of a steel substrate–boride coating composite. Int J Plast 25(11):2225–2248. https://doi.org/10.1016/j.ijplas.2009.01.001

37. Balokhonov RR, Romanova VA, Schmauder S, Martynov SA, Kovalevskaya ZhG (2014) Mesomechanical analysis of plastic strain and fracture localization in a material with a bilayer coating. Compos: Part B: Eng 66:276–286. https://doi.org/10.1016/j.compositesb.2014.05.020

38. Nayebpashaee N, Seyedein SH, Aboutalebi MR, Sarpoolaky H, Hadavi SMM (2016) Finite element simulation of residual stress and failure mechanism in plasma sprayed thermal barrier

coatings using actual microstructure as the representative volume. Surf Coat Technol 291:103–114. https://doi.org/10.1016/j.surfcoat.2016.02.028

39. Balokhonov RR, Romanova VA, Panin AV, Kazachenok MS, Martynov SA (2018) Strain localization in titanium with a modified surface layer. Phys Mesomech 21(1):32–42

40. Needleman A (1990) An analysis of tensile decohesion along an interface. J Mech Phys Solids 38(3):289–324. https://doi.org/10.1016/0022-5096(90)90001-K

41. Needleman A, Ortiz M (1991) Effect of boundaries and interfaces on shear-band localization. Int J Solids Struct 28(7):859–877. https://doi.org/10.1016/0020-7683(91)90005-Z

42. Rabinovich VL, Sarin VK (1996) Modelling of interfacial fracture. Mater Sci Eng, A 209(1–2):82–90. https://doi.org/10.1016/0921-5093(95)10141-1

43. Needleman A, Rosakis AJ (1999) The effect of bond strength and loading rate on the conditions governing the attainment of intersonic crack growth along interfaces. J Mech Phys Solids 47(12):2411–2449. https://doi.org/10.1016/S0022-5096(99)00012-5

44. Chandra N, Ghonem H (2001) Interfacial mechanics of push-out tests: theory and experiments. Compos A Appl Sci Manuf 32(3–4):575–584. https://doi.org/10.1016/S1359-835X(00)00051-8

45. Chiu Z-C, Erdogan F (2003) Debonding of graded coatings under in-plane compression. Int J Solids Struct 40(25):7155–7179. https://doi.org/10.1016/S0020-7683(03)00360-3

46. Wu X-F, Jenson RA, Zhao A (2014) Stress-function variational approach to the interfacial stresses and progressive cracking in surface coatings. Mech Mater 69(1):195–203. https://doi.org/10.1016/j.mechmat.2013.10.004

47. Guan K, Jia L, Kong B, Yuan S, Zhang H (2016) Study of the fracture mechanism of NbSS/Nb5Si3 in situ composite: based on a mechanical characterization of interfacial strength. Mater Sci Eng, A 663:98–107. https://doi.org/10.1016/j.msea.2016.03.110

48. Dehm G, Jaya BN, Raghavan R, Kirchlechner C (2018) Overview on micro- and nanomechanical testing: new insights in interface plasticity and fracture at small length scales. Acta Mater 142:248–282. https://doi.org/10.1016/j.actamat.2017.06.019

49. Meyers MA, Murr LE (1981) Shock waves and high-strain-rate phenomena in metals. Plenum Press, New York

50. Dudarev EF, Kornienko LA, Bakach GP (1991) Effect of stacking-fault energy on the development of a dislocation substructure, strain hardening, and plasticity of fcc solid solutions. Russ Phys J 34:207–216

51. Kozlov EV, Teplykova LA, Koneva NA, Gavrilyuk VG, Popova NA (1996) Role of solid solution hardening and interactions in dislocation ensemble in formation of yield stress of austenite nitrogen steel. Russ Phys J 39:211–229

52. Nemat-Nasser S, Guo W-G (2005) Thermomechanical response of HSLA-65 steel plates: experiment and modeling. Mech Mater 37(2–3):379–405. https://doi.org/10.1016/j.mechmat.2003.08.017

53. Beukel AVD, Kocks UF (1982) The strain dependence of static and dynamic strain-aging. Acta Metall 30(5):1027–1034. https://doi.org/10.1016/0001-6160(82)90211-5

54. Kubin LP, Estrin Y, Perriers C (1992) On static strain aging. Acta Metall Mater 40(5):1037–1044. https://doi.org/10.1016/0956-7151(92)90081-O

55. Deryugin EE, Panin VE, Schmauder S, Storozhenko IV (2001) Effects of deformation localization in Al-based composites with Al_2O_3 inclusions. Phys Mesomech 4(3):35–47

56. Casarotto L, Tutsch R, Ritter R, Weidenmüller J, Ziegenbein A, Klose F, Neuhäuser H (2003) Propagation of deformation bands investigated by laser scanning extensometry. J Comput Mater Sci 26:210–218. https://doi.org/10.1016/S0927-0256(02)00401-9

57. Nagornih SN, Sarafanov GF, Kulikova GA, Daneliya GV, Tsypin MI, Sollertinskaya ES (1993) Plastic deformation instability in cooper alloys. Russ Phys J 36(2):112–117

58. Toyooka S, Madjarova V, Zhang Q, Suprapedi (2001) Observation of elementary process of plastic deformation by dynamic electronic speckle pattern interferometry. Phys Mesomech 4(3):23–27

59. Klose FB, Ziegenbein A, Weidenmüller J, Neuhäuser H, Hähner P (2003) Portevin-LeChatelier effect in strain and stress controlled tensile tests. Comput Mater Sci 26:80–86. https://doi.org/10.1016/S0927-0256(02)00405-6

60. McCormick P, Ling CP (1995) Numerical modeling of the Portevin-Le Chatelier effect. Acta Metall Mater 43(5):1969–1977. https://doi.org/10.1016/0956-7151(94)00390-4
61. Kok S, Barathi MS, Beaudoin AJ, Fressengeas C, Ananthakrishna G, Kubin LP, Lebyodkin M (2003) Spatial coupling in jerky-flow using polycrystal plasticity. Acta Mater 51(13):3651–3662. https://doi.org/10.1016/S1359-6454(03)00114-9
62. Hähner P, Rizzi E (2003) On the kinematics of Portevin-Le Chatelier bands: theoretical and numerical modeling. Acta Mater 51(12):3385–3397. https://doi.org/10.1016/S1359-6454(03)00122-8
63. Balokhonov RR, Romanova VA, Schmauder S, Makarov PV (2003) Simulation of meso–macro dynamic behavior using steel as an example. Comput Mater Sci 28:505–511
64. Balokhonov RR, Romanova VA, Martynov SA, Schwab EA (2013) Simulation of deformation and fracture of coated material with account for propagation of a Lüders- Chernov band in the steel substrate. Phys Mesomech 16(2):133–140
65. Kachanov LM (1974) Fundamentals of fracture mechanics. Nauka, Moscow
66. Balokhonov RR, Romanova VA, Kulkov AS (2020) Microstructure-based analysis of deformation and fracture in metal-matrix composite materials. Eng Fail Anal 110:104412. https://doi.org/10.1016/j.engfailanal.2020.104412
67. Ravnikar D, Dahotre NB, Grum J (2013) Laser coating of aluminum alloy EN AW 6082-T651 with TiB2 and TiC: Microstructure and mechanical properties. Appl Surf Sci 282:914–922
68. Grigorieva IS, Meilihova EZ (eds) (1991) Physical values. Reference book. Energoatomizdat, Moscow
69. Richtmyer RD, Morton KW (1967) Difference methods for initial-value problems. Wiley, Hoboken (New Jersey)
70. Wilkins ML (1999) Computer simulation of dynamic phenomena. Springer, Berlin
71. Mal AK, Singh SJ (1990) Deformation of elastic solids. Pearson College Div, London

Formation of a Nanostructured Hardened Surface Layer on the TiC-(Ni-Cr) Metal-Ceramic Alloy by Pulsed Electron-Beam Irradiation

Vladimir E. Ovcharenko, Konstantin V. Ivanov, and Bao Hai Yu

Abstract The efficiency and service life of products made from metal-ceramic tool alloys and used as cutting tools and friction units are determined by a combination of physical and strength properties of their surface layers with a thickness of up to 200 µm. Therefore, much attention is paid to their improvement at the present time. An effective way to increase the operational properties of the metal-ceramic alloy products is to modify the structure and the phase composition of the surface layers by forming multi-scale internal structures with a high proportion of low-dimensional (submicro and nano) components. For this purpose, surfaces are treated with concentrated energy fluxes. Pulse electron-beam irradiation (PEBI) in an inert gas plasma is one of the most effective methods. This chapter presents results of theoretical and experimental studies of this process. An example is the nanostructured hardened surface layer on the TiC-(Ni-Cr) metal-ceramic alloy (ratio of components 50:50) formed by PEBI in the plasma of argon, krypton, and xenon. Its multi-level structure, phase composition, as well as tribological and strength properties are shown.

Keywords Metal-ceramic composite · Pulsed electron-ion beam irradiation · Inert gas plasma · Structure · Wear resistance

V. E. Ovcharenko (✉) · K. V. Ivanov
Institute of Strength Physics and Materials Science of Siberian Branch of Russian Academy of Sciences, Tomsk, Russia
e-mail: ove45@mail.ru

K. V. Ivanov
e-mail: ikv@ispms.tsc.ru

B. H. Yu
Institute of Metal Research, Chinese Academy of Sciences, Shenyang, China
e-mail: bhyu@imr.ac.cn

© The Author(s) 2021
G.-P. Ostermeyer et al. (eds.), *Multiscale Biomechanics and Tribology of Inorganic and Organic Systems*, Springer Tracts in Mechanical Engineering,
https://doi.org/10.1007/978-3-030-60124-9_18

1 Introduction

Processing materials and products with concentrated energy fluxes to modify the structure, physical and strength characteristics of their surface layers is one of the effective ways to improve operational properties of metals and alloys. High-frequency [1–7], plasma [8–15], laser [16–23], ion- and electron-beam treatment [24–29] are widely used. The technological advantages of these methods include locality and non-contact heating of the surface layers. Each of these methods has its own features determined by the depth and the maximum temperature of the heated surface layer, heating and cooling rates, as well as duration and frequency of pulses. The purpose of the above surface hardening methods is to form multi-scale structures and phase compositions in the surface layer that determine a higher level of physical and strength properties. These structures and the phase compositions enable to decrease local plastic deformations and contribute to a more uniform distribution of elastic stresses in the material volume [30]. As a result, an energy level required to form stress concentrators in the nanostructured surface layer increases, and the probability of defects decreases. Also, a transition zone (instead of an interface) appears between the surface layer and the base material (they smoothly pass into each other). This determines the damping properties of the surface layer with respect to the base material under external shock mechanical and thermal loads. The transition zone prevents premature nucleation and propagation of microcracks from the surface into the material volume which cause its failure. The high-strength nanostructured surface layer, even having multiple micro-cracks, reduces the probability of the crack formation due to increased ductility.

Pulsed electron-beam irradiation (PEBI) has particular advantages for the modification of the structure and the phase composition of the surface layers of metals and alloys. It makes possible to treat a rather large surface area per one pulse with a power density of more than 10^6 W/cm^2. Also, PEBI has some another benefits such as pulse frequency up to $10\,\mathrm{s}^{-1}$, the low reflection coefficient from the irradiated metal surfaces (less than 10%) and high efficiency (more than 90%), as well as possibility to modify the surface layers up to several tens of micrometers in depth.

One of the first comprehensive studies of the metals treated by PEBI has been carried out by scientists of the Tomsk Scientific Center of the Siberian Branch of the USSR Academy of Sciences in the mid-1970s [31]. A little bit later, in the 1980s and subsequent years, some investigations have been done related to the impact of the high-energy electron beams on inorganic materials (Kurchatov Institute of Atomic Energy, Russia; D.V. Efremov Research Institute of Electrophysical Equipment, Russia; The Chernyshev Moscow Machine-Building Enterprise, Russia [30–34], and Sandia National Laboratory, USA [35, 36] have been devoted to amorphization and surface hardening of the materials. However, many important relationships and mechanisms of the formation of the phase composition and the substructure in the surface layer of the metals and the alloys after PEBI are still unclear. In particular, the modes that make possible to initiate the dynamic recrystallization and the subsequent formation of nano- and submicroscale multiphase structures have not been established.

The mentioned possibilities to modify the surface layer of the metal and the alloys by PEBI are in many ways superior to other hardening technologies, such as deposition of wear-resistant coatings, ion irradiation and implantation, pulsed plasma and laser surface treatment [37–39]. The results of preliminary experimental studies have enabled to make conclusions that the development of an industrial technology for the modification of the surface layers is especially relevant for the manufacturing of metal-ceramic alloy tools. These alloys, based on refractory and high-hardness chemical compounds (carbides, nitrides, carbonitrides, and oxides) with a metal binder, are up-to-date dispersion-strengthened composite materials. They have high mechanical and tribological properties, such as strength, hardness, fracture toughness, wear resistance, and thermal stability. In this regard, the metal-ceramic alloys are widely used in various industries as functional materials for heavily loaded tribological units, machine parts, cutting elements operating under conditions of abrasive action, high temperatures and aggressive environments.

World-wide market analysis results show that about 67% of the total number of the metal-ceramic alloys is used for metal cutting tools, 13% is for drilling and rock cutting, 11% is for woodworking, and 9% is for chipless material-removal. At the same time, global consumption of the metal-ceramic alloy tools has doubled over the past 5 years.

Nowadays, the main direction of improving the operational properties of the metal-ceramic alloys is the development of technologies for sintering particles of highly hard components with a size of less than 1 μm (down to 100 nm). This increases the dispersion of the structure, hardness and strength of the alloys [40, 41]. Despite these advantages of the nanostructured hard alloys, their widespread implementation in the industries is hampered by the lack of effective technologies for mass production by the powder metallurgy method. Therefore, the widely used materials for the critical heavily loaded mechanisms are conventional metal-ceramic materials with particle sizes ranging from units to several tens of micrometers.

The operational efficiency of the metal-ceramic alloys in the mentioned conditions is determined by a set of mechanical properties of the surface layer with a thickness of 100–200 μm (the maximum permissible wear of a tool in the industries). Therefore, recent approaches to increase their lifecycle are based on the improvement of the surface layer characteristics. The main challenge is to eliminate powder sintering defects, including thermally activated micro- and macro-discontinuities of the solid structure. This is achieved by the interfacial interaction of the components under non-equilibrium thermodynamic conditions when heating to critical temperatures and then cooling at ultrahigh rates. The interaction of the metal and ceramic components of the alloys occurs at the peak of heating in the pulsed mode and, as a result, non-equilibrium structures are formed having new physical and strength properties. They remain in the surface layer after cooling at ultrahigh rate. A facility, designed and implemented in the Institute of High Current Electronics (Tomsk, Russia), makes it possible to modify the surface by PEBI [42]. The material is heated at a rate of up to 10^9 K/s and, then, cooled at a rate of $10^4 - 10^9$ K/s forming temperature gradients

Table 1 Atomic mass and ionization energy of the plasma-forming gases

Gas	Ionization energy (kJ/mol)	Atomic mass (g/mol)
Ar	1520.6	39.948
Kr	1350.0	83.798
Xe	1170.0	131.290

in the surface layer up to $10^7 - 10^8$ K/m [43]. To date, in collaboration with the Institute of Strength Physics and Materials Science SB RAS (Tomsk, Russia), a new approach has been developed to increase the depth of the surface layer on the metal-ceramic alloys which has a high content of nanoscale components. This is achieved by simultaneous pulsed irradiation with electron and ion beams in a single cycle. Inert gases (krypton and xenon), having higher atomic masses and lower ionization energy levels than argon (Table 1), are used as plasma-forming gas. Heavy ions interact with the surface of the irradiated material, increase the effective energy density in the electron beam, form a high level of radiation defects in the surface layer, and accelerate the processes of mass transfer and atom redistribution [43, 44].

This chapter presents the results of the studies of the nanostructured hardened surface layer on the TiC-(Ni-Cr) metal-ceramic alloy (ratio of components 50:50) formed by PEBI in the plasma of argon, krypton, and xenon. Its multi-level structure, phase composition, as well as tribological and strength properties are shown.

2   Temperature Fields in the Surface Layer under Pulsed Electron-Beam Irradiation

Temperature fields in the surface layer of the TiC-(Ni-Cr) metal-ceramic alloy heated by irradiation with high-energy beams of a rather large diameter can be quantified with a high degree of reliability using a one-dimensional model [45]. The model is based on PEBI of a cylindrical sample of radius r and length X from the end side. Denote electron-beam energy density E_s, pulse duration t_i, and pause between pulses t_0. An approximation has been made in the model that the energy density distribution over the cross section of the electron beam is uniform. Therefore, the sample heating by the electron beam is determined by the equation [46]:

$$c\rho \frac{\partial T}{\partial t} = \lambda \frac{\partial^2 T}{\partial x^2} - \frac{\chi}{x_0}(T - T_0) - \frac{\delta}{x_0}(T^4 - T_0^4),\ 0 \leq x \leq X, \tag{1}$$

where T is temperature; t is time; c, λ are heat capacity and thermal conductivity, respectively, in the solid material as a function of temperature; ρ is density; χ is the convective heat transfer coefficient; δ is the radiation heat transfer coefficient; T_0 is ambient temperature; x is longitudinal coordinate.

According to the additivity law, the heat capacity, thermal conductivity and density values of composite materials are calculated as:

$$c = v_{Ni-Cr}c_{Ni-Cr} + v_{TiC}c_{TiC}, \lambda = v_{Ni-Cr}\lambda_{Ni-Cr} + v_{TiC}\lambda_{TiC},$$

$$\rho = v_{Ni-Cr}\rho_{Ni-Cr} + v_{TiC}\rho_{TiC}, c_{Ni-Cr} = (c_{Ni}a_{Ni} + c_{Cr}a_{Cr})/(a_{Ni} + a_{Cr}),$$

$$c_{TiC} = (c_{Ti}a_{Ti} + c_{C}a_{C})/(a_{Ti} + a_{C}),$$

$$\lambda_{Ni-Cr} = (\lambda_{Ni}a_{Ni} + \lambda_{Cr}a_{Cr})/(a_{Ni} + a_{Cr}); \lambda_{TiC} = (\lambda_{Ti}a_{Ti} + \lambda_{C}a_{C})/(a_{Ti} + a_{C}),$$

$$\rho_{Ni-Cr} = (\rho_{Ni}a_{Ni} + \rho_{Cr}a_{Cr})/(a_{Ni} + a_{Cr}); \rho_{TiC} = (\rho_{Ti}a_{Ti} + \rho_{C}a_{C})/(a_{Ti} + a_{C}),$$

$$(2)$$

where c_j, λ_j, ρ_j, v_j are the heat capacity, thermal conductivity, density and relative mass fraction of the j-th component of the composite (the TiC-(Ni-Cr) metal-ceramic alloy in the investigated case).

For a single electron pulse, boundary conditions on the irradiated surface ($x = 0$) are represented as:

$$-\lambda\frac{\partial T}{\partial x} = \begin{cases} E_s/t_i > 0, t \in [(n-1)(t_i + t_0), (n-1)t_0 + nt_i] \\ 0, t \notin [(n-1)(t_i + t_0), (n-1)t_0 + nt_i] \end{cases}, x = X: \frac{\partial T}{\partial x} = 0$$

$$(3)$$

where n is the number of pulses.

The initial data for the TiC-(Ni-Cr) metal-ceramic alloy having the ratio of the components 50:50 are [47–49]: $c_{NiCr} = 452$ J/kgK, $c_{TiC} = 408$ J/kgK, $\lambda_{NiCr} = 88.5$ J/sKm, $\lambda_{TiC} = 42$ J/sKm, $\rho_{NiCr} = 8800$ kg/m^3, $v_{NiCr} = v_{TiC} = 0.5$, $\chi = 10$ J/sKm^2, $\delta = 3 \times 10^{-7}$ J/sK^{4}m^2, $T_0 = 300$ K, $r = 0.01$ m, $x = 0.001$ m.

An increase in pulse duration causes a significant decrease in temperature of the surface layer and an increase in its depth. When pulse duration is 200 μs, the surface layer temperature achieves 550 K and its depth slightly increases. The electron-beam energy density values have a strong effect on the temperature profile in the surface layer. Both temperature and depth of the surface layer rise simultaneously with an increase in electron-beam energy density. For example, the surface layer temperature rises from 760 to 1250 K when electron-beam energy density enhances from 5 to 10 J/cm^2. Its further increase up to 40 J/cm^2 raises the surface layer temperature up to 4250 K. This effect becomes stronger with increasing pulse duration.

At a constant energy density value, an increase in the number of pulses up to 300 enhances the heated surface layer depth by more than two times. Then, upon reaching a certain number of pulses, the surface layer depth remains constant rising only with an increase in electron-beam energy density. At a frequency of 1 s^{-1}, an increase in the number of pulses up to 50 has almost no effect on the surface layer temperature which increases only after 100 pulses. The temperature profile in the surface layer changes little with increasing pulse frequency from 1 to 20 s^{-1}.

The temperature gradient under PEBI significantly affects structural and phase transformations in the surface layer. The calculated dependences of the temperature gradient in the surface layer of the TiC-(Ni-Cr) metal-ceramic alloy from pulse frequency and duration, the number of pulses, as well as electron-beam energy density show that an increase in pulse duration sharply decreases the temperature gradient while, on the contrary, an increase in electron-beam energy density enhances it. Rising pulse frequency and the number of pulses do not affect the temperature gradient in the surface layer. Thus, the main patterns of the structural and phase transformations in the surface layer of the TiC-(Ni-Cr) metal-ceramic alloy under PEBI are determined primarily by electron-beam energy density, as well as the number of pulses and pulse duration. A weak correlation of the temperature distribution in the surface layer with the number of pulses and pulse frequency means that the temperature profiles are formed during the first pulse and remains almost unchanged with an increase in these parameters. Raising the number of pulses enhances duration of the surface layer existence in non-equilibrium temperature–time conditions. It also determines the interfacial interaction of the surface layer components and, as a result, the formed structure.

An analysis of the temperature profiles and the surface layer depth enables to estimate the ranges of the experimental parameters (electron-beam energy density, the number of pulses and pulse duration) for the surface layer structure modification by PEBI. The main criteria are the modified surface layer depth (from 100 to 200 μm), an increase in its temperature during the first pulse (up to 3000 K), and a minimum temperature gradient. Electron-beam energy density of $40 - 50\,\mathrm{J/cm^2}$ and pulse duration of 100–200 μs correspond to these criteria. Taking into account that the temperature profiles in the surface layer are formed precisely during the first pulse and remain almost unchanged then, an increase in the number of pulses changes only the duration of the interfacial interaction of the components in the TiC-(Ni-Cr) metal-ceramic alloy under non-equilibrium temperature–time conditions for the specified electron-beam energy density and pulse duration values. Thus, the results of the surface layer temperature profile calculations enable to verify their compliance with the general patterns. However, an increase in the content of the ceramic component in the TiC-(Ni-Cr) metal-ceramic alloy enhances the maximum heating temperature in the surface layer, as well as the temperature gradient in the surface layer.

3 The Effect of Pulsed Electron-Beam Irradiation in Different Plasma-Forming Gases on the Surface Layer Structure and Properties

3.1 Material and Experimental Methods

Samples ($10 \times 10 \times 4$ mm plates) were made from the TiC-(Ni-Cr) metal-ceramic alloy. The flat sample surfaces were polished to a mirror shine before irradiation. PEBI was performed using a plasma cathode setup [50, 51]. Argon, krypton, and xenon were used as a plasma-forming gas; their atomic mass and ionization energy

Table 2 PEBI parameters

No.	Plasma-forming gas	Pulse duration (μs)	Electron-beam energy density (J/cm^2)	Number of pulses
1	Ar	200	40	15
2	Kr	100	40	15
3	Xe	100	40	15
4	Ar	150	60	15
5	Kr	200	60	15
6	Xe	150	60	15

are presented in Table 1. The values of electron-beam energy density, pulse duration, and frequency are presented in Table 2. The wide-aperture electron beam covered the entire sample surfaces.

Specimens for the microstructure analysis were prepared by the focused ion beam method using an 'Jeol EM-09100IS' setup (accelerating voltage 8 kV, ion beam incidence angle 4°).

Three-point bending tests were carried out using an 'Instron-3369' machine under continuous loading at a speed of 0.2 mm/min in accordance with ISO 3327-82. Transverse bending strength was calculated without taking into account any possible plastic deformations by the formula $\sigma = 3Pl/2h^2b$, where P was force corresponding to the sample failure; l was the distance between the pillars; h was the sample height (the size coincided with the force direction); b was the sample width (perpendicular to height).

X-ray phase analysis of the surface layers of the samples having the minimum and maximum values of transverse bending strength after PEBI in different plasma-forming gases was carried out using a 'Shimadzu XRD 6000' X-ray diffractometer in Cu radiation (accelerating voltage 40 kV, current 30 mA).

Microhardness values on the surface layers were measured using a 'DM 8B' microhardness tester in accordance with ISO 14577-1:2015. Nanohardness and Young's modulus of the surface layers were investigated using a 'CSEM Nano Hardness Tester' facility in accordance with ISO 14577-1:2015.

Wear resistance of the surface layers and the friction coefficients on the sample surfaces were studied using a 'CSEM Tribometer High Temperature S/N 07-142' tribometer (CSEM Instruments, Switzerland). The samples were rotated on the stationary counterpart (a diamond cone) during the tests. Load on the diamond cone was 5 N, the number of the sample revolutions was 2500. Friction force was continuously measured using a 'Micromesure System STIL' micrometer system (Science et Techniques Industrielles de la Lumere. STILS. A, France). Then, absolute friction coefficient values were calculated. The surface profiles (the depth and the cross-section area of grooves made with the diamond cone) were identified using a 'MICRO MEASURE 3D station (STIL)' profiler after the end of the tests.

Also, abrasive wear resistance of the sample surfaces was investigated in accordance with ASTM G65. The flat samples were pushed down to a rotated rubber lined wheel with a diameter of 218 mm. Load was 36 N, rotating speed was 200 revolutions per minute. An abrasive powder (13A electrocorundum with a particle size of 200 − 250 μm) was supplied to the friction surface. The abrasive flow rate was 270 − 280 g/min.

3.2 Changes in the Structure and the Properties of the Surface Layer after Pulsed Electron-Beam Irradiation

The study of the TiC-(Ni-Cr) metal-ceramic alloy microstructure by scanning electron microscopy (SEM) showed that titanium carbide particles in the initial state (after sintering) had pronounced unequal shapes (Fig. 1a). The average particle size was 2.7 ± 1.2μm, the maximum size was 7.6 μm (Fig. 1b). The alloy in the initial state had three main structural components:

(1) high-strength titanium carbide particles with a size of 1–10 microns;
(2) interparticle layers of the metal binder;
(3) "particles–binder" transition zones with a width of about four microns.

Fracture surfaces of the samples had a brittle intergranular appearance (Fig. 2). Figure 3 shows SEM images of the alloy structure in the initial state, as well as titanium carbide particles and the nickel-chrome binder.

After PEBI, the shapes of titanium carbide particles smoothed and became more rounded (especially of small ones) in the surface layer (Fig. 4a). In addition, their average sizes changed (Fig. 4b). As electron-beam power density (W_S) increases, their average and maximum sizes changed along curves with a maximum at 3.2×10^5 W/cm^2 (Fig. 5, curve 1 and 2, respectively).

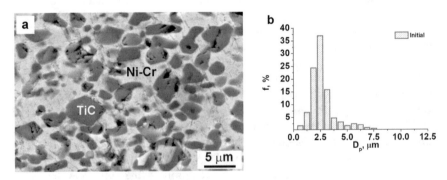

Fig. 1 SEM images of the structure of the metal-ceramic alloy polished surface in the initial state (**a**) and the element distribution in the TiC-(Ni-Cr) metal-ceramic alloy at the interface between titanium carbide particles and the nickel-chrome binder (**b**)

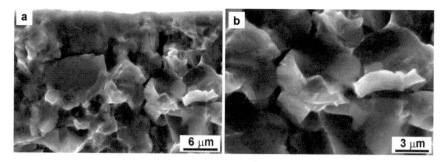

Fig. 2 The fracture surface of the TiC-(Ni-Cr) metal-ceramic alloy in the initial state

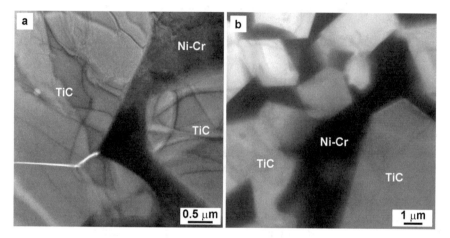

Fig. 3 TEM images of the structure of the TiC-(Ni-Cr) metal-ceramic alloy in the initial state

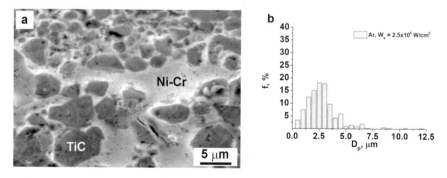

Fig. 4 The structure of the TiC-(Ni-Cr) metal-ceramic alloy after PEBI $\left(W_s = 2.5 \times 10^5 \, \text{W/cm}^2\right)$ in the argon plasma (**a**) and the size distribution of titanium carbide particles (**b**, $D = 2.8 \pm 1.7 \, \mu\text{m}$)

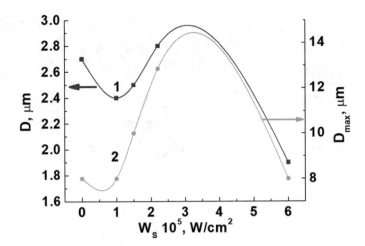

Fig. 5 The average (curve 1) and maximum (curve 2) titanium carbide particle sizes on the irradiated surface versus electron-beam power density

PEBI caused high-rate melting and subsequent solidification of the nickel-chrome binder in the surface layer. As a result, titanium carbide particles were refined by partially dissolving in the molten nickel-chrome binder and simultaneous large particle cracking due to high thermal stresses (Figs. 6 and 7).

In addition to microcracking, PEBI changed the structure and the phase composition of the surface layer. The next process options occurred depending on the PEBI parameters: the nickel-chrome binder melted and solidified, or the primary titanium carbide particles partially dissolved and the secondary titanium carbide precipitated, or large carbide particles cracked and these microcracks were filled with the molten nickel-chrome binder. Figures 8 and 9 show microphotographs of the surface layer after PEBI that demonstrate the results of the mentioned processes.

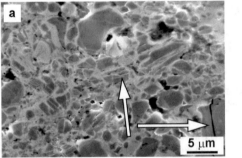
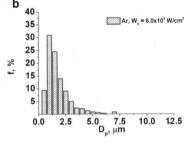

Fig. 6 The structure of the TiC-(Ni-Cr) metal-ceramic alloy after PEBI $\left(W_S = 6.0 \times 10^5 \text{ W/cm}^2\right)$ in the argon plasma (**a**), and the distribution of the titanium carbide particle sizes (**b**, $D = 1.9 \pm 1.2\,\mu\text{m}$)

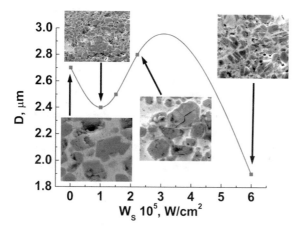

Fig. 7 Failure types of titanium carbide particles under thermo-elastic stresses versus electron-beam power density after irradiation in the argon plasma

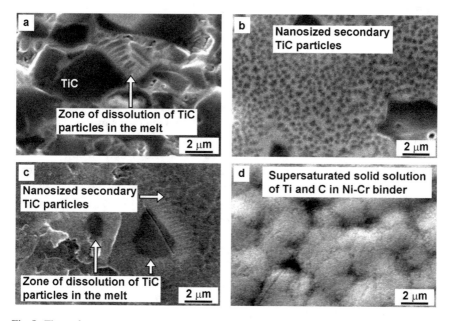

Fig. 8 The surface structure of the TiC-(Ni-Cr) metal-ceramic alloy after PEBI in the argon plasma at an electron-beam power density from 2×10^5 up to 10×10^5 W/cm^2 : $a - E_S = 40.0$ J/m^2, $t_i = 200.0$ μs, $W_S = 2 \times 10^5$ W/cm^2; $b - E_S = 40.0$ J/cm^2, $t_i = 80.0$ μs, $W_S = 4 \times 10^5$ W/cm^2, $c - E_S = 40.0$ J/cm^2, $t_i = 50.0$ μs, $W_S = 8 \times 10^5$ W/cm^2; $d - E_S = 2.5$ J/cm^2, $t_i = 2.5$ μs, $W_S = 10 \times 10^5$ W/cm^2

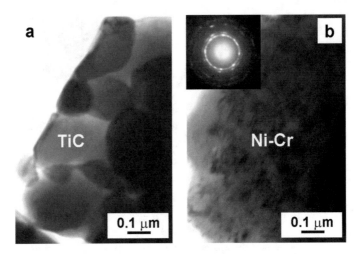

Fig. 9 TEM images of the structure of the TiC-(Ni-Cr) metal-ceramic alloy after PEBI $\left(E_S = 40\,\text{J/cm}^2,\ t_i = 50\,\mu\text{s},\ V = 1\,\text{s}^{-1},\ n = 15\,\text{pulses}\right)$: **a**—titanium carbide particles; **b**—the nickel-chrome binder

The various stages of the interaction of titanium carbide particles with the molten nickel-chrome binder are illustrated by the micrographs in Fig. 8. The initial stage of the titanium carbide particle melting was at $2 \times 10^5\,\text{W/cm}^2$ (Fig. 8a). Then, the sizes of the primary titanium carbide particles and their volume fraction decreased with increasing electron-beam power density (Fig. 8b, c). Finally, the process completed with the partial dissolution of titanium carbide particles in the molten nickel-chrome binder and the formation of a nanostructure in the surface layer (Fig. 8d).

The partial dissolution of the primary titanium carbide particles and the formation of titanium carbide dissolution zones in the molten nickel-chrome binder occurred in the surface layer at an electron-beam power density of $2 \times 10^5\,\text{W/cm}^2$ (Fig. 8a). Rising electron-beam power density up to $4 \times 10^5\,\text{W/cm}^2$ caused an increase in dissolution rate of the primary titanium carbide particles in the molten nickel-chrome binder, the formation of the extensive titanium carbide dissolution zones, and the release of the secondary titanium carbide nanoparticles inside them (Fig. 8b). A subsequent increase in electron-beam power density up to $8 \times 10^5\,\text{W/cm}^2$ resulted in the almost complete dissolution of the primary titanium carbide particles in the surface layer, and the subsequent formation of a supersaturated titanium-carbon solid solution in the nickel-chromium binder at $10 \times 10^5\,\text{W/cm}^2$ (Fig. 8c, d).

A glass-like structure formed in the surface layer. In this structure, titanium carbide particles were almost completely dissolved in the molten nickel-chrome binder (Fig. 9). Microdiffraction studies showed that the glass-like structure consisted of the submicron (200–250 nm) titanium carbide particles and the nanostructured nickel-chrome binder (crystallite size was in the range of 10–30 nm).

Figure 10 shows typical surface layer structures after PEBI using various pulse durations with constant other energy parameters (electron-beam energy density was

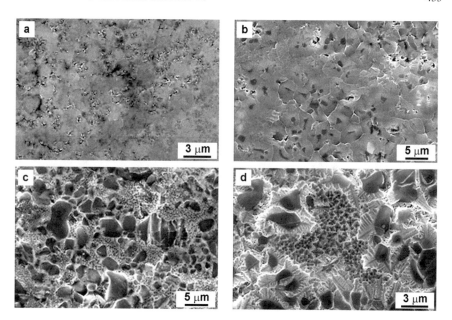

Fig. 10 SEM images of the structure of the TiC-(Ni-Cr) metal-ceramic alloy after PEBI $(E_S = 40 \, J/cm^2, \, v = 1 \, s^{-1}, \, n = 15 \, pulses)$: **a** $t_i = 50 \, \mu s$, **b** $t_i = 100 \, \mu s$, **c** $t_i = 150 \, \mu s$, **d** $t_i = 200 \, \mu s$

$40 \, J/cm^2$, pulse frequency was $1 \, s^{-1}$, the number of pulses was 15). Almost complete dissolution of the primary titanium carbide particles occurred in the molten nickel-chrome binder, and a titanium-carbon solid solution was formed in the surface layer at a pulse duration of 50 μs. An increase in pulse duration up to 100 μs caused the formation of a fine-grain structure with residues of the primary titanium carbide particles in the central part of each grain. The uniform structure was formed at a pulse duration of 150 μs. It included the primary titanium carbide microparticles and interlayers of the nickel-chrome binder saturated with the secondary titanium carbide nanoparticles.

The dependence of the dimension of the titanium carbide particles on electron-beam power density is shown in Fig. 11. At low electron-beam power density values, large titanium carbide particles were refined due to thermal shock and partially dissolved in the molten nickel-chrome binder. Then, this effect intensified to a large extent with an increase in electron-beam power density. As a result, a supersaturated titanium-carbon solid solution in the nickel-chrome binder was formed at electron-beam power density of $8 \times 10^5 \, W/cm^2$ and higher.

Finally, the secondary titanium carbide nanoparticles precipitated in the nickel-chrome binder upon cooling (Fig. 12).

It was found by the local X-ray spectral analysis that the titanium concentration in the nickel-chrome binder interlayers varied within 19–29 wt% and increased

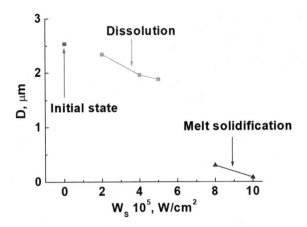

Fig. 11 Average titanium carbide particle sizes versus electron-beam power density

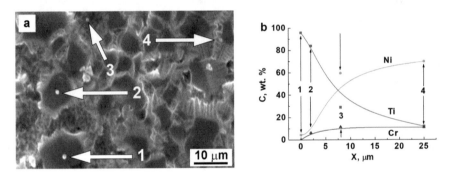

Fig. 12 The structure of the TiC-(Ni-Cr) metal-ceramic alloy after PEBI $\left(\mathbf{a}, E_S = 40\,\mathrm{J/cm}^2,\ t_i = 200\,\mu\mathrm{s},\ v = 1\,\mathrm{s}^{-1},\ n = 15\,\mathrm{pulses}\right)$, and the effect of the distance from titanium carbide particles on the nickel, titanium, and chromium content (**b**, numbers indicate the measurement points and the results of local quantitative X-ray spectral analysis)

significantly near the interface between titanium carbide particles and the nickel-chrome binder. The values were maximum at the surfaces of titanium carbide particles (Fig. 13). These results indicated the intensive dissolution of titanium carbide particles in the molten nickel-chrome binder and its excessive saturation with titanium and carbon, as well as proved the inevitability of the formation of the multiphase composition in the surface layer during its high-rate cooling after electron-beam pulses. In this regard, it was suggested that one of the reasons for the nanostructure (Fig. 10a) formation in the surface layer after PEBI $\left(E_S = 40\,\mathrm{J/cm}^2,\ t_i = 50\,\mu\mathrm{s}\ \text{and 15 pulses}\right)$ was the supersaturation of the nickel-chromium binder with carbon and titanium.

These features were revealed by the results of the SEM investigations (Fig. 13). Dendritic structures typical for high-rate solidification were formed at the mentioned PEBI parameters. However, there were some differences depended on the PEBI mode due to various heat input rates characterized by electron-beam power density.

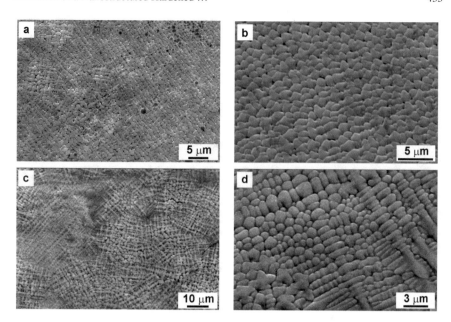

Fig. 13 SEM images of the structure of the TiC-(Ni-Cr) metal-ceramic alloy surface after PEBI:
a and **b** − $E_S = 40\,\text{J/cm}^2$, $t_i = 50\,\mu\text{s}$, $n = 15$ pulses; c and $d - E_S = 40\,\text{J/cm}^2$, $t_i = 200\,\mu\text{s}$, $n = 15$ pulses

In the formed structure, titanium carbide particles with a size of $0.5 - 1.5\mu\text{m}$ had an almost equiaxial shape at a pulse duration of 50 μs (Fig. 13a, b). An increase in pulse duration up to 200 μs decreased cooling rate of the surface layer. Therefore, a dendritic structure with clearly defined first and second order axes was formed during solidification (Fig. 13c, d). Carbide particle shapes varied from an isotropic type with a size of $0.5 - 1.5\,\mu\text{m}$ to a highly anisotropic one having a length of $1.5 - 6\,\mu\text{m}$ and a width of $0.5 - 0.8\mu\text{m}$.

Thus, the phase composition, the defective substructure state, and the solid solution in the gradient structure formed by PEBI depended on the location of the initial structure with respect to titanium carbide particles. The whole thickness of the modified surface layer increased from 13 to $40\,\mu\text{m}$ with rising pulse duration from 50 to 200 μs, but the thickness of its glass-like part continuously decreased down to zero at a pulse duration of 200 μs.

Modification of the structure and the phase composition in the surface layer changed the mechanism of its failure. Fractography analysis of the surface layer after PEBI with a pulse duration of 200 μs and an electron-beam energy density of $40\,\text{J/cm}^2$ (Fig. 14) showed that failures had occurred by the ductile-brittle mechanism (numerous plastic deformation steps were on the facets) while the initial sample fracture type was brittle.

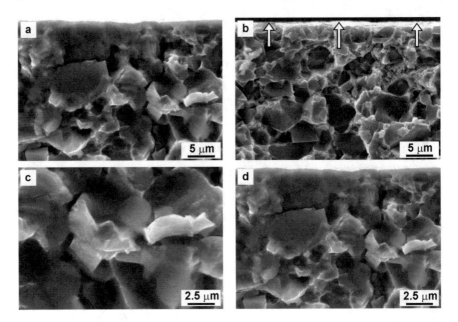

Fig. 14 The fracture surface of the TiC-(Ni-Cr) metal-ceramic alloy in the initial state (a and b, the arrows indicate the irradiation surface) and after PEBI $\left(c \text{ and } d, E_S = 40 \text{ J/cm}^2, t_i = 200 \,\mu\text{s}, n = 15 \text{ pulses}\right)$

The revealed diversity of the structures and the phase composition in the surface layer after PEBI was due to a high level of temperature gradients and the distribution of the alloying elements. The basic indication of the non-equilibrium state achievement in the surface layer was the formation of the modified structure included titanium carbide submicroparticles and the nanostructured nickel-chrome binder. The nickel-chrome binder nanostructuring changed the fracture mechanism from the brittle type to the ductile-brittle one. In this case, the brittle component was determined by the titanium carbide particle failures, and the ductile one was driven by the failure of the nanostructured nickel-chrome binder.

3.3 Theoretical Assessment of the Effect of Plasma-forming Gases on the Pulsed Electron-Beam Irradiation Process

An integral part of the PEBI process is the plasma formation by the inert gas ionization in a vacuum chamber. The inert gas plasma is also an electron source in the case of using a plasma-cathode setup. Additionally to electrons, inert gas ions interact with the irradiated material surface and enhance the surface layer modification. Argon is typically used as a plasma-forming gas for PEBI. However, krypton and xenon are also of significant interest. They differ significantly from argon in terms of atomic mass [52] and ionization energy [53] (Table 1).

It can be assumed that atomic mass and ionization energy of a plasma-forming gas affects the result of the surface layer modification by PEBI at the same pressure in the electron-beam setup chamber. In order to verify or disproof this, PEBI of the end surface of a round plate TiC-(Ni-Cr) metal-ceramic alloy sample with a radius of r and a thickness of X is considered. When the energy density distribution in the electron-beam cross section is uniform, the thermal-conductivity equation is similar to the Eq. (1).

For a single electron pulse, boundary conditions on the irradiated surface ($X = 0$) are represented as:

$$-\lambda \frac{\partial T}{\partial x} = \begin{cases} E_S(g)/t_i > 0, \ 0 < t \le t_i \\ 0, \qquad\qquad t > t_i \end{cases}, \tag{4}$$

and at the end ($x = X$)

$$-\lambda \frac{\partial T}{\partial x} = \chi(T - T_0) + \varepsilon(T^4 - T_0^4) \tag{5}$$

In Eqs. (4) and (5), the following notations are used: t_i is pulse duration; $E_S(g)$ is effective electron-beam energy density in an inert gas plasma. In order to determine $E_S(g)$, it was assumed that an additional ion flux, formed in the inert gas plasma due to electron-beam energy exposure, increases effective (total) radiation density. It is presented in the following approximated form:

$$E_S(g) \approx E_S(Ar) + \Delta E(g) \tag{6}$$

where $E_S(Ar)$ is electron-beam energy density in the argon plasma, $\Delta E(g)$ is electron-beam energy density increment due to additional ionization in the krypton or xenon plasma ($g = Kr$ or Xe). It is also suggested that a decrease in ionization energy of the inert gas increases the number of ions in the electron-plasma flow and, accordingly, the value of $\Delta E(g)$. Hence, effective electron-beam energy density increases with decreasing gas ionization energy and increasing $\Delta E(g)$. Since $E_{iAR} > E_{iKr} > E_{iXe}$ (according to Table 1), the next condition $0 < \Delta E(Kr) < \Delta E(Xe)$ is satisfied, which can be rewritten as

$$E_s(Ar) < E_s(Kr) < E_s(Xe) \tag{7}$$

Numerical calculations were carried out using the system of the Eqs. (2)–(5) and the relations (6) and (7). Heat capacity and thermal conductivity of the TiC-(Ni-Cr) metal-ceramic alloy were taken constant. The values of thermo-physical properties are represented in paragraph 2 and in [54–58].

In order to facilitate the analysis of the effect of the plasma-forming gas on the possibility and intensity of the structure and phase transformations in the surface

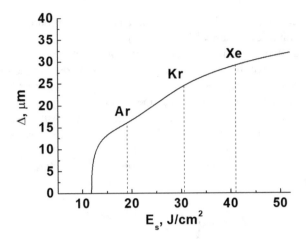

Fig. 15 Depth of the TiC-(Ni-Cr) metal-ceramic alloy surface layer heated up to a temperature of 1500 K by PEBI (argon, krypton, and xenon plasma, $t_i = 150\,\mu s$, $n = 1$ pulse) versus energy density

layer under PEBI, the characteristic temperature index T^* has been used. Structural and phase transformations are negligible when $T < T^*$, and are quite intense when $T \geq T^*$. Figure 15 shows the dependence of the surface layer depth heated up to $T^* = 1500$ K under single electron-beam pulse with a duration of 150 µs on effective electron-beam energy density for argon, krypton and xenon used as a plasma-forming gas. It increases with the change of the plasma-forming gas from argon to krypton, and, then, to xenon at constant pulse duration and electron-beam energy density. Accordingly, there is also a high probability of an increase in the intensity of the interfacial interaction of the TiC-(Ni-Cr) metal-ceramic alloy components in the surface layer and its depth.

3.4   The Effect of the Plasma-Forming Gases on the Structure and the Properties of the Modified Surface Layer

The results of the SEM studies of the modified surface layer structure after PEBI in the different plasma-forming gases showed their patterns in terms of heterogeneity and diversity in the quantitative structure parameters. All modified layers consisted of a set of sublayers. The top sublayer, directly adjacent to the irradiated surface, had a nanosized metal-ceramic columnar structure oriented perpendicular to the outer sample surface (Figs. 16 and 17, sublayer 1). Below, a thicker sublayer was formed with a coarser columnar metal-ceramic structure (Figs. 16 and 17, sublayer 2). Columns were mostly oriented perpendicular to the outer sample surface. The next sublayer had a dendritic type structure. It was a transition layer to the initial metal-ceramic structure (Figs. 16 and 17, sublayer 3). The thicknesses of the mentioned

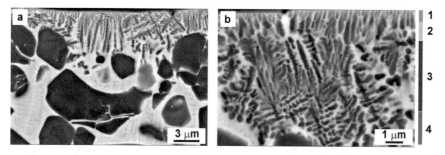

Fig. 16 SEM images of the structure of the TiC-(Ni-Cr) metal-ceramic alloy surface layer after PEBI (argon plasma, $E_S = 60$ J/cm^2)

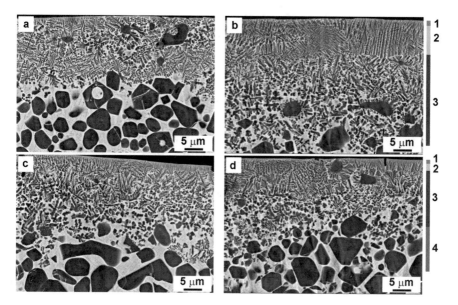

Fig. 17 SEM images of the structure of the TiC-(Ni-Cr) metal-ceramic alloy surface layer after PEBI: **a** and **b**—krypton plasma, $E_S = 40$ J/cm^2; **c** and **d**—xenon plasma, $E_S = 60$ J/cm^2

sublayers depended on the PEBI parameters and the plasma-forming gas. Figure 18 shows the dependences of the thickness of the top sublayer and the whole modified surface layer on pulse duration at electron-beam energy densities of 40 and 60 J/cm^2 using argon, krypton, and xenon as a plasma-forming gas. The effect of the electron-beam energy density on the whole modified surface layer thickness was minimal. However, it significantly increased in the krypton plasma, and even more in the xenon one. The range of the maximum thicknesses of the modified surface layer varied from 35 μm in the argon plasma to 40 μm in the krypton one, and 47 μm in the xenon plasma at an electron-beam energy density of 60 J/cm^2 (Fig. 18). An increase in pulse duration caused an increase in the thickness of both the top sublayer and the whole

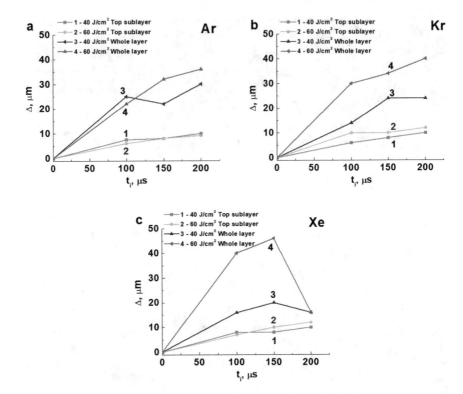

Fig. 18 The thickness of the upper part of the surface layer having the columnar nanostructure (curves 1 and 2) and the whole thickness of the TiC-(Ni-Cr) metal-ceramic alloy surface layer having the modified structure (curves 3 and 4) versus pulse duration (argon, krypton, and xenon plasma, $E_S = 40$ and $60 \, J/cm^2$, $n = 15$ pulses)

modified surface layer (Figs. 16 and 17). However, the top sublayer thickness was almost independent of both the plasma-forming gas and the electron-beam energy density values.

Figure 19 shows histograms of the size distribution of titanium carbide particles in the top, middle, and bottom sublayers modified in the plasma of argon, krypton, and xenon at an electron-beam energy density of $40 \, J/cm^2$. The size distribution of titanium carbide particles varied with enhancing distance from the surface to the sample depth. In addition, their average size became larger. It should be noted that this effect was minimal after PEBI in the xenon plasma. The dispersion of the titanium carbide particles almost did not change in the top and middle sublayers, but became a little bit larger in the bottom one. In addition, the greatest thickness of the modified layer was also in the xenon plasma.

Changes in the structure and the phase composition in the surface layer were due to the dissolution of titanium carbide particles in the molten nickel-chrome binder and the formation of a supersaturated with titanium and carbon dissolution at the

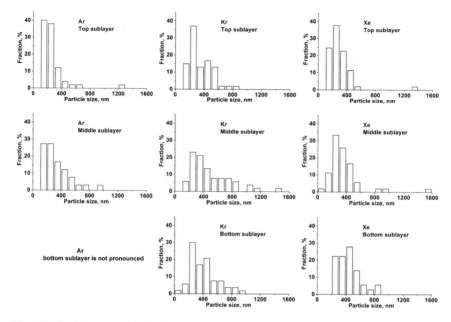

Fig. 19 The histograms of the titanium carbide particle size distribution in the upper, middle and lower parts of the TiC-(Ni-Cr) metal-ceramic alloy modified surface layer after PEBI (argon, krypton, and xenon plasma, $E_S = 40\,\mathrm{J/cm^2}$)

maximum temperature. Also, the secondary titanium carbide nanoparticles precipitated in zones of the primary titanium carbide highest concentration at the interfaces between primary titanium carbide particles and the molten nickel-chrome binder. As a result, a metal-ceramic structure oriented perpendicular to the irradiated surface was formed during solidification under conditions of high temperature gradients. It is shown in Fig. 20 the red line. The primary titanium carbide particles on the surface, where the secondary titanium carbides nanoparticles were formed, are shown by arrows below the red line.

A layer having a metal-ceramic dendrite structure was between the top nanostructured layer and the primary titanium carbide particles partially dissolved in the molten nickel-chrome binder. Elongated titanium carbide particles oriented in different directions were in it (Fig. 21).

The same results were obtained by transmission electron microscopy (TEM). Large titanium carbide particles dispersed to a nanoscale level in the surface layer after PEBI in different plasma-forming gases are shown in Fig. 22. The dispersion of the primary titanium carbide particles occurred by dissolving them in the molten nickel-chrome binder, as well as releasing titanium carbide nanoparticles from a supersaturated with titanium and carbon solid solution in the molten nickel-chrome binder during PEBI in the plasma of argon (a) and krypton (b). The titanium and carbon concentrations in the molten nickel-chrome binder were maximum at the

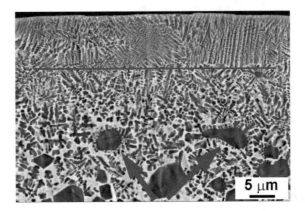

Fig. 20 SEM images of the structure of the TiC-(Ni-Cr) metal-ceramic alloy surface layer after PEBI. Nanosized titanium carbide particles at the interface between the primary titanium carbide particles and the nickel-chrome binder are shown by the arrows

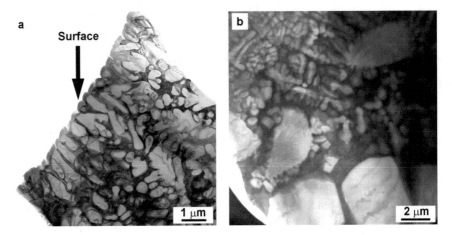

Fig. 21 The dendrite structure of the TiC-(Ni-Cr) metal-ceramic alloy surface layer after PEBI (krypton plasma, $E_S = 60\,\text{J/cm}^2$, $t_i = 150\,\mu\text{s}$, $n = 15$ pulses)

surfaces of the primary titanium carbide particle. Therefore, the number of titanium carbide nanoparticles was also maximum at the interface between the primary titanium carbide particles and the nickel-chrome binder after solidification (Fig. 22, shown by arrows). As their number increases in the nickel-chrome binder interlayers, the dendritic structure included titanium carbide nanoparticles oriented perpendicular to the irradiated surface that had formed by the temperature field (Fig. 20). The dispersion of the primary titanium carbide particles, as the mentioned mechanism of the titanium carbide nanoparticle formation, apparently occurred by mechanical refining as a result of thermal shock, followed by filling of the discontinuities with the molten nickel-chrome binder upon PEBI in the xenon plasma (Fig. 22c, shown

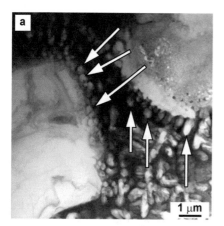
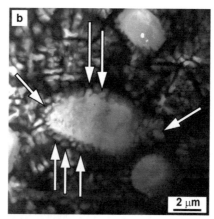
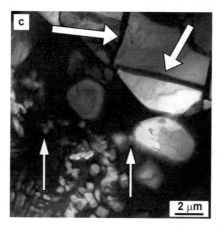

Fig. 22 TEM images of dispersed titanium carbide particles in the initial structure of the TiC-(Ni-Cr) metal-ceramic alloy surface layers after PEBI: **a**—argon plasma, $E_S = 60\,\text{J/cm}^2$, $t_i = 150\,\mu\text{s}$, $n = 15$ pulses; **b**—krypton plasma, $E_S = 60\,\text{J/cm}^2$, $t_i = 150\,\mu\text{s}$, $n = 15$ pulses; **c**—xenon plasma, $E_S = 60\,\text{J/cm}^2$, $t_i = 150\,\mu\text{s}$, $n = 15$ pulses. Nanosized titanium carbide particles formed from a supersaturated solid solution of carbon and titanium on the surface of the primary titanium carbide particles are shown by small arrows. The mechanical failures of the primary carbide particles are shown by the large arrows

by the arrows). Thus, it can be concluded that the heterophase nanostructure was formed in the surface layer of the TiC-(Ni-Cr) metal-ceramic alloy under PEBI in the plasma of light (argon) and heavy (krypton and xenon) inert gases. The modified layer depth increased with the change of the plasma-forming gas from argon to krypton and xenon.

Let us consider X-ray diffraction data. From Table 3, it follows that the phase composition of the material in the initial state (the ceramic to metal component ratio) corresponded to the specified values for the sintered titanium carbide and

Table 3 The parameters of the structure and the phase composition in the TiC-(Ni-Cr) metal-ceramic alloy surface layer in the initial state

Phase type	The relative content (wt%)	The lattice parameter (nm)	The size of the coherent scattering regions (nm)	The lattice micro-distortion, 10^{-3}	Texture
(Ni-Cr)	37.34	0.35711	49.24	4.685	–
TiC	62.66	0.43122	42.56	2.635	–

nickel-chrome powder mixture. There was no texture in the surface layer of the sample.

PEBI significantly changed the ratio of the ceramic and metal components in the surface layer, as well as formed the (002) texture (Table 4). The presented data were verified by the results of the above numerical estimation of the relative component content in the surface layer. Also, there were a decrease in the nickel-chrome binder content with a corresponding increase in the titanium carbide content, as well as an increase in the lattice parameters of the nickel-chrome binder and titanium carbide. The reason was the mutual solubility of the components when the surface layer was heated under PEBI.

Tables 5 and 6 show diffraction patterns and tables of the structure and phase composition parameters of the surface layers after PEBI in the krypton plasma at electron-beam energy densities of 40 and 60 J/cm^2 The main features of the structure and the phase composition were a decrease in the titanium carbide content from 90.08 down to 82.51 wt% and corresponding increase in the nickel-chrome binder

Table 4 The parameters of the structure and the phase composition in the TiC-(Ni-Cr) metal-ceramic alloy surface layer (argon plasma, $E_S = 60$ J/cm^2, $t_i = 150\,\mu$s, $n = 15$ pulses)

Phase type	The relative content (wt%)	The lattice parameter (nm)	The size of the coherent scattering regions (nm)	The lattice micro-distortion, 10^{-3}	Texture
(Ni-Cr)	9.39	0.35900	48.90	3.688	(200)
TiC	90.61	0.43137	59.14	1.181	–

Table 5 The parameters of the structure and the phase composition in the TiC-(Ni-Cr) metal-ceramic alloy surface layer (krypton plasma, $E_S = 40$ J/cm^2, $t_i = 200\,\mu$s, $n = 15$ pulses)

Phase type	The relative content (wt%)	The lattice parameter (nm)	The size of the coherent scattering regions (nm)	The lattice micro-distortion, 10^{-3}	Texture
(Ni-Cr)	9.92	0.35772	–	–	(220)
TiC	90.08	0.43037	187.79	2.936	–

Table 6 The parameters of the structure and the phase composition in the TiC-(Ni-Cr) metal-ceramic alloy surface layer (krypton plasma, $E_S = 60\,\text{J/cm}^2$, $t_i = 200\,\mu\text{s}$, $n = 15$ pulses)

Phase type	The relative content (wt%)	The lattice parameter (nm)	The size of the coherent scattering regions (nm)	The lattice micro-distortion, 10^{-3}	Texture
(Ni-Cr)	17.49	0.35778	33.57	3.725	(200)
TiC	82.51	0.43152	20.31	1.953	–

content from 9.92 up to 17.49 wt%. In addition, the titanium carbide lattice parameter increased from 0.43037 up to 0.43152 nm. Also, the size of the coherent scattering regions of the ceramic component decreased from 187.79 down to 20.31 nm. These features became more pronounced after PEBI in the xenon plasma. This can be concluded from the diffraction patterns, as well as the tables of the structure and phase composition parameters presented in Tables 7 and 8 (electron beam energy densities was 40 and 60 J/cm², respectively). An increase in electron-beam energy density from 40 up to 60 J/cm² reduced the ceramic content from 93.33 down to 65.00 wt%. This value almost corresponded to the level of its content in the initial state. The lattice parameters increased from 0.35798 up to 0.35920 nm for the nickel-chrome binder and from 0.42844 up to 0.43157 nm for titanium carbide. In addition, the coherent scattering regions for the nickel-chrome binder decreased from 39.01 to 14.36 nm. The lattice microdistortions also decreased from 7.242×10^{-3} down to 1.242×10^{-3}; texture of the surface layer was the same (002).

Table 7 The parameters of the structure and the phase composition in the TiC-(Ni-Cr) metal-ceramic alloy surface layer (xenon plasma, $E_S = 40\,\text{J/cm}^2$, $t_i = 150\,\mu\text{s}$, $n = 15$ pulses)

Phase type	The relative content (wt%)	The lattice parameter (nm)	The size of the coherent scattering regions (nm)	The lattice micro-distortion, 10^{-3}	Texture
(Ni-Cr)	6.67	0.35798	39.01	7.280	(200)
TiC	93.33	0.42844	–	–	–

Table 8 The parameters of the structure and the phase composition in the TiC-(Ni-Cr) metal-ceramic alloy surface layer (xenon plasma, $E_S = 60\,\text{J/cm}^2$, $t_i = 150\,\mu\text{s}$, $n = 150$ pulses)

Phase type	The relative content (wt%)	The lattice parameter (nm)	The size of the coherent scattering regions (nm)	The lattice micro-distortion, 10^{-3}	Texture
(Ni-Cr)	35.0	0.35920	14.36	1.242	(200)
TiC	65.0	0.43157			

Summarized results of the X-ray phase analysis are presented in Figs. 23 and 24. PEBI in the argon plasma caused a significant decrease in the nickel-chrome binder content in the surface layer, but increased this value by changing the plasma-forming gas to krypton. The nickel-chrome binder content in the surface layer was about the initial state level when the heaviest xenon plasma had been used (Fig. 23a).

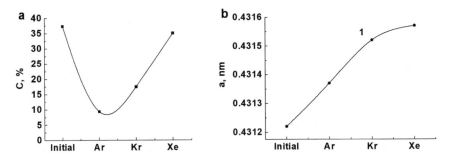

Fig. 23 The relative content of the nickel-chrome binder in the TiC-(Ni-Cr) metal-ceramic alloy surface layer (**a**) and the crystal lattice parameter of titanium carbide (**b**) after PEBI $\left(E_S = 60 \, \text{J/cm}^2, t_i = 150 \ldots 200 \, \mu\text{s}, n = 15 \text{ pulses}\right)$ versus plasma-forming inert gas: 1—the initial state, 2—argon, 3—krypton, 4—xenon

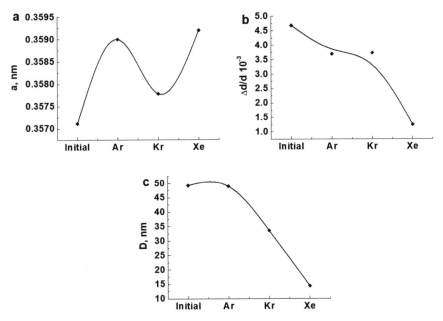

Fig. 24 The lattice parameter (**a**), the lattice micro-distortion (**b**), and the size of the coherent scattering regions (**c**) in the TiC-(Ni-Cr) metal-ceramic alloy surface layer after PEBI $\left(E_S = 60 \, \text{J/cm}^2, t_i = 150 \ldots 200 \, \mu\text{s}, n = 15 \text{ pulses}\right)$ versus plasma-forming inert gas: 1—the initial state, 2—argon, 3—krypton, 4—xenon

A feature of the structure and the phase composition of the surface layer after PEBI was an increase in the crystal lattice parameter. This effect intensified with increasing atomic mass of the plasma-forming gas. For example, the titanium carbide crystal lattice parameter reached 0.43157 nm after PEBI in the xenon plasma. This corresponded to the $C/Ti \approx 0.65$ ratio at the maximum titanium carbide hardness.

Figure 24 shows the effect of the plasma-forming gas on the crystal lattice parameters, the lattice microdistortions, and the size of the coherent scattering regions in the nickel-chrome binder. The crystal lattice parameters increased with an increase in atomic mass of the plasma-forming gas as a result of doping of the nickel-chrome binder during its interaction with titanium carbide particles. However, the lattice microdistortions and the sizes of the coherent scattering regions in the nickel-chrome binder decreased with increasing atomic mass of the plasma-forming gas.

3.5   The Effect of the Plasma-Forming Gases on the Nano- and Microhardness, and Wear Resistance of the Modified Surface Layer

Figure 25 shows dependences of nanohardness values in the surface layer after PEBI in the plasma of argon, krypton, and xenon from electron-beam energy density for pulse durations of 100, 150, and 200 μs. Based on their comparison, it can be concluded that the plasma-forming gas had a significant effect on the nanohardness values in the surface layer. They decreased with increasing electron-beam energy density up to 40 J/cm^2, but then enhanced significantly with rising electron-beam energy density up to 50 J/cm^2 (regardless of the plasma-forming gas). However, this effect was greater after PEBI in the plasma of krypton or xenon. The maximum nanohardness values were after PEBI in the xenon plasma.

A similar dynamics of the change in the nanohardness values was more concise with a change in pulse duration when electron-beam energy densities were 40 and 60 J/cm^2 (Fig. 26). Nanohardness values decreased with an increase in pulse duration up to 150 μs (regardless of the plasma-forming gas). Then, they enhanced slightly as pulse duration increased up to 200 μs.

Figure 27 shows dependences of microhardness values in the surface layer after PEBI from pulse duration in the plasma of argon, krypton, and xenon for various values of electron-beam energy density. They have common patterns with the same number of pulses. Initially, the microhardness values enhanced with increasing electron-beam energy density.

However, they change along a curve with a maximum at 150 μs as pulse duration increased, regardless of the plasma-forming gas (Fig. 28). The maximum microhardness values were after PEBI in the plasma of the lightest inert gas (argon). They decreased as the plasma-forming gas changed to krypton and xenon.

The presented data on the effect of atomic mass of the plasma-forming gases on the microhardness values in the surface layer after PEBI enabled to make

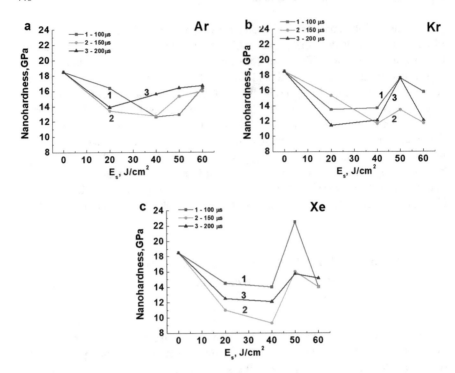

Fig. 25 Nanohardness of the TiC-(Ni-Cr) metal-ceramic alloy surface layer versus electron-beam energy density (argon, krypton, and xenon plasma, $t_i = 100,\ 150$ and $200\,\mu s$)

conclusions that the best results were obtained at electron-beam energy densities of 40, 50 and 60 J/cm² (Fig. 29).

Figure 30 presents dependences of the groove depth from electron-beam energy density after PEBI in the plasma of argon, krypton, and xenon. Modification of the surface layer greatly increased its wear resistance regardless of the plasma-forming gas. However, it increased with enhancing electron-beam energy density. The plasma-forming gas also had a significant effect. The maximum wear resistance values were after PEBI in the xenon plasma at an electron beam energy density of 60 J/cm². It decreased with a change in the plasma-forming gas to krypton and reached the minimum values after PEBI in the argon plasma (Fig. 30).

The presented results were verified by the abrasion test results in accordance with ASTM G65. In the initial period, wear resistance of the samples after PEBI was several times higher than that of the samples in the initial state. The maximum wear resistance was after PEBI in the xenon plasma. However, the modified layer became thinner as the number of revolutions of the abrasive disk increased. As a result, the dendritic structure began to interact with the counterpart and wear resistance of the surface layer decreased to the level of the sample in the initial state.

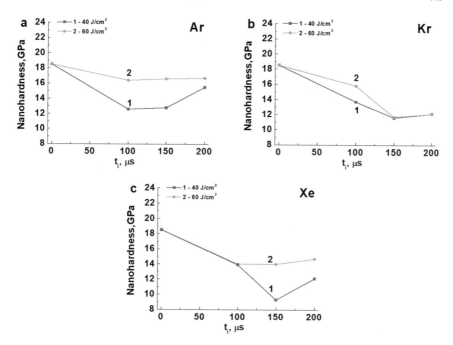

Fig. 26 Nanohardness of the TiC-(Ni-Cr) metal-ceramic alloy surface layer versus pulse duration (argon, krypton, and xenon plasma, $E_S = 40$ and 60 J/cm^2)

Figure 31 shows dependence of the friction coefficient on pulse duration. The main feature of these dependences was a sharp strong decrease in the friction coefficient after PEBI for all studied parameters (regardless of the plasma-forming gas) as compared with the sample in the initial state.

At a pulse duration of 150 μs, the general pattern of dependences shown in Fig. 32 was the invariance of the friction coefficient with changes in electron-beam energy density and pulse duration in all the plasma-forming gases studied. This means the invariance of the type of the structure and the phase composition of the top part of the surface layer with a change in the PEBI energy parameters.

One of the important parameters of the effect of the surface layer nanostructuring on its tribological properties was the dependence of the friction coefficient on temperature. Figure 33 shows the temperature dependences of the friction coefficient in the initial state and after PEBI by pulses of various durations in the plasma of argon, krypton, and xenon. Modification of the surface layer significantly reduced the friction coefficient over the entire studied temperature range up to 600 °C (especially at room temperature). The friction coefficient increased with rising temperature for all samples after PEBI. The friction coefficient of the sample in the initial state changed along a curve with a maximum at 200 °C. However, it was larger than that of the samples after PEBI in the entire studied temperature range. The samples after PEBI in the xenon plasma had the smallest values among others.

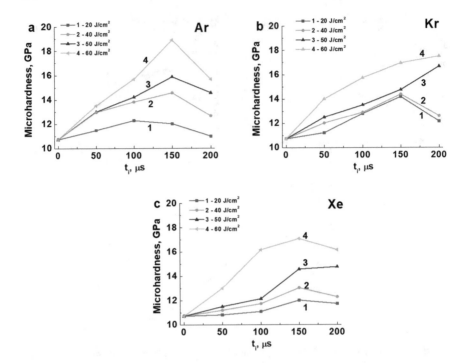

Fig. 27 Microhardness of the TiC-(Ni-Cr) metal-ceramic alloy surface layer versus pulse duration (argon, krypton, and xenon plasma, $E_S = 20,\ 40,\ 50$ and $60\,J/cm^2$, $n = 15$ pulses)

3.6 The Effect of the Nanostructured Surface Layer on Transverse Bending Strength

It follows from the distribution of the titanium carbide particle sizes, that the top part of the surface layers had the structure included columnar titanium carbide nanoparticles oriented perpendicular to the irradiated surface and the nickel-chrome binder interlayers. It is obvious that an increase in the nanostructuring level in the top part of the surface layer lowered the friction coefficient, and also increased its wear resistance and ductility. As a result, transverse bending strength of the samples increased under loading from the side of the irradiated surface. Figure 34 shows a dependence of transverse bending strength from pulse duration after PEBI. Electron-beam energy density was 40, 50 and $60\,J/cm^2$, the number of pulses was 15, the plasma-forming gases were argon, krypton, and xenon. Modification of the surface layer increased transverse bending strength in all investigated cases. The maximum values (indicated by the ellipses in the graphs) were after PEBI in the xenon plasma with the following combinations of electron-beam energy density and pulse duration: $40\,J/cm^2$ and $100\,\mu s$, $50\,J/cm^2$ and $200\,\mu s$, as well as $60\,J/cm^2$ and $150\,\mu s$.

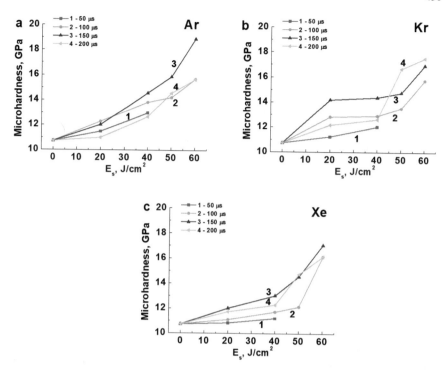

Fig. 28 Microhardness of the TiC-(Ni-Cr) metal-ceramic alloy surface layer versus electron-beam energy density (argon, krypton, and xenon plasma, $t_i = 50, 100, 150$ and $200\,\mu s$, $n = 15$ pulses)

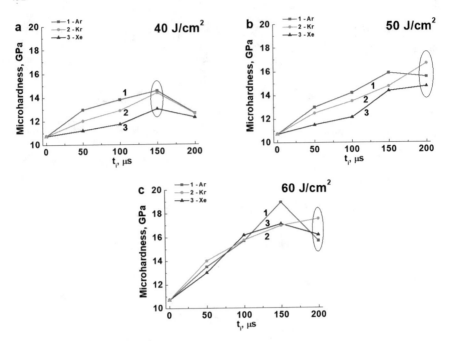

Fig. 29 Microhardness of the TiC-(Ni-Cr) metal-ceramic alloy surface layer versus pulse duration (argon, krypton, and xenon plasma, $E_S = 20$, 40 and 60 J/cm^2, $n = 15$ pulses). The ovals highlight the areas of greatest values

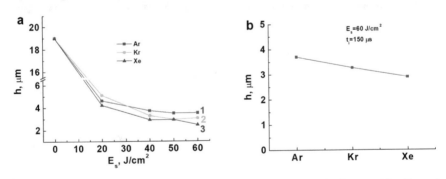

Fig. 30 Depth of the track on the surface of the TiC-(Ni-Cr) metal-ceramic alloy cut with a diamond counterpart after PEBI (argon, krypton, and xenon plasma, $t_i = 150\,\mu s$, $n = 15$ pulses) versus energy density

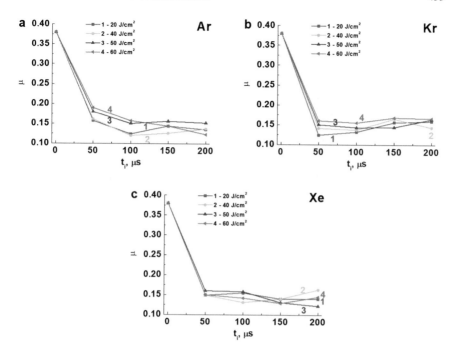

Fig. 31 The friction coefficient of the TiC-(Ni-Cr) metal-ceramic alloy surface after PEBI (argon, krypton, and xenon plasma, $E_S = 20, 40, 50$ and $60\,\text{J/cm}^2$, $n = 15$ pulses) versus pulse duration

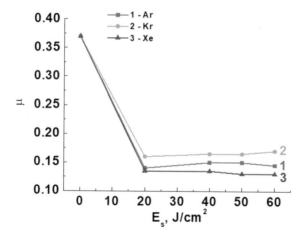

Fig. 32 The friction coefficient of the TiC-(Ni-Cr) metal-ceramic alloy surface after PEBI (argon, krypton, and xenon plasma, $t_i = 150\,\mu\text{s}$, $n = 15$ pulses) versus electron-beam energy density

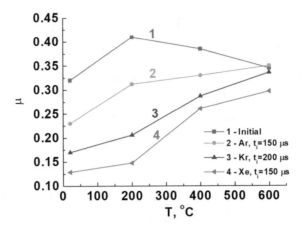

Fig. 33 The friction coefficient of the TiC-(Ni-Cr) metal-ceramic alloy surface after PEBI (argon, krypton, and xenon plasma, $E_S = 60\,\text{J/cm}^2, t_i = 150$ and $200\,\mu s, n = 15\,\text{pulses}$) versus temperature

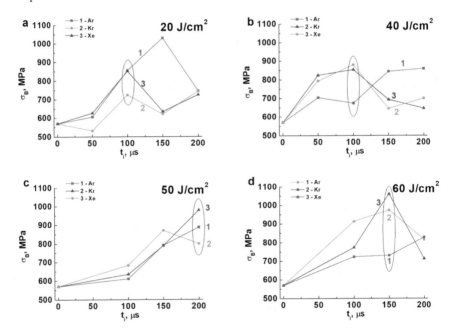

Fig. 34 Bending strength of the TiC-(Ni-Cr) metal-ceramic alloy after PEBI (argon, krypton, and xenon plasma, $E_S = 20, 40, 50$ and $60\,\text{J/cm}^2$, $n = 15\,\text{pulses}$, loading to the irradiated side) versus pulse duration

4 Conclusions

The results of the studies of the structure and the phase composition, as well as the physical, mechanical, and tribological properties of the surface layer of the TiC-(Ni-Cr) metal-ceramic alloy after PEBI the inert gas plasma enabled to draw the following conclusions.

1. PEBI in the plasma of the light (argon) and heavy (krypton, xenon) inert gases formed the nanostructured heterophase structure in the surface layer.
2. The modified layer depth increased with the change of the plasma-forming gas from argon to krypton and xenon.
3. Analysis of the results of the electron microscopic studies of the surface layer microstructure showed various mechanisms of the surface layer nanostructuring with an increase in atomic mass of the plasma-forming gas. Formation of titanium carbide nanoparticles directly at the interface of the carbide particles and the molten nickel-chromium binder was supplemented by the mechanical dispersion of titanium carbide particles and filling gaps with the molten nickel-chromium binder.
4. The change in the plasma-forming gas had the significant effect on the nanohardness values in the surface layer after PEBI. They increased greatly with rising atomic mass of the inert gas.
5. The nanohardness values decreased with increasing electron-beam energy density up to $40\,J/cm^2$, but then enhanced significantly with rising electron-beam energy density up to $50\,J/cm^2$ (regardless of the plasma-forming gas). However, this effect was greater after PEBI in the plasma of krypton or xenon. The maximum nanohardness values were after PEBI in the xenon plasma.
6. Typically, the microhardness values changed along the curve with the as pulse duration increased, regardless of the plasma-forming gas. The maximum microhardness values were after PEBI in the plasma of the lightest inert gas (argon). They decreased as the plasma-forming gas changed to krypton and xenon.
7. PEBI in the plasma of the heavy inert gases formed the thicker modified surface layers having lower microhardness values and, accordingly, higher ductility.
8. Modification of the surface layer greatly increased its wear resistance regardless of the plasma-forming gas. However, it increased with enhancing electron-beam energy density. The plasma-forming gas also had a significant effect. The maximum wear resistance values were after PEBI in the xenon plasma at an electron beam energy density of $60\,J/cm^2$.
9. The friction coefficient of the nanostructured modified surface layers decreased after PEBI for all studied parameters (regardless of the plasma-forming gas).
10. The general pattern was the invariance of the friction coefficient with changes in electron-beam energy density and pulse duration in all the plasma-forming gases studied. This means the invariance of the type of the structure and the phase composition of the top part of the surface layer with a change in the PEBI energy parameters.

11. Modification of the surface layer by PEBI in the plasma of the inert gases enhanced transverse bending strength of the samples under loading from the side of the irradiated surface.

12. The maximum values of the transverse bending strength were after PEBI in the xenon plasma with the following combinations of electron-beam energy density and pulse duration: 40 J/cm^2 and 100 μs, 50 J/cm^2 and 200 μs, as well as 60 J/cm^2 and 150 μs.

13. The maximum values of the transverse bending strength were after PEBI in the xenon plasma when electron-beam energy density was 60 J/cm^2. Transverse bending strength increased from 570 MPa in the initial state up to 1061 after PEBI (almost doubled).

References

1. Kidin IN (1950) Heat treatment of steel during induction heating. Metallurgizdat, Moscow (in Russian)
2. Turlygin SY (1959) Some issues of high-frequency heating of steel for hardening. Gosenergoizdat, Moscow (RU), 167 (in Russian)
3. Nemkov VS, Polevodov VS (1980) High frequency library, volume 15, Mashinostroenie, Leningrad (RU) (in Russian). (Mathematical computer modelling of high-frequency heating installations)
4. Kuvaldin AV (1988) Induction heating of ferromagnetic steel. Energoatomizdat, Moscow (RU) (in Russian)
5. Volodin VL, Sarychev VD (1990) The effect of pulsed magnetic fields on the structure and properties of metal alloys, News of the universities. Ferrous Metallurgy 10:99–104 (in Russian)
6. Pustovoyt VN, Rusin PI, Kudryakov OV (1991) Features of the organization of the steel structure as a result of processing by a concentrated energy flow during heating of the high-frequency current. Metallurgy Heat Treatment Metals 2:112–116 (in Russian)
7. Schukin VG, Marusin VV (1990) Thermophysics of high-frequency pulsed hardening of steel parts. Institute of Thermophysics SB RAS, Novosibirsk (RU) (in Russian)
8. Borisov YuS, Borisova AL (1986) Plasma powder coatings. Tekhnika, Kiew (RU) (in Russian)
9. Kudinov VV (1977) Plasma coatings. Nauka, Moscow (RU) (in Russian)
10. Pfender E (1988) Thermal plasma processing in the nineties. Pure Appl Chem 60(5):591–606
11. Ushio M (1988) Recent advances in thermal plasma processing. In: Proceedings of Japanese symposium on plasma chemistry, Tokyo, 28–29 July 1988, vol 1, pp 187–194
12. Spiridonov NV, Kobyakov OS, Kupriyanov IL (1988) Plasma and laser methods for hardening machine parts. Vysshaya shkola/High School, Minsk (RU) (in Russian)
13. Yoshida T (1990) The future of thermal plasma processing. Mater Trans, JIM 31(1):1–11
14. Zhukov MF, Solonenko OP (1990) High-temperature dusty jets in powder processing. Institute of Thermophysics SB RAS, Novosibirsk (RU) (in Russian)
15. Kudinov VV, Pekshev PJ, Belashchenko VE (1990) Plasma coating deposition. Nauka, Moscow (RU) (in Russian)
16. Ready JF (1971) Effects of high-power laser radiation. Academic Press, New York (US)
17. Rykalin NN, Uglov AA, Kokora AN (1975) Laser processing of materials: handbook. Mashinostroenie, Moscow (RU) (in Russian)
18. Duley WW (1983) Laser processing and analyzing of materials. Plenum Press, New York (US)
19. Prokhorov ASh, Konov VI, Ursu I, Mikheilesku IN (1988) Interaction of laser radiation with metals. Nauka, Moscow (RU) (in Russian)

20. Solntsev YuP (1988) Metallurgy and metal technology. Metallurgia, Moscow (RU) (in Russian)
21. Andriyakhin VM (1988) Laser welding and heat treatment processes. Nauka, Moscow (RU) (in Russian)
22. Sadovskiy VD, Schastlivtsev VM, Tabatchikova TI, Yakovleva IL (1989) Laser heating and steel structure: atlas of microstructures. UR O AN USSR (Ural branch of the Russian Academy of Sciences), Sverdlovsk (RU). (in Russian)
23. Leont'ev PA, Chekanova NT, Khan MG (1986) Laser surface treatment of metals and alloys. Metallurgia, Moscow (RU) (in Russian)
24. Mesyats GA (1986) Pulsed high-current electron-beam devices and their application in technology.In: Proceedings of the international conference on electron beam technologies. Publishing House of the Bulgarian Academy of Sciences, Sofia, pp 144–150 (in Russian)
25. Poate JM, Foti G, Jacobson DC (1983) Surface modification and alloying by laser, ion and electron beams. Plenum Press, New York (US)
26. Kadyrzhanov KK, Komarov FF, Pogrebnyak AD, Rusakov VS, Turkebaev TE (2005) Ion-beam and ion-plasma modification of materials. Publishing House of the Moscow State University, Moscow (RU) (in Russian)
27. Ivanov YuF, Koval NN (2007) The Structure and properties of promising metallic materials. Tomsk (RU): Publishing house NTL, 345–382 (Low-energy electron beams of submillisecond duration: production and some aspects of application in the field of materials science. Ch. 13) (in Russian)
28. Xu Y, Zhang Y, Hao SZ, Perroud O, Li MC, Wang HH, Grosdidier T, Dong C (2013) Surface microstructure and mechanical property of WC-6%Co hard alloy irradiated by high current pulsed electron beam. Appl Surf Sci 279:137–141
29. Hao S, Xu Y, Zhang Y, Zhao L (2013) Improvement of surface microhardness and wear resistance of WC/Co hard alloy by high current pulsed electron beam irradiation. Int J Refractory Metals Hard Mater 41:553–557
30. Psakhie SG, Ovcharenko VE, Knyazeva AG, Shilko EV (2011) The formation of a multiscale structure in the surface layers and the resistance of a cermet alloy under mechanical stress. Phys Mesomechanics 14(6):23–34 (in Russian)
31. Baksht RB, Mesyats GA, Proskurovskiy DI et al (1976) Development and application of sources of intense electron beams. Novosibirsk (RU): Nauka, pp 141–153 (The impact of a powerful short-term electron flow on metal) (in Russian)
32. Vasil'ev VYu, Demidov BA, Kuz'menko TG et al (1982) The formation of an amorphous structure in iron-based alloys during surface treatment by a high-current electron beam. *Reports from the Academy of Sciences of the USSR (DAN of the USSR)*, vol 268, issue 3, pp 605–607 (in Russian)
33. Shulov VA, Paykin AG, Belov AB et al (2005) Mechanisms for the redistribution of elements in the surface layers of parts made of heat-resistant materials when they are irradiated with high-current pulsed electron beams. Phys Chem Mater Process 2(32–41):607 (in Russian)
34. Shulov VA, Paykin AG, Teryaev AD et al (2009) Structural changes in the surface layers of parts made of titanium alloys VT6 and VT9 under exposure of pulsed electron beams. Hardening Technol Coatings 1:29–31 (in Russian)
35. Follstaedt DM (1982) Laser and Electron-beam interactions with solids. In: Proceedings of the Materials Research Society Annual Meeting, November 1981, Boston Park Plaza Hotel, Boston, Massachusetts, U.S.A., vol 4. Elsevier, New York (US), pp 377–388 (Metallurgy and microstructures of pulse melted alloys)
36. Knapp JA, Follstaedt DM (1982) Laser and Electron-beam interactions with solids: Proceedings of the Materials Research Society Annual Meeting, November 1981, Boston Park Plaza Hotel, Boston, Massachusetts, U.S.A., vol 4. Elsevier, New York (US), pp 407–412 (Pulsed electron beam melting of Fe)
37. Zhang C, Cai J, Lv P, Zhang Y, Xia H, Guan Q (2017) Surface microstructure and properties of Cu-C powder metallurgical alloy induced by high-current pulsed electron beam. J Alloys Compounds 697:96–103

38. Yu B-H, Ovcharenko VE, Ivanov KV, Mokhovikov AA, Zhao Y-H (2018) Effect of surface layer structural-phase modification on tribological and strength properties of a TiC–(Ni–Cr) Metal Ceramic Alloy. Acta Metallurgica Sinica (English Letters) 31:547–551

39. Ovcharenko VE, Lapshin OV, Ivanov KV, Klimenov VA (2018) Effectiveness of inert plasma gases in formation of modified structures in the surface layer of a cermet composite under pulsed electron irradiation. Int J Refractory Metals Hard Mater 77:31–36

40. Zhao J, Holland T, Unuvar C, Munir ZA (2009) Sparking plasma sintering of nanometric tungsten carbide. Int J Refractory Metals Hard Mater 27:130–139

41. Fang ZZ, Wang X, Ryu T, Hwang KS, Sohn HY (2009) Synthesis, sintering, and mechanical properties of nanocrystalline cemented tungsten carbide—a review. Int J Refractory Metals Hard Mater 27(2):288–299

42. Koval NN, Shchanin PM, Devyatkov VN, Tolkachev VS, Vintizenko LG (2005) A facility for metal surface treatment with an electron beam. Instruments Exp Techniques 48:117–121

43. Gavarini S, Millard-Pinard N, Garnier V, Gherrab M, Baillet J, Dernoncourt L, Peaucelle C, Jaurand X, Douillard T (2015) Elaboration and behavior under extreme irradiation conditions of nano- and micro-structured TiC. Nuclear Instruments Methods Phys Res Section B Beam Interactions Mater Atoms 356–357:114–128

44. Ovchinnikov VV, Goloborodsky BYu, Gushchina NV, Semionkin VA, Wieser E (2006) Enhanced atomic short-range ordering of the alloy Fe-15 at.% Cr caused by ion irradiation at elevated temperature and thermal effects only. Appl Phys A 83:83–88

45. Ovcharenko VE, Lapshin OV (2008) Calculation of the temperature field in the surface layer of a cermet with electron-pulsed irradiation. Metal Sci Heat Treatment 50(5–6):238–241

46. Lykov AV (1967) Thermal conductivity theory. Vysshaya shkola, Moscow (RU) (in Russian)

47. Smithells CI (1992) Metals: reference book. Butterworth-Heinemann, London (UK)

48. Samsonov GV, Vinnitskiy IM (1976) Refractory compounds: reference book. Metallurgia, Moscow (RU) (in Russian)

49. Varavka VN, Brover GI, Magomedov MG, Brover AV (2001) Thermophysical peculiarities of process of tool steel pulsed laser treatment. Vestnik DGTU 1(7):56–63 (in Russian)

50. Devyatkov VN, Koval NN, Schanin PM, Grigoryev VP, Koval TB (2003) Generation and propagation of high-current low-energy electron beams. Laser Particle Beams 21(2):243–248

51. Devyatkov VN, Koval NN, Schanin PM, Tolkachev LG, Vintizenko LG (2004) Installation for treatment of metal surfaces by low energy electron beam. In: Proceedings of the 7th Intern. conference on modification of materials with particle beams and plasma flows, Tomsk, Russia, 25–30 July 2004. Tomsk (RU): Publisher of the IAO SB RAS, pp 43–46

52. Wieser ME, Holden N, Coplen TB, Böhlke JK, Berglund M, Brand WA, De Bièvre P, Gröning M, Loss RD, Meija J, Hirata T, Prohaska T, Schönberg R, O'Connor G, Walczyk T, Yoneda S, Zhu X-K (2013) Atomic weights of the elements. Pure Appl Chem 85(5):1047–1078

53. Huheey JE, Keiter EA, Keiter RL (1993) Inorganic chemistry: principles of structure and reactivity, 4th edn. HarperCollins College Publishers, New York (US)

54. Laby TH, Kaye GWC (1995) Tables of physical and chemical constants. Longman Sc & Tech, New York (US)

55. Lide DR (1998) Chemical rubber company—CRC handbook of chemistry and physics 79th edition: a ready-reference book of chemical and physical data. CRC Press, Boca Raton (US)

56. Dean JA (1999) Lange's handbook of chemistry, 15th edn. McGraw-Hill Professional, New York (US)

57. James AM, Lord MP (1992) Macmillan's chemical and physical data. Macmillan Publishers, London (UK)

58. Cox JD, Wagman DD, Medvedev VA (1989) CODATA key values for thermodynamics. Hemisphere Publishing Corp, New York (US)

Adhesion of a Thin Soft Matter Layer: The Role of Surface Tension

Valentin L. Popov

Abstract We consider an adhesive contact between a thin soft layer on a rigid substrate and a rigid cylindrical indenter ("line contact") taking the surface tension of the layer into account. First, it is shown that the boundary condition for the surface outside the contact area is given by the constant contact angle—as in the case of fluids in contact with solid surfaces. In the approximation of thin layer and under usual assumptions of small indentation and small inclination angles of the surface, the problem is solved analytically. In the case of a non-adhesive contact, surface tension makes the contact stiffer (at the given indentation depth, the contact half-width becomes smaller and the indentation force larger). In the case of adhesive contact, the influence of surface tension seems to be more complicated: For a flat-ended punch, it increases with increasing the surface tension, while for a wedge, it decreases. Thus, the influence of the surface tension on the adhesion force seems to be dependent on the particular geometry of the contacting bodies.

Keywords Adhesion · Capillarity · Surface tension · Winkler foundation · Contact angle

1 Introduction

Classical contact mechanics as represented by the works of Hertz [1] or Bussinesq [2], see also [3], neglects the surface tension of the contacting solids. In reality, all three surfaces of bodies in contact (Fig. 1a) can be characterized by their specific surface energies γ_1, γ_2 and γ_{12}. Depending on their values, one can distinguish several cases. If the specific surface energy of the surface of elastic body outside the contact area can be neglected, then we have an adhesive contact with specific work of separation $w = \gamma_2 - \gamma_{12}$. This case was first considered in the classic work by Johnson, Kendall and Roberts [4]. If the surface energy of the elastic body outside the contact area is finite, $\gamma_1 \neq 0$, but the work of adhesion, which in the general case is equal to

V. L. Popov (✉)
Technische Universität Berlin, 10623 Berlin, Germany
e-mail: v.popov@tu-berlin.de

© The Author(s) 2021
G.-P. Ostermeyer et al. (eds.), *Multiscale Biomechanics and Tribology of Inorganic and Organic Systems*, Springer Tracts in Mechanical Engineering,
https://doi.org/10.1007/978-3-030-60124-9_19

461

$$w = \gamma_1 + \gamma_2 - \gamma_{12} \tag{1}$$

is zero, then we have a non-adhesive contact with surface tension. One can consider such system as an elastic body coated with a stressed membrane. The corresponding theory was first developed in [5]. The general case is when both work of adhesion and surface tension of the "free surface" are finite. This leads to a general adhesive contact with surface tension. The latter attracted much interest in the last two decades in the context of indentation of soft matter (e.g. gels or biological tissues) [6–8]. Let us also note that another contact problem with adhesion and surface tension represents a contact of an elastic solid with a *fluid* [9]. This class differs significantly from contact of elastic bodies with surface tension and will not be discussed here.

2 Model

In the present paper, we consider an adhesive contact between a thin soft layer on a rigid substrate and a rigid cylindrical indenter ("line contact") taking the surface tension of the layer into account, Fig. 1a (left). Without consideration of the surface tension, this problem has been solved in [10]. Here we extend the study carried out in [10] to include the effect of surface tension of contacting bodies. This contact problem can be treated asymptotically exactly, but only under strong assumptions. In particular, we assume that the following conditions are fulfilled: $d \ll h$, $h \ll a$, where d is the indentation depth, h the thickness of the layer, and a the half-width of the contact (definitions see in Fig. 1a (left)). Additionally, it is assumed that the slope of the profile of contacting bodies and of the free surface outside the contact

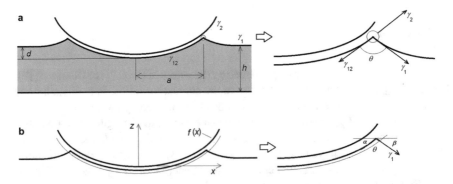

Fig. 1 a left: principal sketch of the system consisting of a rigid indenter in contact with a thin elastic layer (grey) having the initial thickness h. **a** right: Detailed view of the boundary of the contact with three surface forces corresponding to the tree surfaces meeting at the boundary. **b** left: Definitions of coordinates and profile of the rigid indenter as well as free body diagram (of the system over the thin gray line). **b** right: Detailed picture of a part of the free body diagram showing the surfaces forces acting on the contact boundary

is much smaller than unity. Further conditions, if necessary, will be specified later in this paper.

Under the above conditions, the elastic layer is deformed uniaxially, independently in each point and the layer can be considered as a two-dimensional elastic foundation with effective modulus [11]

$$\tilde{E} = \frac{E(1 - \nu)}{(1 + \nu)(1 - 2\nu)}, \tag{2}$$

where E is elastic modulus and ν Poisson number. Due to local uniaxial deformation, the layer can be considered as a two-dimensional elastic foundation composed of independent springs placed with separation Δx and Δy correspondingly, while each spring has the stiffness

$$\Delta k = \tilde{E} \frac{A}{h}, \tag{3}$$

with $A = \Delta x \Delta y$. When a rigid profile $f(x)$ (Fig. 1b (left)) is indented into this elastic foundation by a depth d, then the vertical displacements of the springs in contact are equal to

$$u_z(x) = d - f(x), \quad |x| \le a \tag{4}$$

(Note that while the axis z for the definition of the profile shape is directed upwards, the positive direction of the displacement $u_z(x)$ is accepted to be downwards).

The shape of the surface outside the contact area, is governed by the equation

$$\gamma_1 \frac{\partial^2 u_z(x)}{\partial x^2} = \frac{\tilde{E}}{h} u_z(x), \tag{5}$$

which simply equates the elastic stress to the stress produced by the tensioned surface (surface tension γ_1 multiplied with the surface curvature $\partial^2 u_z(x)/\partial x^2$). Solution of Eq. (5) reads

$$u_z(x) = -C \exp(-x/l), \tag{6}$$

where

$$l = \sqrt{\frac{\gamma_1 h}{\tilde{E}}}. \tag{7}$$

This length plays the role of the "elastocapillary length" in the present problem.

3 Boundary Condition at the Contact Boundary

Equation (5) must be completed through boundary conditions at the boundary of
the contact area. To derive this boundary condition, consider a small part of the
boundary encircled in Fig. 1a (right) by a gray circle. The sum of all forces acting
on the boundary line parallel to the surface of the rigid body, should vanish, if the
boundary friction is neglected:

$$\gamma_{12} + \gamma_1 \cos\theta - \gamma_2 = 0. \tag{8}$$

The elastic force can be neglected in this equation as it vanishes if the size of the
circle tends towards zero (while the surface tensions remain constant). This equation
is the same as in the case of a contact of a liquid with a solid. Karpitschka et al.
come to the same conclusion by performing minimization of the complete energy
functional [12].

4 The Force Acting on the Rigid Indenter

As we consider the line contact (no dependency on the coordinate perpendicular to
the plane (x, z) (not shown in Fig. 1)), it is convenient to use instead of the normal
force the normal force per length, P_N and also normalize all other forces per unit
length. The elastic force per unit length acting on the rigid indenter from the elastic
layer is simply:

$$P_{el} = \frac{\tilde{E}}{h} \int_{-a}^{a} u_z(x)\mathrm{d}x = \frac{\tilde{E}}{h} \int_{-a}^{a} (d - f(x))\mathrm{d}x. \tag{9}$$

Apart from this elastic force, there is an additional force acting on the indenter by
the surface of the elastic body outside the contact area, which is equal to (see Fig. 1b
(right))

$$P_{\mathrm{surf}} = -2\gamma_1 \sin\beta. \tag{10}$$

Under assumption that all slopes in the considered system are small, the total force acting on the rigid indenter from the elastic body is equal to

$$P_N = 2\frac{\tilde{E}}{h} \int_0^a (d - f(x))dx - 2\gamma_1 \left.\frac{\partial u_z(x)}{\partial x}\right|_{x=a+0}. \tag{11}$$

In this equation, the half-width of the contact, a, is still not defined.

5   Contact Half-Width

The boundary condition for the surface shape can be written as

$$\left.\frac{\partial f(x)}{\partial x}\right|_{x=a-0} + \left.\frac{\partial u_z(x)}{\partial x}\right|_{x=a+0} + \theta = \pi. \tag{12}$$

Note that we assume that all slopes are small, so that the contact angle should be almost equal to π. Smaller contact angles can be realized physically but they cannot be treated in the approximation of small slopes, which is used in the present model (see, however, the next section for a more detailed discussion of the area of applicability).

At the boundary of the contact area two equations have to be fulfilled:

$$d - f(a) = -C \exp(-a/l), \tag{13}$$

and

$$\left.\frac{\partial f(x)}{\partial x}\right|_{x=a-0} + \frac{C}{l} \exp(-a/l) + \theta = \pi. \tag{14}$$

Substituting (13) as well as the solution of Eq. (8),

$$\pi - \theta \approx \sqrt{\frac{2w}{\gamma_1}}, \tag{15}$$

into (14) gives

$$f(a) = d + \sqrt{\frac{2wh}{\tilde{E}}} - \sqrt{\frac{\gamma_1 h}{\tilde{E}}} \cdot \left.\frac{\partial f(x)}{\partial x}\right|_{x=a-0}. \tag{16}$$

Here

$$w = \gamma_1 + \gamma_2 - \gamma_{12} \tag{17}$$

is the work of adhesion.

In the limit $\gamma_1 = 0$ (vanishing surface tension), (16) reduces to

$$f(a) = d + \sqrt{\frac{2wh}{\tilde{E}}}, \tag{18}$$

which, according to [10], is the correct result for an arbitrary profile if the surface tension is neglected.

6 Area of Applicability of Eq. (16)

Note that all above considerations are valid under the assumptions listed in Sect. 2. In particular, the thin layer approximation can only be used if all slopes are small. For our problem, this implies that the angle $\pi - \theta$ should also be small. From Eq. (15) it then follows that the current approximation is strictly valid only if $2w \ll \gamma_1$. This means that the limit $\gamma_1 \to 0$ is not covered by the present theory. However, Eq. (18), obtained in the limit $\gamma_1 = 0$, reproduces the exact solution of the corresponding problem without surface tension. This suggests that Eq. (16) can be used as an approximate solution in the whole range of values of $0 \le \gamma_1 \le \infty$.

7 Case Studies

Case study 1: rigid plane In this case, $f(x) = 0$ and Eq. (16) takes the form

$$0 = d + \sqrt{\frac{2wh}{\tilde{E}}}. \tag{19}$$

For negative indentation depth (which correspond to the adhesion case), this equation is fulfilled for one single value of the distance between the rigid plane and the elastic layer:

$$|d_c| = -d_c = \sqrt{\frac{2wh}{\tilde{E}}}. \tag{20}$$

If the distance becomes larger, the contact shrinks and disappears; if it becomes smaller, then it spreads to infinity. At exactly the critical value, the contact is in an indefinite equilibrium at any contact size. These properties are the same as in the case of adhesive contact without tension.

Case study 2: flat punch with half-width a As is clear from the Case study 1, a flat ended punch will detach at once at the critical distance, given by Eq. (20). The surface shape outside the contact is given by

$$u_z(x) = -\sqrt{\frac{2wh}{\tilde{E}}} \exp\left(\frac{-x+a}{l}\right). \tag{21}$$

Equation (11) for the normal force now gives

$$P_N = -2a\sqrt{\frac{2w\tilde{E}}{h}} - 2\sqrt{2w\gamma_1} = -2^{3/2}w^{1/2}\left(a\sqrt{\frac{\tilde{E}}{h}} + \sqrt{\gamma_1}\right). \tag{22}$$

The force of adhesion is minus the normal force acting on the rigid indenter:

$$P_A = 2^{3/2}w^{1/2}\left(a\sqrt{\frac{\tilde{E}}{h}} + \sqrt{\gamma_1}\right). \tag{23}$$

We see that, at the given work of separation w, the surface tension leads to an increase of the force of adhesion. The contact configuration is illustrated in Fig. 2.

Case study 3: wedge shape If the shape of the rigid indenter is given by $f(x) = |x|\tan\delta$ (Fig. 3a), then Eq. (16) takes the form

$$a\tan\delta = d + \sqrt{\frac{2wh}{\tilde{E}}} - \sqrt{\frac{\gamma_1 h}{\tilde{E}}} \cdot \tan\delta \tag{24}$$

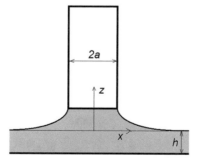

Fig. 2 Adhesive contact with surface tension of a flat rigid indenter with a thin elastic layer

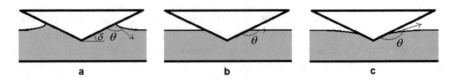

Fig. 3 Contact of a wedge shaped indenter for three different contact angles. Case (**b**) correspond to the vanishing adhesion force

and the half-width of the contact area is given by

$$a = (\tan \delta)^{-1} \left(d + \sqrt{\frac{2wh}{\tilde{E}}} \right) - \sqrt{\frac{\gamma_1 h}{\tilde{E}}}. \tag{25}$$

The shape of the surface outside the contact can be written as

$$u_z(x) = \left(-\sqrt{\frac{2wh}{\tilde{E}}} + \sqrt{\frac{\gamma_1 h}{\tilde{E}}} \cdot \tan \delta \right) \exp\left(\frac{-x+a}{l} \right). \tag{26}$$

For the normal force, Eq. (11), we get:

$$P_N = 2\frac{\tilde{E}}{h} \left(ad - \frac{a^2}{2} \tan \delta \right) + 2\left(-\sqrt{2w\gamma_1} + \gamma_1 \cdot \tan \delta \right). \tag{27}$$

Solving Eq. (24) with respect to d,

$$d = a \tan \delta - \sqrt{\frac{2wh}{\tilde{E}}} + \sqrt{\frac{\gamma_1 h}{\tilde{E}}} \cdot \tan \delta \tag{28}$$

and inserting this result into (27) gives the force per length as a function of the contact half-width a:

$$P_N = 2\frac{\tilde{E}}{h} \left(\frac{a^2}{2} \tan \delta - a\left(\sqrt{\frac{2wh}{\tilde{E}}} - \sqrt{\frac{\gamma_1 h}{\tilde{E}}} \cdot \tan \delta \right) \right) + 2\left(-\sqrt{2w\gamma_1} + \gamma_1 \cdot \tan \delta \right). \tag{29}$$

Minimizing with respect to a, gives the adhesion force

$$F_A = \left| P_{N,\min} \right| = \frac{2w}{\tan \delta} - \gamma_1 \tan \delta. \tag{30}$$

Surface tension leads to a decrease of the force of adhesion. The adhesion force vanishes when

$$\tan \delta = \sqrt{\frac{2w}{\gamma_1}}, \tag{31}$$

or, under consideration of (15) and assuming $\tan \delta \approx \delta$,

$$\delta + \theta = \pi. \tag{32}$$

This equation has a very simple physical interpretation: The adhesion force disappears if the slope of the indenter and the contact angle allow a horizontal non-deformed surface outside the contact area, as illustrated in Fig. 3b. Under condition (31) the whole dependence of the force on the contact half-width, Eq. (29), is reduced to that for the non-adhesive contact without surface tension:

$$P_N = \frac{\tilde{E}}{h} a^2 \tan \delta. \tag{33}$$

This is because, in this case, the surface tension force is directed horizontally and does not contribute to the normal force.

Case study 4: parabolic shape Let us consider the special case of a parabolic profile

$$f(x) = \frac{x^2}{2R}. \tag{34}$$

Equation (16) now takes the form

$$\frac{a^2}{2R} = d + \sqrt{\frac{2wh}{\tilde{E}}} - \sqrt{\frac{\gamma_1 h}{\tilde{E}}} \frac{a}{R} \tag{35}$$

Its solution with respect to a reads

$$a = -\sqrt{\frac{\gamma_1 h}{\tilde{E}}} + \sqrt{\frac{\gamma_1 h}{\tilde{E}} + 2R\left(d + \sqrt{\frac{2wh}{\tilde{E}}}\right)} \tag{36}$$

In the case of vanishing surface tension, $\gamma_1 = 0$, Eq. (36) reduces to $a = \sqrt{2R\left(d + \sqrt{\frac{2wh}{\tilde{E}}}\right)}$ meaning that the contact boundary is defined by cutting the profile at the height $d + \sqrt{\frac{2wh}{\tilde{E}}}$ which coincides with the result of paper [10] for the corresponding problem with vanishing surface tension.

8 Non-adhesive Contact

Let us consider the limiting case of non-adhesive contact with tension separately. Under "non-adhesive" contact we will understand the contact of surfaces with vanishing work of separation, $w = 0$. From the definition (1), it follows that in this case

$$\gamma_1 = \gamma_{12} - \gamma_2. \tag{37}$$

From (8), it the follows that

$$\cos\theta = \frac{\gamma_2 - \gamma_{12}}{\gamma_1} = -1 \tag{38}$$

and $\theta = \pi$. This means that the slope is continuous at the boundary of the contact. Equation (14) now takes the form

$$\left.\frac{\partial f(x)}{\partial x}\right|_{x=a-0} + \frac{C}{l}\exp(-a/l) = 0. \tag{39}$$

Taking (13) into account, we come to the equation

$$d - f(a) = l \cdot \left.\frac{\partial f(x)}{\partial x}\right|_{x=a-0} = \sqrt{\frac{\gamma_1 h}{\tilde{E}}} \cdot \left.\frac{\partial f(x)}{\partial x}\right|_{x=a-0}. \tag{40}$$

Consider as an example a contact of a parabolic indenter $f(x) = x^2/(2R)$. Equation (40) takes the form

$$a^2 + 2al - 2Rd = 0 \tag{41}$$

For the contact radius we thus get

$$a = -l + \sqrt{2Rd + l^2} \tag{42}$$

This means that the surface tension leads to a decrease of the contact width compared with the non-adhesive contact without surface tension.
   For the total normal force we get according to (11)

$$P_N = \frac{2\tilde{E}}{3Rh}\left[\left(2Rd + l^2\right)^{3/2} - l^3\right] = \frac{2\tilde{E}}{3Rh}\left[\left(2Rd + \frac{\gamma_1 h}{\tilde{E}}\right)^{3/2} - \left(\frac{\gamma_1 h}{\tilde{E}}\right)^{3/2}\right] \tag{43}$$

For given d, the normal force with surface tension is larger than that without surface tension.

9 Conclusion

In the present paper, we considered a general adhesive contact of a thin elastic body with a rigid indenter. An important conclusion is that at the boundary of the contact area, the surface of the elastic layer meets the surface of the rigid indenter under a fixed contact angle, which is determined uniquely by the specific surface energies of the rigid body, the elastic body and the interface. The solution obtained for very small surface tension (compared with the work of adhesion) seems to provide a good approximation for arbitrary values of the surface tension. In the case of a non-adhesive contact, surface tension makes the contact stiffer (at the given indentation depth, the contact half-width becomes smaller and the indentation force larger). In the case of adhesive contact, the influence of surface tension seems to be more complicated: For a flat-ended punch, it increases with increasing the surface tension, while for a wedge, it decreases. Thus, the influence of the surface tension on the adhesion force seems to be dependent on the particular geometry of the contacting bodies.

Acknowledgements The author is thankful to Qiang Li, Weike Yuan and Iakov Lyashenko for helpful discussions and Nikita Popov for proofreading. This work was financially supported by the DFG (project number PO 810/55-1).

References

1. Hertz H (1882) Über die Berührung fester elastischer Körper. Journal für die reine und angewandte Mathematik 92:156–171
2. Boussinesq VJ (1885) Application des Potentiels a L'etude de L'Equilibre et du Mouvement des Solides Elastiques. Gautier-Villars, Paris
3. Popov VL, Heß M, Willert E (2019) Handbook of contact mechanics exact solutions of axisymmetric contact problems. Springer, Heidelberg, 347p. https://doi.org/10.1007/978-3-662-587 09-6
4. Johnson KL, Kendall K, Roberts AD (1971) Surface energy and the contact of elastic solids. R Soc Publ Londn A 324:301–313. https://doi.org/10.1098/rspa.1971.0141
5. Hajji MA (1978) Indentation of a membrane on an elastic half space. ASME J Appl Mech 45(2):320–324. https://doi.org/10.1115/1.3424295
6. Cao Z, Stevens MJ, Dobrynin AV (2014) Elastocapillarity: adhesion and wetting in soft polymeric systems. ACS Publ Macromolecules 47:6515–6521. https://doi.org/10.1021/ma5 013978
7. Carrillo J-MY, Dobrynin AV (2012) Contact mechanics of nanoparticles. ACS Publ Langmuir 28:10881–10890. https://doi.org/10.1021/la301657c
8. Style RW, Hyland C, Boltyanskiy R et al (2013) Surface tension and contact with soft elastic solids. Nature Commun 4:2728–2733. https://doi.org/10.1038/ncomms3728
9. Makhovskaya YY, Goryacheva IG (1999) The combined effect of capillarity and elasticity in contact interaction. Tribol Int 32:507–515. https://doi.org/10.1016/S0301-679X(99)00080-8
10. Li Q, Popov VL (2019) Adhesive contact between a rigid body of arbitrary shape and a thin elastic coating. Springer Link Acta Mechanica 230:2447–2453. https://doi.org/10.1007/s00 707-019-02403-0

11. Popov VL (2017) Contact mechanics and friction. Physical principles and applications. Springer, Berlin. https://doi.org/10.1007/978-3-662-53081-8
12. Karpitschka S, van Wijngaarden L, Snoeijer JH (2016) Surface tension regularizes the crack singularity of adhesion. Soft Matter 12:4463. https://doi.org/10.1039/C5SM03079J

Adhesion Hysteresis Due to Chemical Heterogeneity

Valentin L. Popov

Abstract According the JKR theory of adhesive contact, changes of the contact configuration after formation of the adhesive neck and before detaching are completely reversible. This means, that after formation of the initial contact, the force-distance dependencies should coincide, independently of the direction of the process (indentation or pull-off). In the majority of real systems, this invariance is not observed. The reasons for this may be either plastic deformation in the contacting bodies or surface roughness. One further mechanism of irreversibility (and corresponding energy dissipation) may be chemical heterogeneity of the contact interface leading to the spatial dependence of the specific work of adhesion. In the present paper, this "chemical" mechanism is analyzed on a simple example of an axisymmetric contact (with axisymmetric heterogeneity). It is shown that in the asymptotic case of a "microscopic heterogeneity", the system follows, during both indentation and pull-off, JKR curves, however, corresponding to different specific surface energies. After the turning point of the movement, the contact area first does not change and the transition from one JKR curve to the other occurs via a linear dependency of the force on indentation depth. The macroscopic behavior is not sensitive to the absolute and relative widths of the regions with different surface energy but depends mainly on the values of the specific surface energy.

Keywords Adhesion · Hysteresis · Energy dissipation · JKR theory · MDR · Specific surface energy · Heterogeneity

1 Introduction

Johnson, Kendall and Roberts published 1971 their famous work on adhesive contact of elastic parabolic bodies [1]. Contrary to the non-adhesive contact, an adhesive contact shows a hysteresis: the dependencies of force on approach depend on whether the bodies are brought into contact or pulled off. The area enclosed in the hysteresis

V. L. Popov (✉)
Technische Universität Berlin, 10623 Berlin, Germany
e-mail: v.popov@tu-berlin.de

© The Author(s) 2021
G.-P. Ostermeyer et al. (eds.), *Multiscale Biomechanics and Tribology of Inorganic and Organic Systems*, Springer Tracts in Mechanical Engineering, https://doi.org/10.1007/978-3-030-60124-9_20

473

loop is the energy, which is irreversible dissipated during one complete "cycle" of an adhesive contact. According to the JKR theory, after the first contact, an adhesive neck of finite radius appears. If we now would try to pull off the bodies, they remain in contact even for negative values of the indentation depth up to the point of instability where the contact is lost at once. The mechanical energy is irreversibly lost only in such points of instabilities [2]. Both before and after the instability, the processes of approach and detachment are reversible, which is obvious if we remember that the JKR theory is based on the principle of virtual work, which assumes absence of static frictional forces [3]. However, experiments show that adhesive contacts show often pronounced hysteresis even after the formation of initial contact [4–6]. Some authors attribute this to plastic deformation [5]. However, it was shown in [6, 7] that such hysteresis could be seen also in pure elastic contacts between rough bodies. The mechanism of this hysteresis is very simple: The energy dissipation occurs in each act of instable movement of the contact boundary. If the contact area has a complicated shape, the movement of the boundary can proceed in a series of jumps [8] leading to energy dissipation, which means that the boundary feels a dissipative force (see also the supplementary video [9]). Another mechanism of the adhesive hysteresis in the already formed contact state could be the chemical heterogeneity of the contact interface. In the present paper, we analyze this mechanics on a simple example when both the shape of contacting bodies and the chemical heterogeneity have axial symmetry.

2 Problem Statement and Model Description

Consider an adhesive contact between an axissymmetrical rigid body $z = f(r)$, where z is the coordinate in the normal to interface direction and r polar radius in the contact plane, and an elastic half space. It is assumed that the specific work of adhesion assumes two constant values γ_1 and γ_2 in the alternating rings having the widths h_1 and h_2 (Fig. 1). In the framework of the Method of Dimensionality Reduction (MDR) [10, 11], it is possible to map a three-dimensional axisymmetric contact problem to a contact of the modified plane shape

$$g(x) = |x| \int\limits_0^{|x|} \frac{f'(r)}{\sqrt{x^2 - r^2}} dr .$$ (1)

and a one-dimensional elastic foundations consisting of independent springs (Fig. 2) having the stiffness

$$\Delta k = E^* \Delta x.$$ (2)

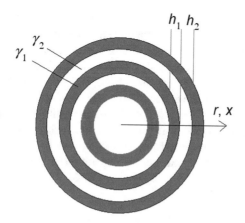

Fig. 1 Schematic representation of the chemical heterogeneity in the considered system. The specific work of adhesion take two constant values γ_1 and γ_2 in the alternating rings having the widths h_1 and h_2

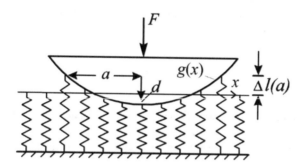

Fig. 2 Scheme of the MDR-representation of an adhesive contact. The equivalent profile $g(x)$ given by (1) is brought into contact with elastic foundation defined by Eq. (2). The contact radius is defined by the condition that the elongation of the springs at the boundary of the contact is given by the rule of Heß, Eq. (5). The normal force, contact radius and the indentation depth in this MDR-model are the same as in the initial three-dimensional contact problem

Here Δx is the space between two adjacent springs, and

$$E^* = E/(1 - \nu^2),\tag{3}$$

with E the Young modulus and ν the Poisson number of the elastic half-space.

In the MDR, it can be shown [10, 11] that the indentation depth, the contact radius and the normal force calculated as a sum of forces of all springs in contact:

$$F(a) = 2E^* \int_0^a (d - g(\tilde{a}))\mathrm{d}\tilde{a}\tag{4}$$

coincide with their values in the original three-dimensional problem. The radius of the adhesive contact is determined from the requirement of the minimum of the total energy of the system. This means that if detachment of two springs on both sides of the contact is leading to a decrease of the total energy (elastic energy plus surface energy) then it will detach. On the other hand, if the formation of contact for the springs adjacent to those at the edge of the contact, leads to a decrease of energy, the contact will spread further. Detachment of two springs leads to a decrease of elastic energy by $E^* \cdot \Delta x \cdot \Delta l^2$, where Δl is the elongation which a spring has in the attached state (Fig. 2). When it detaches, a free surface having the area $2\pi a \Delta x$ is formed, which increases the energy by the work of separation $2\pi a \Delta x \gamma$. The boundary is in equilibrium if these two energies are equal and thus

$$\Delta l = \sqrt{\frac{2\pi a \gamma}{E^*}}. \tag{5}$$

This equation, which is equivalent to the Griffith criterion for crack equilibrium [12], was first found first found by Heß [13] and is known as *rule of Heß* [14]. Using the relation $u(x) = d - g(x)$, where $u(x)$ is the vertical displacement at the position x, we can rewrite (5) in the form

$$d = g(a) - \sqrt{\frac{2\pi a \gamma}{E^*}}. \tag{6}$$

This equation connects the indentation depth with the equilibrium contact radius, a. In the following, for simplicity, we will assume that the contact is realized by a very stiff system, which means that the indentation d can be considered as controlling parameter.

3 Attachment and Detachment of a Chemically Heterogeneous Body

Consider the system with specific surface energy depending on the position as shown in Fig. 1. Assume $\gamma_1 < \gamma_2$. During the indentation, there are three repeating stages in the movement (Fig. 3).

(1) If at some moment of time the contact radius a_1 coincides with the inner edge of the ring having the surface energy γ_1, then at this moment

$$d_1 = g(a_1) - \sqrt{\frac{2\pi a_1 \gamma_1}{E^*}}. \tag{7}$$

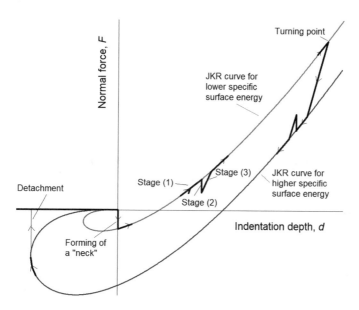

Fig. 3 Processes of approach, formation of contact and pull-off for a heterogeneous contact. During the Stage (1), the boundary moves reversibly along the ring with lower specific surface energy. During the Stage (2) it jumps over the ring with higher specific surface energy. During the Stage (3) it "sticks" in this position until the force reaches the JKR curve. This movement occurs at a constant contact radius and is thus linear. After that, this quasi-periodic process is repeated (the repetitions are not shown in the Figure). If the direction of movement is changed to the opposite ("turning point"), the contact radius first remains constant causing a linear dependency of the normal force on approach. After the force have reached the JKR curve corresponding to the higher specific surface energy, the process consisting of reversible propagation inside the rings with high specific surface energy, jumps over the rings with low specific surface energy and linear returns to the JKR curve

If the indentation depth increases, the contact radius will also increase (exactly accordingly to the corresponding JKR curve with surface energy γ_1, Fig. 3, Stage (1)) unless it reaches the outer edge of the ring having the surface energy γ_1. At this moment

$$d_2 = g(a_2) - \sqrt{\frac{2\pi a_2 \gamma_1}{E^*}}, \quad a_2 = a_1 + h_1. \tag{8}$$

(2) Further increasing of indentation depth leads to a jump-like increase of the surface energy. Therefore, the contact boundary will jump over the whole width of the ring with higher surface energy (at the given indentation depth (8)) and stop at the edge of the ring having lower surface energy. At this point, the configuration is given by the pair

$$(d_2, a_2 + h_2). \tag{9}$$

This jump in the contact area will lead to a (negative) jump in the force (see Fig. 3, Stage (2)).

(3) During further indentation, the contact radius will remain constant and the force will therefore increase linearly with indentation depth until it reaches again the JKR curve.

After that, we are again in the repetition of the Stage (1), and the movement occurs along the JKR curve to the next jump, and so on.

We see that in the phase of indentation the system follows the JKR curve corresponding to the *lower surface energy*, with periodic negative jumps and linear returns to the JKR curve. The maximum amplitude of a jump corresponds to the "distance" between the JKR curves for γ_1 and γ_2. In the following, we assume that the amplitude of jumps is small compared to this "distance". Under this assumption, the indentation occurs practically along the JKR curve for smaller surface energy with small variations.

If at some point the indentation stops and reversed movement starts, then the system first remains stuck in this point. This is because in spreading, the contact area is pinned by the areas with lower specific surface energy while in detaching, it is pinned by the areas with higher specific surface energy. Therefore, the transition from indentation to pulling off first leads to the "switching of the criterion for propagation" which leads to pinning the boundary to the position at the beginning of reverse motion. As the contact area remains constant, the force-distance dependency is in this stage linear until the force reaches the JKR curve corresponding to the higher specific surface energy. In the following, it moves along the JKR curve corresponding to the higher specific surface energy until the boundary reaches the ring with lower energy. At this point, the whole ring with low surface energy detaches at once causing a (positive) jump in the normal fore. After that, the contact area remains constant and the normal force depends linearly on the indentation depth until this linear dependency reaches the JKR-curve (Fig. 3). Thus, the back movement is very similar to the indentation with the only difference that now the systems moves along the JKR curve corresponding to the higher specific surface energy. The hysteresis and the corresponding energy dissipation is solely due to instable stages (jumps). This mechanism of energy dissipation is very similar to that described by Prandtl [15, 16].

4 Complete Cycle of Attachment and Detachment

The attachment-detachment process becomes especially simple if we assume that the thickness of the rings with different values of specific surface energy are so small that they are not "seen" from the macroscopic point of view. It is easy to give mathematical form to this condition.

In the state (8), the normal force is given by

$$F_2 = 2E^* \int_0^{a_1+h_1} (d_2 - g(\tilde{a})) d\tilde{a} \tag{10}$$

and in the state (9) by

$$F_3 = 2E^* \int_0^{a_1+h_1+h_2} (d_2 - g(\tilde{a})) d\tilde{a}. \tag{11}$$

The jump of the force is estimated as

$$F_2 - F_3 = 2E^* \left[\int_0^{a_2} (d_2 - g(\tilde{a})) d\tilde{a} - \int_0^{a_2+h_2} (d_2 - g(\tilde{a})) d\tilde{a} \right]$$

$$= -2E^* \left[\int_{a_2}^{a_2+h_2} (d_2 - g(\tilde{a})) d\tilde{a} \right] \approx 2E^* h_2 \Delta l(a_2) = h_2 \sqrt{8\pi E^* a_2 \gamma_1}. \tag{12}$$

For a rough estimation let us introduce a "characteristic value" of the specific surface energy, γ (e.g. the average of γ_1 and γ_2), the "characteristic value" of the ring width as h and the "characteristic value" of contact radius, a, e.g. the critical value at the neck formation,

$$a = \left(\frac{9\pi R^2 \gamma}{8E^*} \right)^{1/3}. \tag{13}$$

Then the characteristic value of a jump in the force will be $\Delta F_{\text{Jump}} \approx h\sqrt{8\pi E^* a\gamma}$ and that of the "distance" between the two JKR curves $\Delta F_0 \approx (3/2)\pi R\gamma$, [11]. The condition that the jumps are small compared with ΔF_0 can now be written as

$$\frac{\Delta F_{\text{Jump}}}{\Delta F_0} \approx \frac{h}{R} \sqrt{\frac{32 E^* a}{9\pi \gamma}} << 1 \tag{14}$$

or

$$h^3 << R^2 \frac{\gamma}{E^*}. \tag{15}$$

This form is applicable only for indenters with parabolic shape. Written in the form

$$h << a, \tag{16}$$

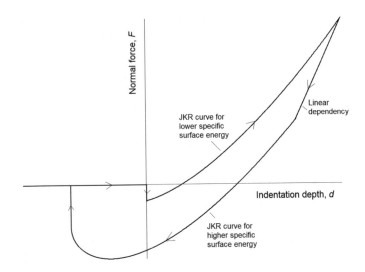

Fig. 4 A complete cycle of the force-indentation dependence for indentation and detachment for the case that the thickness of rings is "microscopic" so that the jumps are not seen on the macroscopic level. In this case, approach occurs along the JKR-curve corresponding to lower specific surface energy and pull-off along the JKR curve for higher surface energy. The are connected by a linear part following the turning point

it can be applied to any shapes. The criterion just means that the characteristic size of heterogeneity should be much smaller than the contact area.

If this condition is fulfilled, we will only see the averaged macroscopic behavior. As was shown in the Sect. 3, the contact configurations and the force-indentation dependencies will follow the JKR solution corresponding to the lower surface energy, γ_1. On the return, the system follows the JKR curve with higher surface energy, γ_2. The transition from one curve to the other at the turning point occurs via a linear force-displacement dependency at a constant contact radius (Fig. 4).

Adhesion cycles of the shape qualitatively very similar to that presented in Fig. 4 are often observed in experiment. As an example, in Fig. 5 results are shown, which have been obtained experimentally in the papers [17, 18]. Experiments were carried out by indenting a glass ball against a plane PDMS substrate and subsequent pulling it off. The main features of the behavior are the same as predicted theoretically: Both during loading and during unloading, the system moves along the JKR curves, however, corresponding to different specific surface energies. By turning, a transition from one curve to the other occurs.

Another example of loading-unloading curves showing very clearly the linear transition region after turning from loading to unloading is shown in Fig. 6. The contact area was observed and recorded by a video camera placed beneath the rubber sheet. In the videos (which are not part of this publication), it is clearly seen that after changing the direction of loading the contact area first remained unchanged. During

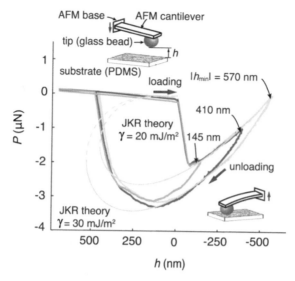

Fig. 5 Experimental loading-unloading curves (adapted from the paper [17]). According to [17], curves were measured using AFM for contacts between a glass sphere and a PDMS substrate. The glass sphere was of diameter $\approx 50\ \mu$m. The gray dashed curves are the fit of the loading and unloading branches of the measured P–h data to the JKR theory. Comparison with theoretical curves in Fig. 4 shows that the contact behavior in experiment is, at least qualitatively very similar to that predicted theoretically: In the loading phase the system moves along a JKR curve corresponding to a lower specific energy. During unloading the transition from one JKR curve to the other one can be clearly identified

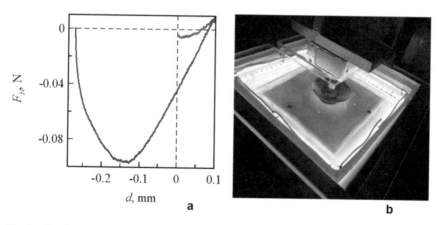

Fig. 6 a Loading-unloading curve for an adhesive contact between a spherical steel indenter (radius of curvature $R = 33$ mm) and a layer of soft transparent rubber TARNAC CRG N3005 (thickness 5 mm). Part of experimental setup is shown in subplot (**b**). In the subplot (**a**), it is clearly seen that after changing the direction of motion, a long linear part of the force-distance dependency is observed. Observation of the contact area via a video camera from beneath the layer shows that during the linear part of this curve the contact area remains constant

this "sticking phase", the dependency of the normal force on approach follows a linear dependency, which can be easily identified in Fig. 6.

5 Conclusions

We considered a simple adhesive contact with axially symmetric chemical heterogeneity of the interface. In this case, the system follows the JKR curves both during the indentation and detachment phases. However, there exist two different specific surface energies – one for forming the contact (during indentation) and the other one for its destruction (pull off). During the indentation, the system follows the JKR curve corresponding the lower specific surface energy, and during retraction the JKR curve corresponding to the higher value. If the chemical heterogeneity can be considered as "microscopic" (that means that the characteristic wavelength of heterogeneity fulfils the criterion (15) or (16)) this result does not depend on the absolute and relative thicknesses of the regions with different specific surface energies, but depends solely on the values of surface energy itself.

The main conclusions of this paper seem to be very generic. The predicted features are often observed in experimental systems not fulfilling the simple assumptions of the present model. The reason for such generality maybe just that the chemical heterogeneity leads to appearance of a force of friction for the moving contact boundary. From the macroscopic, phenomenological point of view, it is not important what is the physical mechanism leading to microscopic instabilities and thus friction in the boarder line. This can be regular heterogeneity as in the present paper or irregular heterogeneity (which also leads to local instabilities in movement of the contact boundary) or roughness. Macroscopically, the appearance of the force of friction of boundary line is equivalent to existence of two surface energies—for closing and for opening the contact. Thus, the phenomenological appearance may be the same independently of particular mechanism leading to the boundary line friction.

It would be interesting to prove whether this main conclusion will remain valid for non-axially symmetric cases and what are then the governing parameters determining the effective surface energies.

Acknowledgements The author is grateful to I. A. Lyashenko for providing experimental results presented in Fig. 6. This work was partially supported by the Deutsche Forschungsgemeinschaft (DFG, PO 810-55-1).

References

1. Johnson KL, Kendall K, Roberts AD (1971) Surface energy and the contact of elastic solids. Proc R Soc Lond Ser A 324:301–313
2. Popov VL (2019) Adhesive contribution to friction. AIP Conf Proc 2167(1):020286

3. Popova E, Popov VL (2018) Note on the history of contact mechanics and friction: interplay of electrostatics, theory of gravitation and elasticity from Coulomb to Johnson–Kendall–Roberts theory of adhesion. Phys Mesomechanics 21(1):1–5
4. Lyashenko IA, Popov VL (2019) Mechanics of adhesive contacts: experiment and theory. In: AIP Conference Proceedings, 2167, 020201. https://doi.org/10.1063/1.5132068
5. Hassenkam T, Skovbjerg LL, Stipp SLS (2009) Probing the intrinsically oil-wet surfaces of pores in North Sea chalk at subpore resolution. Proc Natl Acad Sci 106(15):6071–6076
6. Dalvi S, Gujrati A, Khanal SR, Pastewka L, Dhinojwala A, Jacobs TDB (2019) Linking energy loss in soft adhesion to surface roughness. Proc Natl Acad Sci 116(51):25484–25490. https://doi.org/10.1073/pnas.1913126116
7. Li Q, Pohrt R, Popov VL (2019) Adhesive strength of contacts of rough spheres. Front Mech Eng 5(7). https://doi.org/10.3389/fmech.2019.00007
8. Popov VL, Pohrt R, Li Q (2017) Strength of adhesive contacts: influence of contact geometry and material gradients. Friction 5(2):308–325
9. Supplementary video to the paper [8]: https://www.youtube.com/watch?v=aV2W91d8vwQ
10. Popov VL, Heß M (2015) Method of dimensionality reduction of contact mechanics and friction. Springer, Berlin
11. Popov VL, Heß M, Willert E (2019) Handbook of contact mechanics. Exact solutions of axisymmetric contact problems, Springer, Berlin, p 347p
12. Griffith AA (1921) The phenomena of rupture and flow in solids. Philos Trans R Soc A Math Phys Eng Sci 221:582–593
13. Heß M (2011) Über die exakte Abbildung ausgewählter dreidimensionaler Kontakte auf Systeme mit niedrigerer räumlicher Dimension. Cuvillier Verlag, Göttingen
14. Popov VL (2017) Contact mechanics and friction: physical principles and applications, 2nd edn. Springer, Berlin
15. Prandtl L (1928) Ein Gedankenmodell zur kinetischen Theorie der festen Körper. Zeitschrift für angewandte Mathematik und Mechanik 8:85–106
16. Popov VL, Gray JAT (2012) Prandtl-Tomlinson model: history and applications in friction, plasticity, and nanotechnologies. ZAMM—J Appl Math Mech 92:683–708
17. Deng W, Kesari H (2019) Depth-dependent hysteresis in adhesive elastic contacts at large surface roughness. Sci Rep 9:1639. https://doi.org/10.1038/s41598-018-38212-z
18. Kesari H, Doll JC, Pruitt BL, Cai W, Lew AJ (2010) Role of surface roughness in hysteresis during adhesive elastic contact. Philos Mag Lett 90(12):891–902

Theoretical Study of Physico-mechanical Response of Permeable Fluid-Saturated Materials Under Complex Loading Based on the Hybrid Cellular Automaton Method

Andrey V. Dimaki and Evgeny V. Shilko

Abstract We give a brief description of the results obtained by Prof. Sergey G. Psakhie and his colleagues in the field of theoretical studies of mechanical response, including fracture, of permeable fluid-saturated materials. Such materials represent complex systems of interacting solid and liquid phases. Mechanical response of such a medium is determined by processes taking place in each phase as well as their interaction. This raised a need of developing a new theoretical approach of simulation of such media—the method of hybrid cellular automaton that allowed describing stress-strain fields in solid skeleton, transfer of a fluid in crack-pore volume and influence of fluid pressure on the stress state of the solid phase. The new method allowed theoretical estimation of strength of liquid-filled permeable geomaterials under complex loading conditions. Governing parameters controlling strength of samples under uniaxial loading and shear in confined conditions were identified.

Keywords Hybrid cellular automaton · Poroelasticity · Strength · Permeability · Fluid-saturated materials

1 Introduction

Many natural and man-made materials and media, such as permeable rocks (including coal and oil bearing strata) [1–4], bone tissue [5], filtering materials [6], endoprostheses [7] and other, are fluid-saturated porous or cracked porous media. The mechanical response of fluid-saturated permeable materials exhibits some features that differentiate them from solid composite materials in which phases also have different elastic and rheological characteristics. These features are associated with the ability of a liquid/gaseous phase to redistribute in cracks and pores of an enclosing material. As a result of the redistribution, mean stress can either level off in different regions of the material or strongly oscillate due to dilatancy and filling of new discontinuities or due to fluid exchange with the surrounding medium.

A. V. Dimaki (✉) · E. V. Shilko
Institute of Strength Physics and Materials Science SB RAS, Tomsk, Russia
e-mail: dav18@yandex.ru

© The Author(s) 2021
G.-P. Ostermeyer et al. (eds.), *Multiscale Biomechanics and Tribology of Inorganic and Organic Systems*, Springer Tracts in Mechanical Engineering,
https://doi.org/10.1007/978-3-030-60124-9_21

At present, numerous computational methods do exist describing the mechanical behavior of a continuous medium within a certain scale level (finite difference, boundary element, and cellular automata methods [8], etc.). The most popular among them are the finite element method and different variations of the particle method (molecular dynamics, discrete elements, movable cellular automaton (MCA) [9–11] etc.). Distribution of gas and liquid in various porous media is simulated using such modern and intensively developing methods as the lattice Boltzmann method [12] and methods based on the solution of the Navier–Stokes equations on a finite-difference grid. However, the description of a multiscale contrast medium containing interacting solid, liquid, and gas phases within a common approach meets problems connected with description of interrelation between solutions of equations describing the behavior of each phase. Solution of this problem can be found in development of new methods and approaches giving explicit consideration to the multiscale and behavioral peculiarities of studied objects. The most effective among them are methods based on the principles of physical mesomechanics that considers a solid as a multilevel system [13].

Professor Sergey Psakhie founded the method of hybrid cellular automaton (HCA) and was the leader of numerous pioneering works devoted to theoretical study of multiscale and multiphase media using this method. His ideas allowed for solution of many fundamental and practical problems connected with behavior of geological media under complex dynamic loading. Among these media are coal beds filled with gas, porous permeable materials filled with liquid, weakly connected boundaries (shear bands) in geomaterials with gradients of permeability under shear loading conditions.

2 Brief Description of the Hybrid Cellular Automaton Method

One possible theoretical approach to the study of fluid-saturated permeable materials is to use coupled models that account for the following important aspects of such systems behavior: (1) relation of solid skeleton deformation with volume and porosity variation in the pore and crack volume; (2) relation between pore pressure and stress in the enclosing volume of a solid; (3) fluid redistribution in the pore and crack volume. The most famous representatives of this approach are analytical macroscopic models of poroelasticity whose theoretical basis was first discussed by Biot [14, 15]. They were further developed by taking into account a range of scales in real materials, damage accumulation, dilatancy and their influence on the skeleton elastic properties and pore fluid pressure [16–22].

The computational methods based on a discrete representation of the medium are widely used to describe media where fracture on different scales is a factor determining a mechanical response. A well-known representative of this family of models is the discrete element method (DEM) in which the modeled material is represented as an ensemble of interacting finite-size particles [23–32]. Its major advantages result from the ability of discrete elements to change their surroundings, which is crucial for simulation of complex phenomena such as contact interaction, cracking and fragmentation of solids, flow of granulated media and other.

Various explicit DEMs use different approximations to describe strain distribution within the discrete element volume and the influence of element shape/geometry on its kinematics and interaction with the surroundings [29]. A common approach to description of an element shape is an approximation of an equivalent circular disc or sphere [23, 28, 29]. Further we consider equiaxed or nearly equiaxed elements in approximation of equivalent circular discs and use the term "discrete element" to mention the given simplified representation of an element shape. The given approach has a simple mathematical formulation and apparent advantages in modelling deformation and fracture.

Further development of the DEM formalism expanded its application to a wider range of spatial and structural scales [33–38] and yielded a coupled numerical method for studying permeable media on the meso- and macroscale levels. This coupled numerical method was called the hybrid cellular automaton method. In the HCA method the mechanical response of the enclosing solid is described by the MCA method [39, 40]; fluid filtration and diffusion in cracks and pores of the solids (taken into account implicitly) are described by a finite difference method.

The numerical HCA method is based on separation of the problem into two parts: (1) description of the mechanical behavior of the enclosing solid (skeleton), and (2) description of a fluid transfer in the filtration volume of the solid represented by a system of connected channels, pores, cracks and other discontinuities. Depending on structural features of a considered permeable material and on the simulated scale, the dimensions of discrete elements can be much larger than the linear dimensions of discontinuities in the solid, can be comparable to them or smaller. The influence of "micropores" in the solid skeleton (i.e., pores, channels and other discontinuities whose typical size is smaller than discrete element size) on the mechanical properties and response of a discrete element is accounted for explicitly. Additionally, the MCA method is used to solve the problem of filtration fluid transfer in the network of connected "micropores" of the enclosing solid. Fluid mass transfer between "micropores" and "macropores", which are considered as regions between spaced and noninteracting discrete elements, is calculated using a finer grid embedded in a laboratory coordinate system (Fig. 1a). The same grid is used to calculate "macropore" volumes (Fig. 1b).

Within the MCA formalism local failure is modelled by changing the state of a pair of interacting elements from linked (or bonded) to unlinked. In simulations described below we apply the modified Drucker-Prager failure criterion taking into account local pore fluid pressure:

$$\sigma_{DP} = 0.5(\lambda + 1)\sigma_{eq} + 1.5(\lambda - 1)\left(\sigma_{mean} + bP_{pore}\right) = \sigma_c, \qquad (1)$$

where $\lambda = \sigma_c/\sigma_t^i$ is the ratio of compressive σ_c to tensile σ_t strength, σ_{eq}—equivalent stress, b is the dimensionless coefficient determining a contribution of a fluid pore

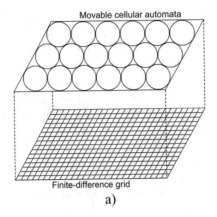
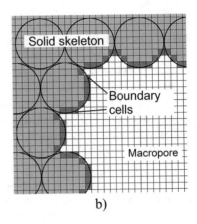

Fig. 1 Layers of discrete elements (movable cellular automata) and finite-difference grid (**a**); grid cells at a boundary between solid skeleton and macropores (**b**)

pressure into mean stress σ_{mean} and P_{pore} is the fluid pore pressure. A comprehensive description of details of the numerical implementation of hybrid cellular automaton method is given in [41, 42].

3 Strength of Porous Fluid-Filled Samples Under Uniaxial Loading: A Competition Between Compression and Fluid Filtration

The developed model was applied to study a mechanical response of porous elastic-brittle samples with water-filled filtration volume under uniaxial compression. The samples were fixed between a matrix (at the bottom) and a punch (at the top) that moved downwards and compressed the sample with constant velocity V_y. The compression direction coincided with the vertical sample axis. The problem was solved in a 2D statement in the plane stress approximation. The sample structure was assumed to be homogeneous, without pores and inclusions. The sample height was $H = 0.1$ m and width was $W = 0.05$ m. Numerical experiments were performed using the following parameters of the model material: $K = 37.5$ GPa, $K_s = 107$ GPa, $G = 5.77$ GPa, $\rho = 2000$ kg/m^3, $\lambda = 7$, $\sigma_c = 70$ MPa, $K_{fl} = 2.2$ GPa (see details in [41, 42]). Initial value of material porosity was $\varphi_0 = 0.1$.

Two hypotheses of micropore distribution in the solid skeleton were considered with regard to the influence of pore pressure on sample strength.

1. Micropores are distributed homogeneously. Their size is much smaller than the characteristic size of damages and cracks formed during fracture. In this case, we assume that the effect of pore fluid on skeleton strength is governed by the porosity

value and is taken into account by subsequent determination of coefficient b in failure criterion (1): $b = \varphi_0$.

2. Micropores are distributed inhomogeneously, and damages in the material are formed through coalescence of several micropores. In this case, the effect of pore fluid on skeleton strength is directly defined by the pore pressure value: $b = 1$.

The simulation results revealed that pore fluid pressure exerts a strong influence on the mechanical response of brittle porous samples. Other factors that significantly influence the sample strength are the loading rate, characteristic filtration channel diameter (this quantity, along with open porosity, determines material permeability) and geometrical dimensions of the sample.

Fluid pressure in pores of a solid under uniaxial compression is governed by two competing processes: (1) solid skeleton deformation accompanied by pore volume reduction and by pore pressure increase (and, consequently, increasing influence of fluid on the stress state of the skeleton); (2) fluid discharge to the environment through the lateral faces of the sample, due to which pore pressure decreases and the influence of fluid on the stress state of the skeleton also decreases.

A balance of the mentioned two processes is governed by permeability coefficient. At low sample permeability fluid outflow from the sample does not compensate fluid density increase during pore deformation. As a result, fluid pressure in the filtration volume constantly increases in the course of deformation, due to which the effective sample strength decreases. At large permeability, the rate of fluid outflow from the sample is sufficient to reduce pore pressure to zero. The influence of fluid on the stress state of such samples is nearly absent and their strength tends to the strength of a "dry" sample. Between these "limiting" cases the rate of fluid pressure decrease caused by fluid outflow from the sample is comparable to the rate of fluid pressure increase due to solid skeleton compression.

Sample strength is also determined by fluid pressure distribution in the pore volume across the sample cross section. This distribution depends on the ratio of sample width W to its height H. The material volume which is adjacent to the lateral face and through which fluid escapes to the environment decreases with the increasing sample width. Correspondingly, the specific amount and pressure of the fluid retained in pores at the beginning of fracture increases with the growing W/H ratio. This leads to an unexpected conclusion: other factors being equal, the strength of a water-saturated sample of larger width appears to be lower than the strength of a "narrower" sample.

As above, the strength of fluid-saturated permeable samples is governed by the competition of mechanical deformation under applied external load and fluid discharge from the pore space to the environment. To reveal general mechanisms of this competition, we studied the effect of loading rate on uniaxial compressive strength of fluid-saturated samples. Uniaxial compression of fluid-saturated samples with different values of d_{ch} was modeled for different loading rates V_y. The initial pore fluid pressure P_{init} was assumed to be zero (pore volume of samples was completely filled with fluid at atmospheric pressure).

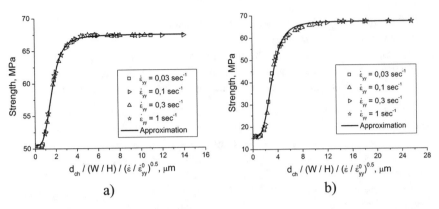

Fig. 2 Generalized dependence of water-saturated sample strength on filtration channel diameter at different loading rates: **a** approximation of homogeneous micropore distribution ($b = \varphi = 0.1$); **b** approximation of relatively large micropores whose evolution leads to macrocrack formation ($b = 1$). Initial fluid pressure in solid skeleton pores is $P_{init} = P_0$. In all calculations the aspect ratio of samples was $W/H = 0.5$

Figure 2 gives the dependences of sample strength on the characteristic filtration channel diameter at different strain rates. The value of parameter b that determines the pore pressure contribution to failure criterion (1) strongly affects the strength of fluid-saturated samples. For example, within the approximation of homogeneous micropore distribution in the solid skeleton ($b = \varphi = 0.1$, Fig. 2a) the pore pressure contribution is rather low (maximum decrease in sample strength does not exceed 25%). Within the second approximation ($b = 1$, Fig. 2b) the strength of water-saturated samples at low permeability values can decrease several-fold. The strength dependence flattens out in this case due to a stronger influence of residual fluid in the pore space on sample strength.

Analysis of the obtained dependences of sample strength on the effective filtration channel diameter d_{ch} at different loading rates and different W/H ratios revealed that they can be reduced to a single dependence of strength on the reduced effective filtration channel diameter:

$$d_{ch} / \left((W/H) \sqrt{\dot{\varepsilon}_{yy} / \dot{\varepsilon}_{yy}^0} \right), \tag{2}$$

where $\dot{\varepsilon}_{yy} = V_y/H$ is the strain rate, and $\dot{\varepsilon}_{yy}^0$ is the scale multiplier that has the dimension of strain rate. The curves shown in Fig. 2 are plotted in these variables ($\dot{\varepsilon}_{yy}^0$ was assumed to be equal to 1 s^{-1}).

It is known that the processes whose occurrence is governed by the competition of several factors or phenomena (e.g., biological population growth, etc.) are often described by a logistic function [43]. Based on the above assumption about the decisive role of the competition between pore pressure increase and fluid outflow from the sample, we used the following logistic function to approximate the dependence of strength of uniaxially compressed water-saturated samples on the reduced effective filtration channel diameter:

$$\sigma_c(d_{ch}, \dot{\varepsilon}_{yy}) = \sigma_c^{\min} + \frac{\sigma_c^0 - \sigma_c^{\min}}{1 + \left(d_{ch} / \left(d_0(W/H)\sqrt{\dot{\varepsilon}_{yy}/\dot{\varepsilon}_{yy}^0}\right)\right)^p} \tag{3}$$

where σ_c^0 is the sample strength under uniaxial compression in the absence of fluid in the pore space, $\sigma_c^{\min}$ is the water-saturated sample strength in the absence of fluid mass transfer, d_0 is the parameter of the approximating function having the dimension of distance, $\dot{\varepsilon}_{yy}$ is the axial strain rate of the sample, and p is the exponent. The parameters of Eq. (3) are defined by the elastic moduli of the solid skeleton and fluid, fluid viscosity, porosity value and so on. As one can see, logistic function (3) allows approximation the numerically calculated data given in Fig. 2 with good accuracy (at $d_0 = 1.62\ \mu m$, $p = 3.7$ for the curve in Fig. 2a and at $d_0 = 3.07\ \mu m$, $p = 4$ for the curve in Fig. 2b).

In order to generalize the results it is useful to take reduced material permeability k as the parameter determining fluid filtration rate:

$$\frac{d_{ch}}{\sqrt{\frac{\dot{\varepsilon}_{yy}}{\dot{\varepsilon}_{yy}^0} \frac{W^2}{H^2}}} \rightarrow \sqrt{\frac{k}{\dot{\varepsilon}_{yy}} \frac{H^2}{W^2} \frac{\dot{\varepsilon}_{yy}^0}{\varphi_0}}. \tag{4}$$

Within this formulation, parameter (2) and approximating function (3) take a more general meaning and can be applied to permeable materials with different structure of filtration volume. Complex relations between the parameters characterizing the mechanical response of the solid skeleton, physical and mechanical properties of fluid and its filtration redistribution dynamics in the system of pores define the nonlinear dependence of sample strength on a combination of these parameters and necessitates the application of numerical methods to study the mechanical response of fluid-saturated porous materials.

4 Influence of Pore Fluid Pressure and Material Dilation on Strength of Shear Bands in Fluid-Saturated Rocks

A range of laboratory and full-scale geological and geophysical research suggests that irreversible deformation in rock samples and rock massifs is strongly localized in shear bands at different scales, the largest of which are tectonic faults [44–46]. These narrow zones not only determine the compliance of rocks in the form of localized relative shear displacements of structural blocks, but control seismic activity of rock massifs. The latter explains the current interest in the mechanical properties of fault zones and the rapid increase in the number of published works in this area.

A. V. Dimaki and E. V. Shilko

One of the key mechanical properties is the maximum (or peak) strength of the fault zone under given stress and confinement conditions. Reaching the maximum strength corresponds to a change in the response of the shear band from pervasive strain (strain hardening stage) to strain localization (strain softening stage) [47, 48]. Maximum shear strength estimation of fault zones is both a fundamental and practical problem widely discussed in fault and rock mechanics [49–52].

The conditions for onset of pervasive inelastic strain and subsequent reaching of the maximum strength of shear bands (including faults) are mainly affected by the pore structure and pore fluid pressure. The pore pressure dynamics is controlled by two interrelated processes [53–63]: (1) fluid flow and (2) pore volume change. The pervasive inelastic deformation of rock is often accompanied by its dilatancy [64–66]. The volume of the connected crack-pore space increases during pervasive shear deformation of the shear band, that leads to decrease of local pore pressure. The reduction of fluid pressure reduces the intensity of the relaxation processes associated with the formation of new discontinuities and coalescence of existing ones. This effect is called dilatancy hardening [54]. In turn, fluid inflow is able to compensate for the pore pressure drop and reduce the effect of strain hardening [63, 67]. The ratio of fluid flow rate to strain rate (which governs the dilation rate) determines the specific value of shear strength of shear bands. Note that the influence of the competition between dilatancy and fluid flow on shear strength is strongly pronounced for shear bands surrounded by material blocks with a similar permeability to that of the shear band gouge. This particularly corresponds to healed (consolidated) faults where the difference in the porosity and permeability of the principal slip zone (of width 1–10 cm) and surrounding periphery zone (up to several meters wide) is much less pronounced than in faults with a mature zone of unconsolidated gouge [68].

Conventionally, the effect of pore fluid on the maximum shear strength (hereinafter referred to as strength) of shear bands, including fault zones, has been studied for limiting modes of deformation (very slow and very fast) that correspond to drained and undrained hydrological conditions [69] due to the fact that limiting modes correspond to the long-term creep and short-term dynamic modes of deformation of fault regions. Numerous experimental and theoretical works, starting with the classical work of Brace and Martin [69], show that the strength of permeable rock samples can increase significantly (up to 30–50%) in transition from the drained to undrained condition [67, 70, 71]. This is explained by the limitation of fluid inflow to the increasing pore space of the incipient shear band and corresponding inhibition of pore pressure drop recovery.

There is still no unambiguous understanding of how the strength of a shear band in the depth of constrained permeable rock massif changes in the transition region between the undrained and drained conditions. We studied the nature and functional form of the dependence of the shear band strength on the ratio of shear strain rate to fluid flow rate under constrained conditions corresponding to faults in rock massifs. The study was performed by numerical modelling of the shear deformation of a fluid-saturated permeable shear band using the discrete element method.

Consider a model sample consisting of two blocks separated by an interfacial layer (shear band) in the plane strain approximation (Fig. 3). The blocks mimic regions of

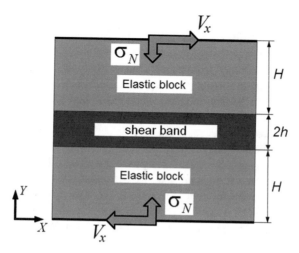

Fig. 3 Sample structure and loading scheme for modelling of constrained shear of a porous fluid-saturated shear band surrounded by porous fluid-saturated blocks

the medium adjacent to the shear band, which are less damaged than the shear band and therefore deform elastically under the considered loading conditions. The shear band of width $2h$ is a layer of an elastic-plastic dilatant material, which simulates the layer of consolidated gouge in principal slip zones of faults [47, 72]. The width of the model blocks was $H = 20h$. We used the following reference values of widths of the shear band and blocks: $2h \approx 1.5$ cm, $H = 15$ cm. The shear band and blocks were assumed to be permeable and fluid saturated.

The model shear band with surrounding fragments of blocks was numerically simulated by the discrete element method using a fully coupled macroscopic model of fluid-saturated porous brittle materials [33–41]. Within the model discrete elements simulating parts of the shear band and surrounding blocks are treated as porous and permeable. The effect of the fluid contained in the crack-pore volume of a discrete element on its stress state is described based on Biot linear model of poroelasticity [14, 15]. The inelastic behavior of the permeable brittle material of the discrete element is described using a plastic flow model of rocks with a non-associated flow law and the Mises–Schleicher yield criterion (Nikolaevsky's model) [16]. The elastic characteristics of the shear band and blocks were assumed to be similar and corresponded to typical values for sandstones with a porosity of 10–15% (Young's modulus $E = 15$ GPa, Poisson's ratio $v = 0.3$). The material of discrete elements modelling the blocks was treated as elastic-brittle and high-strength. The material of discrete elements modelling the shear band was a model elastic-plastic material with linear hardening with the following plasticity and strength parameters: $\beta = 0.57$, $\Lambda = 0.36$, $Y = 10.84$ MPa (this corresponds to a yield stress of 28 MPa under uniaxial compression), strain hardening modulus $\Pi = 515$ MPa, uniaxial compressive strength $\sigma_c = 40$ MPa, uniaxial tensile strength $\sigma_t = 13.33$ MPa ($\lambda = 3$). The calculations were carried out at an initial mean stresses σ_{mean}^0 below the brittle-ductile transition threshold (in this case, at $\sigma_{mean}^0 < 40$ MPa).

The initial values of porosity ($\phi_0 = 0.1$) and permeability k_0 of the shear band and the blocks were assumed to be equal. This approximation is consistent with the low gradients of porosity and permeability in central zones of healed (consolidated) faults. Initially, all interacting elements were linked that simulates a consolidated shear band.

We modelled constrained shear of the sample in the horizontal plane along the X axis (Fig. 3). Periodic boundary conditions were specified on the lateral faces in the horizontal direction to simulate an infinitely long shear band. The sample was loaded in two stages. At the first stage, a normal load σ_N was applied to the upper and lower sample faces. The initial fluid concentration in the pore space of the sample was chosen so as to create the specified pore pressure P_{pore}^0 in the sample. There was no plastic deformation in the sample by the end of the first loading stage. The stress and pore pressure distributions were homogeneous. At the second stage, the sample was subject to simple shear by applying constant tangential velocity V_x and zero normal velocity (along the Y axis) to the upper and lower faces to fulfil the constrained shear condition. The sample deformation proceeded until crack initiation in the shear band.

The described 2D system models a horizontal cross section of a healed fault between structural blocks of a rock massif at a certain depth. Note that the initial "horizontal" (in the XY plane) stresses in the given formulation of the problem exceed the "vertical" ones. This is consistent with experimental data indicating that horizontal stresses are considerably higher than vertical ones in regions with high deformation activity [73]. We used isolated conditions on the external surfaces of the sample (hydraulically isolated sample) that correspond to the hydrological conditions in the central regions of fault zones in the bulk of low permeable host rocks.

The simulation results showed that at high strain rates the magnitude of shear strength tends toward the upper limit, and at low strain rates—toward the minimal value. Such regularity was first observed by Brace and Martin [69], and then reported in numerous experimental and theoretical studies of confined compression of rock samples and shear loading of model fault zones (see e.g. [67]).

The reduction of shear strength from the upper to the lower limit with reduction of strain rate is not monotonous. At a certain intermediate strain rate the shear strength reaches a local minimum. Further reduction of strain rate leads to an *increase* of shear strength up to a local maximum. At even smaller strain rates the shear strength again *decreases* down to the lower limit. This result quantitatively agrees with recent experimental studies [74–76].

We varied the shear strain rate $\dot{\varepsilon}_{xy} = V_x/(H + h)$, the initial permeability of the blocks and the shear band k_0 (at fixed $\phi_0 = 0.1$), the dynamic fluid viscosity η, and system size ($2H + 2h$) within several orders of magnitude: $\dot{\varepsilon}_{xy}$ from 5×10^{-4} s^{-1} to 1 s^{-1}, k_0 from 10^{-18} m^2 to 10^{-13} m^2, η from 2×10^{-4} Pa s to 2×10^{-2} Pa s (dynamic viscosity of water at room temperature is about 10^{-3} Pa s), ($2H + 2h$) from 15 to 150 cm. We found that the parameter combination

$$A_{xy} = \frac{\dot{\varepsilon}_{xy}\eta(H+h)^2}{k_0} \qquad (5)$$

unambiguously determines the value of shear strength of the shear band for a given initial mean stress σ_{mean}^0, pore pressure P_{pore}^0, and ratio h/H. In other words, shear band zones have the same shear strength if they are characterized by the same value of A_{xy}, (even if the specific values of the parameters k_0, η, $\dot{\varepsilon}_{xy}$, and h differ by orders of magnitude). The parameter A_{xy} means the relation of strain rate to fluid flow rate.

Figure 4 shows a typical dependence of the shear strength τ_c of the modelled shear band on the parameter A_{xy} for a hydraulically isolated system. Each point of the curve corresponds to a separate calculation at given values of k_0, η, $\dot{\varepsilon}_{xy}$ and h (at $h/H = const$, $\sigma_N = const$, $P_{pore}^0 = const$). The region $A_{xy} \rightarrow \infty$ (region I in Fig. 4) corresponds to combinations of k, η, $\dot{\varepsilon}_{xy}$ and h where the fluid flow rate is extremely low compared to the rate of pore pressure change caused by pore volume variation. This corresponds to the hydrological conditions close to the undrained condition of the shear band. The region of low A_{xy} values (region III in Fig. 4) corresponds to low shear rate, low fluid viscosity or high permeability of the blocks. In this region the fluid flow rate is relatively high, and the hydrological conditions for the shear band approach a fully drained (the pore pressure distribution in the sample is close to homogeneous during the entire course of deformation).

The curve shown in Fig. 4 has three characteristic regions where the change of strength is monotonous, that implies a presence of a dominant mechanism determining the direction of the change.

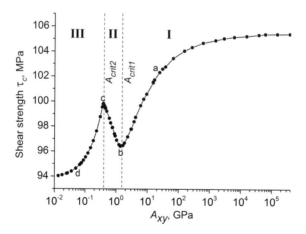

Fig. 4 A typical dependence of the shear strength τ_c of the shear band on the parameter A_{xy} for a hydraulically isolated sample. Roman numerals I–III mark the curve regions corresponding to different behaviour modes of the fluid-saturated sample under shear loading. The values A_{crit1} and A_{crit2} correspond to the local minimum and maximum shear strength. The top and bottom faces of the sample are fixed in vertical direction. Lower A_{xy} imply faster fluid flow or lower strain rate, higher A_{xy} imply slower fluid flow or higher strain rate

In region I the dominating mechanism lies in the decrease of the linear dimensions of the blocks due to fluid outflow to the shear band (poroelastic contraction). Decrease in the value of A_{xy} is accompanied by the inflow of a large amount of fluid into the shear band and hence by reduction of the constraint imposed on the shear band by the compressed blocks (effective normal stiffness of the blocks decreases). This mechanism determines the decrease of the shear strength in region I as A_{xy} decreases.

In region II the trend-determining mechanism is tied to the increase of the dilation rate of the shear band with decreasing value of A_{xy} due to slowing of pore pressure reduction and maintaining nonzero pore pressure during most of the shear process. This mechanism provides an increase in the absolute value of effective mean stress in the sample and hence increase in the strength of the shear band in region II as A_{xy} decreases.

In region III the trend-determining mechanism is linked to the fact that pore pressure in the shear band remains non-zero during the entire shear process. In this region pore pressure in the shear band is higher in the samples characterized by lower values of A_{xy}. Because of this fact an absolute value of effective mean stress in a shear band is also lower in the samples with lower A_{xy}. Decrease of an absolute value of effective mean stress leads to gradual decrease in the shear strength down to the absolute minimum at $A_{xy} \to 0$.

The described three parts of the curve $\tau_c(A_{xy})$ have sigmoid profiles that is the result of the competition between shear band dilatancy and poroelastic contraction of the blocks due to fluid outflow. Analysis of the obtained result allowed formulating the following general dependence of the shear strength of constrained shear band zones on the parameter A_{xy}. The dependence is expressed as a sum of a constant and three sigmoid contributions:

$$\tau_c = \tau_0 + \frac{\tau_1}{1 + \left(c_1 A_{xy}\right)^{-p_1}} + \frac{\tau_2}{1 + \left(c_2 A_{xy}\right)^{p_2}} - \frac{\tau_3}{1 + \left(c_3 A_{xy}\right)^{p_3}}, \qquad (6)$$

where τ_1, τ_2 and τ_3 are the amplitudes of contributions of the three above-mentioned mechanisms to the shear strength, c_1, c_2 and c_3 are the inverse positions of the sigmoid midpoints, the exponents p_1, p_2 and p_3 determine the steepness of the sigmoid functions and τ_0 is a constant contribution independent of the fluid flow dynamics.

The amplitudes of the contributions have the following physical meanings. The constant contribution τ_0 is the strength of the shear band in the absence of fluid flow (or under high strain rate) at a constant value of normal force σ_N applied to the upper and lower sample faces during the entire course of shear. Another three contributions are concerned with squeezing of the blocks due to shear band dilation in a mechanically constrained sample.

The specific parameter values of Eq. (6) depend on mechanical characteristics of the skeleton of shear band and blocks, bulk modulus of fluid, the h/H ratio, initial mean stress σ_{mean}^0, fluid content in the blocks and hydrological conditions.

5 Conclusion

We have presented the results of application of the hybrid numerical technique for theoretical study of deformation and fracture of fluid-saturated permeable materials and media. In the framework of this technique the simulated medium is considered as a superposition of two interdependent layers. One layer is represented by an ensemble of particles (simply deformable discrete elements), and the other—by a finite-difference grid. This approach combined with Biot's model of poroelasticity is suitable for studying complex and interrelated processes of solid skeleton deformation and fracture and fluid redistribution (mass transfer) in the pore volume.

Strength of water-saturated elastic-brittle samples under uniaxial compression significantly depends on fluid pressure in the pore space as well as on solid permeability, physical and mechanical properties of fluid, strain rate and sample dimensions. The numerically simulated dependences of strength of water-saturated samples are well described by a logistic function. This bears witness to the decisive role of the competition of two processes, such as pore pressure increase and fluid outflow to the environment, in permeable brittle materials in the course of mechanical loading.

It was shown that strength of a saturated shear band is directly connected with shear rate, fluid viscosity and permeability of the shear band zone and surrounding massif. The governing combination of these parameters A_{xy}, together with the obtained empirical dependence $\tau_c(A_{xy})$ allows prediction of shear band strength under given loading conditions. The latter may be especially important for estimation of transition point of the shear mode of consolidated fault segments from stick to dynamic slip.

The reported results demonstrate broad potentials of the developed DEM-based coupled model of a poroelastic medium and show the importance of numerical modeling application to the study of the mechanical properties (including strength) of dynamically loaded fluid-saturated materials.

References

1. Volfkovich YuM, Filippov AN, Bagotsky VS (2014) Structural properties of porous materials and powders used in different fields of science and technology. Springer, London
2. Doyen PM (1988) Permeability, conductivity, and pore geometry of sandstone. J Geophys Res-Solid Earth 93(B7):7729–7740
3. Dong T, Harris NB, Ayranci K, Twemlow CE, Nassichuk BR (2015) Porosity characteristics of the Devonian Horn River shale, Canada: insights from lithofacies classification and shale composition. Int J Coal Geol 141–142:74–90
4. Carey JW, Lei Z, Rougier E, Mori H, Viswanathan H (2015) Fracture-permeability behavior of shale. J Unconv Oil Gas Resour 11:27–43
5. Taylor D (2007) Fracture and repair of bone: a multiscale problem. J Mater Sci 42:8911–8918
6. Fernando JA, Chung DDL (2002) Pore structure and permeability of an alumina fiber filter membrane for hot gas filtration. J Porous Mater 9:211–219

7. Azami M, Samadikuchaksaraei A, Poursamar SA (2010) Synthesis and characterization of hydroxyapatite/gelatin nanocomposite scaffold with controlled pore structure for bone tissue engineering. Int J Artif Organs 33:86–95
8. Wolfram S (1986) Theory and applications of cellular automata. World Scientific Publishing Co., Inc., New Jersey
9. Psakhie SG, Smolin AY, Korostelev SY, Dmitriev AI, Shilko EV, Alekseev SV (1995) Investigation of establishment of steady-state deformation of solids by movable cellular automata method. Pisma Zh Tekh Fiz 21(20):72–76
10. Psakhie SG, Ostermeyer GP, Dmitriev AI, Shilko EV (2000) Method of movable cellular automata as a new trend of discrete computational mechanics. I. Theoretical description. Phys Mesomech 3(2):5–12
11. Psakhie SG, Horie Y, Ostermeyer GP, Korostelev SYu, Smolin AYu, Shilko EV, Dmitriev AI, Blatnik S, Špegel M, Zavšek S (2001) Movable cellular automata method for simulating materials with mesostructure. Theoret Appl Fract Mech 37(1–3):311–334
12. Sukop MC, Thorne DT (2007) Lattice Boltzmann modeling: an introduction for geoscientists and engineers. Springer, Berlin
13. Panin VE (ed) (1998) Physical mesomechanics of heterogeneous media and computer-aided design of materials. Cambridge International Science Publishing Ltd., Cambridge
14. Biot MA (1941) General theory of three-dimensional consolidation. J Appl Phys 12:155–164
15. Biot MA (1957) The elastic coefficients of the theory of consolidation. J Appl Phys 24:594–601
16. Garagash IA, Nikolaevskiy VN (1989) Nonassociative flow rules and localization of plastic deformation. Adv Mech 12(1):131–183 (in Russian)
17. Hamiel Y, Lyakhovsky V, Agnon A (2004) Coupled evolution of damage and porosity in poroelastic media: theory and applications to deformation of porous rocks. Geophys J Int 156(3):701–713
18. Lyakhovsky V, Hamiel Y (2007) Damage Evolution and Fluid Flow in Poroelastic Rock. Izv Phys Solid Earth 43(1):13–23
19. Meirmanov AM (2007) Nguetseng's two-scale convergence method for filtration and seismic acoustic problems in elastic porous media. Sib Math J 48(3):645–667
20. Hörlin NE, Göransson P (2010) Weak, anisotropic symmetric formulations of Biot's equations for vibro-acoustic modelling of porous elastic materials. Int J Numer Meth Eng 84(12):1519–1540
21. Hörlin NE (2010) A symmetric weak form of Biot's equations based on redundant variables representing the fluid, using a Helmholtz decomposition of the fluid displacement vector field. Int J Numer Meth Eng 84(13):1613–1637
22. Bocharov OB, Rudiak VI, Seriakov AV (2014) Simplest deformation models of a fluid-saturated poroelastic medium. J Min Sci 50(2):235–248
23. Cundall PA, Strack ODL (1979) A discrete numerical model for granular assemblies. Géotechnique 29(1):47–65
24. Mustoe GGW (1992) A generalized formulation of the discrete element method. Eng Comput 9(2):181–190
25. Shi G-H (1992) Discontinuous deformation analysis—a new numerical model for statics and dynamics of block systems. Eng Comput 9(2):157–168
26. Lisjak A, Grasseli G (2014) A review of discrete modeling techniques for fracturing processes in discontinuous rock masses. J Rock Mech Geotech Eng 6(4):301–314
27. Munjiza AA (2004) The combined finite-discrete element method. Wiley, Chichester
28. Bićanić N (2004) Encyclopedia of computational mechanics. Volume 1: Fundamentals. Wiley, Chichester (Discrete element methods)
29. Jing L, Stephansson O (2007) Fundamentals of discrete element method for rock engineering: theory and applications. Elsevier, Amsterdam
30. Williams JR, Hocking G, Mustoe GGW (1985) The theoretical basis of the discrete element method. NUMETA 1985 Rotterdam—Numerical methods of engineering, theory and applications, pp 897–906

31. Potyondy DO, Cundall PA (2004) A bonded-particle model for rock. Int J Rock Mech Min Sci 41(8):1329–1364

32. Tavarez FA, Plesha ME (2007) Discrete element method for modelling solid and particulate materials. Int J Numer Meth Eng 70(4):379–404

33. Psakhie SG, Shilko EV, Grigoriev AS, Astafurov SV, Dimaki AV, Smolin AY (2014) A mathematical model of particle–particle interaction for discrete element based modeling of deformation and fracture of heterogeneous elastic–plastic materials. Eng Fract Mech 130:96–115

34. Shilko EV, Psakhie SG, Schmauder S, Popov VL, Astafurov SV, Smolin AY (2015) Overcoming the limitations of distinct element method for multiscale modeling of materials with multimodal internal structure. Comput Mater Sci 102:267–285

35. Hahn M, Wallmersperger T, Kröplin B-H (2010) Discrete element representation of discontinua: proof of concept and determination of material parameters. Comput Mater Sci 50:391–402

36. Dmitriev AI, Osterle W, Kloß H (2008) Numerical simulation of typical contact situations of brake friction materials. Tribol Int 41(1):1–8

37. Psakhie S, Ovcharenko V, Yu B, Shilko E, Astafurov S, Ivanov Y, Byeli A, Mokhovikov A (2013) Influence of features of interphase boundaries on mechanical properties and fracture pattern in metal-ceramic composites. J Mater Sci Technol 29(11):1025–1034

38. Psakhie SG, Ruzhich VV, Shilko EV, Popov VL, Astafurov SV (2007) A new way to manage displacements in zones of active faults. Tribol Int 40(6):995–1003

39. Psakhie SG, Shilko EV, Smolin AYu, Dimaki AV, Dmitriev AI, Konovalenko IS, Astafurov SV, Zavshek S (2011) Approach to simulation of deformation and fracture of hierarchically organized heterogeneous media, including contrast media. Phys Mesomech 14(5–6):224–248

40. Zavšek S, Dimaki AV, Dmitriev AI, Shilko EV, Pezdič J, Psakhie SG (2013) Hybrid cellular automata method. Application to research on mechanical response of contrast media. Phys Mesomech 16(1):42–51

41. Psakhie SG, Dimaki AV, Shilko EV, Astafurov SV (2016) A coupled discrete element-finite difference approach for modeling mechanical response of fluid-saturated porous material. Int J Numer Meth Eng 106(8):623–643

42. Shilko EV, Dimaki AV, Psakhie SG (2018) Strength of shear bands in fluid-saturated rocks: a nonlinear effect of competition between dilation and fluid flow. Sci Rep 8(1):1428

43. Zwietering MH, Jongenburger I, Rombouts FM, van't Riet K (1990) Modeling of the bacterial growth curve. Appl Environ Microbiol 56(6):1875–1881

44. Ben-Zion Y, Sammis C (2010) Mechanics, structure and evolution of fault zones. Birkhauser Verlag AG, Basel

45. Fossen H, Schultz RA, Shipton ZK, Mair K (2007) Deformation bands in sandstone: a review. J Geol Soc 164:755–769

46. Agard P, Augier R, Monie P (2011) Shear band formation and strain localization on a regional scale: evidence from anisotropic rocks below a major detachment (Betic Cordilleras, Spain). J Struct Geol 33(2):114–131

47. Marone C, Scholz CH (1989) Particle-size distribution and microstructures within simulated fault gouge. J Struct Geol 11(7):799–814

48. Marone C (1998) Laboratory-derived friction lows and their application to seismic faulting. Annu Rev Earth Planet Sci 26:643–696

49. Marone C (1995) Fault zone strength and failure criteria. Geophys Res Lett 22(6):723–726

50. Scuderi MM, Carpenter BM, Johnson P, Marone C (2015) Poromechanics of stick-slip frictional sliding and strength recovery on tectonic faults. J Geophys Res: Solid Earth 120(10):6895–6912

51. Duarte JC, Schellart WP, Cruden AR (2015) How weak is the subduction zone interface? Geophys Res Lett 42(8):2664–2673

52. Weiss J, Pellissier V, Marsan D, Arnaud L, Renard F (2016) Cohesion versus friction in controlling the long-term strength of a self-healing experimental fault. J Geophys Res: Solid Earth 121(12):8523–8547

53. Hubbert MK, Rubey WW (1959) Role of fluid pressure in mechanics of overthrust faulting. Geol Soc Am Bull 70(2):115–166
54. Rice JR (1975) On the stability of dilatant hardening for saturated rock mass. J Geophys Res 80(11):1531–1536
55. Hardebeck JL, Hauksson E (1999) Role of fluids in faulting inferred from stress field signatures. Science 285(5425):236–239
56. Rudnicki JW (2001) Coupled deformation-diffusion effects in the mechanics of faulting and failure of geomaterials. Appl Mech Rev 54(6):483–502
57. Chambon G, Rudnicki JW (2001) Effects of normal stress variations on frictional stability of a fluid-infiltrated fault. J Geophys Res: Solid Earth 106(B6):11353–11372
58. Hamiel Y, Lyakhovsky V, Agnon A (2005) Rock dilation, nonlinear deformation, and pore pressure change under shear. Earth Planet Sci Lett 237(3–4):577–589
59. Paterson MS, Wong TF (2005) Experimental rock deformation—the brittle field. Springer, Berlin
60. Rozhko AY, Podladchikov YY, Renard F (2007) Failure patterns caused by localized rise in pore-fluid overpressure and effective strength of rocks. Geophys Res Lett 34(22). https://doi.org/10.1029/2007gl031696
61. Ougier-Simonin A, Zhu W (2013) Effects of pore fluid pressure on slip behaviors: an experimental study. Geophys Res Lett 40(11):1–6
62. Ougier-Simonin A, Zhu W (2015) Effects of pore pressure buildup on slowness of rupture propagation. J Geophys Res: Solid Earth 120(12):7966–7985
63. Scuderi MM, Collettini C (2016) The role of fluid pressure in induced vs. triggered seismicity: insights from rock deformation experiments on carbonates. Sci Rep 6:24852. https://doi.org/10.1038/srep24852
64. Nur A (1975) A note on the constitutive law for dilatancy. Pure Appl Geophys 113:197–206
65. Germanovich LN, Salganik RL, Dyshkin AV, Lee KK (1994) Mechanisms of brittle fracture of rock with pre-existing cracks in compression. Pure Appl Geophys 143:117–149
66. Main IG, Bell AF, Meredith PG, Geiger S, Touati S (2012) The dilatancy-diffusion hypothesis and earthquake predictability. Geol Soc London Spec Publ 367:215–230
67. Duda M, Renner J (2013) The weakening effect of water on the brittle failure strength of sandstone. Geophys J Int 192(3):1091–1108
68. Matthai SK, Fischer G (1996) Quantitative modeling of fault-fluid-discharge and fault-dilation-induced fluid-pressure variations in the seismogenic zone. Geology 24(2):183–186
69. Brace WF, Martin RJ (1968) A test of the law of effective stress for crystalline rocks of low porosity. Int J Rock Mech Min Sci Geomech Abstr 5(5):415–426
70. Atkinson C, Cook JM (1993) Effect of loading rate on crack propagation under compressive stress in a saturated porous material. J Geophys Res: Solid Earth 98(B4):6383–6395
71. Samuelson J, Elsworth D, Marone C (2011) Influence of dilatancy on the frictional constitutive behavior of a saturated fault zone under a variety of drainage conditions. J Geophys Res: Solid Earth 116(B10). https://doi.org/10.1029/2011jb008556
72. Sibson RS (2003) Thickness of the seismic slip zone. Bull Seismol Soc Am 93(3):1169–1178
73. Zoback MD, Barton CA, Brudy M, Castillo DA, Finkbeiner T, Grollimund BR, Moos DB, Peska P, Ward CD, Wiprut DJ (2003) Determination of stress orientation and magnitude in deep wells. Int J Rock Mech Min Sci 40(7–8):1049–1076
74. Chung SF, Randolph MF, Schneider JA (2006) Effect of penetration rate on penetrometer resistance in clay. J Geotech Geoenviron Eng 132(9):1188–1196
75. Quinn TAC, Brown MJ (2011) Effect of strain rate on isotropically consolidated kaolin over a wide range of strain rates in the triaxial apparatus. In: Proceedings of the fifth international symposium on deformation characteristics of geomaterial, 1–3 Sept 2011, Seoul, Korea. IOS Press, pp 607–613
76. Robinson S, Brown MJ (2013) Rate effects at varying strain levels in fine grained soils. In: Proceedings of the 18th international conference on soil mechanics and geotechnical engineering, Paris, 2–6 Sept 2013, pp 263–266

Transfer of a Biological Fluid Through a Porous Wall of a Capillary

Nelli N. Nazarenko and Anna G. Knyazeva

Abstract The treatise proposes a model of biological fluid transfer in a dedicated macropore with microporous walls. The distribution of concentrations and velocity studies in the capillary wall for two flow regimes—convective and diffusive. The largest impact on the redistribution of concentration between the capillary volume and its porous wall is made by Darcy number and correlation of diffusion coefficients and concentration expansion. The velocity in the interface vicinity increases with rising pressure in the capillary volume or under decreasing porosity or without consideration of the concentration expansion.

Keywords Capillary · Diffusion · Peclet number · Convective and diffusive flow regimes

1 Introduction

Contemporary medicine widely implements agents for tissue culture, delivery systems for pharmaceuticals, implants, bandages, arterial conduits, etc. The efficacy of all synthesized materials depends on their structure, including the structure of the pore space, which largely controls the kinetics of biochemical processes. For example, an implant should possess a strictly determined pore size promoting the formation of blood vessels during tissue growth. The structure of biological porous media is multiscale. Along with macroscopic pores, there are a lot of capillaries. Pore walls, in turn, consist of several layers, each of which has its own structure. The properties of the surfaces of pores and capillaries also affect the flow of biological fluids.

The main biological fluids of a person include blood, tissue fluid and lymph. The first performs mainly a transport function. Essentials for life enter the cells through the tissue fluid from the blood into the cells. The main function of the third fluid

N. N. Nazarenko (✉) · A. G. Knyazeva
Institute of Strength Physics and Materials Science, Siberian Branch of the Russian Academy of Sciences, Tomsk 634021, Russia
e-mail: nnelli@ispms.tsc.ru

© The Author(s) 2021
G.-P. Ostermeyer et al. (eds.), *Multiscale Biomechanics and Tribology of Inorganic and Organic Systems*, Springer Tracts in Mechanical Engineering, https://doi.org/10.1007/978-3-030-60124-9_22

is protective. Lymph destroys pathogens and ensures the return of tissue fluid into the bloodstream. The blood vessels through which blood flows from the heart form the arterial system, and the vessels that collect blood and carry it to the heart form the venous system. The metabolism between the blood and body tissues is carried out using capillaries that penetrate the organs and most tissues. The main functions of the blood and circulatory system are to connect organs and cells to ensure their vital functions—in the delivery of oxygen, nutrients, hormones, excretion of decomposition products, maintaining a constant body temperature, and protection from harmful microbes [1, 2]. All this suggests the need to study the flow of biological fluids in the system of vessels and capillaries, taking into account the features of the structure and rheological properties and the development of appropriate models [3]. The rheological properties of blood are mainly due to the processes of hydrodynamic interaction of erythrocytes with plasma, which contribute to the formation and decay of aggregates, rotation and deformation of red blood cells, their redistribution, and the corresponding orientation in the flow [4]. Blood is a heterogeneous and multiphase physical and chemical system. It can be represented as a suspension and non-Newtonian fluid with complex rheological properties. In addition to modeling blood flow in large blood vessels [2, 4–8] there are a number of works in the literature in which the flow in capillaries is simulated.

For example, work [9] has analyzed three variants of mathematical models describing the flow of a viscous incompressible fluid in a long cylindrical capillary with its internal surface covered by a permeable porous layer. The authors have shown that for thin weakly permeable porous layers on the capillary walls, the Brinkman model is not applicable; one just can use the Navier slip condition. If the porous layer is thick and/or is weakly permeable, it is not allowed to neglect the effect of the flow in it on the total fluid flow rate through the capillary, and an adequate description of the filtration process should be made using the Brinkman model.

In [10, 11], the authors have studied the blood flow through porous blood vessels taking into account an electromagnetic field. They have suggested a blood flow model in an artery with porous walls within the model of a non-Newtonian fluid in the presence of electromagnetic field. In these works, the viscosity of the non-Newtonian fluid depended on the temperature and magnetic field and was calculated by the models of Reynolds and Vogel or was assumed to be constant.

There are a lot of works devoted to modeling of blood flow in capillaries [12–14]. The authors investigate the influence of diverse parameters on the fluid motion in capillaries: vessel curvature [15], capillary radius and shape [16–18], dynamics of oxygen transportation [19], hemodynamics of vascular prostheses and implants [20, 21].

Important applications of biomedical systems, such as biological tissues, require taking into account the flow, heat and mass exchange through a porous medium [22]. The theory of transfer in a porous medium on the basis of various models, such as Darcy-Brinkman model of momentum transfer and local thermal equilibrium for energy transfer, were analyzed by the authors and can be particularly useful in describing different biological applications.

In general case, biological fluids possess specific rheological properties. As a rule, biological fluids are non-Newtonian fluids that are described by various rheological models. In the literature, models of viscoelastic, viscoplastic, pseudo-plastic and dilatant fluids are widely used. All the models, with due consideration of complex rheology, are non-linear. The non-linear effects also manifest when accounting the dependence of properties (for instance, viscosity) on concentration.

Current work suggests a model of biological fluid transfer in a selected macropore with microporous walls. Unlike [10, 11], we assume isothermal conditions; however, we assume some the state equations for pressure in the fluid to be differential [23, 24], which yields a non-linear coupled model.

2 General Equations

Let us formulate the problem on the transfer of a biological fluid (or a pharmaceutical) in a selected cylindrical macropore with radius R_1 having microporous walls (Fig. 1). Area 1 is the macropore, area 2 is the porous layer with thickness $\delta = R_2 - R_1$.

To construct the model we use the continuity equation:

$$\frac{\partial \rho}{\partial t} + \nabla \cdot (\rho \mathbf{v}) = 0, \tag{1}$$

balance equation for species

$$\rho \left(\frac{\partial C_k}{\partial t} + \mathbf{v} \nabla C_k \right) = -\nabla \cdot \mathbf{J}_k, \tag{2}$$

and motion equation

$$\rho \frac{\partial \mathbf{v}}{\partial t} + \rho \mathbf{v} \nabla (\mathbf{v}) = -\nabla \cdot \boldsymbol{\sigma} + \rho \mathbf{F}, \tag{3}$$

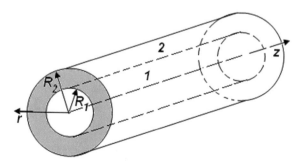

Fig. 1 Cylindrical pore with radius R_1 having porous walls

where ρ is the density, $\mathbf{v}$ is the velocity of centre of mass; C_k—species (component) concentrations; $\mathbf{J}_k$ is diffusion flux of this component; σ is stress tensor; $\mathbf{F}$ is the mass force vector; $\nabla \ldots \equiv grad \ldots$; $\nabla \cdot \ldots \equiv div \ldots$.

We will describe the flow in the macropore (area 1) using Navier-Stokes equations. The microporous medium (area 2) will be modeled as Brinkman medium. In a first approximation, the biological fluid is assumed incompressible. Navier-Stokes equations follows from (3) when

$$\sigma_{ij} = -P\delta_{ij} + 2\mu e_{ij}, \tag{4}$$

and $\mathbf{F} = \mathbf{F}_1 = -\frac{1}{\rho}\nabla(gz)$. Here p is hydrodynamic pressure and e_{ij} is the tensor of strain rates,

$$e_{ij} = \frac{1}{2}\left(\frac{\partial V_i}{\partial x_j} + \frac{\partial V_j}{\partial x_i}\right),$$

V_i are components of the velocity vector.

Brinkman medium appears when we assume

$$\mathbf{F} = \mathbf{F}_1 + \mathbf{F}_2, \tag{5}$$

where $\mathbf{F}_2$ is the force of internal friction depending on filtration velocity, $\mathbf{w}$. Then $\mathbf{v} = \mathbf{w}/a$, $a = S_p/S$, and S_p is the area occupied by pores in the section S.

If the fluid is incompressible (which is usually accepted for slow flows), instead continuity equations (1) will remain:

$$\nabla \cdot \mathbf{v}_i = 0; \quad i = 1, 2. \tag{6}$$

As a result we obtain for area inside the capillary:

$$0 < r < R_1: \quad \rho_1\left(\frac{\partial C_k}{\partial t} + \mathbf{v}_1\nabla C_k\right) = -\nabla \cdot \mathbf{J}_k, \tag{7}$$

$$\rho_1\left(\frac{\partial \mathbf{v}_1}{\partial t} + \mathbf{v}_1\nabla \mathbf{v}_1\right) = -\nabla p_1 + \rho_1 gz + \nabla \cdot (\mu_1\nabla \mathbf{v}_1), \tag{8}$$

and for the porous walls:

$$R_1 < r < R_2: \quad \rho_2\left(\frac{\partial C_k}{\partial t} + \mathbf{v}_2\nabla C_k\right) = -\nabla \cdot \mathbf{J}_k, \tag{9}$$

$$\rho_2\left(\frac{\partial \mathbf{v}_2}{\partial t} + \mathbf{v}_2\nabla \mathbf{v}_2\right) = -\nabla p_2 + \rho_2 gz + \left(\nabla \cdot (\mu_2\nabla \mathbf{v}_2) - \mu_2' m \frac{\mathbf{v}_2}{k_f}\right). \tag{10}$$

Here $\mathbf{v}_1$, $\mathbf{v}_2$, ρ_1, ρ_2 are the vectors of velocities and densities of the liquid in areas 1 and 2, C_k is the concentration of the k-th component, $\mathbf{J}_k = -D_k \rho_i \nabla C_k$ is the diffusion flux of the k-th component; P_1, P_2, μ_1, μ_2 are the pressure and viscosity of the fluid in areas 1 and 2, μ_2' is the viscosity in the Darcy's law, in general case it differs from μ_2; g is the force of gravity; k_f is the permeability of the porous medium; m is the porosity of pore walls.

In the case of slow (crawl) flow, the second summands in the left brackets of the motion equations for porous walls can be neglected.

We should add the state equation connecting the pressure with temperature and fluid composition. For constant temperature, we can write [23, 25]

$$dp = -\rho \beta_T^{-1} d\gamma + \sum_{k=1}^{n} p_k dC_k \qquad (11)$$

where $p_k = \alpha_k \beta_T^{-1}$, α_k is concentration expansion coefficients, β_T is isothermal compressibility coefficient, $\beta_T^{-1} = K$; K is bulk module for fluid. Then for incompressible fluid with constant properties for each area we have

$$p_2 - p_{20} = 3K\alpha(C_2 - C_{20}) \quad \text{and} \quad p_1 - p_{10} = 3K\alpha(C_1 - C_{10}), \qquad (12)$$

where C_{10} and C_{20} is preset zero approximation, p_{10} and p_{20}—is initial pressures values in areas.

3 Stationary Model

From (7)–(10) we obtain stationary model for individual pore The hydrodynamic part of the problem will include equations:

$$0 < r < R_1 : \quad \rho_1 V_1 \frac{dV_1}{dr} = -\frac{dp_1}{dr} + \frac{1}{r}\frac{d}{dr}\left(r\mu_1(C_1)\frac{dV_1}{dr}\right), \qquad (13)$$

$$R_1 < r < R_2 : \quad \frac{dp_2}{dr} - \left(\frac{1}{r}\frac{d}{dr}\left(r\mu_2(C_2)\frac{dV_2}{dr}\right) - \mu_2'(C_2)m\frac{V_2}{k_f}\right) = 0, \qquad (14)$$

where V_k, $k = 1,2$ are radial components of velocity for areas.

The boundary conditions for the hydrodynamic part of the problem will be as follow. In the point $r = 0$ we have the symmetry condition

$$V_1 = 0. \qquad (15)$$

In the interface between two areas, mass velocities and stress tensor components are equal

$$r = R_1: \quad \rho_1 V_1 = m\rho_2 V_2, \quad -p_1 + \mu_1(C_1)\frac{dV_1}{dr} = m\left(-p_2 + \mu_2(C_2)\frac{dV_2}{dr}\right).$$
(16)

We can assume that the outer wall of the capillary ($r = R_2$) is free of load, on

$$\sigma_{rr} = -p_2 + \mu_2\frac{dV_2}{dr} = 0$$
(17)

or (other case)

$$V_2 = 0.$$
(18)

The viscosities linearly depend on concentration:

$$\mu_1(C_1) = \mu_{10} + \mu_{11}C_1, \quad \mu_2(C_2) = \mu_{20} + \mu_{21}C_2.$$
(19)

For the diffusion part of the problem, we have:

$$0 < r < R_1: \quad \rho_1 V_1 \frac{dC_1}{dr} = \frac{1}{r}\frac{d}{dr}\left(D_1\rho_1 r \frac{dC_1}{dr}\right),$$
(20)

$$R_1 < r < R_2: \quad \rho_2 V_2 \frac{dC_2}{dr} = \frac{1}{r}\frac{d}{dr}\left(D_2\rho_2 r \frac{dC_2}{dr}\right),$$
(21)

$$r = 0: \quad \frac{dC_1}{dr} = 0,$$
(22)

$$r = R_1: \quad C_1 = mC_2, \quad \rho_1\left[D_1\frac{dC_1}{dr} - V_1 C_1\right] = \rho_2 m\left[D_2\frac{dC_2}{dr} - V_2 C_2\right],$$
(23)

$$r = R_2: \quad \rho_2\left(D_2\frac{dC_2}{dr} - V_2 C_2\right) = \Omega,$$
(24)

where D_1, D_2 are the diffusion coefficients in areas 1 and 2.

The condition (22) is symmetry condition; first of (23) follows from chemical potential continuity, second of (23) is the equality of the total mass flows; condition (24) contains the mass sink on the outer wall of the capillary Ω.

Taking into account the connection between pressure and concentration (12), from (13), (14), (16), (17) we obtain

$$\rho_1 V_1 \frac{dV_1}{dr} = -3K\alpha\frac{dC_1}{dr} + \frac{1}{r}\frac{d}{dr}\left(r\mu_1(C_1)\frac{dV_1}{dr}\right);$$

$$3K\alpha\frac{dC_2}{dr} - \left(\frac{1}{r}\frac{d}{dr}\left(r\mu_2(C_2)\frac{dV_2}{dr}\right) - \mu_2'(C_2)m\frac{V_2}{k_f}\right) = 0;$$

$$- (3K\alpha(C_1 - C_{10}) + p_{10}) + \mu_1(C_1)\frac{dV_1}{dr}$$

$$= m\left(-(3K\alpha(C_2 - C_{20}) + p_{20}) + \mu_2(C_2)\frac{dV_2}{dr}\right);$$

$$\sigma_{rr} = -(3K\alpha(C_2 - C_{20}) + p_{20}) + \mu\frac{dV_2}{dr} = 0.$$

In this model we assumed that viscosity of fluid and diffusion coefficients in pore and in porous wall are different, that connect with special structure of porous space affecting the fluid mobility. These problem is coupling in general case.

4 Special Case

The simplest stationary diffusion model for individual pore can be analyzed for the case when the pressure gradient along the macro pores is given, and the fluid composition in pore is fixed (we neglect the gravitational force):

$$\nabla p_1 = \omega = const, \quad C_1 = C_{10}. \tag{25}$$

Because in interface $\rho_1 V_1 = m\rho_2 V_2$ and the pressure is proportional to concentration, then we do not mistake if assume:

$$\nabla p_1 = const = \beta\omega \sim \nabla p_2, \tag{26}$$

where ω, β is some constants. In this case the hydrodynamical part of the problem turns to

$$\omega - \frac{1}{r}\frac{d}{dr}\left(r\mu_1(C_{10})\frac{dV_1}{dr}\right) = 0; \tag{27}$$

$$\omega\beta - \left(\frac{1}{r}\frac{d}{dr}\left(r\mu_2(C_2)\frac{dV_2}{dr}\right) - \frac{V_2}{k_f}\right) = 0; \tag{28}$$

$$r = 0: \quad V_1 = 0; \quad r = R_2; \quad V_2 = 0; \tag{29}$$

$$r = R_1: \quad \rho_1 V_1 = m\rho_2 V_2; \quad \mu_1(C_{i0})\frac{dV_1}{dr} = m\mu_2(C_2)\frac{dV_2}{dr}. \tag{30}$$

For the case of constant viscosity, the exact analytical solution of this problem is presented in [26].

Diffusion part of the problem takes a place only for the area $R_1 \leq r \leq R_2$:

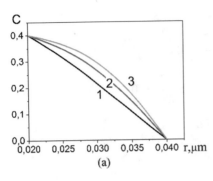
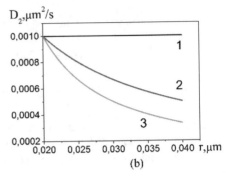

Fig. 2 Concentration distribution in the wall of pore (**a**) for different diffusion coefficients (**b**). $D_{20} = 10^{-3}$ μm/s, 1—$\alpha_{p2} = 0$; 2—$\alpha_{p2} = 6$; 3—$\alpha_{p2} = 10$

$$\rho_2 V_{r2}(r; C_{10}, C_2, m, \omega)\frac{dC_2}{dr} = \frac{1}{r}\frac{d}{dr}\left(D_2\rho_2 r\frac{dC_2}{dr}\right);$$

$$r = R_1 : \quad C_{10} = mC_2;$$

$$r = R_2 : \quad C_2 = 0.$$

We assume that velocity distribution in the walls is given and does not depend on concentration. It is obviously, when the velocity is equals to zero, and diffusion coefficient is constant value, we come to concentration distribution coinciding with the exact analytical solution (Fig. 2a, b—lines 1):

$$C_2 = \frac{C_{10}}{m}\frac{\ln(r/R_2)}{\ln(R_1/R_2)}.$$

This solution does not contain density and diffusion coefficient.

If diffusion coefficient depends on space coordinate (that could be connected with the change of pore structure, for example, using the equation $D_2 = D_{20}/(1 + \alpha_{p2}(r - R_1))$, then concentration distribution changes (2 and 3 curves correspondingly) in this figure.

The positive value of given filtration velocity effects on concentration distribution similarly (Fig. 3). However, the type of the velocity distribution is not essential concentration (Fig. 4).

The concentration distribution in Fig. 4b is given for the velocity functions $V_2(r)$, presented in the Fig. 4a. The change of the velocity with the coordinate could be connects with the complex structure of porous space, with structure of pore surface, with their specific tortuosity, the close pores availability leading to inhibition of concentration distribution along pore walls.

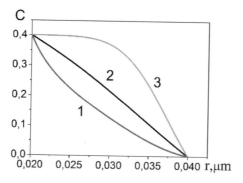

Fig. 3 Concentration distribution for given filtration rate. 1—$V_2 = 0.1$; 2—$V_2 = 0$; 3—$V_2 = 0.2$ $\mu m/s$, $D_{20} = 10^{-3}$ $\mu m/s$

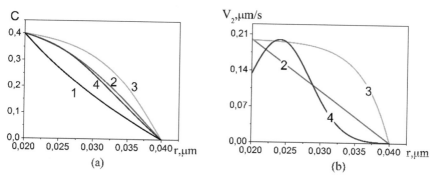

Fig. 4 Concentration distribution (**a**) and liquid velocity in pore wall (**b**). Black line—is exact analytical solution for $V_2 = 0$; the colors of the lines to the left correspond to colors of the lines to the right. $D_{20} = 10^{-3}$ $\mu m/s$

5 Dimensionless Variables and Parameters in Total Stationary Model

Let us introduce the following dimensionless variables:

$$\xi = \frac{r}{R_2}, \quad \overline{V}_i = \frac{V_i}{\mu_{10}/\rho_1 R_2}, \quad \bar{p}_i = \frac{p_i}{\mu_{10}^2/\rho_1 R_2^2}.$$

Then the equations and boundary conditions in dimensionless variables will be rewritten as

$$0 < \xi < 1 - \Delta : \quad \overline{V}_{r1}\frac{d\overline{V}_1}{d\xi} = -\frac{d\bar{p}_1}{d\xi} + \frac{1}{\xi}\frac{d}{d\xi}\left(\xi\bar{\mu}_1(C_1)\frac{d\overline{V}_1}{d\xi}\right), \quad (31)$$

$$Pe_D\overline{V}_1\frac{dC_1}{d\xi} = \frac{1}{\xi}\frac{d}{d\xi}\left(\xi\frac{dC_1}{d\xi}\right), \tag{32}$$

$$1 - \Delta < \xi < 1: \quad Da\frac{d\bar{p}_2}{d\xi} - \left(\frac{Da}{\xi}\frac{d}{d\xi}\left(\xi\,\bar{\mu}_2(C_2)\frac{d\overline{V}_2}{d\xi}\right) - \bar{\mu}_2'm(C_2)\overline{V}_2\right) = 0, \tag{33}$$

$$Pe_D\overline{V}_2\frac{dC_2}{d\xi} = \overline{D}\frac{1}{\xi}\frac{d}{d\xi}\left(\xi\frac{dC_2}{d\xi}\right), \tag{34}$$

$$\xi = 0: \quad \overline{V}_1 = 0, \quad \frac{dC_1}{d\xi} = 0, \tag{35}$$

$$\xi = 1 - \Delta: \quad \overline{V}_{r1} = m\bar{\rho}\overline{V}_{r2}. \quad -\bar{p}_1 + \bar{\mu}_1(C_1)\frac{d\overline{V}_{r1}}{d\xi} = m\left(-\bar{p}_2 + \bar{\mu}_2(C_2)\frac{d\overline{V}_{r2}}{d\xi}\right), \tag{36}$$

$$C_1 = mC_2, \quad \frac{dC_1}{d\xi} - Pe_D\overline{V}_1C_1 = m\bar{\rho}\left(\overline{D}\frac{dC_2}{d\xi} - Pe_D\overline{V}_2C_2\right), \tag{37}$$

$$\xi = 1: \quad \bar{\mu}_2\frac{d\overline{V}_{r2}}{d\xi} = \bar{p}_{20} + K_{bez}(C_2 - C_{20}), \tag{38}$$

$$\overline{D}\frac{dC_2}{d\xi} - Pe\overline{V}_2C_2 = \overline{\Omega}, \tag{39}$$

where $\bar{\mu}_1(C_1) = 1 + \alpha_1C_1$, $\bar{\mu}_2(C_2) = \beta + \alpha_2C_2$, $\bar{\mu}_2' = \bar{\mu}_2$, $\bar{p}_2 - \bar{p}_{20} = K_{bez}(C_2 - C_{20})$ and $\bar{p}_1 - \bar{p}_{10} = K_{bez}(C_1 - C_{10})$.

Stationary model contains following dimensionless parameters:

$$Pe_D = \frac{V_*R_2}{D_1}, \quad Da = \frac{k_f}{R_2^2}, \quad K_{bez} = \frac{K\alpha R_2^2\rho_1}{\mu_{10}^2}, \quad \overline{D} = \frac{D_2}{D_1}, \quad \Delta = \frac{\delta}{R_2},$$

$$\bar{\rho} = \frac{\rho_2}{\rho_1}, \quad m, \quad \alpha_1 = \frac{\mu_{11}}{\mu_{10}}, \quad \alpha_2 = \frac{\mu_{21}}{\mu_{10}}, \quad \beta = \frac{\mu_{20}}{\mu_{10}}, \quad \overline{\Omega} = \frac{\Omega R_2}{\rho_2 D_1}.$$

Diffusion Peclet number Pe_D includes the velocity $V_* = \mu_{10}/\rho_1 R_2$ and characterizes the relation between convective and diffusion forces; Darcy number Da, relation of elastic and viscous forces K_{bez} and diffusion coefficients $\overline{D}$ together with Pe_D determine the nature of the flow. parameters $\overline{\Omega}$, $\bar{\rho}$, $\alpha_1, \alpha_2, \beta$ are not so significant.

To assess the dimensionless parameters, the following physical values were used that characterize the diffusing fluid (blood) and physical parameters of the capillary [27–31]: $\mu_{10} = 4.5 \times 10^{-3}$ Pa s, $\rho_1 = 1064$ kg/m^{-3}, $K = 2.2 \times 10^9$ Pa, $\alpha = 0.3$, $R_1 = 3 \times 10^{-6}$ m, $R_2 = 4 \times 10^{-6}$ m, $D_1 = 2.1 \times 10^{-10}$ m^2/s, $k_f = 0.5 \times 10^{-12}$ m^2. By using data above we can assess the region of alteration of dimensionless complexes: $0.38 \leq Pe_D \leq 1.43 \times 10^3$, $K_{bez} = 5.54 \times 10^4$. In the calculations, the following dimensionless parameters were varied: m, Pe_D, $\overline{D}$, Da and p_{10}, p_{20}. The rest of the parameter were fixed: $C_{10} = 0.1$, $C_{20} = 0$, $\xi_1 = 0.75$, $\xi_2 = 1$, $\bar{\rho} = 1$, $\alpha_1 = 0.1$, $\alpha_2 = 0.1$, $\beta = 1$, $\overline{\Omega} = 0$.

The stationary problem for the porous wall (31)–(39) was solved numerically. In differential Eqs. (31), (32) and (34), the convective summand is approximated by the difference against the flow [32]. Such difference provides approximation of the convective summand for any direction of the flow velocity and yields stable algorithm. The initial distribution of velocities and concentrations is specified first. Then the differential equations for concentration and velocity are solved by the double-sweep method. Obtained distributions are used as initial for next iteration. The interface between media is distinguished explicitly. In the direct marching, the coefficients are found with a special approximation of boundary conditions in point at the interface. During reverse marching, we first find C_2 at the interface and then, by using first condition (37) in the point in interface, we find C_1. The same operations are applied to velocities. The process is repeated until a special condition is met. The calculation is carried out until a special condition is fulfilled—until a solution with a given accuracy is obtained. The variation of spatial steps changes the results no more than 1–5% in the wide region of varying model parameters.

6   Analysis of Results

The Peclet number which characterizes two flow regimes-convective ($Pe_D > 1$) and diffusive one ($Pe_D < 1$)—presents the main interest in the study of fluid transfer through a capillary with porous wall.

At small Peclet numbers, the main contribution to the distribution of concentrations in a capillary is made by diffusion. Since diffusion is a slow process, a smaller amount of a diffusant gets into the porous capillary wall (line 3 in Fig. 5a). With

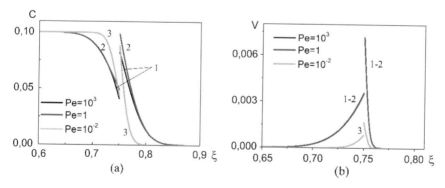

Fig. 5 Distribution of concentration (**a**) and velocity (**b**) along the radius at different Peclet numbers 1—$Pe = 10^3$, 2—$Pe = 1$, 3—$Pe = 10^{-2}$, $\overline{D} = 1$, $m = 0.3$, $Da = 0.01$, $p_{10} = 2$, $p_{20} = 0.2$, $K_{bez} = 5.54 \times 10^{-2}$, $\Delta = 0.25$, $\overline{\Omega} = 0$

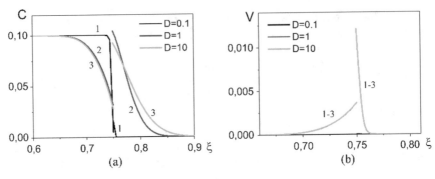

Fig. 6 Distribution of concentration (**a**) and velocity (**b**) along the radius under convective mass transfer at different values of parameter $\overline{D}$, 1—$\overline{D} = 0.1$, 2—$\overline{D} = 1$, 3—$\overline{D} = 10$, $Pe_D = 10^3$, $m = 0.3$, $Da = 0.01$, $p_{10} = 2$, $p_{20} = 0.2$, $K_{bez} = 5.54 \times 10^{-2}$, $\Delta = 0.25$, $\overline{\Omega} = 0$

growing Peclet number ($Pe_D > 1$), the contribution of convective diffusion becomes dominating; the flow velocity is higher in both areas (lines 1 in Fig. 5b). This leads to increased amount of the diffusant permeating the porous wall of the capillary (lines 1 and 2 in Fig. 5a). Line 2 in Fig. 5 corresponds to the transition regime, a convective-diffusive mass transfer. The main changes to concentration and velocity are observed in the vicinity of the interface $\xi = \Delta$.

The redistribution of the diffusant concentration between the materials is appreciably affected by parameter $\overline{D}$ (relation of the diffusion coefficient in area 2 to the diffusion coefficient in area 1). At $\overline{D} > 1$ the fraction of the diffusant in the capillary wall increases (Fig. 6a), while at $\overline{D} < 1$ the diffusant is almost absent in area 2. It has almost zero effect on the velocity distribution (Fig. 6b). The character of concentration distribution both at convective (Fig. 6) and diffusive mass transfer is qualitatively similar under variation of $\overline{D}$. The difference is only in the velocity values.

Increased capillary wall thickness decreases the concentration and velocity at the interface in both phases both under convective and diffusive mass transfer. This was demonstrated in Table 1, because it was difficult to demonstrate in a figure.

Increased wall porosity decreases the concentration of the diffusant in the second area near the interface; however, the diffusant permeates deeper into the capillary wall under both convective (Fig. 7a, c) and diffusive (Fig. 7b, d) regimes. This is

Table 1 Concentration and velocity at interface for different capillary wall thickness under convective mass transfer

$\Delta = \xi_2 - \xi_1$	$C_1(\xi_1)$	$C_2(\xi_1)$	$V_1(\xi_1)$	$V_2(\xi_1)$
0.25	0.05056	0.16855	0.00368	0.012284
0.3	0.05050	0.16834	0.00367	0.012249
0.35	0.05041	0.16803	0.00366	0.012247
0.4	0.05032	0.16775	0.00367	0.012243
0.45	0.05021	0.16738	0.00367	0.012239

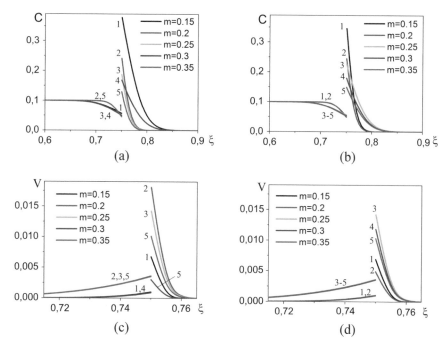

Fig. 7 Distribution of concentration (**a, b**) and velocity (**c, d**) along radius under convective (**a, c**) and diffusive (**b, d**) mass transfer and different values of parameter m, 1—$m = 0.15$, 2—$m = 0.2$, 3—$m = 0.25$, 4—$m = 0.3$, 5—$m = 0.35$, $\overline{D} = 1$, $Da = 0.01$, $p_{10} = 2$, $p_{20} = 0.2$, $K_{bez} = 5.54 \times 10^{-2}$, $\Delta = 0.25$; $\overline{\Omega} = 0$

explained by increased volume of porous space and the diffusant moves more freely in the second area. The variation of porosity hardly affects the concentration distribution in the first area at any Peclet number. The concentration negligibly reduces only near the interface. The velocity behaves ambiguously (Fig. 7c, d) which is due to the interdependence of contrary physical mechanisms.

Decreased permeability (decreased Darcy number) negligibly reduces the concentration and reduces the velocity in both regions at any Peclet number.

The major impact on the velocity distribution is caused by the pressure gradient. An increase in the initial pressure in the first area (not shown) augments the velocity in both areas, while it has almost no effect on the concentration distribution; the concentration drops in both areas only in the interface vicinity. Similar behavior is observed under the diffusive regime of mass transfer. An increase in the initial pressure in the second area has no effect on the concentration distribution, while the velocity in both areas negligibly reduces in any regime.

All previous calculations were made with due regard to the concentration expansion. Parameter α which is included into dimensionless complex $K_{bez} = K\alpha R_2^2 \rho_1/\mu_{10}^2$. This excites interest in the comparison of the concentration distribution and velocities in the porous wall with and without due consideration of this

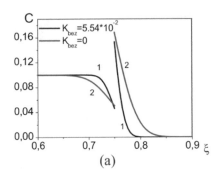 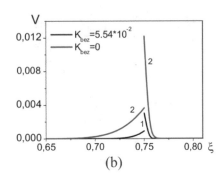

(a)   (b)

Fig. 8 Distribution of concentration (**a**) and velocity (**b**) along the radius under convective mass transfer and for different values K_{bez}, 1—$K_{bez} = 5.54 \times 10^{-2}$, 2—$K_{bez} = 0$, $Pe_D = 10^3$, $\overline{D} = 1$, $m = 0.3$, $Da = 0.01$, $p_{10} = 2$, $p_{20} = 0.2$, $\Delta = 0.25$; $\overline{\Omega} = 0$

effect (Fig. 8). Evidently, without consideration of the concentration expansion, the velocity in the interface vicinity increases, which is valid for both sides of the interface. The concentration expansion causes more diffusant to permeate into the porous capillary wall. Such considerable difference is observed even at small values of coefficient K_{bez}.

The effect of viscosity versus concentration on concentration and velocity distributions is illustrated in the Fig. 9. An increase in viscosity with concentration leads to an increase in the fraction of diffusant in the capillary wall in the convective flow regime (Fig. 9a, c), and in the diffusion mode to an insignificant decrease (Fig. 9b, d). In this case, the velocity increases in the convective mode, and decreases in the diffusion mode.

For all figures above it was accepted $\overline{\Omega} = 0$.

The mass flow affects the nature of the concentration distribution in the diffusion mode (Fig. 10b, d red lines), but in the convective mode, no effect is detected (Fig. 10a, c red lines). A smaller amount of diffusant remains in the capillary wall (Fig. 10b) when mass flow is taken into account, the speed also decreases (Fig. 10c).

7 Conclusions

The work suggested a model of biological fluid transfer in a selected macropore with microporous walls with due account for concentration expansion phenomena appearing in stet equation. For two flow regimes—convective ($Pe_D > 1$) and diffusive ($Pe_D < 1$)—we have studied the concentration distribution in the capillary wall. It was shown that the largest impact on the redistribution of concentration between the capillary volume and its porous wall is made by Darcy number Da and correlation of diffusion coefficients. The concentration of the diffusant in the porous layer increases with growing parameter $\overline{D}$ or decreasing porosity or permeability under diffusive mass transfer. The velocity in the interface vicinity increases with rising

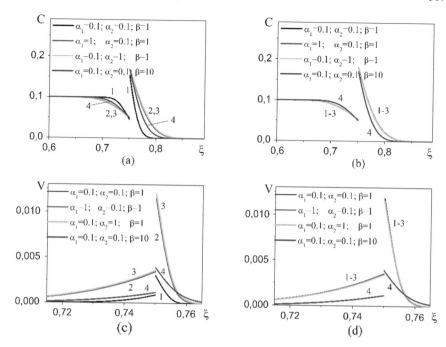

Fig. 9 Distribution of concentration (**a, b**) and velocity (**c, d**) along radius under convective (**a, c**) and diffusive (**b, d**) mass transfer and different values of viscosity of liquid; $\overline{D} = 1$, $Da = 0.01$, $p_{10} = 2$, $p_{20} = 0.2$, $K_{bez} = 5.54 \times 10^{-2}$, $m = 0.3$, $\Delta = 0.25$; $\overline{\Omega} = 0$

pressure in the capillary volume or under decreasing porosity at any Peclet number. It was discovered that the concentration expansion appreciably affects the distribution of velocity and concentration. The ambiguous impact of model parameters on different flow regimes is connected with the interrelation between contrary physical mechanisms. Described model contains practically significant parameters allowing understanding how the concentration distribution changes with flow type variation.

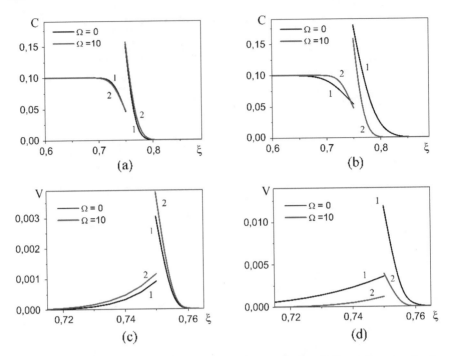

Fig. 10 Distribution of concentration (**a**, **b**) and velocity (**c**, **d**) along radius under convective (**a**, **c**) and diffusive (**b**, **d**) mass transfer and different values of parameter $\overline{\Omega}$, 1—$\overline{\Omega} = 0$, 2—$\overline{\Omega} = 10$, $\overline{D} = 1$, $Da = 0.01$, $p_{10} = 2$, $p_{20} = 0.2$, $K_{bez} = 5.54 \cdot 10^{-2}$, m = 0.3, $\Delta = 0.25$

Acknowledgements The work was performed within the Program of Fundamental Scientific Research of the State Academies of Science for 2013–2020, number III.23.2.5.

References

1. Caro CG, Pedley TJ, Schroter RC, Seed WA (2012) The mechanics of the circulation. Cambridge University Press, Cambridge
2. Pedley TJ (1980) The fluid mechanics of large blood vessels Cambridge monographs on mechanics and applied mathematics. Cambridge University Press, Cambridge
3. Petrov IB (2009) Mathematical modeling in medicine and biology based on models of continuum mechanics. Process MIRT 1(1):5–16 (in Russian)
4. Gupta AK, Agrawal SP (2015) Computational modeling and analysis of the hydrodynamic parameters of blood through stenotic artery. Procedia Comput Sci 57:403–410
5. Astrakhantseva EV, Gidaspov VYU, Reviznikov DL (2005) Mathematical modeling of hemodynamics of large blood vessels. Matematicheskoe modelirovanie 17(8):61–80 (in Russian)
6. Parshin VB, Itkin GP (2005) Biomechanics of blood circulation. Publishing of MGTU them. N.E. Bauman, Moscow (in Russian)
7. Selmi M, Belmabrouk H, Bajahzar A (2019) Numerical study of the blood flow in a deformable human aorta. Appl Sci 6(9):1216–1227. https://doi.org/10.3390/app9061216

8. Ku DN (1997) David Blood flow in arteries. Annu Rev Fluid Mech 29:399–434
9. Filippov AN, Khanukaeva DY, Vasin SI, Sobolev VD, Starov VM (2013) Liquid flow inside a cylindrical capillary with walls covered with a porous layer (Gel). Colloid J 75(2):214–225
10. Rahbari A, Fakour M, Hamzehnezhadd A, Vakilabadi MA, Ganji DD (2017) Heat transfer and fluid flow of blood with nanoparticles through porous vessels in a magnetic field: a quasi-one dimensional analytical approach. Math Biosci 283:38–47. https://doi.org/10.1016/j.mbs.2016.11.009
11. Ghasemi SE, Hatami M, Sarokolaie AK, Ganji DD (2015) Study on blood flow containing nanoparticles through porous arteries in presence of magnetic field using analytical methods. Physica E 70:146–156. https://doi.org/10.1016/j.physe.2015.03.002
12. Jafari A, Zamankhan P, Mousavi SM, Kolari P (2009) Numerical investigation of blood flow. Part II: in capillaries. Commun Nonlinear Sci Numer Simul 14(4):1396–1402
13. Pries AR, Secomb TW (2008) Handbook of physiology: section 2, the cardiovascular system, vol IV, Microcirculation, 2nd edn. Academic Press, San Diego, pp 3–36 (Blood flow in microvascular networks)
14. Xiong G, Figueroa CA, Xiao N, Taylor ChA (2011) Simulation of blood flow in deformable vessels using subject-specific geometry and spatially varying wall properties. Int J Numer Methods Biomed Eng 27:1000–1016
15. Overko VS, Beskrovnaya MV (2013) Modeling a blood flow in pathologically curved vessels. Visnik NTU "KhP1" 5(979):211–220 (in Russian)
16. Shabrykina NS, Vistalin NN, Glachaev AG (2004) Modeling the influence of a blood capillary shape on filtration and read sorption processes. Russ J Biomech 8(1):67–75 (in Russian)
17. Hammecker C, Mertz JD, Fischer C, Jeannette D (1993) A geometrical model for numerical simulation of capillary imbibition in sedimentsry rocks. Transp Porous Media 12:125–141
18. Koroleva YO, Korolev AV (2019) Herschel-bulkley model of blood flow through vessels with rough walls. Colloq J 15(39). https://doi.org/10.24411/2520-6990-2019-10460
19. Kislyakov YY, Kislyakova LP (2000) Mathematical modeling of O2 transport dynamics in red cells and blood plasma in a capillary. Sci Instrum Eng 10(1):44–51 (in Russian)
20. Schiller NK, Franz T, Weerasekara NS, Zilla P, Reddy BD (2010) A simple fluid–structure coupling algorithm for the study of the anastomotic mechanics of vascular grafts. Comput Methods Biomech Biomed Eng 13(6):773–781. https://doi.org/10.1080/10255841003606124
21. Dobroserdova TK, Olshanskii MA (2013) A finite element solver and energy stable coupling for 3d and 1d fluid models. Comput Methods Appl Mech Eng 259:166–176
22. Khaled RA, Vafai K (2003) The role of porous media in modeling flow and heat transfer in biological tissues. Int J Heat Mass Transf 46:4989–5003
23. Knyazeva AG (2009) One-dimensional models of filtration with regard to thermal expansion and volume viscosity. In: Proceedings of the XXXVII summer school–conference advanced problems in mechanics (APM, St. Petersburg 2009), pp 330–337
24. Knyazeva AG (2006) Thermodynamic model of a viscous heat conductive gas and its application in modeling of combustion processes. Math Model Syst Process 14:92–108 (in Russian)
25. Knyazeva AG (2018) Pressure diffusion and chemical viscosity in the filtration models with state equation in differential form. J Phys: Conf Ser 1128–1132
26. Filippov AN, Khanukaeva DYu, Vasin SI, Sobolev VD, Starov BM (2013) Modeling of flow of multi component biological fluid in macropore with microporous walls. Colloide J 75(2):237–249
27. Nazarenko NN, Knyazeva AG, Komarova EG, Sedelnikova MB, Sharkeev YuP (2018) Relationship of the structure and the effective diffusion properties of porous zinc- and copper-containing calcium phosphate coatings. Inorg Mater: Appl Res 9(3):451–459. https://doi.org/10.1134/S2075113318030243
28. Sevriugin VA, Loskutov VV (2009) Influence of geometry on self-diffusion of liquid molecules in porous media in long time regime. J Porous Media 12(1):29–41
29. Grigoriev IS, Radzig AA (1997) Handbook of physical quantities. CRC Press, Boca Raton

30. Virgilyev YS et al (1975) Carbon-based structural materials, on the interconnection of permeability with some physical properties of carbon material, Metallurgiya, Moscow, pp 136–139 (in Russian)
31. Ortega JM, Poole WG Jr (1981) Numerical methods for differential equations. Pitman, London
32. Roache PJ (1972) Fundamentals of computational fluid dynamics. Hermosa Pub., New Mexico

Failure Mechanisms of Alloys with a Bimodal Graine Size Distribution

Vladimir A. Skripnyak, Evgeniya G. Skripnyak, and Vladimir V. Skripnyak

Abstract A multi-scale computational approach was used for the investigation of a high strain rate deformation and fracture of magnesium and titanium alloys with a bimodal distribution of grain sizes under dynamic loading. The processes of inelastic deformation and damage of titanium alloys were investigated at the mesoscale level by the numerical simulation method. It was shown that localization of plastic deformation under tension at high strain rates depends on grain size distribution. The critical fracture stress of alloys depends on relative volumes of coarse grains in representative volume. Microcracks nucleation at quasi-static and dynamic loading is associated with strain localization in ultra-fine grained partial volumes. Microcracks arise in the vicinity of coarse and ultrafine grains boundaries. It is revealed that the occurrence of a bimodal grain size distributions causes increased ductility, but decreased tensile strength of UFG alloys. The increase in fine precipitation concentration results not only strengthening but also an increase in ductility of UFG alloys with bimodal grain size distribution.

Keywords Bimodal distribution of grain sizes · Computational plasticity · High strain rates · Titanium alloys · Magnesium alloys

1 Introduction

Hexagonal close-packed (HCP) metals, such as Ti, Mg, Zn and Zr, are of interest for engineering and medical applications due to their unique combination of high ductility and strength. In recent years, intensive efforts have been focused on investigations of the physical and mechanical properties of alloys, whose grain structure is formed by surface rolling, severe plastic deformation, selective laser melting, or selective laser sintering, and friction stir processing [1–5].

Recently Long et al. showed a bimodal microstructure formation in the ultrafine-grained Ti–6Al–4V alloy produced by means of the spark plasma sintering of

V. A. Skripnyak (✉) · E. G. Skripnyak · V. V. Skripnyak
National Research Tomsk State University (TSU), 36 Lenin Avenue, 634050 Tomsk, Russia
e-mail: skrp2006@yandex.ru

© The Author(s) 2021
G.-P. Ostermeyer et al. (eds.), *Multiscale Biomechanics and Tribology of Inorganic and Organic Systems*, Springer Tracts in Mechanical Engineering, https://doi.org/10.1007/978-3-030-60124-9_23

a mixture of ball-milled and unmilled powders [6]. The additive manufacturing technology of HCP metals opens new possibilities for obtaining parts from alloys combining high strength, ductility, and fracture toughness.

Usually, the formation of nano-sized and ultra-fine grained structures in metal alloys leads to increased strength and decreased ductility. The low ductility of ultra-fine (UFG) alloys limits their applications as advanced engineering materials. In recent years it was found that the alloys with a bimodal grain size distribution may exhibit increased yield strength without reducing ductility [7, 8]. Guo showed the ductility of fine-grained (FG) Zr with an average grain size of 2–3 μm was 17% higher than that of coarse-grained alloy [2]. Pozdnyakov [9], Malygin [10], Ulacia [11] showed that the bimodal grain distribution can be formed in light alloys by means of severe plastic deformation and follow-up heat treatment. It was revealed that hexagonal close-packed (HCP) alloys with a bimodal grain size distribution exhibit a number of anomalies in mechanical behavior [12, 13]. Therefore, the optimization of strength and ductility requires a profound knowledge of volume fraction and distribution of ultra-fine grains and coarse grain in multi-modal alloys. Research of the influence of the grain structure on the strength and ductility of the alloys under cyclic loadings, dynamic loadings, quasistatic loadings are carried out using experimental methods and by computer simulation [14–17].

Berbenni proposed a theoretical micromechanical model taking into account the grain size distribution and representing the accommodation of grain deformation in HCP metals (Zirconium-α) [7]. Raeisinia proposed a model to examine the effect of unimodal and bimodal grain size distributions on the uniaxial tensile behavior of a number of polycrystals [12]. The model demonstrated that bimodal grain size distributions have enhanced macro mechanical properties as compared with their unimodal counterparts. The authors of this article proposed a multiscale model of UFG metals with bimodal grain size to predict the localization of plastic flow in UFG light alloys under dynamic loads with respect to the ratio of volume concentrations of small and large grains. Results of calculations have shown increased dynamic ductility of UFG titanium and magnesium alloys, when the specific volume of coarse grains is greater than 30%.

Zhu proposed a micromechanics-based model to investigate the mechanical behavior of polycrystalline dual-phase metals with a bimodal grain size distribution, and fracture by means of nano/microcracks generation during plastic deformation. Results have shown that the volume fraction of coarse grains controls the strength and ductility of metals [18–21]. Clayton developed a continuum model of crystal plasticity for the interpretation of experimental data on shock wave propagation and spall fracture in polycrystalline aggregates [22]. The model accounted for complex features of the mechanical response of alloys: plastic anisotropy, large volumetric strain, heat conduction, thermos-elastic heating, rate- and temperature-dependent flow stress. Magee proposed a multiscale modeling technique to simulate the microstructural deformation of an alloy with bimodal grain size distribution. The simulation has shown that intergranular cracks nucleate at the coarse grains/ultra-fined grains interfaces [23]. Although much research is carried out on the mechanical properties of alloys with a bimodal grain size distribution, a number of issues remains poorly understood. The effect of the bimodal grain structure of alloys on the ultimate strain to failure at high strain rates is poorly studied [24]. The laws of localization of plastic deformation and the laws of damage accumulation in HCP alloys with a bimodal grain size distribution under dynamic impacts are not well studied. These problems are of great practical importance and are associated with the transition to new technologies for digital design and production of technical and medical products.

In this paper, we develop a multilevel approach to study the mechanical behavior of HCP alloys in a wide range of strain rates.

2 Computational Model

Titanium, zirconium, and magnesium alloys with a hexagonal closed-packed (HCP) structure have a significant low crystal symmetry compared to the face-centered cubic (FCC) and body-centered cubic (BCC) crystal structures of engineering alloys. The mechanical behavior of HCP alloys under quasistatic and dynamic loading at temperatures T/T_m less than 0.5 is determined by dislocation mechanisms and twinning [25–29]. Authors used modifications of the constitutive equations developed in the framework of the micro-dynamical approach and taking into account the thermally activated dislocation mechanisms [15–17, 30–32].

The grain sizes and the grain-boundary phase structure affect the glide of dislocations and the formation of dislocation substructures during the plastic flow [27].

Mechanical response of polycrystalline alloys can be described by parameters of states averaging over the model representative volume element (RVE). Therefore, it is required for model volume to represent not only a given grain size distribution but also realistic values of the physical properties of the system.

We use the approach of a finite mixture model for the estimation of average sizes in multimodal grain size distribution [7, 17, 33].

The multimodal grain size distribution described by probability density function $g(d_g)$. Distribution $g(d_g)$ is a mixture of k component distributions g_1, g_2, g_3 of ultra-fined grains (UFG), fine grains (FG), coarse grains (CG), respectively:

$$g(d_g) = \sum_{k=1}^{m} \lambda_k g_k(d_g),$$ (1)

where λ_k are the mixing weights, $\lambda_k > 0$, $\sum_{k=1}^{3} \lambda_k = 1$, $k = 1, 2, 3$.

This method allows us to determine the multimodal grain size distribution function and the range of grain size groups (UFG, FG, CG). In this paper, we present simulation results for unimodal and bimodal (UFG and CG) grain distribution. We used experimental data of grain size distribution of alloys after severe plastic deformation to calibrate the computational model.

The RVE can be created using the experimental data on grain structures obtained by the analysis of Electron Backscatter Diffraction (EBSD) based on scanning electron microscopy (SEM) [34–36]. This analysis gives quantitative information about sizes and shapes of grains, Euler angles, angles of misorientation at the grain/subgrain boundaries with angular resolution ~0.5°. The shape coefficient was determined by the relation of minimal grain size to maximal size $\left(\xi_k = d_g^{\min}/d_g^{\max}\right)_k$, where k is the number of grain size group.

The analysis of grain structure distributions has shown that there are several types of grains structure characterized by unimodal (near log-normal) distribution, and bimodal grain size distribution, multimodal grain size distribution [2, 7, 8, 35].

The specific volume of ultrafine grains (with grain size $100\,\text{nm} < d_g < 1\,\mu\text{m}$), microstructural grains $\left(1 < d_g < 10\,\mu\text{m}\right)$, and coarse grains $\left(10\,\mu\text{m} < d_g < d_g\max\right)$ was estimated using the probability density function $g(d_g)$ of grain size distribution:

$$C_{UFG} = \int_{d_g^{\min}}^{1\,\mu\text{m}} g_1(x)dx, \quad C_{FG} = \int_{1\,\mu\text{m}}^{10\,\mu\text{m}} g_2(x)dx, \quad C_{CG} = \int_{10\,\mu\text{m}}^{d_g^{\max}} g_3(x)dx, \quad (2)$$

where C_{UFG}, C_{FG}, C_{CG} are specific volumes of UFG, FG, and CG, g_1, g_2, g_3 are the probability density functions of UFG, FG, and CG systems, respectively.

The relative volume of coarse and ultra-fine grains in model RVE was determined in accordance with the probability density function of the grain size distribution.

Figure 1 shows a 3D RVE model of a titanium alloy with a bimodal grain size distribution. An RVE with dimensions of $14 \times 8 \times 1\ (\mu\text{m})^3$ was used.

Computational domains were meshed with eight-node linear bricks and reduced integration together with hourglass control.

The kinematic boundary conditions correspond to macroscopic tension. The scheme of boundary conditions is shown in Fig. 1.

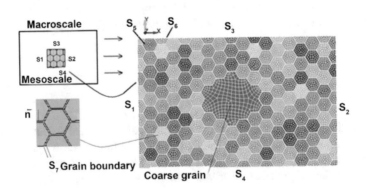

Fig. 1 Scheme of boundary conditions

$$u_x(x_k, t) = 0, \quad x_k \in S_1$$
$$u_x(x_k, t) = v_x, \quad x_k \in S_2$$
$$\sigma_{22} = 0, \quad x_k \in S_3, \quad x_k \in S_4$$
$$u_z = 0, \quad x_k \in S_5, \quad x_k \in S_6$$
$$u_k^A - u_k^B = 0, \quad \sigma_n^A = -\sigma_n^B, \quad k = 1, 2, 3, \quad x_k \in S_7 \quad (3)$$

where u_k^A, u_k^B are the projections of the displacement rate onto the external normal to the boundary S_7 of the grain and the grain boundary phase at the boundary points, respectively, σ_n^A, σ_n^B are the components of the surface forces along the external normal $\vec{n}$ to the boundary S_7, respectively, x_k is the Cartesian coordinate.

Mechanical behavior at the mesoscale level is described within the approach of the damaged elastic-plastic medium. The system of equations includes:

Conservation equations (4);
Kinematic relations (5);
Constitutive relations (8);
Equation of State (9);
Relaxation equation for the deviatoric stress tensor (10).

$$\frac{d\rho}{dt} = \rho \frac{\partial u_i}{\partial x_i}, \quad \frac{\partial \sigma_{ij}}{\partial x_j} = \rho \frac{du_i}{dt}, \quad \rho \frac{dE}{dt} = \sigma_{ij}\dot{\varepsilon}_{ij}, \quad (4)$$

where ρ is the mass density, u_i is the components of the particle velocity vector, x_i is the Cartesian coordinates, $i = 1, 2, 3$, E is the specific internal energy, σ_{ij} are the components of the effective stress tensor of the damaged medium, $\dot{\varepsilon}_{ij}$ are the components of strain rate tensor.

Kinematics of the medium was described by the local strain rate tensor:

$$\dot{\varepsilon}_{ij} = (1/2)[\partial u_i/\partial x_j + \partial u_j/\partial x_i)], \quad \dot{\omega}_{ij} = (1/2)[\partial u_i/\partial x_j - \partial u_j/\partial x_i)], \quad (5)$$

where $\dot{\varepsilon}_{ij}$, $\dot{\omega}_{ij}$ are the components of strain rate tensor and the bending-torsion tensor, u_i is the component of particles velocity vector.

The components of strain rate tensor are expressed by the sum of elastic and inelastic terms:

$$\dot{\varepsilon}_{ij} = \dot{\varepsilon}_{ij}^e + \dot{\varepsilon}_{ij}^p, \quad \dot{\varepsilon}_{ij}^p = \dot{e}_{ij}^p + \delta_{ij}\dot{\varepsilon}_{kk}^p/3, \quad (6)$$

where $\dot{\varepsilon}_{ij}^e$ are the components of the elastic strain rate tensor, $\dot{\varepsilon}_{ij}^p$ are the components of the inelastic strain rate tensor.

The bulk inelastic strain rate is described by relation:

$$\dot{\varepsilon}_{kk}^p = \dot{f}_{growth}/(1 - f), \quad (7)$$

where f is the damage parameter, the substantial time derivative is denoted via dot notation.

The bulk inelastic strain rate ε_{kk}^p is equal to zero only when the material is undamaged.

$$\sigma_{ij} = \sigma_{ij}^{(m)}\varphi(f), \quad \sigma_{ij}^{(m)} = -p^{(m)}\delta_{ij} + S_{ij}^{(m)}, \tag{8}$$

$$p^{(m)} = p_x^{(m)}(\rho) + \Gamma(\rho)\rho E_T, \quad E_T = C_p T,$$
$$p_x^{(m)} = \frac{3}{2}B_0 \cdot \left((\rho_0/\rho)^{-7/3} - (\rho_0/\rho)^{-5/3}\right)\left[1 - \frac{3}{4}(4 - B_1) \cdot \left((\rho_0/\rho)^{-2/3} - 1\right)\right] \tag{9}$$

$$DS_{ij}^{(m)}/Dt = 2\mu(\dot{\varepsilon}_{ij}^e - \delta_{ij}\dot{\varepsilon}_{kk}^e/3), \quad \dot{e}_{ij}^p = \lambda\partial\Phi/\partial\sigma_{ij}, \tag{10}$$

the function $\varphi(f)$ establishes a relation between the effective stresses of the damaged medium and the stresses in the condensed phase, Γ is the Grüneisen coefficient, ρ_0 is the initial mass density of the condensed phase of the alloy, γ_R, ρ_R, n, B_0, B_1 are the material's constants, C_p is the specific heat capacity, $D(\cdot)/Dt$ is the Jaumann derivative, μ is the shear modulus, $\dot{f}_{growth}$ is the void growth rate, f is the void volume fraction in the damaged medium, λ is the plastic multiplier derived from the consistency condition $\dot{\Phi} = 0$, and Φ is the plastic potential. The plastic potential was described using the Gurson–Tvergaard model (GTN) [32, 37–39].

The Grüneisen coefficient Γ was equal to 1.42 and 1.09 for Mg–3Al–1Zn and Ti–5Al–2.5Sn, respectively.

The function $\varphi(f)$ takes the form of $(1 - f)$ for pressure and is implicitly defined for the deviatoric stress tensor [40].

The temperature rise associated with energy dissipation during plastic flow can be evaluated by relation [25, 32]:

$$T = T_0 + \int_0^{\varepsilon_{eq}^p} (\beta/\rho C_p)\sigma_{eq}d\varepsilon_{eq}^p, \tag{11}$$

where T_0 is the initial temperature and $\beta \sim 0.9$ is the parameter representing a fraction of plastic work converted into heat.

The specific heat capacity for Ti–5Al–2.5Sn titanium was calculated by the phenomenological relations within the temperature range 293–1115 K [32]:

$$C_p = 248.389 + 1.53067T - 0.00245T^2 \text{ (J/kg K)} \quad \text{for } 0 < T < T_{\alpha\beta} = 1320\,\text{K}, \tag{12}$$

The temperature dependence of the shear modulus for alpha titanium alloy was described by the equation:

$$\mu(T) = 48.66 - 0.03223T \text{ (GPa)}, \quad (273K < T < 1200\,\text{K}). \tag{13}$$

The flow stress was described by equation:

$$\sigma_s = \sigma_{s0} \exp\left\{C_1\sqrt{(1 - T/T_m)}\right\} + C_2\sqrt{1 - \exp\{-k_0\varepsilon_{eq}^p\}}$$
$$\exp\{-C_3T\} \exp\{C_4T \ln(\dot{\varepsilon}_{eq}/\dot{\varepsilon}_{eq0})\}, \tag{14}$$

where $\dot{\varepsilon}_{eq} = [(2/3)\dot{\varepsilon}_{ij}\dot{\varepsilon}_{ij}]^{1/2}$, $\dot{\varepsilon}_{eq0} = \gamma_1 \exp\{-T/\gamma_2\} + \gamma_3$, $\varepsilon_{eq}^p = \int_0^t \dot{\varepsilon}_{eq}^p dt$ is the equivalent plastic strain $\gamma_1 = 2115.08615\,\text{s}^{-1}$, $\gamma_2 = 38.26589\,\text{K}$, $\gamma_3 = 9.82388 \times 10^{-5}\,\text{s}^{-1}$ and T_m is the melting temperature, $\dot{\varepsilon}_{eq0} = 1.0\,\text{s}^{-1}$.

$$\sigma_s = \sigma_{s0} + C_5(\varepsilon_{eq}^p)^{n_1} + k_{hp}d_g^{-1/2} - C_2 \exp\{-C_3T + C_4T \ln(\dot{\varepsilon}_{eq}/\dot{\varepsilon}_{eq0})\}, \tag{15}$$

where $\sigma_{s0}, C_5, n_1, k_{hp}, C_2, C_3, C_4$ are the material parameters, d_g is the average grain size.

The impact of grain size distribution on the stress of hcp alloys with average grain sizes in single-mode and bimodal microstructures was taken into account in Eq. (15) by analogy with the Hall-Petch relation.

Material parameters of Ti–5Al–2.5Sn and Mg–3Al–1Zn are shown in Tables 1 and 2, respectively.

The influence of damage on the flow stress was taken into account using the Gurson–Tvergaard model [32, 36, 38]:

$$(\sigma_{eq}^2/\sigma_s^2) + 2q_1 f^* \cosh(-q_2 p/2\sigma_s) - 1 - q_3(f^*)^2 = 0, \tag{16}$$

where σ_s is the yield stress and q_1, q_2 and q_3 are the model parameters, $\sigma_{eq} = \sqrt{\frac{3}{2}\sigma_{ij}\sigma_{ij} - \frac{1}{2}\sigma_{kk}\sigma_{kk}}$.

$$\dot{f} = \dot{f}_{nucl} + \dot{f}_{growth},$$
$$\dot{f}_{nucl} = \dot{\varepsilon}_{eq}^p (f_N/s_N) \exp\{-0.5[(\varepsilon_{eq}^p - \varepsilon_N)/s_N]^2\},$$
$$\dot{f}_{growth} = (1 - f)\dot{\varepsilon}_{kk}^p, \tag{17}$$

Table 1 Material parameters of Eq. (14)

Parameter	σ_{s0} (GPa)	C_1	C_2 (GPa)	C_3 (K^{-1})	C_4 (K^{-1})	k_0	T_m (K)
Ti–5Al–2.5Sn	0.02	3.85	0.56	0.0016	0.00009	8.5	1875

Table 2 Material parameters of Eq. (15)

Coefficients	σ_{s0} (GPa)	k_{hp} (Pa nm$^{1/2}$)	C_2 (GPa)	C_3 (K^{-1})	C_4 (K^{-1})	C_5 (GPa)	n_1
Mg–3Al–1Zn	0.141	6.2	0.315	0.0029	0.000389	0.505	0.2514

Table 3 Dimensionless parameters for the Gurson–Tvergaard–Needleman (GTN) model for alpha titanium and magnesium alloys

	q_1	q_2	q_3	f_0	f_N	f_C	f_f	ε_N	S_N
Ti–5Al–2.5Sn	1.3	1	1.69	0.00	0.2	0.035	0.4	0.28	0.1
Mg–3Al–1Zn	1.5	1	2.25	0.00	0.4	0.073	0.128	0.1	0.2

where ε_N and S_N are the average nucleation strain and the standard deviation, respectively. The amount of nucleating voids is controlled by the parameter f_N

$$f^* = f \ \text{for} \ f \leq f_c;$$
$$f^* = f_c + (\bar{f}_F - f_c)/(f_F - f_c) \ \text{for} \ f > f_c, \quad (18)$$

where $\bar{f}_F = (q_1 + \sqrt{q_1^2 - q_3})/q_3$.

The final stage in ductile fracture comprises the voids coalescence [38]. This causes softening of the material and accelerated growth rate of the void fraction f^*.

The model parameters for Ti–5Al–2.5Sn and Mg–3Al–1Zn were determined using numerical simulation. These parameters are given in Table 3.

We use the ductile fracture criteria for alloys at the room and elevated temperatures owing to relatively low melting temperature [31, 38, 41].

Finite elements are removed from the grid model and free boundary conditions are introduced at the formed boundary, when the local fracture criterion is met.

Specific features of mechanical properties of nanostructured materials are connected to distinctions of matter properties in a crystalline phase of grains and in the boundary of grains. A decrease in the mass density of nanostructured materials is caused by increased defect's density and relative volume of grain boundary phase.

3 Results and Discussion

Figure 2 shows the fields of equivalent plastic strain under tension at $v_x = 2.3$ m/s. Damages were localized near the grain boundary of coarse grains.

Damage nucleation in alloys with a bimodal grain size distribution occurs in shear bands and zones of their intersection.

These results agree with experimental observations of strain localization and fracture of titanium alloys by Sharkeev [42], authors of this work [32, 43], Valoppi [44] and Zheng [45]. It is significant that increases in fine precipitates concentration in alloys caused the increase in resistance to plastic flow within both coarse and ultrafine grains.

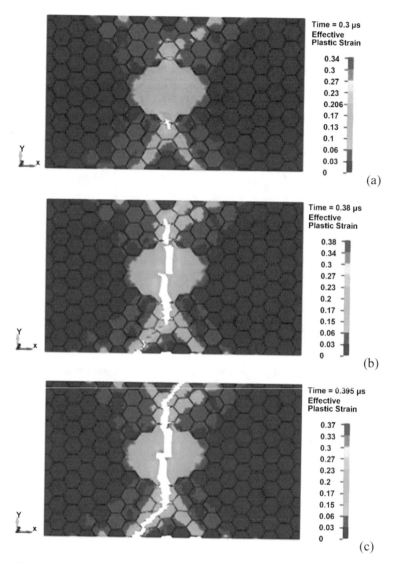

Fig. 2 Effective plastic strain field at time, **a** 0.332 μs, **b** 0.3895 μs, **c** 0.398 μs in Ti–5Al–2.5Sn alloy with bimodal grain structure

Thus, the fracture of alloys with bimodal grain structures is caused by the damage nucleation at the boundary of coarse grains with an ultrafine-grained structure and further growth of damage in mesoscopic bands of localized plastic deformation.

Thus, the process of fracture of alloys with bimodal grain structures is associated with the formation of mesoscopic bands of localized plastic deformation.

The segregation of impurity atoms in the grain-boundary phase affects the formation of plastic shear bands [14, 46].

The mechanical properties of the grain boundary phase were varied to take into account the effect of segregation of impurity atoms on the yield stress in the simulation.

Figure 2b shows the field of equivalent strains, indicating that the formation of cracks in large grains can be accompanied by a change in the orientation of the shear localization bands at the mesoscopic level. This occurs as a result of a change in the orientation of the plane of maximum shear stresses during the evolution of a triaxial stress state near cracks. A macroscopic crack obtained by modeling the tension of flat samples has the same configuration as the cracks observed in experimental studies by Verleysen [24] and authors of this work [31, 32].

The averaged strain along the axis of tension of the computational domain at the moment of crack crossed the representative volume was interpreted as the ultimate strain to fracture of the alloy.

Figure 3 shows the dependence of strain to fracture of titanium and magnesium alloys on the logarithm of strain rates. Experimental data reported by Ulacia for coarse-grained Mg–3Al–1Zn alloy are marked by filled square symbols [11]. The experimental data for the Ti–5Al–2.5Sn alloy are shown by filled triangular symbols [32, 47].

Strains to fracture behave nonmonotonically and nonlinearly with increasing strain rate in the range from 0.001 to 1000 1/s as shown in Fig. 3 (Curves 1 and 3). In coarse-grained HCP alloys, this is related to more intense twinning under dynamic loading.

Curve 2 is obtained by the approximation of calculated values of the strains to fracture under tension of a magnesium alloy with a bimodal grain structure with a specific volume of large grains of 70%. With an increase in the concentration of micron and

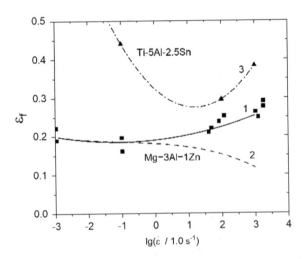

Fig. 3 The strain to fracture versus logarithm of strain rates for magnesium and titanium alloys with a unimodal (curve 2, 3) and a bimodal grain sizes distribution (curve 1, the concentration of coarse grains is 70%)

submicron grains in the alloy, the strain to failure under tension decreases nonlinearly. The simulation results agree with the available experimental data [11, 45–50].

The strain to fracture of alloy with bimodal grain size distribution versus specific volume of coarse grains under quasi-static tension can be described by the relation:

$$\varepsilon_f^n = 0.01 \exp(C_{cg}/0.363), \tag{19}$$

where ε_f^n is the strain to fracture under quasi-static tension, C_{cg} is the specific volume of coarse grains.

Equation (19) describes the ductility of alloys with a bimodal grains distribution versus the specific volume of coarse grains. The increase in ductility of HCP alloys under quasi-static tension occurs when a specific volume of coarse grains is greater than 30%.

4 Conclusions

The multiscale approach was used in the computer simulation of fracture of magnesium and titanium alloys at high strain rates. Structured RVEs were proposed to predict the mechanical properties of alloys taking into account the grain size distribution and the segregation impurity atoms in the grain-boundary.

The results of computer simulation showed that damage nucleation in alloys with a bimodal grain size distribution occurs in the shear bands and their intersection zones. Damage arises at the boundary between coarse- and ultrafine-grained structures. Further damage growth occurs in the mesoscopic bands of localized plastic deformation.

Thus the computer simulation can be used to estimate the influence of grains size distribution on the dynamic strength and ductility of HCP alloys.

Localization of plastic flow in HCP alloys with bimodal grain size distribution under tension at high strain rates depends on the ratio between volume concentrations of fine and coarse grains. As a result, the strain to fracture of hcp alloys with bimodal grain size distribution varies nonlinearly with tensile strain rate in the range from 0.001 to 1000 1/s.

The dynamic ductility of HCP alloys with bimodal grain size distribution is increased when a specific volume of coarse grains is greater than 30%.

References

1. Fall A, Monajati H, Khodabandeh A, Fesharaki MH, Champliaud H, Jahazi M (2019) Local mechanical properties, microstructure, and microtexture in friction stir welded Ti-6Al-4V alloy. Mater Sci Eng: A 749:166–175. https://doi.org/10.1016/j.msea.2019.01.077

2. Guo D, Zhang Z, Zhang G, Li M, Shi Y, Ma T, Zhang X (2014) An extraordinary enhancement of strain hardening in fine-grained zirconium. Mater Sci Eng, A 591:167–172

3. Jiang L, Pérez-Prado MT, Gruber PA, Arzt E, Ruano OA, Kassner ME (2008) Texture, microstructure and mechanical properties of equiaxed ultrafine-grained Zr fabricated by accumulative roll bonding. Acta Mater 56(6):1228–1242

4. Toth LS, Gu C (2014) Ultrafine-grain metals by severe plastic deformation Mater. Mater Charact 92:1–14

5. Wang YL, Hui SX, Liu R, Ye WJ, Yu Y, Kayumov R (2014) Dynamic response and plastic deformation behavior of Ti-5Al-2.5Sn ELI and Ti-8Al-1Mo-1V alloys under high-strain rate. Rare Met 33:127–133. https://doi.org/10.1007/s12598-014-0238-y

6. Long Y, Wang T, Zhang HY, Huang XL (2014) Enhanced ductility in a bimodal ultrafine-grained Ti–6Al–4V alloy fabricated by high energy ball milling and spark plasma sintering. Mater Sci Eng, A 608:82–89

7. Berbenni S, Favier V, Berveiller M (2007) Impact of the grain size distribution on the yield stress of heterogeneous materials. Int J Plast 23:114–142. https://doi.org/10.1016/j.ijplas.2006. 03.004

8. Chong Y, Deng G, Gao S, Yi J, Shibata A, Tsuji N (2019) Yielding nature and Hall-Petch relationships in Ti-6Al-4V alloy with fully equiaxed and bimodal microstructures. Scripta Mater 172:77–82. https://doi.org/10.1016/j.scriptamat.2019.07.015

9. Pozdnyakov VA (2007) Ductility of nanocrystalline materials with a bimodal grain structure. Tech Phys Lett 33:1004–1006

10. Malygin GA (2008) Strength and plasticity of nanocrystalline metals with a bimodal grain structure. Phys Solid State 50:1032–1038

11. Ulacia I, Salisbury CP, Hurtado I, Worswick MJ (2001) Tensile characterization and constitutive modeling of AZ31B magnesium alloy sheet over wide range of strain rates and temperatures. J Mater Process Technol 211:830–839

12. Raeisinia B, Sinclair CW, Poole WJ, Tome CN (2008) On the impact of grain size distribution on the plastic behaviour of polycrystalline metals. Model Simul Mater Sci Eng 16(2):025001. https://doi.org/10.1088/0965-0393/16/2/025001

13. Revil-Baudard B, Cazacu O, Flater P, Chandola N, Alves JL (2016) Unusual plastic deformation and damage features in titanium: experimental tests and constitutive modeling. J Mech Phys Solids 88:100–122. https://doi.org/10.1016/j.jmps.2016.01.003

14. Aksyonov DA, Lipnitskii AG, Kolobov YuR (2013) Grain boundary segregation of C, N and O in hcp titanium from first-principles. Model Simul Mater Sci Eng 21:075009. https://doi.org/ 10.1088/0965-0393/21/7/075009

15. Skripnyak VA (2012) Mechanical behavior of nanostructured and ultrafine-grained materials under shock wave loadings. Experimental data and results of computer simulation. Shock compression of condensed matter. AIP Conf Proc 1426:965–970. https://doi.org/10.1063/1. 3686438

16. Skripnyak VA, Skripnyak EG (2017) Nanomechanics. Rijeka (Cratia): InTechOpen. (Article: Mechanical behaviour of nanostructured and ultrafine-grained metal alloy under intensive dynamic loading. Chapter 2: Nanotechnology and Nanomaterials). https://doi.org/10.5772/ intechopen.68291. Print ISBN 978-953+-51-3181-6

17. Skripnyak VA, Skripnyak NV, Skripnyak EG, Skripnyak VV (2017) Influence of grain size distribution on the mechanical behavior of light alloys in wide range of strain rates. AIP Conf Proc 1793:110001. https://doi.org/10.1063/1.4971664

18. Zhai J, Luo T, Gao X, Graham SM, Baral M, Korkolis YP, Knudsend E (2016) Modeling the ductile damage process in commercially pure titanium. Int J Solids Struct 91:26–45. https:// doi.org/10.1016/j.ijsolstr.2016.04.031

19. Zhao QY, Yang F, Torrens R, Bolzoni L (2019) In-situ observation of the tensile deformation and fracture behaviour of powder-consolidated and as-cast metastable beta titanium alloys. Mater Sci Eng, A 750:45–59. https://doi.org/10.1016/j.msea.2019.02.037

20. Zhu L, Shi S, Lu K, Lu J (2012) A statistical model for predicting the mechanical properties of nanostructured metals with bimodal grain size distribution. Acta Mater 60:5762–5772

21. Zhu L, Lu J (2012) Modelling the plastic deformation of nanostructured metals with bimodal grain size distribution. Int J Plast 30–31:166–184
22. Clayton JD (2005) Dynamic plasticity and fracture in high density polycrystals: constitutive modeling and numerical simulation. J Mech Phys Solids 53:261–301
23. Magee AC, Ladani L (2015) Representation of a microstructure with bimodal grain size distribution through crystal plasticity and cohesive interface modeling. Mech Mater 82:1–12
24. Verleysen P, Peirs J (2017) Quasi-static and high strain rate fracture behaviour of Ti6Al4V. Int J Impact Eng 108:370–388. https://doi.org/10.1016/j.ijimpeng.2017.03.001
25. Armstrong RW, Zerilli FJ (1994) Dislocation mechanics aspects of plastic instability and shear banding. Mech Mater 17(2–3):319–327. https://doi.org/10.1016/0167-6636(94)90069-8
26. Döner M, Conrad H (1975) Deformation mechanisms in commercial Ti-5Al-2.5Sn (0.5 At. pct Oeq) alloy at intermediate and high temperatures (0.3-0.6Tm). Metall Trans A 6:853–861. https://doi.org/10.1007/BF02672308
27. Frost HJ, Ashby MF (1982) Deformation-mechanism maps. Pergamon Press, Oxford
28. Salem AA, Kalidindi SR, Doherty RD (2003) Strain hardening of titanium: role of deformation twinning. Acta Mater 51:4225–4237. https://doi.org/10.1016/s1359-6454(03)00239-8
29. Wang Q, Ren J, Wang Y, Xin C, Xiao L, Yang D (2019) Deformation and fracture mechanisms of gradient nanograined pure Ti produced by a surface rolling treatment. Mater Sci Eng, A 754:121–128. https://doi.org/10.1016/j.msea.2019.03.080
30. Herzig N, Meyer LW, Musch D, Halle T, Skripnyak VA, Skripnyak EG, Razorenov SV, Krüger L (2008) The mechanical behaviour of ultrafine grained titanium alloys at high strain rates. In: Proceedings of 3rd international conference on high speed forming—11–12 Mar 2008, Dortmund, Germany, pp 141–150. https://doi.org/10.17877/de290r-8660
31. Skripnyak VV, Kozulyn AA, Skripnyak VA (2019) The influence of stress triaxiality on ductility of α titanium alloy in a wide range of strain rates. Mater Phys Mech 42:415–422. https://doi.org/10.18720/MPM.4242019_6
32. Skripnyak VV, Skripnyak EG, Skripnyak VA (2020) Fracture of titanium alloys at high strain rates and under stress triaxiality. Metals 10(3):305. https://doi.org/10.3390/met10030305
33. Gao PF, Qin G, Wang XX, Li YX, Zhan M, Li GJ, Li JS (2018) Dependence of mechanical properties on the microstructural parameters of TA15 titanium alloy with tri-modal microstructure. Mater Sci Eng, A 739:203–213. https://doi.org/10.1016/j.msea.2018.10.030
34. Cayron C, Artaud B, Briottet L (2006) Reconstruction of parent grains from EBSD data. Mater Charact 57(4–5):386–401
35. Hémery S, Villechaise P (2019) In situ EBSD investigation of deformation processes and strain partitioning in bi-modal Ti-6Al-4V using lattice rotations. Acta Mater 171:261–274. https://doi.org/10.1016/j.actamat.2019.04.033
36. Shi Y, Li M, Guo D, Ma T (2013) Tailoring grain size distribution for optimizing strength and ductility of multi-modal Zr. Mater Lett 108:228–230
37. Needleman A, Tvergaard V, Bouchaud E (2012) Prediction of ductile fracture surface roughness scaling. J Appl Mech 79(3):031015. https://doi.org/10.1115/1.4005959
38. Springmann M, Kuna M (2005) Identification of material parameters of the Gurson–Tvergaard–Needleman model by combined experimental and numerical techniques. Comput Mater Sci 32:544–552. https://doi.org/10.1016/j.commatsci.2005.02.002
39. Tvergaard V (2015) Study of localization in a void-sheet under stress states near pure shear. Int J Solids Struct 60–61:28–34. https://doi.org/10.1016/j.ijsolstr.2015.08.008
40. Heibel S, Nester W, Clausmeyer T, Tekkayan AE (2017) Failure assessment in sheet metal forming using a phenomenological damage model and fracture criterion: experiments, parameter identification and validation. Procedia Eng 207:2066–2071. https://doi.org/10.1016/j.proeng.2017.10.1065
41. Orozco-Caballero A, Li F, Esqué-de los Ojos D, Atkinson MD, Quinta da Fonseca J (2018) On the ductility of alpha titanium: the effect of temperature and deformation mode. Acta Mater 149:1–10. https://doi.org/10.1016/j.actamat.2018.02.022
42. Sharkeev YP, Vavilov VP, Belyavskaya OA, Skripnyak VA, Nesteruk DA, Kozulin AA, Kim VM (2016) Analyzing deformation and damage of VT1-0 titanium in different structural states

by using infrared thermography. J Nondestr Eval 35(3):42. https://doi.org/10.1007/s10921-016-0349-5

43. Yoo VH (1981) Slip, twinning, and fracture in hexagonal close-packed metals. Metall Trans A 12:409–418

44. Valoppi B, Bruschi S, Ghiotti A, Shivpur R (2017) Johnson–Cook based criterion incorporating stress triaxiality and deviatoric effect for predicting elevated temperature ductility of titanium alloy sheets. Int J Mech Sci 123:94–105. https://doi.org/10.1016/j.ijmecsci.2017.02.005

45. Zheng G, Tang B, Zhou Q, Mao X, Dang R (2020) Development of a flow localization band and texture in a forged near-α titanium alloy. Metals 10:121. https://doi.org/10.3390/met100 10121

46. Raabe D, Herbig M, Sandlöbes S, Li Y, Tytko D, Kuzmina M, Ponge D, Choi P-P (2014) Grain boundary segregation engineering in metallic alloys: a pathway to the design of interfaces. Curr Opin Solid State Mater Sci 18(4):253–261

47. Zhang B, Wang J, Wang Y, Wang Y, Li Z (2019) Strain-rate-dependent tensile response of Ti–5Al–2.5Sn alloy. Materials 12:659. https://doi.org/10.3390/ma12040659

48. Liang H, Pan FS, Chen YM, Yang JJ, Wang JF, Liu B (2011) Influence of the strain rates on tensile properties and fracture interfaces for Mg-Al alloys containing Y. Adv Mater Res 284–286:1671–1677. https://doi.org/10.4028/www.scientific.net/amr.284-286.1671

49. Wei K, Zeng X, Huang G, Deng J, Liu M (2019) Selective laser melting of Ti-5Al-2.5Sn alloy with isotropic tensile properties: The combined effect of densification state, microstructural morphology, and crystallographic orientation characteristics. J Mater Process Technol 271:368–376. https://doi.org/10.1016/j.jmatprotec.2019.04.003

50. Yang L, Zhicong P, Ming L, Yonggang W, Di W, Changhui S, Shuxin L (2019) Investigation into the dynamic mechanical properties of selective laser melted Ti-6Al-4V alloy at high strain rate tensile loading. Mater Sci Eng, A 745:440–449. https://doi.org/10.1016/j.msea.2019.01.010

Self-reproduction Cycles of Living Matter and Energetics of Human Activity

Leonid E. Popov

Abstract In the author's opinion, many global problems that face humanity—in the fields of education, medicine, management etc. can be tackled more effectively if the cyclic nature of self-reproducing systems—including living beings—is taken into account. Summarizing the main physiological findings of the last decades on "adaptation reactions", one can very roughly say that the way of action which is *effective* in the sense of productive activity of people happens at the same time to be *healthy*, and it gives the participants of the process the feeling of *happiness*. The present paper represents a very short overview of the contemporary concepts of the adaptation reactions based on the fundamental understanding of their cyclic nature due to general properties of self-reproducing systems. One interesting feature of self-reproduction cycles is its first "phase of orientation" which was not discussed in detail in the past but plays a key role in the whole cycle.

Keywords Self-reproducing systems · Life · Adaptation reactions · Activation · Training · Stress · Health · Memory · Catabolic phase · Anabolic phase

1 Introduction

Any activity of living systems is cyclic at all levels of the organization of life. The cyclic character of activity of biological systems is associated not only with the periodic nature of the action of external factors on it (such as the rotation of the Earth around its axis, its movement around the Sun or the cyclic change in solar activity). The above mentioned cycles are the result of adaptation of living systems to periodic changes in the environment. However, life is cyclic in itself, by virtue of its very essence. Let us first briefly discuss this fundamental aspect of life which is essential for the following discussion.

How does "life" differ from inert matter? This question was put forward many times in the history of science. In retrospect, we can state that there is one point where

L. E. Popov (✉)
Tomsk State University of Architecture and Construction, Pl. Solyanaya 2, 634003 Tomsk, Russia
e-mail: l.popov@vemn.de

© The Author(s) 2021
G.-P. Ostermeyer et al. (eds.), *Multiscale Biomechanics and Tribology of Inorganic and Organic Systems*, Springer Tracts in Mechanical Engineering,
https://doi.org/10.1007/978-3-030-60124-9_24

many researchers do agree: the essential property of living systems is their "reproduction". Let us consider this property closer. Chilean neuroscientists H. Maturana and F. Varela provide the following explanation: "When we speak of living beings, we presuppose something in common between them; otherwise we wouldn't put them in the same class we designate with the name 'living'. … Our proposition is that living beings are characterized in that, literally, they are continually self-producing. We indicate this process when we call the organization that defines them an *autopoietic organization*.

What is distinctive about… <living beings> … is that their organization is such that their only product is themselves, with no separation between producer and product. The being and doing of an autopoieic unity are inseparable, and this is their specific mode of organization" [1].

So living matter only exists due to continuously reproducing itself. The word "self-reproduction" very precisely reflects the meaning of the basic characteristic of all life: for all interactions with the environment, the living beings remain "themselves", although in the process of interaction parameters of a living system can significantly deviate from their values at steady state. The deviation is followed by a return to the stationary state of "stable non-equilibrium".

Self-reproduction means turning back to oneself. It would seem that there is no other interpretation here. But in the popular, as well as in scientific literature, self-reproduction is very often considered a synonym for reproduction of generations, and when speaking of "self-reproduction cycles", one usually means a cycle of producing a new generation. However, this aspect means the reproduction of population and the biological species as a whole, while each living unit reproduces itself continuously during its whole life. In the present paper, we focus of this—more unusual, but in reality central aspect of living things—their continuous self-reproduction during the course of individual life.

The process of individual self-reproduction is also cyclical in nature: after deviating from a stationary state, the system returns to the same state. Ugolev [2] formulated the principle of cyclic activity among the "principles of natural technologies of biological systems": "At all levels of organization (from cellular to planetary), biological systems (more precisely, processes in them) are partially or completely cyclized… The principle of recurrence is one of the most important principles to ensure maximum efficiency and effectiveness of living systems through the multiple use of the same structures. Cyclization also ensures coordination of all components, implementing a multi-stage process. Many process systems considered by us as linear will be later characterized as cyclic…".

2 Three Phases of the Self-reproduction Cycle

Thus, our life activity is discrete, and each cycle proceeds in phases. Let us first analyze the general structure of each such a cycle. Each behavioral act begins with the appearance in our mind of an *image of the result of our action* [3]. The formation of this image is an elementary event of prognostic activity of the nervous system. We can talk about the prognostic phase of the cycle. The two subsequent steps are obvious. The second phase is the *action* itself, leading to the achievement of the result using the functional system that was formed at the first stage. During this phase, the work is performed to achieve the desired (predicted) result, and this work is associated with the expenditure of free energy of the body. Finally, the third phase is the *return of the organism to the steady state of stable disequilibrium*. During this phase, the body recovers the energy and possibly structural costs related to the work done in the second phase.

We note two points here. Firstly, the formation of an image of the result of an organism's action is associated with the extraction of some information stored in long-term memory in the central nervous system. In addition, the formation of a functional system corresponding to this image requires preparation for the action of executive bodies. That is, the entire first phase of the cycle is associated with the processes of structure formation, with the processes of *anabolism*.

In the second phase, work is performed that is accompanied by the dissipation of free energy of the body, that is, with the cost of reserves of energy and, possibly, structure. Thus, the second phase is a *catabolic* one. The cost of energy and body structure is normally restored *excessively* in the third phase. Moreover, the processes of anabolism occur only in the structures that took part in the action.

Phenomenologically, the existence of the prognostic phase, during which the action is still absent, was of course known for a very long time, in various contexts, for example under the name of "keeping a clear head". "Keeping a clear head" in an acute situation does not mean not acting at all. "Clear head" refers first of all to the initial phase of orientation and forecast, without which the active action turns into "action nowhere." The absence of a phase of the formation of the image of the result of the action would lead to a chaotic, senseless and potentially dangerous activity for the body, the "panic". The physiological reaction characteristics of this first phase of the behavioral cycle have also been repeatedly described: The heart first "stops" (that is, the pulse drops sharply), and only then begins to "beat wildly". This physiological reaction can be found in such idiomatic expressions as a "heart-stopping event" (something that is very impressive or exciting).

However, it was this first phase of the cycle that was last to be discovered and described in the framework of physiological studies. A single cycle of physiological self-reproduction was studied by Arshavsky with collaborators in the laboratory of age-related physiology and pathology of animals and humans at the Institute of Experimental Medicine since 1935. By this time, researchers have already come to the understanding that living time is discrete (although it is perceived as flowing continuously): it is measured by metabolic cycles, which, as it was originally thought, consist of two phases: catabolic and anabolic. In the catabolic phase, motor activity is carried out, and work is done due to the energy of the destruction of cellular substrates. In the anabolic phase, the disturbed structure is restored, and recovery occurs to an extent exceeding the costs in the second phase. This transfers the body to a higher than the initial level of disequilibrium and ordering.

Bearing in mind this excessive anabolism, I. A. Arshavsky called the metabolic cycle associated with a single motor activity a hypercycle.

But only in 1968 Arshavsky experimentally established the presence of another phase of the cycle: the "phase of the starting metabolism". Studies were conducted on the human fetus: through the wall of the abdomen of the mother, the fetal heartbeat was registered. It was expected that with recurrent "spontaneous" increase in activity would be associated increasing of the heart rate (catabolic processes occur with the dominance of the sympathetic nervous system that increases the heart rate), and then of slowing down (anabolism, dominance of the parasympathetic nervous system). It was found, however, that before each motor activity, the heartbeat slowed down (parasympathetic reaction, anabolism). After that, extensive research on puppies was undertaken during the process of prenatal development. On the obtained cardiograms, you can see all three phases associated with motor activity: a few rarer pulses (the starting anabolism phase), a rapid heartbeat during motor activity (catabolic phase) and a decrease in heart rate after the cessation of motor activity (phase of excessive anabolism).

In the fundamental work "Physiological mechanisms and regularities of individual development", where the results of research performed in the laboratory of age physiology and pathology in the 1930–1980s were summarized, I. A. Arshavsky gives the following definition of the adaptive response of the organism:

> Adaptation is the response of physiological-morphological transformation of the organism and its parts resulting in an increase of its structure-energy potential, that is, free energy and potential for action. This response is induced by such irritants of the environment which can be referred to as physiological, although they do demand the expenditure of an amount of energy. We call these irritants physiological because the energy expenditure resulting from them is compensated by its acquisition, that is by the functional induction of excess anabolism, automatically induced by each activity. As a consequence, a spiral-like transition of a developing organism to a higher level of potential working ability takes place. This adaptive response is characterized by three phases: the first phase is anabolic, the second one catabolic, giving the possibility of realizing another activity, and the third phase excessively anabolic. Speaking in terms of thermodynamics, we can interpret the above definition of adaptation as a response resulting in an increase of the free energy of the organism and … the state of stable nonequilibrium… [4].

Note that the phase of excessive anabolism occurs only after activity. If there is no activity, then there is no development (and not even just recovery, since non-functioning cells self-eliminate by means of apoptosis mechanisms; an inactive organism degrades). It is in the phase of excessive anabolism that all development processes take place, in particular, the processes of long-term memory that stores our life experience. "…All motor activity, beginning with a zygote, is a factor of induction of anabolism, no matter if the motion is stimulated endogenously—due to the necessity of satisfying the demand for food, or exogenously—due to an irritant action of physiological stress. The purpose of this anabolism is not just recovery of the initial state, but necessarily excessive recovery. This transfers the system… to a new level, at which its internal energy increases… In the case of nervous cells it is manifested in trace hyperpolarization and … in an increase in RNA content. For muscle cells it is trace hyperpolarization, trace hyperrelaxation, and an increase in protein content. It was found that … anabolic excess is realized only due to activity. Experimental blocking of activity causes, in spite of continuing food intake, to a temporary stop or inhibition of growth and development, or even its stasis" [4].

Though I. A. Arshavsky, speaking of growth and development, means a growing organism, including the prenatal stage of its development, all of the above concerning the obligatoriness of function for excessive anabolism refers to any living creature independent of its age. In adult individuals, growth and development continue during their entire life in the form of expansion of the volume of long-term memory. The factors controlling the development of the nervous system of a growing organism and the processes within which this development is realized, are concentrated in the brain structures in adulthood, where new distributed multilevel systems of local neuron networks are formed (neocortex columns) in response to a sequence of signals about subjectively new realities of the environment. At present, the complete unity of the biochemical processes of development of the nervous system and of memory formation has been proven.

3 Graphical Representation of Self-reproduction (Adaptation) Cycles

Following [5], let us illustrate the above graphically. When we set ourselves the task of graphically representing the normal physiological adaptive response as a cyclic process, the value of structural energy of the organism (F) is naturally assumed to be on the ordinate axis. It is suitable to quote here a comment made by I. A. Arshavsky concerning this quantity:

> …the most essential qualitative feature of living systems is their state of nonequilibrium, steadily supported by the work of the structures comprising it—the work directed against their transition to the equilibrium state. The chemical energy of food substances entering the living systems can not directly transform into work. The energy, first and foremost, is used for producing free energy in the structure of the living system. In living systems, in contrast to nonliving ones, neither thermal nor chemical energy is the source of the work being performed by them, as is the case with thermo- and chemodynamical engines; the structural energy of the organism is responsible for that. The latter, in contrast to mechanical, thermal, chemical, electromagnetic, gravitational and nuclear energy is a specific form of energy inherent only in living systems [4].

Figure 1a illustrates the changes of the free structural energy during one typical cycle of self-reproduction. Point A corresponds to the moment when an action on the organism requiring its response took place, point B—the time of completion of the response, point C—completion of the (excessively anabolic) phase 3 of the adaptive response. The dashed line shows the minimum value of free energy of the organism when it performs the work related to its adaptive response to the irritant. The figure illustrates the case when this limit is not exceeded: the free energy of the organism at point B (F_B) is in excess of F_{min} ($F_B > F_{min}$). In phase 3, the organism returns

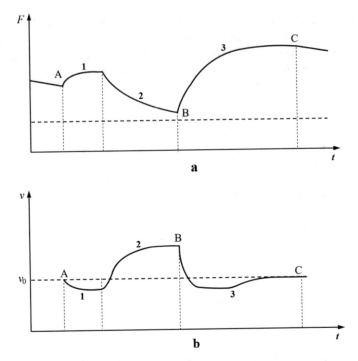

Fig. 1 a The change in the structural energy of an organism (F) with time (t) during an adaptive response (single cycle of self-reproduction). **b** The corresponding change in heart rate (v) with time (t)

to a state of rest (point C), with higher free energy F_C as compared with that of the initial state of rest.

Another important parameter of the cycle characterizing the deviation of the organism from the stable steady state is the heart rate (v). One argument in favor of this choice is the circumstance that a change in the performance of the cardiovascular system in the conditions of catabolism and anabolism occurs in opposite ways. In the catabolic phase, the heart rate increases. In contrast, a decrease in the heart rate (vagal bradycardia [6]) is typical of anabolism. "…Vagal bradycardia is a necessary correlate both of the first and of the third anabolic phases of the physiological response of the entire organism. For the heart, the periodically occurring vagal bradycardia plays the role of providing the growth and development of the heart muscle. This is achieved through an increase in the duration of the diastolic pause ensuring an excess of regeneration processes upon every heart diastole" [4].

Facing a new stimulus, the organism first shows a temporary decrease of the heart rate (anabolic reaction associated with the prognostic phase) which is then followed by an increase of the heart rate in the phase of action, Fig. 1b. After achieving the result, the heart rate falls enabling excessive anabolism and restoration of the structural energy and finally comes to the equilibrium value—however, already with an increased level of structural energy.

Arshavsky's adaptive response results in the increase of structural energy of the organism:

$$\Delta F = F_C - F_A > 0$$

and, consequently, an increase in its ability of performing work under following irritants of a similar (but not only) kind. I. A. Arshavsky referred to phase 3 of a normal physiological adaptive response as a phase of "excessive anabolism". This phase starts after performing a certain function at phases 1 and 2 by the organism and only in those structures which were involved in the realization of the function [4]. A dominating component of the adaptive response at these stages is motor activity both at the level of the whole organism (skeletal muscles), and at the cellular level [7].

4 Hyper Cycles of Self-reproduction

The cyclic nature becomes more visible if the free energy of the system is plotted as function of the heart rate, v.

In Fig. 2, Arshavsky's adaptive response is shown schematically in the coordinate system free energy (F)—heart rate (v). Due to excessive character of anabolic stage, this is an open cycle ("hyper cycle" according to I. A. Arshavsky's terminology).

In Fig. 2, v_0 is the heart rate in the state of rest of a wakeful organism; point A corresponds to the time of irritant action; point B to the completion of necessary

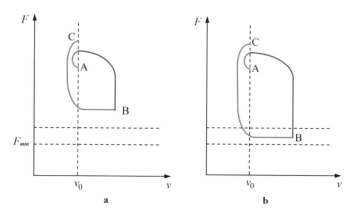

Fig. 2 Schematic of the adaptation response due to I. A. Arshavsky: F—structural energy, v—heart rate. **a** Normal physiological response, **b** a critical physiological response

responses; point C to the completion of the third, anabolic, phase of the response and thereby the entire adaptive response.

Point A is the point at which we are caught up in the need to act. Immediately there is a decrease in heart rate and an increase in free energy in the prognostic phase, the formation of a functional system that will provide the subsequent action. Then this functional system acts (the region indicated by the red line and ending at point B). The action is followed by recovery, but the recovery is excessive, so that the body reaches a higher level of free energy than at the beginning. F_{min} is the energy, below which one should not go. Why? The reason is that during the "action" phase, carbohydrates are used as energy supply. If one continues to act, ignoring the capabilities of the body, the body will have to switch from carbohydrate to lipid metabolism, and then to protein metabolism. Horizontal dashed lines mark the area where lipid-protein metabolism is starting, (this is actually a certain transition region rather than a boundary). Thus, a healthy organism can sometimes use a part of its necessary structural resources (as shown in Fig. 2b) but restore this damage completely during the following excessive anabolic stage.

Depending on the strength of external influences, this cycle can be realized in several forms. For example, if the "problem to solve" turned out to be very simple and a solution was found with a minimum expenditure of resources, then at the completion of the action, the energy will not fall below the initial one (Fig. 3a). After this, however, the recovery phase still follows. All this happens within carbohydrate metabolism. Stronger impacts have already been discussed above.

And finally, if the irritant is too strong than the body is forced to enter the sphere of lipid-protein energetics (Fig. 3b). At the end of the action, in this case, the pulse drops, but not to the equilibrium value, and after that it very slowly returns to normal. Only after this the reverse transition to carbohydrate metabolism proceeds. After this, a

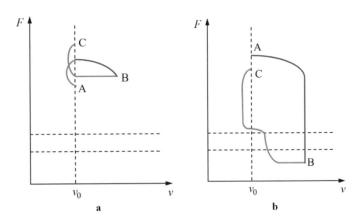

Fig. 3 Schematic of the adaptation response due to I. A. Arshavsky: F—structural energy, v—heart rate. **a** Response to weak stimuli (training reaction), **b** overcritical stimuli (pathogenic reaction—stress)

rapid decrease in pulse occurs, a transition to the stage of anabolism, which, however, may not be enough to completely restore the wasted structural energy.

5 Non-specific Adaptive Responses

The described cycles of self-reproduction are *specific* for each irritant. The particular nature of the irritant determines the image of the result of the action as well as the structures of the organism comprising the functional system formed in the phase 1 and the duration of the activity needed to achieve the result. However, not always the result can be achieved in one cycle so that several cycles may be required. In conditions of prolonged focused activity, a general *non-specific adaptive response* of the body is formed, controlled by the hypothalamus-hypophysis-adrenal gland system [4, 8, 9]. Depending on the strength of the stimulus, it may be a "training reaction" [8], an "activation reaction" [8] or an enhanced activation reaction [10]. Finally, in a situation where efforts to achieve the result are required that exceed the physiological capabilities of the body, a pathological response develops—stress [8, 11].

Apparently, one of the listed adaptation answers corresponds to the dominance of the long-term purposeful activity of each of the options for the elementary reproduction cycle described above.

The four non-specific adaptation reactions are summarized in Fig. 4, which shows the power of vital activity (consumption of structural energy per unit time) versus time.

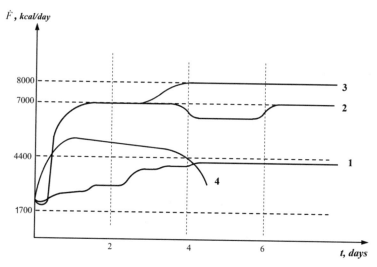

Fig. 4 Metabolic power characteristic for: 1: weak stimuli (training reaction), 2: average stimuli (activation reaction), 3: enhanced activation, 4: stress

Curve 1 corresponds to the case of weak stimuli (Fig. 3a). The maximum metabolic power developed by humans in the state of this "training response" to weak stimuli does not exceed 4400 kcal per day [12].

The second option corresponds to the so-called activation reaction [12]. Being in this state is very becoming to the organism. Changes in the immune system accompanying this response occur in the direction of increasing the organism's resistance, improving the immune systems. Without going into detail about the mechanisms of formation of the activation response ("physiological stress" according to I. A. Arshavsky), we should note that the adaptive responses of the organism are largely controlled by hormonal activity of the hypothalamus–hypophysis–adrenal-gland system. When the organism functions in the regime of the Arshavsky's adaptive response (activation response according to Ukolova [8]), the system releases anabolic agents into the bloodstream: growth-hormone-releasing factor of the hypothalamus, somatotropic hormone of the hypophysis, mineralcorticoids and testosterone of the adrenal gland, etc. In general, the processes of anabolism, recovery, and development dominate in the organism in this case. This is facilitated by the fact that the activation reaction is accompanied by a certain decrease in blood coagulability, which ensures high mobility of changes in blood supply (and therefore nutrition) of all functional tissues of the organism.

During the activation response a human can use up to 7000 kcal per day [12]. For comparison, recall that the base metabolism rate required for sustaining the vital activity of the organism in a state of complete rest is approximately 1600 kcal per day, and in our ordinary vital activity, including laborious activity, we use about 3500 kcal per day.

In the case of prolonged activity, long and intense work is possible within the framework of a reaction of enhanced activation with a power reaching 8000 kcal per day.

6   Stress and Necessity Avoiding It

However, in reality, man often finds himself in situations requiring efforts that exceed the physiological capabilities of the organism. A response to extreme irritants ("strong stimuli" [8, 10]) allows the organism to survive as a living system, but leaves behind injuries and pathological changes. This modus of activity is shown in Fig. 4 as line 4.

The anabolic stage under stress is weak or is entirely absent [4] (Fig. 3b), while the catabolic stage 2 leads to expenditures of structural energy and to structural changes in the organism that exceed its regenerative abilities, the increase in structure energy at stage 3 is lower than the expenditures at the catabolic stage 2: $\Delta F = F_C - F_A < 0$. This adaptive response of animal and human organisms to excessive irritants was discovered earlier than the responses corresponding to normal physiology by Canadian physiologist H. Selye in 1936 [9]. He called the response stress (the term became widely used mostly owing to journalists who used and abused the word). The changes in the hypothalamus–hypophysis–adrenal-gland system controlling the hormonal status of the organism, as Selye [9, 11] and other researchers have shown, are of a mobilizatory character. The hypothalamus, secreting the corticotropic-releasing factor into the blood vessels of the hypophysis triggers the release of adrenocorticotropic (ACTH) hormone by the hypophysis, which in turn induces secretion of corticosteroids by the adrenal cortex, along with the release of catecholamines, adrenaline and noradrenaline, by the adrenal medulla. Both corticosteroids and catecholamines are factors whose function is the emergency mobilization of the organism. Energy is mostly generated on the lipid (adipose) basis. Carbohydrate energy is suppressed. Physiologists even have a proverb: "Fats burn in the carbohydrate fire, but carbohydrates do not burn in the fat fire" [8]. This is a very important circumstance, since the nervous system functions only on the basis of carbohydrate energy. Therefore, while in stress, the cognitive component of vital activity is either suppressed or absent. In other words, stress actions are unacceptable in the educational space and in educational cognitive activity. When designing educational technologies, the pathological adaptive response, stress, should be considered only insofar that it must be eliminated.

Let us note some other features of the pathological adaptive response. An increase in coagulability occurs (a natural evolutionary adaptation: stress situations may result in bloodshed). Blood circulation becomes less dynamic, tissue nutrition is decreased or stopped, including that of functional tissues. One of the characteristic stress symptoms is the formation of multiple bleeding ulcers in the digestive system [8]. Thus, stress never does without bleeding.

The immune system is literally switched off. Its major organ, the thymus, drastically decreases in size, all immune factors are released into the bloodstream, and as a result the organism is relatively resistant to pathogenic influences for three days. But in the stage of exhaustion which sets in on the third or fourth day of stress, the human becomes ill. And although the "weak links" of the organism differ from person to person, in most cases acute tonsillitis occurs, due to the defenselessness of the tonsils, which are lymph-glands, and therefore, a part of the immune system. In any case, if one finds that the mandibular glands are swollen, one should try to remember what trouble happened 3 days ago, and by revealing the stressor, eliminate it, changing the attitude towards it or by restructuring near-term plans.

It should be noted that under stress the water-mineral metabolism changes so much that the weight of a human increases by 2 kg in several hours [12]. Therefore, a person who regularly observes their weight can easily discover stress by weighing.

7 Conclusions

We provided a brief review of adaptation reactions stressing the importance of their cyclic nature as well as the first anabolic phase of a cycle (prognostic phase, when the *image of the result of action and the action itself* is created). The central role in the self-reproducing cycles is played by the anabolic phases. Normally, one of the anabolic phases follows a catabolic phase during which the resources of the organism are expended. However, in exceptional cases, it is possible that the whole activity remains anabolic.

We discuss the importance of avoiding stress reaction—both in terms of health and educational activities (as stress never leads to development of long-term memory).

In the past years, the author discussed some of this topics with S. G. Psakhie in context of creative scientific activity and its organization.

Acknowledgements The author is thankful to V. L. Popov for discussions of the key ideas of the present paper and for his assistance in preparing the manuscript, Elena Komar' for preparing illustrations and M. Popov for proofreading the Chapter.

References

1. Marturana HR, Varela FJ (1987) The tree of knowledge: the biological roots of human understanding. New Science Library/Shambhala Publications, Boston, 263 pp
2. Ugolev AM (1987) Natural technologies of biological systems. Nauka, Leningrad, 318 p (in Russian: 317 c.)
3. Anokhin PK (1968) Biology and neurophysiology of conditional reflex. Medicine, Moscow, p 546 (in Russian)
4. Arshavskii IA (1982) Physiological mechanisms and regularities of individual development. Nauka, Moscow, p 270
5. Popov LE, Postnikov SN, Kolupaeva SN, Slobodskoy MI (2015) Natural resources and technologies in educational activities: education in times of accelerated technological development. Cambridge International Science Publishing Ltd., Cambridge, 133 pp
6. Semenova ST (1993) Vernadsky and Russian Cosmism, Vladimir Vernadsky, Biography. Selecta. Memory Lane of Contemporaries. Descendant Speculations. Sovremennik, Moscow, pp 596–646
7. Penrose R (1994) Shadows of the mind: a search for the missing science of consciousness. Oxford University Press, Oxford, p 457
8. Garkavi LH, Kvakina EB, Ukolova MA (1977) Adaptive responses and resistance of the organism. Rostov University Press, Rostov-on-Don, p 126
9. Selye H (1936) Thymus and adrenals in the response of the organism in injuries and intoxication. Br J Exp Pathol 17(3):234–248
10. Garkavi LH, Kvakina EB, Kuz'menko TS (1998) Antistress responses and activation therapy. IMEDIS, Moscow, p 656
11. Selye H (1972) At the level of the whole organism. Nauka, Moscow, p 122
12. Hammond KA, Diamond JM (1997) Maximum sustained energy budgets in humans and animals. Nature 386:457–462

Seeing What Lies in Front of Your Eyes: Understanding and Insight in Teaching and Research

Elena Popova, Valentin L. Popov, and Alexander E. Filippov

Abstract In the present paper, we considered the phenomena of understanding and discoveries (as a sort of "social understanding") and found that the empirical properties of these phenomena (the critical character and emerging of a new property) have much in common with first-order phase transitions. From this point of view, we discuss both the process of understanding and discoveries and the reasons impeding "seeing what lies in front of our eyes". In our opinion, these ideas can be further studied on the same phenomenological basis, without detailed understanding of the underlying neuronal mechanisms.

Keywords Understanding · Insight · Discovery · Phase transitions · Order parameter · History of science · Friction · Contact · Adhesion

1 Introduction

This paper is devoted to a phenomenon of fundamental importance for many areas of human activity, including research and teaching—the phenomenon of *understanding*. It is clear that teachers must ensure that the material they mediate is *understood* and not only memorized or "only" practiced well. However, this central phenomenon of learning always remains a mystery. The act of understanding can be compared to that of "seeing": When understanding something, we suddenly see "the picture as a whole". However, this is not at all easy—as Goethe once said: "The hardest thing to see is what lies in front of your eyes".

E. Popova (✉) · V. L. Popov
Technische Universität Berlin, Berlin, Germany
e-mail: elena.popova@tu-berlin.de

V. L. Popov
e-mail: v.popov@tu-berlin.de

A. E. Filippov
Donetsk Physico-Technical Institute, Donetsk, Ukraine

© The Author(s) 2021
G.-P. Ostermeyer et al. (eds.), *Multiscale Biomechanics and Tribology of Inorganic and Organic Systems*, Springer Tracts in Mechanical Engineering,
https://doi.org/10.1007/978-3-030-60124-9_25

Based on historical examples, we suggest considering the mechanism of "sudden understanding" as a first-order-phase transition in the cognitive space (individual understanding) or a combination of cognitive and real space (discoveries).

2 Phenomenon of Understanding

The paradoxical truth expressed by Goethe is valid both for the scientific perception of the world and for everyday life. Consider the painting "Gartenlokal an der Havel—Nikolskoe" (Fig. 1).

When people stand in front of this painting, showing a Cafe at one of the lakes in Berlin, they usually do not see anything strange in it—even if they study the painting attentively. There are just people sitting in a cafe and speaking to each other. What goes unnoticed is the fact that none of the chairs in this painting has legs: Everyone is sitting in the air! It is still a matter of debate whether Liebermann just did not have enough time and the painting is incomplete or this was his joke with the observer, but the fact is that the majority of spectators are not at all disturbed by the missing details. They just do not see that they are missing!

In this case, the details we do not see are not important for the correct perception of the painting. However, sometimes it is exactly the details we overlook that are important. In the same way, something that may be hard to understand initially, can

Fig. 1 Max Liebermann, Garden Restaurant on the Havel—Nikolskoe, 1916, Berlin, Nationalgalerie

Prof. Popov | Colloquium Advanved Mechanics | 28.04.2017

Problem 1. Two tractors pull a box with ropes (see picture). The velocities of the tractors are directed along the ropes and are v_1 and v_2. The angle between the ropes is alpha. What is the velocity of the box and how is it directed?

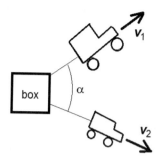

Fig. 2 A task from the "Colloquium Advanced Mechanics" at the TU Berlin

become trivial once seen from the right perspective: The solution lies in front of our eyes, but we do not *see* it.

One of the authors of this paper (V.P.) offers a course at the TU Berlin that is specifically designed to illustrate and to train the ability to understand: the "Colloquium for Advanced Mechanics". The typical feature of the tasks that are handled in the Colloquium is that they are "very difficult problems, that are very easy to solve". The statement is contradictory only at first sight. The tasks of the Colloquium often cannot be solved by students, but only because they do not approach them from the correct perspective.

Here is an example of a task from the Colloquium (Fig. 2). Two tractors pull a box with ropes. The velocities of the tractors are directed along the ropes and are v_1 and v_2. The angle between the ropes is α. What is the velocity of the box and how is it directed? The task sounds simple. It is clear to everybody that it must have a solution: If you pull a box, it will have to move somewhere. And yet it is a task that even scientific collaborators at the University cannot solve immediately. The first idea that comes to mind is to sum up the velocities of the tractors. However, at least after considering the special case of parallel ropes and equal velocities, one can see that this idea is unfortunately completely wrong.

Experience shows that only exceptional personalities can solve this problem quickly and correctly—in spite of the fact that the task is solvable in two lines! There is just nothing to "calculate" in this problem! The only thing one needs is to look at the problem from the correct point of view. Once one understands the underlying principle, one can only say: "Ah!" And that *is* the solution.

The colloquium essentially consists of tasks of this kind, in which one finds the solution not by "working long and hard", but by real *understanding*. As Goethe has said, the hardest thing to see is what lies in front of your eyes!

3 Discoveries in the History of Science as "Seeing the Obvious"

A similar phenomenon of understanding is known in the history of science. History is full of examples where great discoveries were nothing more than seeing what lies in front of your eyes [1].

Let us take fracture theory as an example. In 1921, Alan Griffith put forward an idea why the strength of materials is much lower than the theoretical one and why it depends on the size of engineering parts [2]. His paper was the beginning of the theory of fracture. But what did Griffith really do? He only said that the crack tip is in equilibrium if the change of energy due to a small displacement is zero. This energy change consists of the relaxation of elastic energy and the work of separation of surfaces. One of them is positive and the other negative. If their absolute values are equal then the crack is in equilibrium. This is of course nothing else than the principle of virtual work for mechanical systems. This principle is by no means an invention of Griffith, it was already known to d'Alembert and Leonard Euler. It is a standard, well-known equilibrium condition, which can be found in any introductory textbook on mechanics. New was only that Griffith applied this old principle to the crack. Interestingly, he did not even have to carry out any complicated analytical calculations, because the elastic energy, which is released due to a small movement of a crack tip, was already known. The corresponding problem was solved in 1911 by Inglis [3], and the expression for the work of adhesion is trivial: it is the product of the specific work of adhesion and the new surface area produced by the crack opening. Thus, Griffith equated two contributions, which were both known at that time and thus produced the famous, classical results of fracture theory!

50 years later, Johnson, Kendall and Roberts (JKR) published their famous paper on adhesion, one of the most cited papers in the field of contact mechanics [4]. In their paper, JKR note that their approach is equivalent to that of Griffith. They write: "…the approach followed in this analysis, is similar to that used by Griffith in his criterion for the propagation of a brittle crack." As a matter of fact, JKR realized that the adhesive contact *is* the inverted Griffith crack (in the crack the discontinuity is mostly inside and in the "adhesive contact" outside). After that, JKR applied the principle of energy balance exactly in the way Griffith did. The only difference is in the expressions for the elastic energy. Griffith used the energy of an "internal crack" provided by Inglis and JKR used the expressions provided by Hertz [5] and Boussinesq [6]. The realization of the equivalence of the problem of adhesive contact to the problem of the Griffith crack was the main contribution of JKR, as a matter of fact, their *only* contribution. Already Griffith had all the necessary ingredients for the solution of the adhesive contact problem, he merely had no need to solve it. If he had, we probably would have had the "JKR"-solution already in 1921. It is surprising that 50 years were needed just to recognize the equivalence of these two problems!

But the history of adhesion does not end here. One of the most effective numerical simulation methods in contact mechanics is at the present the Boundary Element Method (BEM), which allows simulation of arbitrary contact configurations [7].

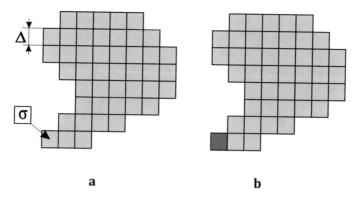

Fig. 3 In each calculation step, stress in each particular discretization element is determined. If the stress σ in a given element at the boundary of the contact area exceeds the critical value (3), it is "detached" and the stress in this element is set zero

However, only in 2015 an idea was put forward on how to simulate adhesive contacts in the framework of BEM: The use of a simple, unmodified Griffith criterion for the BEM detachment condition [8]. Let us explain this in more detail.

In each iteration of the BEM simulation, the stress in each discretized cell is determined and then it is decided whether this cell is still in contact or not. For non-adhesive systems, the answer is very simple: The pressure must be positive. But if you have an adhesive contact then the pressure can become negative and we have to find a criterion for when a particular element will lose the connection to the counter-body. In 2015, two scientists suggested applying the Griffith criterion: Let us assume that the contact of a cell is lost, then the stress in this cell vanishes (Fig. 3). It is easy to calculate analytically what amount of elastic energy is relaxed due to the stress vanishing:

$$\Delta U_{el}(\tau) = \kappa \frac{\sigma^2}{E^*} \Delta^3 \tag{1}$$

with

$$\kappa = \frac{2}{3\pi}\left(1 - \sqrt{2} + \frac{3}{2}\log\left(\frac{\sqrt{2}+1}{\sqrt{2}-1}\right)\right) \approx 0.473201. \tag{2}$$

The element is in the state of indifferent equilibrium if the change of elastic energy is equal to the work of adhesion needed for creating the free surface with the area Δ^2, $\Delta U_{adh} = \gamma_{12}\Delta^2$, or with (1), $\kappa\sigma^2\Delta^3/E^* = \gamma_{12}\Delta^2$, where γ_{12} is the work of adhesion per unit area. For the critical detachment stress they obtained

$$\sigma_c = \sqrt{\frac{E^*\gamma_{12}}{0.473201 \cdot \Delta}}. \tag{3}$$

It is now assumed that if this elastic energy is equal to the work of separation then the element will detach. In other words, as soon as the elastic energy is enough for creating new surfaces, they will be created. This solution is so simple and straightforward that one can only ask why it took another 44 years to apply the Griffith criterion in numerical simulation!

The present paper is not devoted to any detailed substantiation of ideas, let us just mention that the simple criterion presented above reproduces all known analytical solutions exactly [9].

4 Understanding as Changing the Point of View

The above examples illustrate very clearly the idea that the real mechanism behind new theories and paradigms is often a transfer of knowledge from one subject area to another [1]. A similar process in individual perception is *changing the point of view*. But what exactly does that mean? This mechanism reflects fundamental philosophical roots of the term "meaning" that were elaborated in detail in the second half of the twentieth century, owing to the development of system analysis and increased interest in the nature of complicated systems. According to system analysis, properties of a system cannot be understood by looking on it from inside the system [10]. Recall the definition of a system given by R. Ackoff: "A system is a whole which is determined by its function in the system, a part of which it is" [11]. Therefore, understanding of the meaning, purpose, and objectives of any system can be achieved only at the level of the containing system (supersystem). This means that true understanding implies the placing of a given object into a corresponding supersystem where it plays a specific role. Depending on the supersystem, the same object can have different meanings.

The above historical examples have one thing in common: The "seeing" occurs by exchanging the supersystems. Very often, one hears that scientists and engineers have to "think outside the box". From the point of view of system analysis, thinking outside the box becomes a generic, compulsory prerequisite to any true understanding.

Of interest is R. Ackoff's comment on the relationship between the analysis of a part of a system and the supersystem: "Note that an analysis starts with division of a subject into parts and gives knowledge. Synthesis starts with combination of things and gives understanding. Analysis is the way the scientists investigate. Synthetic thinking is manifested in designing... There is one more extremely important aspect of synthetic thinking. The systems we deal with are becoming increasingly sophisticated. Scientists seek the effective ways of treating them. Unfortunately, most of them approach the problem analytically. As a result, they introduce so many variables and relations between them that we cannot cope with them. However, if complexity is dealt with in a synthetic way, by designing, like designing a high-rise building or a city, there is no level of complexity that we cannot effectively cope with" [10, 11].

If we accept the viewpoint of Ackoff, then we have to accept that understanding is the final result of a synthetic activity aimed at a broadening the context in which the given object is considered or at trying a large variety of "contexts" until a "correct" one is found.

There is another potentially important point of this process, which is best illustrated not on examples of personal perception but on the history of discoveries: It is not always clear that the correct supersystem has already been found, so that the final solution may require many attempts. In the history of science, this is seen by the very widespread phenomenon of multiple discoveries.

5 Multiple Discoveries

Very often scientific results are obtained multiple times—"re-discovered" without knowing the predecessors. Take as an example the history of the so-called Method of Dimensionality Reduction [12]. This is a method of representing contact mechanics in a simple way that can be taught even to undergraduate students and can be used by any practical engineer. It was developed in the decade from 2005 to 2015. However, it was invented already much earlier and reinvented many times in the course of history. And each time, people just did not see what was lying in front of their eyes. It was first formulated in a paper by Schubert in 1942 in Germany [13]. The same solution was found by Galin in 1946 in Russia [14] and later by Green and Zerna in 1954 [15]. Sneddon translated the book of Galin from Russian into English in 1950th (see a later publication [16]) and also published 1965 a very influential paper which contains exactly the same solution, since then mostly known as the Sneddon solution [17]. In 1998 Jäger suggested an alternative physical interpretation of the same equations [18]. Finally, they were reformulated as a simple mnemonic rule and generalized to a large variety of contact problems in 2005 to 2015, thus creating MDR [10, 12]. It is striking that it took about 70 years to realize that the solution lies in front of our eyes!

6 Understanding as a Phase Transition

The sudden character of understanding shows that this is a critical phenomenon. A prerequisite for it is a synthetic (design) work, but the final act is often perceived as "instantaneous". Understanding is bringing order in a large number of elements. It can thus be interpreted as a "phase transition" establishing a "long range order" in the corresponding "cognitive space". This transition can be easily followed phenomenologically by considering an impressionist painting from various distances. When considered in the vicinity, the painting looks like a chaotic set of color spots (Fig. 4). Going further from the picture, one suddenly recognizes the picture as a whole. Interestingly, this process is almost reversible so that one can repeat this "act of understanding" or "act of recognition" many times. At this point, we cannot discuss the question of the exact neuronal mechanism of this phase transition. We just use

a   **b**

Fig. 4 Painting "Lisette Sewing in front of the Entrance Door of Marquerol" by Henri Martine: **a** view from vicinity, **b** view of the whole picture. Tel Aviv museum of Art

the finding of the theory of phase transitions, which says that local interactions in a distributed system can lead to a "sudden" establishment of a long-range order in the system when interactions achieve a critical value [19]. This change can lead to the appearance of qualitatively new properties, such as superconductivity. In the human perception we see empirically similar sudden state changes with the emerging of a new property—that of "understanding". It surely would be an intriguing and fruitful task to try to find out what exactly the neuronal mechanisms of this transition are, which could greatly facilitate the development of didactics.

The analogy of understanding and seeing brings another important aspect into play.

7 Interrelation of Personal Understanding and Discoveries

The fastest runner doesn't always win the race,

and the strongest warrior doesn't always win the battle.

The wise sometimes go hungry,

and the skillful are not necessarily wealthy.

And those who are educated don't always lead successful lives.

It is all decided by chance,

by being in the right place at the right time.

Ecclesiastes 9:11

While the acts of personal understanding and scientific discovery have much in common, one has to also see the essential difference between both. The above example of the history of the MDR shows that "understanding" in a historical context does not merely mean that something is seen by an individual scientist. It is important that the act of seeing a scientific result coincides with the act of public recognition of its importance: Understanding "not at the right time" "does not count". Only "collective seeing" is a true discovery. We often do not really know how many times the same scientific truth was re-discovered and intensive historical research is needed to answer the question: "who was the first"? Maybe one can speak of the act of "public understanding", which means understanding reaching a large (relevant) community. However, it is very difficult to know in advance whether a particular result will be broadly accepted by scientific community. That is why it is so difficult to make "discoveries on demand"—in reality, discoveries are in most cases understood as such a posteriori—in rare lucky cases of (later) public recognition.

The analogy with phase transitions leads to the conclusion that "authorship" of particular discoveries is relative. If the medium is far from the critical state, no fluctuations will create the long order parameters, and vice versa, if it is in the critical or overcritical state, then *any* nucleus will lead to the phase transition. Can we blame those researchers who by chance did not work in some field which the future would show to be very important and, on the contrary, praise those who for some reason were "centuries ahead of their time" (in most cases without knowing this)?

For success of a personal scientific career, it is very important to work on problems which may have a public resonance (at least in the relevant scientific community). The highest scientific qualification lies exactly in the general orientation of what problems are "of interest".

However, for the progress of science as a whole, this personal success is of no relevance. The true precondition of phase transitions is the local interaction. Preparation of this local interaction is as important for science as being an "ingenious inventor" (the latter often meaning to be a random "fluctuation" at the right time and at the right place.)

8 What Prevents Us Seeing What Lies in Front of Our Eyes?

The analogy of the phenomenon of understanding with phase transitions suggests a solution to the question of what prevents us from seeing what lies in front of our eyes. The fact that the "act of understanding" often occurs suddenly, implies that the underlying phase transition is a first-order phase transition. This means that the transition always occurs through formation of initially small nuclei, which later expand to the whole phase space. This propagation is connected with some "friction" in the boundary line. A very similar phenomenon takes place in social processes. A society of sufficiently large volume (including the scientific society as well) causes a practically infinite barrier for the immediate acceptance of new knowledge, even if it was already achieved by an individual researcher. Thus, the factor impeding the understanding is a resistance to the change of state. This resistance is a generic property of any first-order-phase transformation. What is sometimes called the "inertia of thinking" is in reality a "friction of thinking". The essential question is of course what factors determine this "friction". Just as for example ferromagnetic materials can have a large or small internal friction for the motion of phase boundaries (corresponding to hard and soft magnetic materials), the resistance to understanding may be higher or lower depending on factors which are not yet understood. Their determination could greatly facilitate didactics.

Similarly, discoveries can be considered as acts of "social understanding". They also occur "suddenly" (on the time scale of historical processes) and thus have the features of first-order transitions. The phase space now contains both "cognitive dimensions" and the real space dimension (in form of a geographic distribution of researchers). Any nucleus of new understanding will have to overcome the frictional force for boundary propagation. In everyday language, one can say that anybody offering a radically new idea will stumble upon a consolidated resistance by the whole system of knowledge accumulated and lovingly ordered on the shelves.

In the theory of first-order transitions it is known that a high boundary friction leads to the necessity of "overheating" the system to initiate the transition. In our analogy with understanding, this means that one has to do much more work and to accumulate more information related to the particular topic in the space of knowledge than it is necessary to perform "act of understanding" itself.

Another property of the first-order transition is the existence of the threshold of "absolute instability". In our analogy, this means that at some point it becomes impossible to ignore new knowledge and the transition becomes unavoidable.

The nuclei of a new phase can also dissolve again if the systems moves away from the critical point. The history of civilization provides many examples of the reverse process when already achieved knowledge "suddenly disappears", if for some reason it looses its social importance.

9 Conclusion

In the present paper, we considered the phenomena of understanding and discoveries (as a sort of "social understanding") and found that the empirical properties of these phenomena (the critical character and emerging of a new property) have much in common with first-order phase transitions. In our opinion, these ideas can be further studied on the same phenomenological basis, without detailed understanding of the underlying neuronal mechanisms—similarly to the famous treatment of the theory of superconductivity in the phenomenological theory of phase transformations [20].

However, just as a statistical physics give a deeper insight into the physical understanding of phase transitions, understanding the neuro-physiological mechanisms of perception and handling of "newness" by our brain would strongly facilitate also the phenomenological view on understanding. Some important findings in this relation can be found in [10].

In philosophy, this picture has already been developed over centuries under the notion of "measure" as a unity of quantity and quality (Hegel) [21]. However, we find that the picture provided by the theory of phase transitions provides more details and more understanding of the underlying processes.

The true understanding is the only quality, which enables a person or a community to make real progress in any branch of science or engineering. The analogy of understanding to a long range order provides an (we hope useful) illustration of what sometimes is called "complete knowledge" [10], meaning that understanding implies seeing the "picture as a whole". The long range order is exactly what provides a picture the "wholeness".

Acknowledgements The authors are grateful to Prof. Leonid Evgenyevich Popov and Emanuel Willert for inspiring and useful discussions and Nikita Popov for proof reading the Chapter.

References

1. Popova E, Popov VL (2018) Note on the history of contact mechanics and friction: Interplay of electrostatics, theory of gravitation and elasticity from Coulomb to Johnson–Kendall–Roberts theory of adhesion. Phys Mesomech 21(1):1–5
2. Griffith AA (1921) The phenomena of rupture and flow in solids. Philos Trans R Soc A: Math Phys Eng Sci 221:582–593
3. Inglis CE (1913) Stresses in a plate due to the presence of cracks and sharp corners. Trans Inst Naval Arch 55:219–241
4. Johnson KL, Kendall K, Roberts AD (1971) Surface energy and the contact of elastic solids. Proc R Soc Lond A 324(1558):301–313
5. Hertz H (1881) Über die Berührung fester elastischer Körper. Journal Für Die Reine Und Angewandte Mathematik 92:156–171
6. Boussinesq J (1885) Applications des Potentiels a l´Etude de l´Equilibre et du Mouvement des Solides Elastiques. Gauthiers-Villars, Paris
7. Pohrt R, Li Q (2014) Complete boundary element formulation for normal and tangential contact problems. Phys Mesomech 17(4):334–340
8. Pohrt R, Popov VL (2015) Adhesive contact simulation of elastic solids using local mesh-dependent detachment criterion in Boundary Elements Method. Facta Univ Ser: Mech Eng 13(1):3–10
9. Popov VL, Pohrt R, Li Q (2017) Strength of adhesive contacts: influence of contact geometry and material gradients. Friction 5(3):308–325
10. Popov LE, Postnikov SN, Kolupaeva SN, Slobodskoy MI (2015) Natural resources and technologies in educational activities: education in times of accelerated technological development. Cambridge International Science Publishing, Cambridge, p 133
11. Ackoff R, Greenberg D (2008) Turning learning right side up: putting education back on track. Wharton School Publishing, New Jersey, p 224
12. Popov VL, Heß M (2015) Method of dimensionality reduction of contact mechanics and friction. Springer, Berlin
13. Schubert G (1942) Zur Frage der Druckverteilung unter elastisch gelagerten Tragwerken. Ingenieur-Archiv 13(3):132–147
14. Galin LA (1946) Three-dimensional contact problems of the theory of elasticity for punches with circular planform. Prikladnaya Matematika I Mekhanika 10:425–448 ((in Russian))
15. Green AE, Zerna W (1954) Theoretical elasticity. Clarendon Press, Oxford

16. Galin LA (1961) Contact problems in the theory of elasticity. North Carolina (USA): Department of Mathematics, School of Physical Sciences and Applied Mathematics, North Carolina State College
17. Sneddon IN (1965) The relation between load and penetration in the axisymmetric Boussinesq problem for a punch of arbitrary profile. Int J Eng Sci 3(1):47–57
18. Jaeger J (1995) Axi-symmetric bodies of equal material in contact under torsion or shift. Arch Appl Mech 65:478–487
19. Landau LD, Lifshitz EM (1980) Statistical physics, 3rd edn. Part 1, vol 5. Butterworth-Heinemann, Oxford. ISBN 978-0-7506-3372-7
20. Ginzburg VL, Landau LD (1950) To the theory of superconductivity. In: Pis'ma v ZhTF, 20, 1064 (in Russian)
21. Hegel GWF (2010) Encyclopedia of the philosophical sciences in basic outline, part 1, science of logic. Cambridge University Press, Cambridge (Cambridge Hegel Translations)

Index